Journal of Neural Transmission

Supplementum 32

JNT

P. Riederer and M.B.H. Youdim (eds.)

Amine Oxidases and Their Impact on Neurobiology

Proceedings of the 4th International Amine Oxidases Workshop, Würzburg, Federal Republic of Germany, July 7–10, 1990

Springer-Verlag Wien New York

Prof. Dr. Peter Riederer
Department of Psychiatry, University of Wurzburg, Federal Republic of Germany
Prof. Dr. Moussa B. H. Youdim
Technion, Israel Institute of Technology, Haifa, Israel

Typeset by Macmillan India Ltd., Bangalore

Printed on acid-free paper

With 110 Figures

ISSN 0303-6995
ISBN-13:978-3-211-82239-5 e-ISBN-13:978-3-7091-9113-2
DOI: 10.1007/978-3-7091-9113-2

Preface

The Amine Oxidase Workshop was the brain child of Brian Callingham and Keith Tipton. The present book is the proceedings of the fourth in the series and in the past we have dedicated our meeting to individuals who have made significant fundamental contributions to the biochemistry, physiology and pharmacology of this interesting group of enzymes. The present meeting is not an exception because we have gathered here in Würzburg to honour an individual whose outstanding work on the autonomic pharmacology has spanned more than 30 years and whose contribution has made a vast impact in our understanding of the kinetic relationship between amine metabolism (that uptake and enzymatic inactivation) and adrenergic neurotransmission. We speak of course of Professor Ullrich Trendelenburg, or Ulli as he likes to be known. Like the amine oxidases, he has a unique place in Pharmacology.

Professor Trendelenburg was born on the eve of the new year of 1922 in the city of Rostock in northern Germany, where his father was Professor of Pharmacology at the Faculty of Medicine of Rostock. His father accepted the offer to take over the corresponding chair in Freiburg im Breisgrau (in Southern Germany). As the youngest of four children, Ulli grew up mainly in the garden of the department, in the animal quarters and in the workshop. Of the various coworkers of his father, one he liked especially was Otto Krayer, the young pharmacologist.

In 1927 his father accepted the call to Berlin – and Otto Krayer went with him. At that time it was customary in Germany that a "professorial apartment" exists in close conjunction with the department. Thus, they lived door to door with the department, once again with many contacts with his father's coworkers. Apart from Otto Krayer, two others should be mentioned – two young ladies who had just finished their medical studies: Edith Bulbring and Marthe Vogt. Edith Bulbring was a close friend of the family and had been a frequent guest in Freiburg when she went to medical school.

In 1929 his father fell ill and in 1931 he died. It was during this sad time that Otto Krayer went out of his way to help the children to adjust to this shock. Ulli was lucky to attend the school the Elector of Brandenburg had founded for the Huguenots when they were persecuted in France (1685). This school was rich in tradition and remarkably free of Nazi influence.

By the time he finished school in 1941 World War II raged. He joined the Luftwaffe, trained as a pilot and was then active as a flying instructor through most of this time. He only saw active service as a fighter pilot in March and

April 1945, was shot down on April 10, 1945 (we are glad that Ulli was shot down and not up, otherwise the discipline of pharmacology would have been different without him) and ended up in a POW camp of the US army in France. For 19 months he stacked the surplus of the US army – then he was discharged to go "home". He went to Göttingen where he spent six years of medical school. He was exceedingly lucky to be invited to the university of Uppsala (Sweden) for one term in 1951, where he came into contact with Ernst Barany, the pharmacologist in Uppsala and thanks to his help, Ulli was able to return to Uppsala twice (again for periods of 6 months each). It was also in Uppsala that he first met Professor J. H. Burn from Oxford who offered to train him in pharmacology: after all, from 1933 onwards, Professor Burn had collaborated with Edith Bulring who had to leave Germany for Oxford in 1933.

In 1952 the British Council gave him a 2-year scholarship. Thus he was able to join probably the best pharmacological department of that time. Moreover, after two years he was offered the position of a Departmental Demonstrator. He spent a total of four years in Oxford (1952–1956). Under the guidance of J. H. Burn, Edith Bulring, Hugh Blaschko, and in close contact with Marthe Vogt and Wilhelm Feldburg, with John Vane and Oleh Hornykiewicz. The atmosphere there must have been incredibly stimulating. Here he fell in love with the autonomic nervous system. The pharmacology of ganglia (i.e. of the superior cervical ganglion ot the cat) was the first topic that he worked on independently. In Oxford he married his utterly charming wife Christel Teschemacher from Tübingen.

In 1956 he returned to Mainz University to work with Professor G. Kuschinsky, but he spent only one rather unhappy year in Mainz. Since at that time he had two offers from the USA (from Acheson and from Otto Krayer), the Trendelenburgs decided to accept the offer from Harvard Medical School, where he spent 11 very happy years. Very soon after his arrival, he had the chance to find coworkers so from then onwards his research always involved a small group. Inspired by his time at Oxford, the group set out to clarify the mechanisms of supersensitivity to norepinephrine (denervation, decentralization, etc. etc.). By about 1962, the problem was solved: denervation causes two types of supersensitivity, of which one is due to the disappearance of an important site of loss (the neuronal uptake), the second one to a slowly developing change in the postsynaptic cells. During the subsequent years, he became more and more interested in these "sites of loss", i.e., in those mechanisms which are responsible for the quick inactivation of the transmitter norepinephrine. Just before he left Harvard, he also became interested in the function of a second (extraneuronal) site of loss for catecholamines. He had very close contact and relations with Otto Krayer, the head of the department. After Barany and Burn, Otto Kreyer was the third teacher who deeply influenced him. In 1966 Otto Krayer retired and Ulli was then acting head of the department for two years. Scientifically, these next two years were very profitable. It is rather sad to learn that the Department of Pharmacology at

Harvard (of which your were once the head) has been merged into the Biochemistry Department. In 1968 the Trendelenburgs moved again back to Germany – to Würzburg. Very soon, he had again collected a group of young people around him and their interest shifted to the "extraneuronal uptake and 0-methylation" of catecholamines. It was due to the urging of these young coworkers such as Professor Graefe, Bönisch and Henseling that they more and more learned to use kinetics as a means to analyse the function of these fascinating systems, in which the catecholamine must be transported to the intracellular enzymes, irrespective of whether they deal with neuronal monoamine oxidase. The group has changed from year to year, but the main direction of their interest has remained the same since 1968, including the family structure.

Ulli has received many honours through the years. They are far too many to be mentioned. He is an honorary member of 4 international pharmacology societies and has also received 4 honorary doctorates of medicine from universities in Finland, Portugal, USA and Poland. He has served on the editorial board to many pharmacology journals. If this were not enough, besides the dedication of this meeting to him a similar symposium was held in 1983 as a celebration of his 60th birthday. Next year a retirement symposium will also be held for him here in Würzburg.

What about Professor Trendelenburg the man? No description of Professor Trendelenburg would be complete without mentioning his family life. He has an extremely happy private life. He married in 1953 and Christel, his wife, has never minded all their travels but was happy to return to Germany in 1968. They were blessed with the birth of their daughter Marie in 1966. Having been a very premature baby, she has a slight spastic handicap, but, with the support of her dedicated parents she has learnt very well to cope with it. The Trendelenburgs efforts to enable her to lead a normal life have been so successful that she is now studying medicine. One of the great delights is a visit to his modest home where drinks and conversation overflow and Christel cooks and most times joins in the ongoing debate. Sometimes in the middle of dinner Ulli will disappear: he goes down to the basement to watch his favourite TV detective series.

I (MBHY) first met Ulli in 1972 at the Sardinia monoamine oxidase meeting organized by Professor Merton Sandler, who is participating in this meeting. We are also delighted to see so many friends from that meeting here including Professors Kopin, Tipton, Oreland, Shih. At the Sardinia meeting one could not but be influenced by Ulli's presentation on the kinetic analysis of amine uptake and metabolism by catechol-0-methyl transferase and monoamine oxidase, in whole organ. This impression remained with me until I got to Oxford where we set up similar kinetic analysis of monoamine oxidase A and B in isolated livers and lungs and brain. Then in 1980 Professor Trendelenburg and Professor Weiner invited Professors Tipton and Finberg and myself to contribute a chapter on "Monoamine Oxidase" to the Handbook of Experimental Pharmacology Catecholamine II. The instructions for it were that it should consist of 45 typed pages, including figures, tables and references. By the time Professors Tipton, Finberg and myself had finished our individual sections, this chapter was not less than 295 typed pages. We did not have the heart to shorten it. Norman Weiner on receiving the paper was shocked by its size. But the most extraordinary response was from Ulli (to whom Norman had sent our paper) who wrote in his letter "the more I read it, the more I like it". In fact it was due to his tenacity for details that our chapter was published almost intact. This was an unusual response from an editor, most of whom always like to delete things. The ensuing exchange of letters between us through the years and the establishment of Germany-Israel Binational Research Grants has certainly brought us closer and has made me even more appreciative of him. This has been especially evident in his quality as a human being and his deep understanding of other peoples feelings. When Professor Riederer moved from Vienna to Würzburg, I received an invitation from Ulli, to visit him. I cannot forget the first time I arrived at his office. His secretary showed me into his room, where he was buried behind hundreds of papers and books. You knew he was there because of the cigarette smoke. What made that visit so vivid was the discussion about relations between Germany and Israel, between Germans and Jews. This conversation which started in his office that afternoon lasted through hours of the next morning at his house. This has been almost the pattern of each visit I have made to him. Sometimes we do break into the realms of science where I have tried to influence him to look at the relationship between MAO A or B, and amine uptake and release. I gather he is doing this now. He wants to understand and contribute to bridge the gap between peoples. We are sure that each one of us who has come to know Ulli has his own similar story to tell. His kindness and helpfulness to young scientists is well known among his associates. This is most evident when refereeing papers of young scientists.

We (MBHY and PR) are also confident that we are speaking on behalf of all "Amine Oxidaseologists" and all those friends and colleagues who could not participate in the 4th Amine Oxidase workshop that it is most fitting to dedicate this meeting to Professor Ulli Trendelenburg.

M. B. H. Youdim

July 1990 P. Riederer

Contents

The state of the art

Trendelenburg, U.: The interaction of transport mechanisms and intracellular enzymes in metabolizing systems 3
Kopin, I. J.: In-vivo quantitative imaging of catecholaminergic nerve terminals in brain and peripheral organs using positron emission tomography (PET) 19
Ackenheil, M.: The mechanism of action of antidepressants revised 29

Monoamine oxidase and its inhibitors

Shih, J. C., Grimsby, J., Chen, K.: The expression of human MAO-A and -B genes 41
Saura Marti, J., Kettler, R., Da Prada, M., Richards, J. G.: Molecular neuroanatomy of MAO-A and MAO-B 49
Oreland, L., Jossan, S. S., Hartvig, P., Aquilonius, S. M., Långström, B.: Turnover of monoamine oxidase B (MAO-B) in pig brain by positron emission tomography using ^{11}C-L-deprenyl 55
Jossan, S. S., Gillberg, P.-G., Karlsson, I., Gottfries, C. G., Oreland, L.: Visualisation of brain monoamine oxidase B (MAO-B) in dementia of Alzheimer's type by means of large cryosection autoradiography: a pilot study 61
Moll, G., Moll, R., Riederer, P., Gsell, W., Heinsen, H., Denney, R. M.: Immunofluorescence cytochemistry on thin frozen sections of human substantia nigra for staining of monoamine oxidase A and monoamine oxidase B: a pilot study 67
Colzi, A., d'Agostini, F., Kettler, R., Borroni, E., Da Prada, M.: Effect of selective and reversible MAO inhibitors on dopamine outflow in rat striatum: a microdialysis study 79
Clement, H.-W., Grote, Chr., Wesemann, W.: In vivo studies on the effect of monoamine oxidase inhibitors on dopamine and serotonin metabolism in rat brain areas 85
Yu, P. H., Davis, B. A.: Some pharmacological implications of MAO-mediated deamination of branched aliphatic amines: 2-Propyl-1-aminopentane and N-(2-propylpentyl)glycinamide as valproic acid precursors 89
May, T., Strauss, S., Rommelspacher, H.: [^{3}H] Harman labels selectively and with high affinity the active site of monoamine oxidase (EC 1.4.3.4) subtype A (MAO-A) in rat, marmoset, and pig 93
Balsa, D., Fernandez-Alvarez, E., Tipton, K. F., Unzeta, M.: Inhibition of MAO by substituted tryptamine analogues 103
Sherry, R. L., Baker, G. B., Coutts, R. T., Mousseau, D. D.: Ring-substituted analogues of tranylcypromine as monoamine oxidase inhibitors 107
McKenna, K. F., Baker, G. B., Coutts, R. T., Rauw, G., Mozayani, A., Danielson, T. J.: Recent studies on the MAO inhibitor phenelzine and its possible metabolites 113

Sablin, S. O., Bachurin, S. O., Tkachenko, S. E.: Stylbasole analogues of MPTP as monoamine oxidase (MAO) substrates ... 119
Colombo, M., Strolin Benedetti, M., Bonsignori, A., Cocchiara, G., Roncucci, R., Dostert, P.: MAO activity, metabolism and anticonvulsant activity of milacemide in rats and mice ... 123
Borbe, H. O., Niebch, G., Nickel, B.: Kinetic evaluation of MAO-B-activity following oral administration of selegiline and desmethyl-selegiline in the rat ... 131
Nickel, B., Borbe, H. O., Szelenyi, I.: Effect of selegiline and desmethylselegiline on cortical electric activity in rats ... 139
White, H. L.: An enzymatic method for detecting MAO-A and MAO-B inhibitors in plasma and its application in studies with reversible MAO-A selective inhibitors 145
Celada, P., Sarrias, M. J., Artigas, F.: Serotonin and 5-hydroxyindoleacetic acid in plasma. Potential use as peripheral measures of MAO-A activity ... 149
Helena Fernandes, M., Soares-da-Silva, P.: Role of monoamine oxidase A and B in the deamination of newly-formed dopamine in the rat kidney ... 155
Caramona, M. M., Cotrim, M. D., Fontes Ribeiro, C., Macedo, T.: Monoamine oxidase activity in blood platelets of migraine patients ... 161
Cesura, A. M., Muggli-Maniglio, D., Lang, G., Imhof, R., Da Prada, M.: Monoamine oxidase inhibition by moclobemide and 2-amino-ethyl carboxamide derivatives: mode of action and kinetic characteristics ... 165
Oxenkrug, G. F., Requintina, P., Yuwiler, A.: Does moclobemide stimulate melatonin synthesis as the other selective MAO-A inhibitors do? ... 171
Hebenstreit, G. F., Loidl, M., Baumhackl, U., Gallhofer, B., Geretsegger, Ch., Radmayr, E., Saletu, M., Schöny, W., Stabl, M.: Efficacy and safety of moclobemide compared with imipramine in the treatment of major depressive disorder. Double-blind Multicenter Study, Austria ... 177
Classen, W., Laux, G.: Psychometric alterations in treatment with MAO-A-inhibitor moclobemide ... 185
Laux, G., Classen, W., Sofic, E., Becker, T., Riederer, P., Lesch, K. P., Beckmann, H.: Clinical, biochemical and psychometric findings with the new MAO-A-inhibitors moclobemide and brofaromine in patients with major depressive disorder ... 189
Fritze, J., Becker, Th., Ziegler, V., Laux, G., Bieck, P., Sofic, E., Riederer, P.: Brofaromine (CGP 11 305 A): estimation of plasma concentrations by a biologic technique as compared to liquid chromatography ... 197
Hinze, C., Harland, D., Zreika, M., Dulery, B., Hardenberg, J.: A double-blind, placebo-controlled study of the tolerability and effects on platelet MAO-B activity of single oral doses of MDL 72.974A in normal volunteers ... 203
Kettler, R., Cesura, A. M., Dingemanse, J., Da Prada, M.: MAO-B inhibition in rabbit tissues and in human platelets by Ro 19-6327 shows similar time course 211

Oxidative stress and autooxidation

Heikkila, R. E., Terleckyj, I., Sieber, B.-A.: Monoamine oxidase and the bioactivation of MPTP and related neurotoxins: relevance to "DATATOP" ... 217
Cohen, G.: Monoamine oxidase and oxidative stress at dopaminergic synapses ... 229
Youdim, M. B. H., Riederer, P., Ben-Shachar, D.: The role of monoamine oxidase, iron-melanin interaction, and intracellular calcium in Parkinson's disease ... 239
Huether, G., Reimer, A., Schmidt, F., Schuff-Werner, P., Brudny, M. M.: Oxidation of the indole nucleus of 5-hydroxytryptamine and formation of dimers in the presence of peroxidase and H_2O_2 ... 249

Sofic, E., Götz, M., Frölich, L., Burger, R., Heckers, S., Riederer, P., Jellinger, K., Beckmann, H.: Reflection of changes in membrane constituents in various regions of Alzheimer brains to differential scanning thermograms 259

Rausch, W.-D., Schallauer, E., Chan, W. W., Riederer, P., Weiser, M.: Effects of L-Deprenyl and Amantadine in an MPTP-Model of Parkinsonism 269

Diamine oxidase – semicarbazide sensitive amine oxidase – polyamine oxidase

Callingham, B. A., Holt, A., Elliott, J.: Some aspects of the pharmacology of semicarbazide-sensitive amine oxidases 279

Sattler, J., Lorenz, W.: Intestinal diamine oxidases and enteral-induced histaminosis: studies on three prognostic variables in an epidemiological model 291

McIntire, W. S., Dooley, D. M., McGuirl, M. A., Cote, C. E., Bates, J. L.: Methylamine oxidase from *arthrobacter* P1 as a prototype of eukaryotic plasma amine oxidase and diamine oxidase 315

Sessa, A., Perin, A.: Chronic ethanol feeding and diamine oxidase activity in rat brain and liver 319

Lizcano, J. M., Escrich, E., Tipton, K. F., Unzeta, M.: Amine oxidase activities in chemically-induced mammary cancer in the rat 323

Fernandez de Arriba, A., Balsa, D., Tipton, K. F., Unzeta, M.: Monoamine oxidase and semicarbazide-sensitive amine oxidase activities in bovine eye 327

Raimondi, L., Pirisoni, R., Banchelli, G., Ignesti, G., Conforti, L., Buffoni, F.: Cultured preadipocytes produce a semicarbazide-sensitive amine oxidase (SSAO) activity 331

Blicharski, J. R. D., Lyles, G. A.: Semicarbazide-sensitive amine oxidase activity in rat aortic cultured smooth muscle cells 337

Lizcano, J. M., Balsa, D., Tipton, K. F., Unzeta, M.: Amine oxidase activities in bovine lung 341

Fogel, W. A.: Mucosal mono- and polyamine oxidase activities in digestive tract are distributed complementary to diamine oxidase 345

Strolin Benedetti, M., Cocchiara, G., Colombo, M., Dostert, P.: Does FAD-dependent polyamine oxidase contribute to the metabolism of milacemide? 351

Nic a Bháird, N., Tipton, K. F.: Behaviour and properties of catechol-0-methyltransferase from human placenta 359

Bertocci, B., Garotta, G., Zürcher, G., Miggiano, V., Da Prada, M.: Monoclonal antibodies recognizing both soluble and membrane bound catechol-0-methyltransferase 369

Zürcher, G., Colzi, A., Da Prada, M.: Ro 40-7592: inhibition of COMT in rat brain and extracerebral tissues 375

Waldmeier, P. C., De Herdt, P., Maitre, L.: Effects of the COMT inhibitor, CGP 28014, on plasma homovanillic acid and 0-methylation of exogenous L-DOPA in the rat 381

Bieck, P. R., Nilsson, E., Antonin, K. H.: Effect of the new selective COMT inhibitor CGP 28 014 A on the formation of 3-0-methyldopa (30MD) in plasma of healthy subjects 387

Uptake processes

Osswald, W.: Oxidative deamination of noradrenaline in human blood vessels ... 395

Finberg, J. P. M., Ari, G., Lavian, G., Hovevey-Sion, D.: Modification of alpha-2 presynaptic receptor activity and catecholamine release following chronic MAO inhibition 405

Bönisch, H., Martiny-Baron, G., Blum, B., Michael-Hepp, J.: Biochemical characterization and purification of the neuronal sodium-dependent noradrenaline transporter ... 413
Graefe, K.-H., Halbrügge, Th., Gerlich, M., Ludwig, J.: The importance of plasma 3,4-dihydroxyphenylglycol (DOPEG) in analyses of the sympathetic nervous system in vivo ... 421
Halbrügge, T., Friedgen, B., Thyen, A., Graefe, K.-H.: The involvement of desipramine-sensitive processes in the extraction of various catecholamines from plasma in the anaesthetized rabbit ... 431
Schömig, E., Babin-Ebell, J., Schönfeld, C.-L., Russ, H., Trendelenburg, U.: Human Caki-1 cells are the first model for extraneuronal transport of noradrenaline (uptake$_2$) which is based on a clonal cell line ... 437
Ludwig, J., Gerlich, M., Halbrügge, T., Graefe, K.-H.: The synaptic noradrenaline concentration in humans as estimated from simultaneous measurements of plasma noradrenaline and dihydroxyphenylglycol (DOPEG) ... 441
Egashira, T., Goto, S., Murayama, F., Yamanaka, Yu: Inhibition of MAO activity, ^{3}H-imipramine binding, ^{3}H-paroxetine binding and ^{3}H-5-HT uptake by human cerebrospinal fluid ... 447
Lesch, K. P., Laux, G., Mueller, T.: α_2-Adrenoceptor responsivity in depression: effect of chronic treatment with moclobemide, a selective MAO-A-inhibitor, versus maprotiline ... 457
Proença, J., Paiva, M. Q., Guimarães, S.: Supersensitivity to catecholamines after inhibition of extraneuronal uptake (uptake$_2$) or 0-methylation ... 463

Sulfation

Kienzl, E., Eichinger, K., Sofic, E., Jellinger, K., Riederer, P., Kuhn, W., Fuchs, G., Laux, G.: Urinary dopamine sulfate: regulations and significance in neurological disorders ... 471
Faraj, B. A., Camp, V. M., Watts, R. L.: Effect of adrenal-caudate transplantation on CSF levels of salsolinol sulfate in patients with Parkinson's disease ... 481
Subject Index ... 485

Listed in Current Contents

The state of the art

J Neural Transm (1990) [Suppl] 32: 3–18

The interaction of transport mechanisms and intracellular enzymes in metabolizing systems*

U. Trendelenburg

Department of Pharmacology and Toxicology, University of Würzburg, Federal Republic of Germany

Summary. The life span of extracellular catecholamines is limited by the combination of uptake and subsequent intracellular metabolism by either monoamine oxidase (MAO) and/or catechol-O-methyl transferase (COMT). Three such "metabolizing systems" are involved in the inactivation of noradrenaline: 1) Neuronal uptake (high-affinity $uptake_1$) in association with neuronal MAO (and vesicular uptake), 2) extraneuronal uptake (low affinity $uptake_2$) in association with intracellular COMT and MAO (in smooth muscles, myocardial cells, glands), and 3) $uptake_1$ of non-neuronal cells in association with intracellular COMT and/or MAO (in vascular endothelium of rat lung).

Such systems function as "pump and leak systems with enzyme(s) inside". The analysis of either uptake or enzyme fails to reveal the characteristics of such systems; they are determined by the interaction of *both* components.

Because of the high activity of these intracellular enzymes, it is unlikely that either COMT or MAO is ever saturated in vivo. However, in vitro saturation of extraneuronal COMT and MAO reveals that extraneuronal COMT is a high-affinity, but extraneuronal MAO a low-affinity enzyme. Hence, membrane-bound COMT appears to be responsible for the extraneuronal O-methylation of noradrenaline.

If intracellular enzymes remain unsaturated, the determination of the rate constants describing the unsaturated enzyme ($K_{ENZYME} = V_{max}/K_m$) is of particular interest. K_{ENZYME} can be determined for metabolizing systems, since this rate constant is not affected by the (usually unknown) fractional size of the metabolizing system.

* Abbreviations: COMT – catechol-O-methyl transferase; DOMA – dihydroxymandelic acid; DOPEG – dihydroxyphenyl-glycol; MAO – monoamine oxidase.

Introduction

After its exocytotic release from cholinergic varicosities, acetylcholine is hydrolysed by acetylcholinesterase localized in the extracellular space. The transmitters of noradrenergic, dopaminergic and serotoninergic neurones, on the other hand, are *not* subject to degradation in the extracellular space: the relevant enzymes are intracellular ones. Therefore, for noradrenaline, dopamine and 5-hydroxytryptamine transport mechanisms are required for the translocation of the extracellularly located transmitter to the intracellular enzyme. This also means that the primary mechanism limiting the life-span of these transmitters in the extracellular space is transport (out of the extracellular space), not an enzyme.

In the following, the discussion will be restricted to the mechanisms responsible for the degradation of noradrenaline (and of other catecholamines) in peripheral organs. The term "metabolizing system" has been proposed for systems in which a transport mechanism functions hand in hand with one or two intracellular enzyme(s). For earlier reviews of this area, the reader is referred to Trendelenburg (1980, 1984a, 1986, 1988).

Three types of metabolizing systems remove catecholamines from the extracellular space of peripheral organs:

1. Neuronal uptake (and re-uptake) is effected by a high-affinity transport mechanism (uptake$_1$) which is highly and specifically sensitive to inhibition by desipramine or cocaine. Uptake$_1$ is arranged in series with the intraneuronal, mitochondrial MAO. About 20 to 30% of the noradrenaline translocated into the neurone are deaminated, while 70 to 80% are stored again in the vesicles (Eisenhofer et al., 1987; Halbrügge et al., 1989).

2. Extraneuronal uptake (into various smooth muscle, myocardial and gland cells) is effected by low-affinity uptake$_2$. This differs from uptake$_1$ with regard to its substrate spectrum (isoprenaline, for instance, is a good substrate of uptake$_2$ but not of uptake$_1$) as well as with regard to inhibitors with high specificity (various corticosteroids and O-methylated catecholamines). Translocation of catecholamines by uptake$_2$ exposes the amines to the intracellular COMT, and in most tissues also to MAO; moreover, it should be realized that extraneuronal cells have *no* storage mechanism for catecholamines.

3. Recent observations indicate that it is no longer correct to equate uptake$_1$ with neuronal uptake, since uptake$_1$ has also been found in certain extraneuronal cell types (glands of rabbit endometrium: Kennedy and de la Lande, 1987; dental pulp of the rabbit: Parker et al., 1987; vascular endothelium of rat lung: Bryan et al., 1988, 1989). In these extraneuronal cells, uptake$_1$ is associated with COMT and (at least in rat lung vascular endothelium) with MAO. Again, there appears to be no storage of catecholamines in these extraneuronal cells.

Most of the experiments discussed here deal with the metabolizing systems (1) and (2). In most studies, one of the two extraneuronal enzymes was inhibited. Hence, such analyses deal with the "neuronal deaminating system", the "extraneuronal O-methylating system" or the "extraneuronal deaminating system". Furthermore, if only *one* enzyme is intact within a metabolizing system, there is just *one* "predominant metabolite" formed: for instance, after the deamination of noradrenaline and adrenaline, the aldehydes are predominantly converted to DOPEG, while there is very little formation of DOMA; the O-methylation of these catecholamines, on the other hand, leads to the formation of normetanephrine and metanephrine, respectively. If necessary, the rather minor formation of DOMA was taken into account in the experiments, but, for brevity's sake, this report will simplify the presentation by concentrating on the "main" metabolite. The reader is referred to the original papers for details.

In this report the term "activity" is used to describe either enzyme or uptake mechanism. It should be realized that this term does not relate to the V_{max}, but rather characterizes the unsaturated enzyme or uptake mechanism. It corresponds to k_{ENZYME} or k_{UPTAKE} ($= V_{max}/Km$, min^{-1}), i.e., the rate constants that characterize the unsaturated enzyme or uptake mechanism. If either enzyme or uptake mechanism is saturated by the substrate, this is then described as a decline of its "efficiency" (since an x-fold increase in substrate concentration is then *not* followed by an x-fold increase in rate of metabolism or uptake).

The term metabolizing system was coined to emphasize that both, transport mechanism and intracellular enzyme, are interacting. Indeed, in the following, an attempt will be made to demonstrate that metabolizing systems have characteristics that can be understood only if *both* components of these systems are considered in conjunction. As a first step in the presentation of the complex systems, a very simple system has to be considered which is equipped with just a transport mechanism (or a "pump").

The pump and leak system

As all biological membranes are (more or less) permeable, a system with just a transport mechanism is known as a "pump and leak" system (Wilbrandt and Rosenberg, 1961). As long as the pump is not saturated (i.e., as long as the pump functions with maximum efficiency), a pump and leak system generates a steady-state inside concentration (Si_{ST-ST}) that is proportional to the activity of the pump and inversely proportional to the magnitude of the leak. Moreover, in the absence of saturation of the pump, any 10-fold increase in pump speed increases Si_{ST-ST} 10 fold; or in other words, the ratio Si_{ST-ST}/So (where So = outside substrate concentration) is then independent of So. However, this changes when So rises so much that the pump becomes saturated (i.e., loses its efficiency). The ratio Si_{ST-ST}/So then declines and approaches unity (with increasing So). The midpoint of the resulting sigmoid curve is obtained at So = Km for the pump (see line with solid points in Fig. 1).

In experiments with isolated tissues, Si cannot be measured directly within the relevant compartment. However, we can determine the "steady-state tissue/medium (T/M_{ST-ST}) ratio" (= tissue content per gram/So; in ml/g). If the relevant compartment has the characteristics of a pump and leak system, then the dependence of T/M_{ST-ST} on So is the same as for the dependence of the ratio Si_{ST-ST}/So on So. Hence, in the following (and in Fig. 1), the term T/M_{ST-ST} will be used.

Functionally, a pump and leak system is unable to sustain a removal of a transmitter from the extracellular space. In steady state, the rate of inward transport equals the rate of leakage; or in other words, there is then no *net* inward pumping.

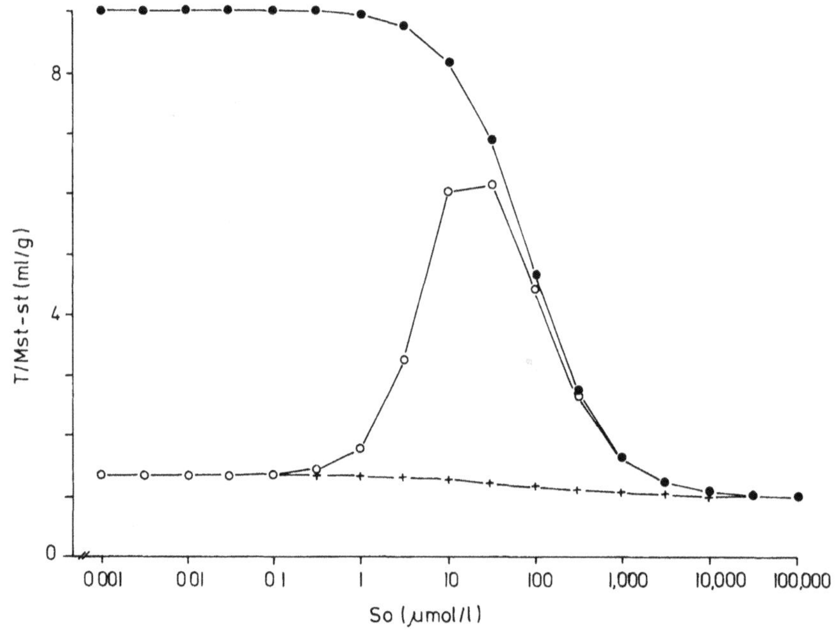

Fig. 1. The dependence of T/M_{ST-ST} on So in a pump and leak system (solid circles), in a pump and leak system with low-Km enzyme inside (open circles), and in a pump and leak system with high-Km enzyme inside (+). Calculations in analogy to those of Kurahashi et al. (1980), but based on equation (2) of Trendelenburg (1988). The pump was assumed to be characterized by $Km_{UPTAKE} = 87$ µmol/l and $V_{max} = 58$ nmol·g^{-1}·min^{-1}, the low-Km_{ENZYME} by $Km_{ENZYME} = 4$ µmol/l and $V_{max} = 2$ nmol·g^{-1}·min^{-1}, the high-Km enzyme by $Km_{ENZYME} = 150$ µmol/l and $V_{max} = 75$ nmol·g^{-1}·min^{-1}. Note that the complex dependence of T/M_{ST-ST} on So is obtained only when Km_{ENZYME} is low relative to Km_{UPTAKE}

The pump and leak system with an enzyme inside

If an unsaturated pump and leak system (that generates a high T/M_{ST-ST}, see above) contains an intracellular enzyme, then the effect of this enzyme on Si is equivalent to a pronounced increase in the size of the leak: T/M_{ST-ST} is the lower the more active the enzyme is.

What happens, if gradual increases in So first lead to saturation of the inside enzyme, and then to saturation of the pump (i.e., if Km_{ENZYME} is low with respect to Km_{UPTAKE})? With the help of the model presented by Kurahashi et al. (1980), such calculations were carried out for $Km_{UPTAKE} = 87$ μmol/l and $Km_{ENZYME} = 4$ μmol/l (i.e., for parameters that correspond to uptake$_2$ of ^{3}H-isoprenaline and its O-methylation by COMT). Figure 1 (open circles) shows that, for non-saturating So, the existence of intracellular COMT lowers T/M_{ST-ST} from about 9 (pump and leak system with no enzyme) to about 1.3 ml/g. As So is increased gradually beyond 0.1 μmol/l, partial saturation of COMT is achieved; hence the enzyme becomes less efficient, and T/M_{ST-ST} begins to rise. Indeed, with increasing So T/M_{ST-ST} more and more approaches the T/M_{ST-ST} obtained when there is *no* enzyme inside. As T/M_{ST-ST} (with *no* enzyme inside) declines, when the pump becomes saturated, T/M_{ST-ST} (*with* COMT inside) first rises and then falls. This curious and rather complex dependence of T/M_{ST-ST} on So very nicely simulates this dependence in experiments with rat hearts perfused with increasing concentrations of ^{3}H-isoprenaline (Fig. 2). Moreover, a very similar dependence of T/M_{ST-ST} on So was very recently observed by Bryan (1990, unpublished observations) in rat lungs perfused with various concentrations of ^{3}H-noradrenaline, where the metabolizing system is restricted to the vascular endothelium (containing uptake$_1$ and COMT – after inhibition of MAO).

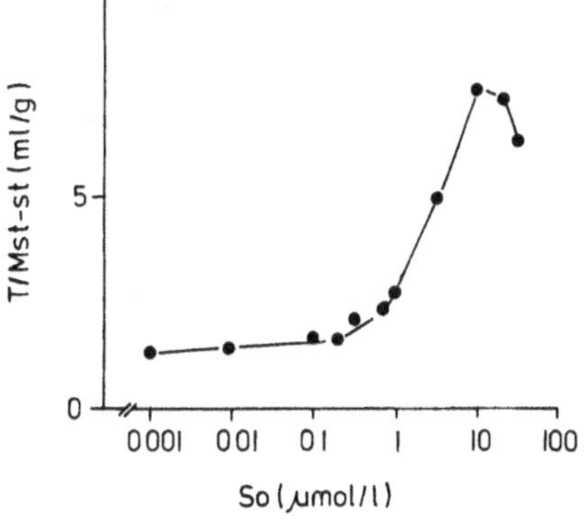

Fig. 2. The dependence of T/M_{ST-ST} on So in rat hearts perfused with ^{3}H-isoprenaline. Note similarity of results with the middle curve of Fig. 1. Results were taken from Kurahashi et al. (1980)

Next, it is of interest to consider what happens when Km_{ENZYME} exceeds Km_{UPTAKE}. Figure 1 (+ — — — — +) shows the result of model calculations based on the entirely hypothetical assumption that the COMT inside the compartment has a high Km (= 150 μmol/l). Figure 1 shows that, when So is gradually increased, saturation of this hypothetical high-Km COMT takes place at such high So that there is hardly any rise of T/M_{ST-ST}; there is simply a gradual decline to a T/M_{ST-ST} of unity. Thus, the curious "hump" (Fig. 2) is observed only if Km_{ENZYME} is low relative to Km_{UPTAKE}.

Such calculations serve to demonstrate that results like those in Fig. 2 can be obtained only if Km_{COMT} for catecholamines is in the low micromolar range. In this context, it is of interest to note this Km for the soluble (cytoplasmic) enzyme is above 100 μmol/l (Guldberg and Marsden, 1975). In recent years more and more evidence has accumulated that there is also a membrane-bound COMT with a substantially lower Km for catecholamines (Rivett and Roth, 1982; Reid et al., 1986). Hence, it is likely that, in the rat heart, the O-methylation of catecholamines is nearly exclusively carried out by membrane-bound COMT.

The lowest curve in Fig. 1 illustrates a second point, for the neuronal deaminating system. As neuronal MAO has a very high activity (see below), axoplasmic noradrenaline levels (corresponding to Si) must be as low as, or lower than, indicated in Fig. 1 for the high-Km_{COMT}. Axoplasmic Si is then a rather constant fraction or multiple of So, over a wide range of So. This explains why Graefe and Eckert (1972) were able to use measurements of rates of ^{3}H-DOPEG formation as a legitimate indicator of initial rates of $uptake_1$ of ^{3}H-noradrenaline. This is possible only if (within the range of concentrations studied) the ratio Si_{ST-ST}/So is virtually independent of So.

As we are unable to directly measure Si, and as the exact fractional size of the metabolizing system is unknown (see below), we are unable to determine the Km of the inside enzyme. However, measurements of the V_{max} of the enzyme are possible, provided $V_{max\,UPTAKE}$ clearly exceeds $V_{max\,ENZYME}$. If the reverse is true, $V_{max\,ENZYME}$ cannot be determined from metabolite formation (as, for instance, for neuronal MAO).

Functionally, a metabolizing system is able to remove the catecholamine from the extracellular space, at least as long as the inside enzyme is not saturated. With increasing saturation of the inside enzyme, the metabolizing system loses this ability, since it then resembles more and more a pump and leak system (see above). Hence, it is of functional interest to determine that So, which half-saturates the inside enzyme (see Grohmann and Trendelenburg, 1985, 1988, for the extraneuronal metabolizing system of the rat heart).

The pump and leak system with two enzymes inside

$Uptake_2$ of noradrenaline in the rat heart translocates the amine into a metabolizing system with *two* enzymes (COMT and MAO). Hence, in the unsaturated system, a rather low T/M_{ST-ST} can be expected. As discussed before, inhibition of one of the two enzymes should increase T/M_{ST-ST}; moreover, this increase should be the more pronounced, the more active the enzyme was prior to inhibition. Figure 3 illustrates that, in the rat heart, inhibition of MAO (by pargyline, P) caused a much smaller increase of T/M_{ST-ST} than did inhibition of COMT (by U-0521, U) (Trendelenburg, 1984b). Thus, as far as the extraneuronal fate of noradrenaline is concerned, unsaturated COMT is more active than is unsaturated MAO.

Enzyme inhibition and rate of metabolite formation

It is tempting to believe that a certain degree of inhibition of an enzyme should lead to a comparable degree of reduction of the formation of the metabolite. However, while this is true after homogenization of the tissue, it is not necessarily true for metabolizing systems (Cassis et al., 1986). The reason is already evident from Fig. 3: inhibition of an active enzyme increases Si, and any increase in Si *must* counteract the effect of partial inhibition of the enzyme. This is best seen in a metabolizing system that generates a very high T/M_{ST-ST}, i.e., in the adrenergic nerve ending.

Vasa deferentia (of rats pretreated with reserpine; COMT inhibited) were exposed to 10 nmol/l ^{3}H-noradrenaline, without or with a prior exposure for 30 min to 10, 20 or 40 μmol/l pargyline (followed by wash-out). (It should be

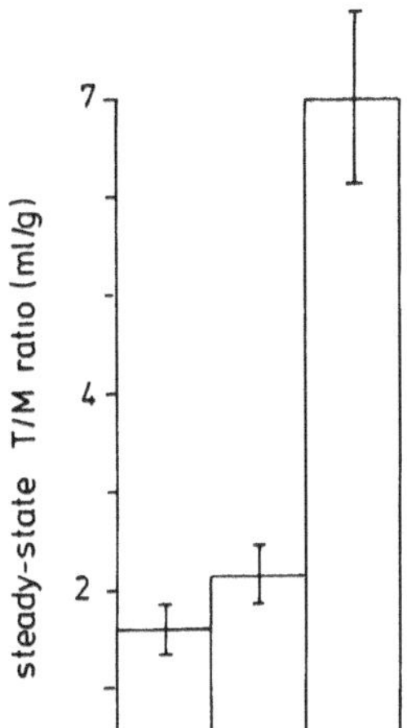

Fig. 3. The effect of inhibition of either extraneuronal MAO (by pargyline, *P*) or of extraneuronal COMT (by U-0521, *U*) on T/M_{ST-ST} in rat hearts perfused with 1 nmol/l ^{3}H-noradrenaline (in the presence of 30 μmol/l cocaine, to block uptake$_1$). *nil* both extraneuronal enzymes uninhibited. Note that such results indicate that, prior to inhibition, extraneuronal COMT was more active than was extraneuronal MAO, when noradrenaline was the substrate. Results taken from Trendelenburg (1984b)

noted that micromolar concentrations of pargyline were used, not the nanomolar ones stated by Cassis et al., 1986). Although tissue homogenates revealed that MAO had been inhibited by about 85%, 94% and > 95% (by 10, 20 and 40 μmol/l pargyline, respectively), Fig. 4A shows that the steady-state rate of ^{3}H-DOPEG formation (which can be equated with the steady-state rate of deamination, since ^{3}H-DOPEG is the only deaminated metabolite of importance) is barely halved after preincubation with 40 μmol/l pargyline (as shown by the slope of the linear part of the lowest curve). The discrepancy between a more than 95% inhibition of MAO and the about 50% reduction of the steady-state rate of deamination is explained by Fig. 4B: the very pronounced inhibition of MAO by 40 μmol/l pargyline went

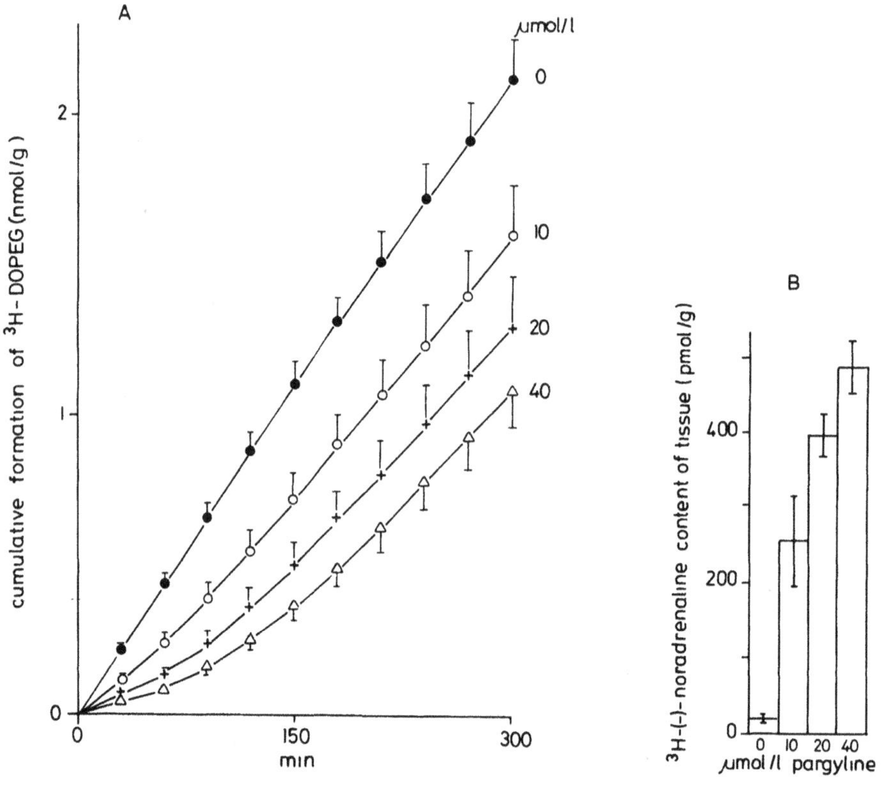

Fig. 4A and B. The effect of partial inhibition of neuronal MAO on the steady-state rate of deamination of ^{3}H-noradrenaline (= steady-state rate of formation of ^{3}H-DOPEG) (in *A*), and on the ^{3}H-noradrenaline content of the tissue (in *B*). Vasa deferentia of reserpine-pretreated rats were exposed for 30 min to 0, 10, 20 or 40 μmol/l pargyline, washed out and then exposed to 10 nmol/l ^{3}H-noradrenaline for 300 min. The steady-state rate of ^{3}H-DOPEG formation (in *A*) is indicated by the slope of the linear part of the curves. After 40 μmol/l pargyline, it was about halved, although the MAO of homogenized tissues was inhibited by about 98% (see text). This discrepancy is due to the about 25-fold increase in ^{3}H-noradrenaline content of tissue (in *B*), after 40 μmol/l pargyline. Results were taken from Cassis et al. (1986)

hand in hand with an about 25-fold increase of Si. Let us assume that 40 μmol/l parglyine resulted in 98% inhibition (or in other words, it reduced the activity of MAO by a factor of 50). If 1/50 of the original enzyme is exposed to a 25-fold increased substrate concentration (and if the enzyme is *not* saturated), then one can expect the steady-state rate of deamination to be 50% of normal. Hence, in metabolizing systems, one *cannot* expect proportionality between degree of enzyme inhibition and degree of reduction of steady-state metabolite formation, and the reason for this discrepancy is found in the fact that inhibition of an inside enzyme inevitably leads to an increase in Si.

Saturation of neuronal MAO inside the metabolizing system

It was emphasized in the preceding section that neuronal MAO was not saturated, in spite of the pronounced increase in Si after inhibition of the enzyme by pargyline. In these experiments, tissues were exposed to a very low concentration of ^{3}H-noradrenaline (10 nmol/l). In other experiments (Stefano and Trendelenburg, 1984), with vasa deferentia of rats pretreated with reserpine and pargyline, loading of the tissues with 10 nmol/l ^{3}H-noradrenaline failed to provide any evidence for saturation of intraneuronal MAO. However, when the loading concentration was increased to 2 μmol/l, then evidence for saturation of neuronal MAO was obtained. Pargyline, as an irreversible inhibitor of MAO, lowers V_{max} of the enzyme without altering the Km for noradrenaline. Thus, in order to saturate MAO, a very high percentage of the enzyme molecules must have been irreversibly inhibited by pargyline; this results in a very high axoplasmic concentration of ^{3}H-noradrenaline. However, saruration of the largely inhibited enzyme is obtained only when the already high axoplasmic concentration of ^{3}H-noradrenaline is further increased by loading with a high concentration of this ^{3}H-amine. This, in turn, indicates that *un*inhibited neuronal MAO is so very active that it can handle whatever noradrenaline is transported into the axoplasm.

This conclusion was supported when it was found that an exposure of the rat vas deferens to 1 mmol/l tyramine (a good substrate of uptake$_1$) failed to saturate neuronal MAO (Langeloh and Trendelenburg, 1987). Saturation of MAO was observed only when the tissue was exposed to 10 mmol/l tyramine. The rate of neuronal uptake of tyramine must be very close to V_{max}, irrespective of whether the outside concentration is 1 or 10 mmol/l. As tyramine is rather lipid soluble (Mack and Bönisch, 1979), it is evident that neuronal MAO was saturated by massive inward *diffusion* of tyramine, not by the inward *transport* of this amine.

In more general terms, it can be stated that the enzymes inside those metabolizing systems that handle noradrenaline, dopamine or 5-hydroxytryptamine are likely to remain unsaturated under most in vivo conditions (unless we deliberately saturate them in our in vitro experiments). This may not be so for uptake$_1$ immediately after the exocytotic release of noradrenaline (which may well reach partly saturating concentrations in the immediate neighbourhood of the varicosity). However, considering plasma levels of circulating catecholamines (even under stress), it is unlikely that the inside enzymes are ever saturated.

The interaction between uptake and inside enzyme

If we consider the fate of extracellular noradrenaline, it is obvious that the rate of uptake (by uptake$_1$ or uptake$_2$) largely determines the rate of intracellular metabolism (by MAO or COMT). However, it is of interest to note that the intracellular enzyme is able to determine the ability of the uptake mechanism to remove noradrenaline from the extracellular space. For this clearance of the extracellular space, *net* uptake is essential. For the neuronal deaminating system of the perfused rabbit heart, it has been shown that *net* uptake of ^{3}H-noradrenaline (= removal from the extracellular space) is constant with time, provided vesicular uptake and MAO are not inhibited. However, net uptake declines with time, if MAO has been inhibited (Graefe et al., 1971); as inhibition of MAO results in a gradual increase in the axoplasmic concentration of noradrenaline, diffusional loss of noradrenaline from the varicosity increases with time, thus reducing *net* uptake in spite of a unidirectional inward transport that remained constant with time. Such observations serve to emphasize the role played by neuronal MAO in keeping the axoplasmic noradrenaline concentration low.

K_{ENZYME} and the fractional size of the meabolizing system

According to the Michaelis-Menten equation

$$v = V_{max} \cdot S/(Km + S). \quad (1)$$

For very low (i.e., non-saturating) substrate concentrations, the equation can be simplified to

$$v = S \cdot V_{max}/Km \quad (2)$$

or

$$v = S \cdot k_{ENZYME}, \quad (3)$$

since

$$k_{ENZYME} = V_{max}/Km. \quad (4)$$

k_{ENZYME} (min^{-1}) is a rate constant characterizing the activity of the unsaturated enzyme. It increases with increasing V_{max} and/or decreasing Km.

As the enzymes of the metabolizing systems discussed here are usually not saturated (see above), determinations of k_{ENZYME} (in intact metabolizing systems) are of great interest.

This is possible, since equation (3) can be adapted to read

$$v_{ST-ST} = Si_{ST-ST} \cdot k_{ENZYME}, \quad (5)$$

where v_{ST-ST} = steady-state rate of metabolite formation and Si_{ST-ST} = steady-state Si.

At this point, it is necessary to discuss the fractional size (f.s.) of the metabolizing systems, i.e. that fraction of the total tissue under study that participates in uptake and metabolism. For instance, autoradiography revealed that ^{3}H-isoprenaline (a good substrate of uptake$_2$ and COMT) distributes (and is O-methylated) in all myocardial and smooth muscle cells of the perfused rat heart (Azevedo et al., 1983). Hence, the f.s. of the extraneuronal metabolizing system of the rat heart is large (i.e. slightly below unity). Moreover, for ^{3}H-noradrenaline it has been shown that extraneuronal COMT and MAO coexist in the *same* extraneuronal compartment (Trendelenburg, 1984b); hence, f.s. may be assumed to be identical for the extraneuronal O-methylating and deaminating systems of the rat heart. For the neuronal metabolizing system, on the other hand, f.s. is very small, since the total volume of all adrenergic varicosities is not more than a very small fraction of total tissue. Hence, f.s. may well be 0.01 or less.

The question then arises whether f.s. of the metabolizing system affects calculations of k_{ENZYME}.

If the f.s. of the metabolizing system is 1.0 (i.e., if *all* cells of the tissue under study participate in the uptake and metabolism of the amine), then the experimentally determined "steady-state amine content of the tissue" is equivalent to Si. Hence, if f.s. = 1.0, k_{ENZYME} can be calculated from the rearranged equation (5):

$$k_{ENZYME} = v_{ST-ST}/Si_{ST-ST}. \quad (6)$$

Let us then assume that we have two isolated tissues (A and B) for which identical steady-state rates of metabolite formation (v_{ST-ST}) and identical tissue contents of the amine under study were obtained (both parameters expressed per gram of total tissue). For tissue A the fractional size (f.s.) of the metabolizing system is assumed to be 1.0 (see above), while the metabolizing system in tissue B has a f.s. of 0.1 (i.e., only *one tenth* of tissue B is responsible for uptake and metabolism of the amine).

If f.s. equals 1.0, the amine content of the tissue amounts to Si_{ST-ST} (see above). Hence, if f.s. equals 1.0, we can write: $k_{ENZYME} = v_{ST-ST}/Si_{ST-ST}$.

However, if f.s. = 0.1, then the amine content is distributed into 1/10 of the tissue, yielding 10 times Si_{ST-ST} within the metabolizing system. Moreover, in that case the steady-state rate of metabolite formation *within the small metabolizing system* equals 10 times v_{ST-ST}. Hence, for a metabolizing system with f.s. = 0.1, we must write: $k_{ENZYME} = 10 \cdot v_{ST-ST}/10 \cdot Si$.

Comparison of these two equations shows that Si_{ST-ST} (inside the metabolizing compartment) and v_{ST-ST} (generated by the metabolizing system) are clearly influenced by the fractional size of the metabolizing system, but k_{ENZYME} is *not* (see Kurahashi et al., 1980, for the mathematical derivation). As a consequence, we can directly compare k_{ENZYME}-values determined in different tissues, for different enzymes and in metabolizing systems of very different f.s.

k_{COMT} and k_{MAO} of the extraneuronal metabolizing system

Rat hearts were perfused with very low, non-saturating concentrations of ^{3}H-catecholamines until the rate of appearance of the relevant ^{3}H-metabolite in the venous effluent reached steady state. $Uptake_1$ was blocked by the presence of 30 µmol/l cocaine (Grohmann, 1987). In these experiments either MAO or COMT was inhibited. Hence, k_{ENZYME} was determined (see equation (6)) as v_{ST-ST}/Si_{ST-ST} for the extraneuronal metabolizing system containing *one* enzyme (either COMT or MAO). Table 1 shows that k_{COMT} declined in the order isoprenaline > adrenaline > dopamine > noradrenaline. It is not known whether such differences are due to differences between Km-values, between V_{max}-values or both. k_{MAO}-values for the extraneuronal enzyme, on the other hand, fell in the order: dopamine > noradrenaline > adrenaline.

It should be noted that, in agreement with the results of Fig. 3, k_{COMT} for noradrenaline was clearly higher than k_{MAO}. For adrenaline, this difference

Table 1. Extraneuronal and neuronal k_{ENZYME}-values (as determined in rat hearts perfused with 50 nmol/l of various ^{3}H-catecholamines)

	A extraneuronal k_{COMT} (min^{-1})	B k_{MAO} (min^{-1})	C neuronal k_{MAO} (min^{-1})
^{3}H-noradrenaline	0.244	0.107	0.349
^{3}H-adrenaline	0.487	0.053	0.107
^{3}H-dopamine	0.312	0.261	0.313
^{3}H-isoprenaline	0.779	–	–

k_{ENZYME} was obtained from V_{ST-ST}/Si_{ST-ST} (i.e., steady-state rate of metabolite formation/steady-state amine content of tissue; see text). Hearts were perfused with 30 µmol/l cocaine (to block $uptake_1$; in A and B) or with 100 µmol/l O-methylisoprenaline (to block $uptake_2$; in C). MAO was blocked by pretreatment with pargyline (in A); COMT was blocked by the presence of 10 µmol/l U-0521 (in B and C). Results were taken from Grohmann (1987).

was even more pronounced; for the dog saphenous vein, Paiva and Guimaraes (1978) demonstrated that the extraneuronal O-methylation of adrenaline greatly exceeds its extraneuronal deamination. Thus, these k_{ENZYME}-values agree well with earlier observations.

It might be tempting to add k_{COMT} and k_{MAO} (for noradrenaline, adrenaline or dopamine), in order to arrive at a k_{ENZYME} for the situation when *both* enzymes are functional. However, if both enzymes (co-existing in *one* compartment) are functional, then MAO deaminates not only noradrenaline but also normetanephrine, while COMT O-methylates noradrenaline, DOPEG and DOMA. Likewise, the aldehyde-metabolizing enzymes then deal with both, the aldehyde and the O-methylated aldehyde. The possible interactions between these multiple substrates of each of the four enzymes (COMT, MAO, aldehyde reductase and aldehyde dehydrogenase) have not yet been explored in detail.

k_{MAO} of the neuronal metabolizing system

Rat hearts were perfused with 50 nmol/l ^{3}H-catecholamines + 100 µmol/l O-methylisoprenaline (to inhibit uptake$_2$); COMT was inhibited (Grohmann, 1987). Under these conditions, v_{ST-ST} for the formation of ^{3}H-metabolites were determined as well as Si_{ST-ST} for the ^{3}H-catecholamines. The resulting (neuronal) k_{MAO}-values are presented in Table 1. k_{MAO}-values for noradrenaline and dopamine are about three times higher than k_{MAO} for adrenaline.

Before the *neuronal* k_{ENZYME}-values are compared with the *extraneuronal* ones (Table 1), a further complication has to be discussed. As pointed out above, extraneuronal tissues are devoid of any ability to "store" catecholamines; hence, these amines may be assumed to be more or less uniformly distributed within the extraneuronal compartment. Varicosities, on the other hand, contain vesicles able to store catecholamines that are then protected from deamination by MAO. The experiments designed to measure k_{ENZYME} were carried out with hearts of reserpine-pretreated rats (to block vesicular uptake). However, even after full block of vesicular uptake by reserpine, ^{3}H-catecholamines are still able to diffuse (from the axoplasm) into storage vesicles where they are then trapped by the low intravesicular pH (Johnson and Scarpa, 1976). According to the analysis of Schömig and Trendelenburg (1987), even after pretreatment with reserpine about 40% of neuronally distributed ^{3}H-noradrenaline is found in the axoplasm (under steady-state conditions), while about 60% are trapped within storage vesicles. This means that a) "total tissue content" greatly overestimates the concentration of noradrenaline in the axoplasmic compartment of the varicosity, and b) k_{MAO} was greatly underestimated. For noradrenaline, the axoplasmic concentration is 2.5 times lower than calculated from "total tissue content"; hence, k_{MAO} is 2.5 times higher than indicated in Table 1. Analogous information about the intraneuronal distribution into the two intraneuronal compartments does not exist for adrenaline and dopamine. However, as these two catecholamines are more lipid soluble than is noradrenaline (Mack and Bönisch, 1979), it is likely that the correction factor may even be somewhat larger than that for noradrenaline.

If the k_{MAO} for noradrenaline is corrected in this way, it amounts to 0.87 min^{-1}; this means that 87% of axoplasmic noradrenaline are deaminated by MAO per minute. This high value supports the view that the activity of neuronal MAO is very high, indeed. As MAO has a rather high Km for noradrenaline (Strolin Benedetti et al., 1983), it is obvious that the V_{max} of neuronal MAO must be very high, exceeding V_{max} of $uptake_1$.

Conclusions

If both, a transport mechanism and an enzyme, are studied in isolation, very little can be predicted about their interactions when they arranged in series in intact cells. However, if their combined effects are interpreted in terms of a "pump and leak system with inside enzyme" (= metabolizing system), then an analysis of the complex interactions is possible. This review may serve to emphasize two main points. 1. In any metabolizing system the (partial or full) inhibition of the enzyme raises Si_{ST-ST}. Hence, proportionality of degree of enzyme inhibition and resulting degree of reduction of metabolite formation cannot be expected. 2. The unknown fractional size of the metabolizing system makes it impossible to determine (in intact cells) Km (and often also V_{max}) of the inside enzyme, but measurements of k_{ENZYME} (= V_{max}/Km) are possible. They are of great interest, since catecholamine-metabolizing enzymes (and especially neuronal MAO) are unlikely to be ever saturated under in vivo conditions.

Acknowledgements

The studies (reported here) of the author's group were supported by the Deutsche Forschungsgemeinschaft (Tr. 96 and SFB 176).

References

Avevedo I, Bönisch H, Osswald W, Trendelenburg U (1983) Autoradiographic study of rat hearts perfused with ^{3}H-isoprenaline. Naunyn-Schmiedebergs Arch Pharmacol 322:1–5

Bryan LJ (1990) Catechol-O-methyltransferase (COMT) in pulmonary endothelial cells of rat perfused lungs has a higher activity than monoamine oxidase (MAO) but is saturated at low concentrations of noradrenaline (NA). J Auton Pharmacol 10:12

Bryan LJ, O'Donnell SR, Westwood NN (1988) The uptake process for catecholamines in endothelial cells in rat perfused lungs is the same as uptake$_1$ in noradrenergic neurones. Br J Pharmacol 95:539P

Bryan LJ, O'Donnell SR, Westwood NN (1989) Further evidence that the uptake process for catecholamines in pulmonary endothelial cells is the same as uptake$_1$ in noradrenergic neurones. Clin Exp Pharmacol Physiol [Suppl] 27

Cassis L, Ludwig J, Grohmann M, Trendelenburg U (1986) The effect of partial inhibition of monoamine oxidase on the steady-state rate of deamination of ^{3}H-catecholamines in two metabolizing systems. Naunyn-Schmiedebergs Arch Pharmacol 333:253–261

Eisenhofer G, Ropchak TG, Stull RW, Goldstein DS, Keiser HR, Kopin IJ (1987) Dihydroxyphenylglycol and intraneuronal metabolism of endogenous and exogenous norepinephrine in the rat vas deferens. J Pharmacol Exp Ther 241:547–553

Graefe K-H, Eckert E (1972) On the stereoselectivity of the neuronal uptake in the cat's nictitating membrane. Naunyn-Schmiedebergs Arch Pharmacol 275:45–68

Graefe K-H, Bönisch H, Trendelenburg U (1971) Time-dependent changes in neuronal net uptake of noradrenaline after pretreatment with pargyline and/or reserpine. Naunyn-Schmiedebergs Arch Pharmacol 271:1–28

Grohmann M (1987) The activity of the neuronal and extraneuronal catecholamine-metabolizing enzymes of the perfused rat heart. Naunyn-Schmiedebergs Arch Pharmacol 336:139–147

Grohmann M, Trendelenburg U (1985) The handling of five catecholamines by the extraneuronal O-methylating system of the rat heart. Naunyn-Schmiedebergs Arch Pharmacol 329:264–270

Grohmann M, Trendelenburg U (1988) The handling of five amines by the extraneuronal deaminating system of the rat heart. Naunyn-Schmiedebergs Arch Pharmacol 337:159–163

Guldberg HC, Marsden CA (1975) Catechol-O-methyl transferase: pharmacological aspects and physiological role. Pharmacol Rev 27:135–206

Halbrügge T, Wölfel R, Graefe K-H (1989) Plasma 3,4-dihydroxyphenylglycol as a tool to assess the role of neuronal uptake in the anaesthetized rabbit. Naunyn-Schmiedebergs Arch Pharmacol 340:726–732

Johnson RG, Scarpa A (1976) Internal pH of isolated chromaffin vesicles. J Biol Chem 251:2189–2191

Kennedy JA, de la Lande IS (1987) Characteristics of the cocaine-sensitive accumulation and O-methylation of ^{3}H-(−)-noradrenaline by rabbit endometrium. Naunyn-Schmiedebergs Arch Pharmacol 336:148–154

Kurahashi K, Rawlow A, Trendelenburg U (1980) A mathematical model representing the extraneuronal O-methylating system of the perfused rat heart. Naunyn-Schmiedebergs Arch Pharmacol 311:17–32

Langeloh A, Trendelenburg U (1987) The mechanism of the ^{3}H-noradrenaline releasing effect of various substrates of uptake$_1$: role of monoamine oxidase and of vesicularly stored ^{3}H-noradrenaline. Naunyn-Schmiedebergs Arch Pharmacol 336:611–620

Mack F, Bönisch H (1979) Dissociation constants and lipophilicity of catecholamines and related compounds. Naunyn-Schmiedebergs Arch Pharmacol 310:1–9

Paiva MQ, Guimaraes S (1978) A comparative study of the uptake and metabolism of noradrenaline and adrenaline by the isolated saphenous vein of the dog. Naunyn-Schmiedebergs Arch Pharmacol 303:221–228

Parker DAS, de la Lande IS, Proctor C, Marino V, Lam NX, Parker I (1987) Cocaine-sensitive O-methylation of noradrenaline in dental pulp of the rabbit:

comparison with the rabbit ear artery. Naunyn-Schmiedebergs Arch Pharmacol 335:32–39

Reid J, Stitzel RE, Head RJ (1986) Characterization of the O-methylation of catechol oestrogens by intact rabbit thoracic aorta and subcellular fractions thereof. Naunyn-Schmiedebergs Arch Pharmacol 334:17–28

Rivett AJ, Roth JA (1982) Kinetic studies on the O-methylation of dopamine by human brain membrane-bound catechol-O-methyltransferase. Biochemistry 21: 1740–1742

Schömig E, Trendelenburg U (1987) Simulation of outward transport of neuronal ^{3}H-noradrenaline with the help of a two-compartment model. Naunyn-Schmiedebergs Arch Pharmacol 336:631–640

Stefano FJE, Trendelenburg U (1984) Saturation of monoamine oxidase by intraneuronal noradrenaline accumulation. Naunyn-Schmiedebergs Arch Pharmacol 328:135–141

Strolin Benedetti M, Boucher T, Fowler CJ (1983) The deamination of noradrenaline and 5-hydroxytryptamine by rat brain and heart monoamine oxidase and their inhibition by cimoxatone, toloxatone and MD 770222. Naunyn-Schmiedebergs Arch Pharmacol 323:315–320

Trendelenburg U (1980) A kinetic analysis of the extraneuronal uptake and metabolism of catecholamines. Rev Physiol Biochem Pharmacol 87:33–115

Trendelenburg U (1984a) Metabolizing systems. In: Fleming WW, Graefe K-H, Langer SZ, Weiner N (eds) Neuronal and extraneuronal events in autonomic pharmacology. Raven Press, New York, pp 93–109

Trendelenburg U (1984b) The influence of inhibition of catechol-O-methyl transferase or of monoamine oxidase on the extraneuronal metabolism of ^{3}H-(−)-noradrenaline in the rat heart. Naunyn-Schmiedebergs Arch Pharmacol 327: 285–292

Trendelenburg U (1986) The metabolizing systems involved in the inactivation of catecholamines. Naunyn-Schmiedebergs Arch Pharmacol 332:201–207

Trendelenburg U (1988) The extraneuronal uptake and metabolism of catecholamines. In: Trendelenburg U, Weiner N (eds) Catecholamines I. Springer, Berlin Heidelberg New York (Handbook exp pharmacol, vol 90/I, pp 279–319)

Wilbrandt W, Rosenberg T (1961) The concept of carrier transport and its corollaries in pharmacology. Pharmacol Rev 13:109–183

Author's address: Prof. Dr. U. Trendelenburg, Department of Pharmacology and Toxicology, University of Würzburg, Versbacher Strasse 9, D-8700 Würzburg, Federal Republic of Germany.

J Neural Transm (1990) [Suppl] 32: 19–27

In-vivo quantitative imaging of catecholaminergic nerve terminals in brain and peripheral organs using positron emission tomography (PET)

I. J. Kopin

NINDS, NIH, Bethesda, Md, U.S.A.

Summary. Positron emission tomogaphy (PET), has made possible quantitative imaging of the origins of positron-emitting isotopes, such as ^{11}C and ^{18}F, in intact animals and in humans. Lack of absolute specificity of enzymes, storage mechanisms and transporters allows ^{11}C-or ^{18}F-labelled "false transmitters" to be formed, stored and released from nerve terminals. Discussed are the assumptions, limitations, and advantages of ^{18}F-6-fluoroDOPA, ^{18}F-6-fluorodopamine, ^{18}F-6-fluorometaraminol, and ^{11}C-N-methyl-metaraminol (m-hydroxyephedrine) for PET imaging of dopaminergic nerve terminals in brain or sympathetic innervation in peripheral organs.

Introduction

Isotopes have been used for over fifty years to trace metabolic pathways and to analyze the kinetics of biochemical and physiological processes. About thirty-five years ago, introduction of liquid scintillation spectrometry simplified assay of beta-emitting isotopes and thus provided a strong impetus to the use of carbon-14 (^{14}C) and tritium (^{3}H) labelled compounds to examine the metabolism and disposition of important endogenous substances and of drugs. Autoradiographic methods have proven useful for quantitative imaging of radioactively labelled compounds in tissues, so that their distribution and sites of retention can be determined. The weak radiations from ^{14}C or ^{3}H penetrate tissues so short a distance that these isotopes can be assayed or imaged only in vitro. With the invention of computer assisted tomography, it became possible to image the sources of high energy, penetrating radiations in intact animals and in humans and to quantify time dependent changes in the concentration of the radiation source in a volume of interest. The radiation resulting from annihilation of a positron originates within a short distance of the decaying atom and consists of two equally intense gamma rays directed at an angle of 180 degrees from each other. Coincidence detection of the two oppositely directed gamma rays enhances spatial resolution; images obtained with positron-emitting isotopes such as carbon-11 (^{11}C), nitrogen-13 (^{13}N), oxygen-15 (^{15}O), and fluorine-18 (^{18}F) are generally superior to those obtained with radionuclides which emit other types of radiations directed randomly in space.

The short half-lives of positron-emitting isotopes, however, create some difficulties which require special attention and place important constraints on their use. The isotopes must be produced within a reasonable distance from the site at which they are to be used. The positron-emitting isotopes that would be most useful for labelling endogenous substances have very short half-lives (^{15}O has a half-life of about 2 min, ^{13}N about 10 min, and ^{11}C about 20 min). Synthesis and purification of labelled compounds has usually required too much time to make practical use of any but the simplest substances (e.g., $H_2{}^{18}O$), but use of ^{11}C is possible because there are available methods for rapid production of several ^{11}C-labelled compounds, particularly those which can be produced using ^{11}C-methyl iodide. Labelling compounds with ^{18}F (half-life of about 2 hours) is more attractive but the biological behavior of the fluorinated analogue must be examined to ensure that it adequately reflects the process being studied. The half-life of the radionuclide in relation to the turnover rate of the process to be examined is another important consideration. It is impractical to attempt to study kinetics of a biologic process that has a half-life of days using a radionuclide having a half-life of minutes because a very large quantity of isotope would be required and the radiation dose would be excessive.

In vivo imaging of appropriately labelled compounds using positron-emission tomography (PET) has developed rapidly and proven extremely useful, but radiopharmaceuticals for quantitative imaging by PET must be selected carefully. As indicated above, the biological behavior of a chemical analogue of the substance being studied must be mimicked by the substitute molecule. Futhermore, because total radioactivity rather than a specific compound is imaged, the chemical identities of the photon source molecules in the region of interest must be established in studies of animal tissues.

Because of the lack of strict specificity of the biosynthetic enzymes, transporters, storage processes and release mechanisms for catecholamines, substitute "false transmitters" can be formed or taken up and stored in various catecholamine-containing tissues (adrenal medulla, noradrenergic and dopaminergic neurons). The concept of false transmitters developed as a result of the search for the mechanism of action of several drugs which influence adrenergic function. Sourkes et al. (1954) described α-methyl-DOPA as a competitive inhibitor of dopa decarboxylase, but the antihypertensive action of this compound (Oates et al., 1960) could not be accounted for by inhibition of this enzyme. Carlsson and Lindquist (1962), showed that α-methyl-DOPA was metabolized to α-methyl analogues of dopamine and norepinephrine which displace the normal transmitters in brain. Day and Rand (1963) proposed that the weaker agonist, α-methylnorepinephrine (α-methyl-NE), replaced and was released instead of the physiological transmitter. This false transmitter hypothesis was confirmed by Muscholl and Maitre (1963) when they showed that α-methyl-NE formed from α-methyl-DOPA was released by sympathetic nerve stimulation. Metaraminol, which is formed after administration of α-methyl-m-tyrosine (Carlsson and Lindquist, 1962) or taken up from the circulation, was also shown to be released by nerve stimulation (Crout et al., 1964). A wide variety of other phenylethylamine derivatives were then shown to be formed or taken up in the vesicles of sympathetic nerve terminals (Musacchio et al., 1965) and released during sympathetic nerve stimulation (Fischer et al., 1965).

Using appropriate false transmitter-like substances labelled with positron-emitting isotopes, it has been possible to image catecholaminergic neurons. It is the purpose of this communication to describe the current status of progress in using PET to image, in intact animals and in humans, the integrity and function of dopaminergic neurons in brain and noradrenergic neurons in the peripheral sympathetic nervous system.

Imaging striatal dopaminergic neurons

Because catecholamines do not penetrate the blood-brain barrier, replacement of dopamine which is deficient in brain of patients with Parkinson's disease requires use of L-DOPA, the amino acid precursor of this neurotransmitter. L-DOPA, usually administered with a peripheral decarboxylase inhibitor, is transported across the blood-brain barrier and is decarboxylated to form dopamine. As indicated above, the decarboxylase is not specific and will decarboxylate a variety of amino acids, including analogues of L-DOPA such as 5- or 6-fluorodopa. Garnett et al. (1978) followed the brain content of ^{18}F by external monitoring of gamma radiation from the head of baboons given ^{18}F-5-fluoroDOPA intravenously. They demonstrated that the ^{18}F accumulation was diminished by reserpine and enhanced by treatment with a monoamine oxidase inhibitor or haloperidol suggesting that ^{18}F-fluoroDOPA was transported into brain, decarboxylated and stored as fluorodopamine. In support of this, they showed that in rats which had unilateral 6-hydroxydopamine lesions of the substantia nigra, fluoroDOPA produced turning behavior similar to that produced by DOPA. Because the 5-fluoro analogue of DOPA is O-methylated more rapidly than the 6-fluoro derivative, they turned to ^{18}F-6-fluorodopa and using PET, visualized the basal ganglia in monkeys (Garnett et al., 1983) and in humans (Garnett et al., 1983). As expected, fluoroDOPA was found to resemble L-DOPA biochemically; it crosses the blood-brain barrier with the similar kinetics, is decarboxylated and the dopamine analogue so formed appears to be stored in vesicles (Firnau et al., 1986, 1987). Furthermore, the rate of O-methylation of ^{18}F-6-fluoro-L-DOPA by catechol-O-methyl transferase was found to be only 1/4 that of L-DOPA. After ^{18}F-6-fluoro-L-DOPA and ^{14}C-L-DOPA were injected simultaneously into rhesus monkeys, the time courses of the changes in striatal concentrations of their metabolites were similar. ^{18}F-6-fluoro-DOPA was metabolized to 6-fluorodopamine (6F-DA), 3,4-dihydroxy-6-fluorophenylacetic acid (6-F-DOPAC), and 6-fluorohomovanillic acid (6F-HVA), corresponding to the metabolites of DOPA. ^{18}F-6-FDA was synthesized more rapidly than was ^{14}C-dopamine and reached a peak concentration at about 20 minutes after its administration. At that time, 6F-DA accounted for the major fraction of the ^{18}F in the striatum and 6F-DOPAC was the major metabolite. After about 90 min, ^{18}F-HVA levels exceeded those of 6F-DA and of 6F-DOPAC. In areas of brain, such as the occipital cortex, in which there is little endogenous dopamine, the major ^{18}F-containing substance was 3-O-methyl-6F-DOPA. This compound is the major metabolite of 6F-DOPA in human plasma, can cross the blood-brain barrier, and, along with the ^{18}F-dopamine metabolites which are not immediately removed into the circulation, contribute to the total ^{18}F content of the striatum (Firnau et al., 1988). The presence of these ^{18}F-labelled compounds precludes precise measurement of the rate of decline in concentration of ^{18}F-dopamine.

Uptake and accumulation of ^{18}F, however, reflects the capacity of the striatum to decarboxylate ^{18}F-6-fluoroDOPA and to store the resulting ^{18}F-fluorodopamine. Although not all the ^{18}F is fluorodopamine, at early times (20–60 min) total radioactivity does appear to reflect integrity of the dopaminergic terminals. Since both 6F-DOPA and its O-methylated derivative are transported into brain by the large neutral amino acid transport system, an increase in the plasma levels of these amino acids will markedly depress accumulation of ^{18}F after administration of ^{18}F-6F-DOPA (Leenders et al., 1986b); clearly control of this factor is of importance when PET studies with ^{18}F-fluoroDOPA are planned.

PET imaging of the striatum with ^{18}F-6-fluoroDOPA has been applied by several groups to demonstrate the degree of striatal dopaminergic neuronal damage in MPTP-treated monkeys (Chiueh et al., 1987; Doudet et al., 1989), in humans exposed to MPTP (Calne et al., 1985), and in patients with Parkinson's disease (Leenders et al., 1986a) or with parkinsonism secondary to manganese intoxication (Wolters et al., 1989) or cyanide poisoning (Roseberg et al., 1989). Since the interval during which uptake, accumulation, and decarboxylation of DOPA and its fluorinated analogue to the corresponding amines is relatively short, ^{11}C-DOPA has recently been introduced to image the striatum (Aquilonius et al., 1989). Use of ^{11}C has the advantages of lower total radiation exposure with better imaging. Furthermore, the physiological dopamine precursor, rather than a substitute analogue, is used.

Imaging peripheral sympathetic innervation

Amine uptake by sympathetic nerve terminals was first discovered over twenty-five years ago when it was found that a major portion of ^{3}H-norepinephrine administered to animals was retained in those tissues which have high endogenous concentrations of norepinephrine. Since sympathetic denervation (Hertting et al., 1961) greatly reduced tissue uptake and retention of ^{3}H-norepinephrine, it was concluded that the sympathetic nerve terminals were the site of amine uptake. Direct autoradiographic demonstration of administered ^{3}H-norepinephrine in sympathetic nerves in the heart (Samorajski et al., 1964) confirmed this conclusion. In addition to norepinephrine uptake by neurons (Uptake$_1$), in most tissues there is a second extraneuronal uptake system (Uptake$_2$) which becomes apparent at high norepinephrine concentrations (Iversen, 1965a). Uptake and retention of tracer amounts of radio-labelled norepinephrine into the sympathetic neurons of various tissues depends upon its delivery via the perfusing blood, its diffusion through the tissue to the region of the nerve terminals and the density of sympathetic nerve terminals (Kopin et al., 1965). Thus, the heart which is both richly perfused and densely innervated by sympathetic nerves accumulates ^{3}H-norepinephrine from the circulation more rapidly and in higher concentrations than any other tissue.

It has long been recognized that the norepinephrine uptake processes are not specific and that many sympathomimetic amines are substrates for Uptake$_1$. It would be expected that these amines, like norepinephrine, would be concentrated in the myocardial sympathetic neurons and in other tissues which are well perfused and densely innervated with sympathetic neurons. Labelling with ^{11}C or ^{18}F of amines which are accumulated rapidly via the norepinephrine uptake system and retained in the storage vesicles might afford the possiblity of quantitative imaging with PET of adrenergic nerve terminals, analysis of the kinetics of the uptake process and assessment of the integrity of sympathetic innervation in peripheral tissues of intact animals and of humans. Several false adrenergic transmitters are released and metabolized, as well as transported and stored, by the same mechanisms as norepinephrine (Kopin, 1968). The rates of decline in levels of such substances reflect norepinephrine turnover and sympathetic activity in addition to providing information about the integrity of the storage mechanisms in the sympathetic terminals.

Metaraminol was one of the first false adrenergic transmitters to be identified (Crout et al., 1964). Because it does not contain a catechol moiety, it will not undergo O-methylation and because of its α-methyl group, it is resistant to deamination. Furthermore, it differs from norepinephrine in its binding affinities. The affinity of metaraminol, as well as other monophenols, for vesicular storage sites is lower than that of catecholamines (Musacchio et al., 1965; Lundborg and Stitzel, 1967), but its affinity for Uptake$_1$ is greater than that of norepinephrine (Iversen, 1965b). Thus the accumulation rate of metaraminol appears to be a better index of $Uptake_1$ than is norepinephrine (Herman and Graefe, 1977).

Wieland et al. (1990) showed that metaraminol labelled with ^{18}F in the 6-position rapidly accumulates in the heart where it is retained avidly for the duration of adequate ^{18}F detection (4 hours). In dog hearts regionally denervated by phenol treatment, uptake and retention of ^{18}F was highly correlated with residual endogenous norepinephrine concentration. Chemical sympathetic denervation with 6-hydroxydopamine, blockade of $Uptake_1$ with desipramine or pretreatment with reserpine markedly diminished the amount of ^{18}F found in the rat ventricle 90 min after injection of ^{18}F-6-fluorometaraminol. These observations provide convincing evidence that the labelled amine is taken up selectively and stored in sympathetic nerve terminals in the heart. Although good myocardial imaging of ^{18}F was obtained in dogs, the relatively low specific activity of ^{18}F required administration of physiologically active quantities of the labelled compound at a dose which might be unsafe for use in cardiac patients (Wieland et al., 1990). As an alternative to ^{18}F-6-fluorometaraminol, the N-methyl analogue of metaraminol, *m*-hydroxyephedrine, which can readily be prepared from metaraminol using ^{11}C-methyl iodide, has been used to image the myocardial sympathetic nerves (Schwaiger et al., 1990). This ^{11}C-labelled compound was obtained with sufficiently high specific activity to circumvent pharmacological effects and appears to be a promising new approach for PET imaging of neuronal amine uptake and storage and thus integrity of sympathetic nerve terminals. The metabolic stability and avid uptake which enhance the value of metaraminol and its derivatives for assessing $Uptake_1$ may, however, limit their ability to reflect the turnover rate of endogenous norepinephrine.

^3H-norepinephrine which is taken up and stored in nerve terminals can be released by nerve stimulation (Hertting and Axelrod, 1961), and ganglionic blockade slows release of the labelled amine from the heart (Hertting et al., 1962). These results suggested that the turnover rate of norepinephrine could be used as an index of sympathetic neuronal function. Studies of the kinetics of ^3H-norepinephrine in tissues of animals have provided a wealth of information about alterations in adrenergic activity during stress, after drug administration or as a result of induced or genetic disorders. It has not been possible, however, to examine directly the kinetics of norepinephrine turnover in human tissues. PET imaging with appropriate labelled compounds, however, may provide opportunities to obtain such information.

Because of their potential use for PET imaging, fluorinated catecholamines were examined as possible false adrenergic transmitters which might reflect the turnover of norepinephrine (Chiueh et al., 1983; Eisenhofer et al., 1989). In these studies it was found that 6-fluorodopamine is taken up into sympathetic nerve terminals where it is largely β-hydroxylated to 6-fluoronorepinephrine and retained in the storage vesicles. The fluorinated catecholamines are released from the perfused cat spleen during sympathetic nerve stimulation. In the rat heart, the rates of decline of tritiated and of fluorinated norepinephrine appear to be almost identical (Chiueh et al., 1983), suggesting that ^{18}F-6 fluoronorepinephrine can provide an index of norepinephrine turnover. In dogs ^{18}F-6 fluorodopamine has been used successfully for PET imaging of sympathetic neurones in the heart, spleen and salivary gland (Goldstein et al., 1990). Sympathetic denervation or pharmacological blockade of neuronal uptake by administration of desipramine markedly reduced uptake of the fluorine tracer, whereas myocardial levels of radio-labelled metaraminol derivatives reach a plateau. After administration of ^{18}F-6-fluorodopamine, myocardial ^{18}F declines significantly during an interval of only a few hours. Since the rate of decline in ^{18}F levels was slowed by ganglionic blockade and accelerated by infusion of tyramine, fluorinated catecholamines appear to hold promise for assessing sympathetic nerve function by PET quantitative imaging.

In conclusion, it is apparent that with advances in technology which permit quantitative imaging of positron-emitting isotopes, it is becoming possible to examine in intact animals and in humans the integrity and function of catecholaminergic neurons. PET imaging of neurotransmitters and in vivo estimation of their turnover rates will provide an important means for examining in humans abnormalities in function of the aminergic nervous system in disease states and in response to drugs.

References

Aquilonius SM, Langstrom B, Tedroff J (1989) Brain dopaminergic mechanisms in Parkinson's disease evaluated by positron emission tomography. Acta Neurol Scand 126:55–59

Calne DB, Langston JW, Martin WR, Stoessl AJ, Ruth TJ, Adam MJ, Pate BD, Schulzer M (1985) Positron emission tomography after MPTP: observations relating to the cause of Parkinson's disease. Nature 317:246–8

Carlsson A, Lindquist M (1962) In-vivo decarboxylation of α-methyl dopa and α-methyl metatyrosine. Acta Physiol Scand 54:87–94

Chiueh CC, Zukowska-Grojec Z, Kirk KL, Kopin IJ (1983) 6-Fluorocatecholamines as false adrenergic transmitters. J Pharmacol Exp Ther 225:529–533

Chiueh CC, Firnau G, Burns RS, Nahmias C, Kopin IJ, Garnett ES (1987) Determination and visualization of damage to striatal dopaminergic terminals in 1-methyl-4-phenyl-1,2,3,6-tetrahydropyridine-induced parkinsonism by [^{18}-F]-labeled 6-fluoro-L-dopa and positron emission tomography. Adv Neurol 45: 167–169

Crout JR, Alpers HS, Tatum EL, Shore PA (1964) Release of metaraminol from the heart by sympathetic nerve stimulation. Science 145:828–829

Day MD, Rand MJ (1963) A hypothesis for the mode of action of α-methyldopa in relieving hypertension. J Pharm Pharmacol 15:221–224

Doudet DJ, Myack H, Finn RT, McLellan CA, Aigner TG, Wan RQ, Adams HR, Cohen RM (1989) 6-^{18}F-L-dopa imaging of the dopamine neostriatal system in normal and clinically normal MPTP-treated rhesus monkeys. Exp Brain Res 78(1):69–80

Eisenhofer G, Hovevey-Sion D, Kopin IJ, Miletich R, Kirk KL, Finn R, Goldstein DS (1989) Neuronal uptake and metabolism of 2- and 6-fluorodopamine. False transmitters for positron emission tomographic imaging of sympathetically innervated tissues. J Pharmacol Exp Ther 248:419–427

Firnau G, Garnett ES, Chirakal R, Sood S, Nahmias C, Schrobilgen G (1986) [^{18}F] fluoro-L-dopa for the in vivo study of intracerebral dopamine. Int J Rad Appl Instrum [A] 37(8):669–675

Firnau G, Sood S, Chirakal R, Nahmias C, Garnett ES (1987) Cerebral metabolism of 6-[^{18}F]fluoro-L-3,4-dihydroxyphenylalanine in the primate. J Neurochem 48(4):1077–1082

Firnau G, Sood S, Chirakal R, Nahmias C, Garnett ES (1988) Metabolites of 6-[^{18}F]fluoro-L-dopa in human blood. J Nucl Med 29(3):363–369

Fischer JE, Horst WD, Kopin IJ (1965) β-Hydroxylated sympathomimetic amines as false neurotransmitters. Br J Pharmacol 24:477–484

Garnett ES, Firnau G, Chan PKH, Sood S, Belbeck LW (1978) [^{18}F] Fluorodopa, an analogue of dopa, and its use in direct external measurements of storage degradation and turnover of intracerebral dopamine. Proc Natl Acad Sci USA 75:464–467

Garnett ES, Firnau G, Nahmias C (1983) Dopamine visualized in the basal ganglia of living man. Nature 205:137–138

Garnett ES, Firnau G, Nahmias C, Chirakal R (1983) Striatal dopamine metabolism in living monkeys examined by positron emission tomography. Brain Res 280:169–171

Goldstein DS, Chang PC, Eisenhofer G, Miletich R, Finn R, Bacher J, Kirk KL, Bacharach S, Kopin IJ (1990) Positron emission tomographic imaging of cardiac sympathetic innervation and function. Circulation 81:1606–1621

Guttman M, Yong VW, Kim SU, Calne DB, Martin WR, Adam MJ, Ruth TJ (1988) Asymptomatic striatal dopamine depletion: PET scans in unilateral MPTP monkeys. Synapse 2:469–473.

Herman W, Graefe KH (1977) Relationship between the uptake of ^{3}H-Metaraminol and the density of adrenergic innervation in isolated rat tissues. Arch Pharm 296:99–110

Hertting G, Axelrod J (1961) Fate of tritiated norepinephrine at the sympathetic nerve ending. Nature 192:172–173

Hertting G, Axelrod J, Kopin IJ, Whitby LC (1961) Lack of uptake of catecholamine after chronic denervation of sympathetic nerves. Nature 189:66

Hertting G, Potter LT, Axelrod J (1962) Effect of decentralization and ganglion blocking agents on the spontaneous release of ^{3}H-norepinephrine. J Pharmacol Exp Ther 136:289–292

Iversen LL (1965a) The uptake of catecholamines at high perfusion concentrations in the rat isolated heart: a novel catecholamine uptake process. Br J Pharmacol 25:18–33

Iversen LL (1965b) Inhibition of noradrenaline uptake by drugs. J Pharm Pharmacol 17:52–64

Kopin IJ (1968) False adrenergic transmitters. Ann Rev Pharmacol 8:377–394

Kopin IJ, Gordon EK, Horst WD (1965) Studies of uptake of L-norepinephrine-^{14}C. Biochem Pharmacol 14:753–759

Leenders KL, Palmer AJ, Quinn N, Clark JC, Firnau G, Garnett ES, Nahmias C, Jones T, Marsden CD (1986a) Brain dopamine metabolism in patients with Parkinson's disease measured with positron emission tomography. J Neurol Neurosurg Psychiatry 49:853–860

Leenders KL, Poewe WH, Palmer AJ, Brenton DP, Frackowiak RS (1986b) Inhibition of L[^{18}F]fluorodopa uptake into human brain amino acids demonstrated by positron emission tomography. Ann Neurol 20:258–262

Lundborg P, Stitzel R (1967) Uptake of biogenic amines by two different mechanisms present in adrenergic granules. Br J Pharmacol 39:342–249

Musacchio JM, Kopin IJ, Weise VK (1965) Subcellular distribution of some sympathomimetic amines and their β-hydroxylated derivatives in the rat heart. J Pharmacol Exp Ther 148:22–28

Muscholl E, Maitre L (1963) Release by sympathetic stimulation of α-methylnoradrenaline stored in the heart after administration of α-methyldopa. Experientia 19:658–659

Oates JA, Gillespie I, Udenfriend S, Sjoerdsma A (1960) Decarboxylase inhibition and blood pressure reduction by α-methyl-3,4-dihydroxy-D,L-phenylalanine. Science 131:1890–1891

Rosenberg NL, Myers JA, Martin WR (1989) Cyanide-induced parkinsonism: clinical, MRI, and 6-fluorodopa PET studies. Neurology 39:142–144

Samorajski T, Marks BN, Webster EJ (1964) An autoradiographic study of the uptake and storage of norepinephrine-^{3}H in tissues of mice treated with reserpine and cocaine. J Pharmacol Exp Ther 143:82–89

Schwaiger M, Hutchins GD, Rosenspire K, Haka MS, Weiland DM (1990) Quantitative evaluation of the sympathetic nervous system in patients with cardiomyopathy. J Nucl Med 31:792

Sourkes TL (1954) Inhibition of dihydroxyphenylalanine decarboxylase by derivatives of phenylalanine. Arch Biochem Biophys 51:444–456

Wieland DM, Rosenspire KC, Hutchins GD, Vandort M, Rothley JM, Mislankar SG, Lee HT, Massin CC, Gildersleeve DL, Sherman PS, Schwaiger M (1990) Neuronal mapping of the heart with 6-[^{18}F]fluorometaraminol. J Med Chem 33:956–964

Wolters EC, Huang CC, Clark C, Peppard RF, Okada J, Chu NS, Adam MJ, Ruth TJ, Li D, Calne DB (1989) Positron emission tomography in manganese intoxication. Ann Neurol 26(5):647–651

Author's address: I.J. Kopin, M.D., NINDS, National Institute of Health, Bethesda, MD 20892, U.S.A.

J Neural Transm (1990) [Suppl] 32: 29–37

The mechanism of action of antidepressants revised

M. Ackenheil

Department of Neurochemistry, Psychiatric Hospital of the University of Munich, Federal Republic of Germany

Summary. The discovery of the clinical efficacy of imipramine and of the MAO-inhibitor iproniazid intensively stimulated biochemical-pharmacological research on the mechanism of action of antidepressants. Due to these investigations, until recently an enhanced activity of the central noradrenergic and/or serotonergic transmitter system was considered essential for the clinical antidepressive action. Such enhancement could be achieved either presynaptically by blocking α_2-adrenergic receptors, or in the synaptic cleft by inhibiting the transmitter reuptake or the main metabolic enzyme, MAO. The common final result, especially of chronic treatment, was the down-regulation of postsynaptic β-receptors, modulated by interaction with the serotonergic system, neuropeptides, and hormones. The delay of clinical response corresponded better with such receptor alterations. However, the introduction of new, more selective antidepressants led to new reflections upon the mechanism of action. On the level of transmitters, α_1-up-regulation, increased activity of the dopaminergic system, an alteration in the balance between the different transmitter systems, are reported and seem to be important. Most promising are recent investigations of the second messenger systems, the adenylate cyclase system and the phosphatidylinositol system. Both systems are modulated by antidepressant drugs including lithium and carbamazepine. These second messengers, in turn, modulate the phosphorylation status of neuronal proteins via protein kinase, which may lead to elevations of the above mentioned receptors and again their transduction systems.

Introduction

The discovery of the antidepressant activity of imipramine (Kuhn, 1957) and of the monoamine oxidase (MAO) inhibitor iproniazide (Crane, 1957) enormously stimulated biological investigations on the pathophysiology of depressive illnesses and the mechanism of action of antidepressants. The observation of the frequent occurence of depressive symptoms after treating hypertension with reserpine, which depletes the vesicles in nerve endings from catecholamine and serotonin (5-HT), lead to the development of the catecholamine (Schildkraut, 1965; Matussek, 1969) and serotonin (Coppen, 1967) hypotheses of depression. It should be mentioned that some other hypotheses, such as the GABA and acetylcholine-hypotheses, are postulated, too. In relation to the mechanism of action of antidepressants this article mainly focusses on the 5-HT, DA and second messenger systems without, however, claiming to be complete. Tricyclic antidepressants reverse the behavioural symptoms such as catalepsy of reserpine in animals and this effect is related to the reuptake inhibition of norepinephrine (NE) and 5-HT into the nerve endings. This reuptake inhibition, which is an important inactivating step, results in an enrichment of the two transmitters NE and 5-HT in the synaptic cleft and thereafter the stimulation of postsynaptic α- and β-adrenergic and 5-HT receptors. A similar effect could be achieved with MAO inhibitors by affecting the catabolism of both NE and 5-HT. For many years this effect was considered to be the precondition for antidepressant activity. Controverse discussions arose only with regard to the relative potency of inhibiting NE or 5-HT reuptake. These controverse discussions were complicated by the fact that tricyclic antidepressants were metabolized, e.g. demethylated in the body and, frequently, the active metabolite had a different potency on these inhibitory effects. However, neither the catecholamine nor the serotonin hypothesis could be confirmed in depressive patients. As reasons for this failure were discussed the heterogeneity of the illness, e.g. neurotic vs. endogenous depressives, and adaptions of the transmitter systems. MHPG and 5-HIAA, the metabolites of the two amines NE and 5-HT, were influenced similarly (Potter et al., 1985) and could neither clarify the discussion. Clinically, the onset of the antidepressant action differs from the biochemical effects. The uptake inhibition occurs suddenly, whereas the therapeutic effect needs two to three weeks.

The development of new antidepressants, e.g. mianserine, having no reuptake inhibition lead to further doubts concerning the validity of this concept. In the meantime Vetulani and Sulser (1975) demonstrated that treatment of animals with tricyclic antidepressants resulted in a decrease of the stimulation of cAMP by isoproterenol, which they termed β-down-regulation. This β-down-regulation was characteristic for all antidepressants known and occurred with a time delay, thus corresponding better with the clinical antidepressant activity. Even the controversy concerning the catecholamine and 5-HT hypotheses could be explained, because β-down-regulation requires an intact 5-HT system (Racagni and Brunello, 1984). The discovery of β-down-regulation together with neuroendocrine findings showing a decreased sensitivity of α_2-receptors (Matussek et al., 1980) and, probably, an increased sensitivity of β-receptors in depressive patients stimulated the interest in transmitter-receptor interaction. The discovery of presynaptic autoreceptors (Langer, 1978) could explain the clinical effectivity of such antidepressants like mianserine showing no reuptake inhibition. Presynaptic autoreceptors regulate the release of transmitters. Therefore, the pharmacological manipulation of such receptors could increase the amount of the transmitter via a blockade of presynaptic α_2-receptors. The discrepancy between the time course of biochemical effects and onset of clinical therapeutic effects could not be explained. Furthermore, attempts to evaluate the β-down regulation in humans had not been successful. The measurement of melatonin, considered to be an indicator of β-receptor sensitivity before and after treatment with different antidepressives, did not demonstrate alterations of melatonin secretion (Checkley, 1988).

Recent development of new antidepressants reanimated the discussion and again induced further progress. The reversible MAO inhibitors moclobemide and brofaromine liberated the principle of MAO inhibition from the manco of being too dangerous drugs with too many side effects. New clinical results could demonstrate that they were as effective as classical antidepressants, even the so-called therapy resistant depressions showed a positive response to such treatment (Schmauss et al., in preparation). So-called specific reuptake inhibitors, such as fluoxetine and fluvoxamine, again stimulated the investigation of the 5-HT system and its importance for antidepressant action. Specific NE reuptake inhibitors, such as (+)-oxaprotiline, respectively non-reuptake inhibitors, such as levoprotiline, gave new impulses. Similarly, the prophylactic antidepressant effect of lithium ions and of carbamazepine as well as the clinical effectiveness of rolipram (Wachtel et al., 1990) focussed interest on the second messenger systems.

Specific 5-HT reuptake inhibitors renewed the view for the importance of the 5-HT system in brain. In the last years, the 5-HT system was intensively investigated and nowadays, on the basis of radioligand assays, six different 5-HT receptors are to be distinguished (review Peroutka, 1987). Most of them, 5-HT_1A, 5-HT_1B, 5-HT_1C, 5-HT_2, are coupled to the G-protein and belong to the so-called *gene protein receptor family* together with e.g. β-receptors (Prichett et al., 1988). The 5-HT_3 receptor is coupled to ion channels (Kilpatrick et al., 1987). A number of physiological functional correlates of these 5-HT receptors have been proposed (review Peroutka, 1987). Some of these physiological functions are considered to be characteristic for antidepressants. Many of the antidepressant treatments potentiate 5-HT functions and these effects persist during chronic treatment. Presynaptically, this effect is reflected in measures of firing rate, synthesis and turnover, which are decreased. However, postsynaptic receptor functions, especially after chronic administration, are enhanced. Therefore, the overall picture is one of enhanced transmission at 5-HT synapses. However, exceptions should be mentioned, too. MAO inhibitors (Olpe and Schellenberg, 1981) were found to reduce 5-HT receptor function. The different 5-HT receptors, however, do not function in an independent manner. As mentioned, increased activity of serotonergic neurones is necessary for β-down regulation. Simultaneously, an increased sensitivity of α_1-receptors, the so-called α-up regulation is observed. 5-HT_1B, somatodendritic autoreceptors, modulate the 5-HT release in the synapses (Engel et al., 1986). Furthermore, there exists interaction between 5-HT_1A- and 5-HT_2 receptors. Indoleamines, such as the hallucinogenic drug LSD, stimulate both receptors, whereas the hallucinogenic drug mescaline, a phenylethylamine, acts only on 5-HT_2 receptors (see Aghajanian et al., 1987). Therefore, 5-HT_2 receptors are considered to be responsible for paranoid-hallucinogenic symptoms, whereas 5-HT_1A receptors seem to be important for antidepressant and for anxiolytic action. The 5-HT_1A agonist ipsapirone e.g., has antidepressant properties in animals (rat swim test) as well as in man. In patients with general anxiety beneficial effects were found, too (own results, in preparation). Furthermore, a balance between 5-HT_1A and 5-HT_2 receptors may be important. The potency of either blocking 5-HT_1 or 5-HT_2 receptors, (Wander et al., 1986) and of the 5-HT reuptake inhibition may account for the antidepressive activity.

Besides these intraserotonergic interactions, influences on other transmitter systems are described, too (see review Meltzer and Lovy, 1987). Some antidepressants act by removing the inhibitory effect of serotonergic on catecholaminergic neurones (Eccleston, 1981). The 5-HT_1A agonist ipsapirone and 8-OH-DPAT increase NE turnover, probably leading to a β-down regulation. There seems to be interaction with the dopaminergic system. An increase of cerebrospinal fluid homovanillic acid (CSF HVA) is described after serotonergic antidepressants. 5-HT_1A agonists increase as well HVA and dihydroxyphenyl acid (DOPAC) in animals. 5-HT has an inhibiting influence on acetylcholine in the striatum (Fuxe et al., 1983) and, finally, interaction with neuropeptides should not be neglected. In summary, 5-HT alone is not the only principle of action for antidepressant activity. This view is supported by antidepressant drugs showing no serotonergic activities, but influencing other systems such as the dopamine system.

Although the euphoric effect of amphetamine, which is most probably related to an increase of dopamine activity, has been well known for many years, the role of the dopamine system with regard to antidepressant effects was neglected. In 1977 already, Randrup and Braestrup described the inhibition of dopamine reuptake by tricyclic antidepressants as well. Nomifensine, with excellent clinical antidepressant properties, inhibits dopamine reuptake, thus leading to an increased dopaminergic activity. Unfortunately, because of severe side effects on the hematopoetic system, it had to be taken from the market. MAO inhibitors increase dopaminergic activity, too, and, as described before, specific serotonergic drugs such as 5-HT_1A agonists modulate the activity of dopaminergic neurones.

New areas of research were opened by the properties of oxaprotiline and its enantiomers. The (+)-enantiomer is a very specific NE reuptake inhibitor inducing β-down regulation, whereas the (–)-enantiomer shows no biochemical effects as produced by the usual antidepressants (Delini-Stula et al., 1983). However, good therapeutic clinical effects have been reported in many clinical trials (Wendt and Binz, 1989). In pharmacological experiments, an enhancement of dopamine related (Mogilnicka et al., 1982; Maj and Wedzony, 1987; Delini-Stula and Mogilnicka, 1987; Delini-Stula et al., 1988), and α_1-receptor related behaviour were demonstrated. Therefore, it may be concluded that the activity of dopamine and NE neurones is increased without pronounced alterations of the biochemical effects. The functional activity of the 5-HT system was increased, too (Delini-Stula and Mogilnicka, 1989). On the level of the "second messenger" phosphatidylinositol system, a slight stimulation was described in blood cells of patients treated with levoprotiline (Schubert et al., 1990, in prep.)

The described controversial results on the level of transmitter turnover and their receptors again stimulated reflections on a common mechanism of action of antidepressants. Further perspectives of such investigations are related to "second" and "third messenger" systems. Most of the receptors, α-, β- and serotonin receptors, are linked via G-protein to the adenylate cyclase system, the phosphatidylinositol system, and the diacylglycerol system. Modulation of receptor sensitivity, either down-regulation or up-regulation, is most probably dependent on the second and third messenger systems. Until now, the prophylactic antidepressant effects of lithium and carbamazepine are not very well understood. The most widely discussed hypothesis is the effect on second messenger systems. Lithium inhibits adenylate cyclase (Dousa and Hechter, 1970; Newman et al., 1983). It further inhibits neuronal inositol phosphatase (Hallcher and Sherman, 1980) and prevents the resynthesis of phosphatidylinositol-2-phosphate. Lithium blocks cholinergic and adrenergic signal transduction by interference with G-proteins (Avissar et al., 1988). In this way, it can damp the overactivity of the adenylatecyclase system and deamplifies the PLC system. Carbamazepine, too, interacts with this second messenger, especially with the adenosine stimulated cAMP (van Calker et al., 1990). The new compound rolipram inhibits the phosphodiesterase (Wachtel, 1983) and, thus, prevents the catabolism of cAMP.

Second messengers, such as cAMP, stimulate cAMP-dependent protein kinase C initiating a phosphorylase cascade. The adenylate cyclase system interacts with the phosphatidylinositol system and both systems influence each other reciprocally. The second messenger dysbalance hypothesis (Wachtel, 1990) assumes a decreased functional availability of cAMP. cAMP and proteinase C dependent kinases are considered to phosphorylate a variety of other proteins which lead to an induction of primary response genes, which are called tetradecanoyl phorbol acetate (TPA) induced sequences or TIS genes. Many transmitters, via their interaction with their specific receptors, initiate a cascade of intracellular events which lead to the induction of such specific primary response genes. Involved in this cascade or events are the TIS genes. One of them, the c-fos protooncogene, is stimulated by dopamine. Other stimulations, such as e.g. kindling in the hippocampus, activate c-fos genes, too. Since antidepressants increase the activity of different transmitters like NE, DA, 5-HT, it can be assumed that this cascade of intracellular events is initiated, too. First results supporting this idea are reported by Perez and colleagues (Perez et al., 1989). 15 days of treatment with desmethylimipramine mediated the cAMP dependent protein kinase and altered the phosphorylation of microtubulae associated proteins (MAP). The best characterized receptor is the β-receptor involved either by stimulating or via feedback in these intracellular processes. For β-down regulation such processes are described, e.g. cAMP regulates the transcription of the β-adrenergic receptor gene (Hadcock and Malbon, 1988). It can be speculated that alterations on the level of the so called third messengers are important for the mechanism of action of antidepressants, too. Today, it is an open question in which precise neuronal processes these genes are regulated. The application of molecular biological methods promises new insights into the mechanism of action of drugs.

References

Aghajanian GK, Sprouse JS, Rasmussen K (1987) Physiology of the midbrain serotonin system. In: Meltzer HY (ed) Psychopharmacology: the third generation of progress. Raven Press, New York, pp 141–127

Arenander AT, de Vellis J, Herschman HR (1989) Induction of c-fos and TIS genes in cultured rat astrocytes by neurotransmitters. J Neurosci Res 24:107–114

Avissar S, Schreiber G, Danon R, Belmaker H (1988) Lithium inhibits adrenergic and cholinergic increase in GTP binding in rat cortex. Nature 331: 440–442

van Calker D, Steber R, Klotz K-N, Greil W (1990) Carbamazepine distinguishes between adenosine receptor syptypes that mediate, respectively, regulation of cyclic AMP accumulation and inhibition of inositolphosphate formation. (Submitted)

Ceccatelli S, Villar MJ, Goldstein M, Hökfelt T (1989) Expression of c-fos immunoreactivity in transmitter-characterized neurons after stress. Proc Natl Acad Sci USA 86:9569–9573

Checkley SA (1988) Monoamines, depression and antidepressant drugs. Pharmacopsychiatry 21:6–8

Checkley SA, Thompson C, Burton S, Francy C, Arendt J (1985) Clinical studies of the effect of (+) and (−)-oxaprotiline upon noradrenaline uptake. Psychopharmacology 87:116–118

Comb M, Hyman SE, Goodman HM (1987) Mechanism of trans-synaptic regulation of gene expression. TINS 10(11):473–478

Coppen A (1967) The biochemistry of affective disorders. Br J Psychiatry 113:1237–1264

Crane GE (1957) Iproniazid (Marsilid) phosphate a therapeutic agent for mental disorders and debilitating diseases. Psychiat Res Rep Am Psychiat Ass 8:142–152

Delini-Stula A, Mogilnicka E (1988) Single treatments with the antidepressant oxaprotiline and its (+) and (–) enantiomers increase behavioural responses to dopaminergic stimulation in the rat. J Neural Transm 71:91–98
Delini-Stula A, Mogilnicka E (1989) Rapid changes in functional responsiveness of the 5-HT system after single-dose and multiple-dose treatment with antidepressants: effects of maprotiline and oxaprotiline and its enantiomers. J Pharmacol 3(1):7–13
Delini-Stula A, Radeke E, van Riezen H (1988) Enhanced functional responsiveness of the dopaminergic system – the mechanism of anti-immobility effects of antidepressants in the behavioural despair test in the rat. Neuropharmacology 27(9):943–947
Dousa T, Hechter O (1970) Lithiuim and brain adenyl cyclase. Lancet i:834–835
Eccelston D (1981) Monoamines in affective illness – is there a place for 5-HT? Br J Psychiatry 138:257–258
Engel G, Göthert M, Hoyer D, Schlicker E, Hillenbrand K (1986) Identity of inhibitory presynaptic 5-hydroxytryptamine (5-HT) autoreceptors in the rat brain cortex with 5-HT_{1B} binding sites. Nauyn-Schmiedebergs Arch Pharmacol 357:1–7
Feuerstein TJ, Hertting G, Jackisch R (1985) Endogenous noradrenaline as modular of hippocampal serotonin (5-HT)-release. Naunyn-Schmiedebergs Arch Pharmacol 329:216–221
Fuxe K, Calza L, Benfenati F, Zini I, Agnati LF (1983) Quantitative autoradiographic localization of [^{3}H]imipramine binding sites in the brain of the rat: relationship to ascending 5-hydroxytryptamine neuron systems. Proc Natl Acad Sci USA 80:3836–3840
Hadrock JR, Malbon CC (1988) Down-regulation of beta-adrenergic receptors: Agonist-induced reduction in receptor mRNA levels. Proc Natl Acad Sci USA 85:5021–5025
Hallcher LM, Sherman WR (1980) The effects of lithium ion and other agents on the activity of myo-inositol-1-phosphate from bovine brain. J Biol Chem 225:10896–10901
Kilpatrick GJ, Jones BJ, Tyers MB (1987) Identification and distribution of 5-HT_3 receptors in rat brain using radioligand binding. Nature 330:746–748
Kuhn R (1957) Über die Behandlung depressiver Zustände mit einem Iminodibenzylderivat (G 22 455). Schweiz Med Wochenschr 35/36: 1135–1140
Langer SZ (1978) Presynaptic receptors and the regulation of transmitter release in the peripheral and central nervous system: physiological and pharmacological significance. In: Catecholamines: basic and clinical frontiers. Proc 4th Int Catecholamine Symposium. Pergamon Press, London, New York
Maj J (1990) Pharmakologie von Levoprotilin. (In preparation)
Maj J, Wedzony K (1988) The influence of oxaprotiline enantiomers given repeatedly on the behavioural effects of d-amphetamine and dopamine injected into the nucleus accumbens. Eur J Pharmacol 145:97–103
Matussek N (1969) Die Catecholamin- und Serotoninhypothese der Depression. In: Hippius H, Selbach H (Hrsg) Das depressive Syndrom. Urban & Schwarzenberg, München Berlin Wien
Matussek N, Ackenheil M, Hippius H, Müller F, Schröder HT, Schultes H, Wasilewski B (1980) Effect of clonidine on growth hormone release in psychiatric patients and controls. Psychiatry Res:25–36
Meltzer HY, Lowry MT (1987) The serotonin hypothesis of depression. In: Meltzer HY (ed) Psychopharmacology: the third generation of progress. Raven Press, New York, pp 513–526

Mogilnicka E, Boissard CG, Waldmeier PC, Delini-Stula A (1983) The effects of single and repeated doses of maprotiline, oxaprotiline and its enantiomers on foot-shock induced fighting in rats. Pharmacol Biochem Behav 19:719–723

Mogilnicka E, Zazula M, Wedzony K (1987) Functional supersensitivity to the α_1-adrenoceptor agonist after repeated treatment with antidepressant drugs is not conditioned by β-down-regulation. Neuropharmacology 26(10):1457–1461

Morgan JI, Curran T (1989) Stimulus-transcription coupling in neurons: role of cellular immediate-early genes. TINS 12(11):459–462

Newman ME, Klein E, Birnmaher B, Feinsod M, Belmaker H (1983) Lithium at therapeutic concentrations inhibits human brain noradrenaline-sensitive cAMP accumulation. Brain Res 278:380–381

Olpe HR, Schellenberg (1981) Rotubule fractions of rat cerebral cortex after prolonged desmethyl-imipramine treatment. Eur J Pharmacol-Mol Pharmacol Section 172:305–316

Perez J, Tinelli D, Brunello N, Racagni G (1989) CAMP-dependent phosphorylation of sulble and crude microtube fractions of rat cerebral cortex after prolonged desmethylimipramine treatment. Eur J Pharmacol-Mol Pharmacol Section 172:305–316

Peroutka SJ (1987) Serotonin receptors. In: Meltzer HY (ed) Psychopharmacology: the third generation of progress. Raven Press, New York, pp 303–311

Potter WZ, Scheinin M, Golden RN (1985) Selective antidepressants and cerebrospinal fluid. Arch Gen Psychiatry 42:1171–1177

Pritchett DB, Bach AW, Wozny M, Taleb O, Dal Toso R, Shih J, Seeburg PH (1988) Structure and functional expression of cloned rat serotonin 5-HT_2 receptor. EMBO J 7:4135–4140

Racagni G, Brunello N (1984) Transsynaptic mechanisms in the action of antidepressant drugs. Trends Pharmacol Sci 5:527

Randrup A, Braestrup C (1977) Uptake inhibition of biogenic amines by newer antidepressant drugs: relevance to the dopamine hypothesis of depression. Psychopharmacology 53:309–314

Robertson HA, Peterson MR, Murphy K, Robertson GS (1989) D1-dopamine receptor agonists selectively activate striatal c-fos independent of rotational behaviour. Brain Res 503:346–349

Schildkraut JJ (1965) The catecholamine hypothesis of affective disorders: a review of supporting evidence. Am J Psychiatry 122:509–522

Vetulani J, Sulser F (1975) Action of various antidepressant treatments reduces reactivity of noradrenergic cyclic AMP-generating system in limbic forebrain. Nature 257:495–497

Wachtel H (1983) Potential antidepressant activity of rolipram and other selective cyclic adenosine 3′,5′-monophosphate phosphodiesterase inhibitors. Neuropharmacology 22:367–372

Wachtel H (1990) The second-messenger dysbalance hypothesis of affective disorders. Pharmacopsychiatry 23:27–32

Waldmeier PC, Baumann PA, Hauser K, Maître L, Storni A (1982) Oxaprotiline, a noradrenaline uptake inhibitor with an active and an inactive enantiomer. Biochem Pharmacol 31(12):2169–2176

Wander TJ, Nelson A, Okazaki H, Richelson E (1986) Antagonism by antidepressants of serotonin S_1 and S_2 receptors of normal human brain in vitro. Eur J Pharmacol 132:115–121

Wendt G, Binz U (1989) Levoprotilin. Therapeutische Wirksamkeit und Verträglichkeit am Beispiel einer Doppelblind-Studie vs. Amitriptylin. Münch Med Wochenschr 131(2):0–0

Author's address: Prof. Dr. M. Ackenheil, Psychiatrische Klinik der Universität München, Nußbaumstrasse 7, D-8000 München 2, Federal Republic of Germany.

Monoamine oxidase and its inhibitors

J Neural Transm (1990) [Suppl] 32: 41–47

The expression of human MAO-A and B genes

J. C. Shih, J. Grimsby, and **K. Chen**

Division of Biological Sciences, School of Pharmacy, University of Southern California, Los Angeles, California, U.S.A.

Summary. Northern analysis using ^{32}P-labeled subfragments of human liver MAO-A and B cDNA clones detected 5Kb and 3Kb transcripts, respectively in fetal tissues and adult brains. The tissue distribution of these transcripts was determined. Small intestine and placenta express, in addition to the MAO-A 5Kb transcript, a 2Kb transcript determined to lack a 3′ flanking region. MAO-A appeared prior to MAO-B in the fetal brain, whereas both MAO-A and B were found in adult brain.

Introduction

Monoamine oxidase A and B function to regulate the levels of biogenic amines in the brain and peripheral tissues by catalyzing the oxidative deamination of these amines (Wyatt and Murphy, 1976; Singer et al., 1979). The two forms of the enzyme have been found to differ in their substrate and inhibitor specificities (Fuller, 1968; Johnston, 1968; Knoll and Magyar, 1972). Earlier studies were directed towards elucidating the molecular nature of these two types of MAO. Immunological studies (Denney et al., 1982), SDS-gel electrophoresis (Weyeler and Salach, 1985), and peptide mapping from enzyme digestion (Cawthon et al., 1981) suggested that the two forms of the enzyme had different primary structures. We have recently isolated two cDNA clones which encode MAO-A and B, respectively, (Bach et al., 1988; Shih et al., 1988) and the functional enzymes were expressed when transfected into COS cells (Lan et al., 1989). The successful cloning of these enzymes allowed us to deduce the amino acid sequence of the two forms of the enzyme. These deduced sequences show 70% identity. Both peptides contain the pentapeptide ser-gly-gly-cys-tyr in the C-terminus region in which the cofactor FAD is covalently bound to cysteine. Additionally, there are conserved sequences in both deduced sequences which reside near the N-terminus and which are presumable binding sites for AMP. The high degree of similarity of these deduced sequences would suggest that they are derived from a common progenitor gene. However, the fact that the identical sequences are randomly distributed over the entire length of the proteins, suggests that MAO-A and B are derived from different genes. This data is conclusive evidence that the two proteins have unique primary structures and their differing catalytic abilities do not arise solely from post transcriptional modifications. In situ hybridization utilizing the MAO-A and B cDNA clones enabled us to localize the position of the two genes to the X chromosomes, Xp11.23 (Lan et al., 1989b). The major concentrations of silver grains were found to be localized on the Xp11.23 snf Xp21 for the MAO-A and one the Xp11.23, Xp21.2, and Xp22.1 regions for the MAO-B. Additionally, we were able to demonstrate that DNA from a patient with Norrie disease lacks both MAO-A and B genes following EcoR1 digestion of the patient's DNA and hybridization of the Southern blot with the MAO-A and B cDNA probes. Coupled with the in situ hybridization, field inversion gel electrophoresis, deletion mapping in this patient with Norrie disease enabled us to demonstrate that the two genes are closely linked to each other and to the DXS7 locus (Xp11.3).

Now having an understanding of some of the basic aspects of MAO genes, i.e. that the two forms of the enzyme are derived from two distinct genes which are closely linked and located on the short arm of the X chromosome, we have studied the tissue specific expression of MAO-A and B.

Materials and methods

Total RNA from various human fetal tissues (19-week old), adult brain subregions and human placenta were prepared as previously described by Chirgwin et al. (1979). Total RNA was fractionated on agarose-formaldehyde gels, transferred to nylon membranes (Bio-Rad), hybridized with ^{32}P-labeled MAO-A or B cDNA subfragments and washed at high stringency. The subfragments were used in order to enhance positive signals. Subfragments used include: the 5′ end region (A_8, nucleotide no. 1–552), the middle region (A_9, nucleotide no. 553–1200) and the 3′ end region (A_{10}, nucleotide no. 1201–1958) of the MAO-A cDNAs, the 5′ end (B_1, nucleotide no. 1–602), the middle region (B_2, nucleotide no. 603–1434), and the 3′ end (B_3, nucleotide no. 1435–2498) of the MAO-B cDNA (Lan et al., 1989a). The hybridization was carried out in 50% formamide/5X SSC at 37 °C washed with 0.1 × SSC containing 0.1% SDS at 65 °C and exposed for two days. Under these conditions, MAO-A and B transcripts do no cross-hybridize.

Results

We found a 5Kb transcript in most of the fetal tissues examined when the MAO-A cDNA subfragments were used (Fig. 1a). The small intestine and placenta had the highest level of 5Kb transcript and an additional 2Kb transcript. For the remainder of the fetal tissues examined, the relative concentrations of 5Kb transcript were as follows: lungs, muscle, kidney > brain, spinal cord, meninges, liver > spleen > adrenal glands. We found no detectable levels of the 5Kb transcript in the thymus and the heart under our experimental conditions. The placenta and small intestine were found to have approximately equal amounts of both the 5Kb and a 2Kb transcript (Fig. 1a). The remainder of the tissues had concentrations of 2Kb transcript which were much less (barely detectable) than that of the 5Kb transcript.

When the MAO-B cDNA subfragments were hybridized with fetal RNA, a 3Kb transcript was detected in the following relative concentrations: small intestine > kidney, liver > adrenal glands > heart > spinal cord > lung. In the fetal brain and muscle the concentrations of the 3Kb transcript were much less than that of the other tissues. No transcript was detectable in the placenta, meninges, spleen, or thymus under our experimental conditions.

We used the MAO-A and B cDNA subfragments to examine regions of the adult brain (Fig. 1b). MAO-A and B transcripts were not found in all the regions examined but were found to be co-expressed in regions with detectable transcripts. It was found that MAO-A was a 5Kb transcript, while that for MAO-B was a 3Kb transcript which was consistent with the other tissues. The relative concentrations were frontal cortex (A10) > locus coeruleus (A6) $\geq$ temporal cortex (A20) > posterior pensylvian cortex-supra marginal gyri (A40) > anterior pensylvian cortex-opercular gyri (A44) $\gg$ hippocampus and thalamus.

The concentrations of RNA were precisely controlled and quantitatively determined using the 28s rDNA probe (Fig. 1a and b). To examine the relationship between the 2Kb and 5Kb MAO-A transcripts, we used a 1Kb Hind III-Sal 1 fragment that lies $\sim$ 1.5 Kb 3′ to the poly $(A)^+$ signaling sequence of a MAO-A genomic clone. When the probe was hybridized against total placental RNA and adult brain RNA, only the 5Kb transcript was seen (Fig. 2, lane 4 and 2 respectively). When the A_8 MAO-A subfragment was used, both the 5Kb and 2Kb transcripts were seen in the placenta (Fig. 2, lane 3).

Discussion

We demonstrated the existence of a 5Kb transcript for MAO-A and a 3Kb transcript of MAO-B in most fetal tissues we examined. Under our experimental conditions, the MAO-A 5Kb transcript was not detected in the thymus and heart while the MAO-B 3Kb transcript was not detected in the meninges and thymus. We were able to exclude the possibility that RNA degradation was causing this finding with use of the 28s rDNA probe. Our finding that the placenta had approximately equal concentrations of both

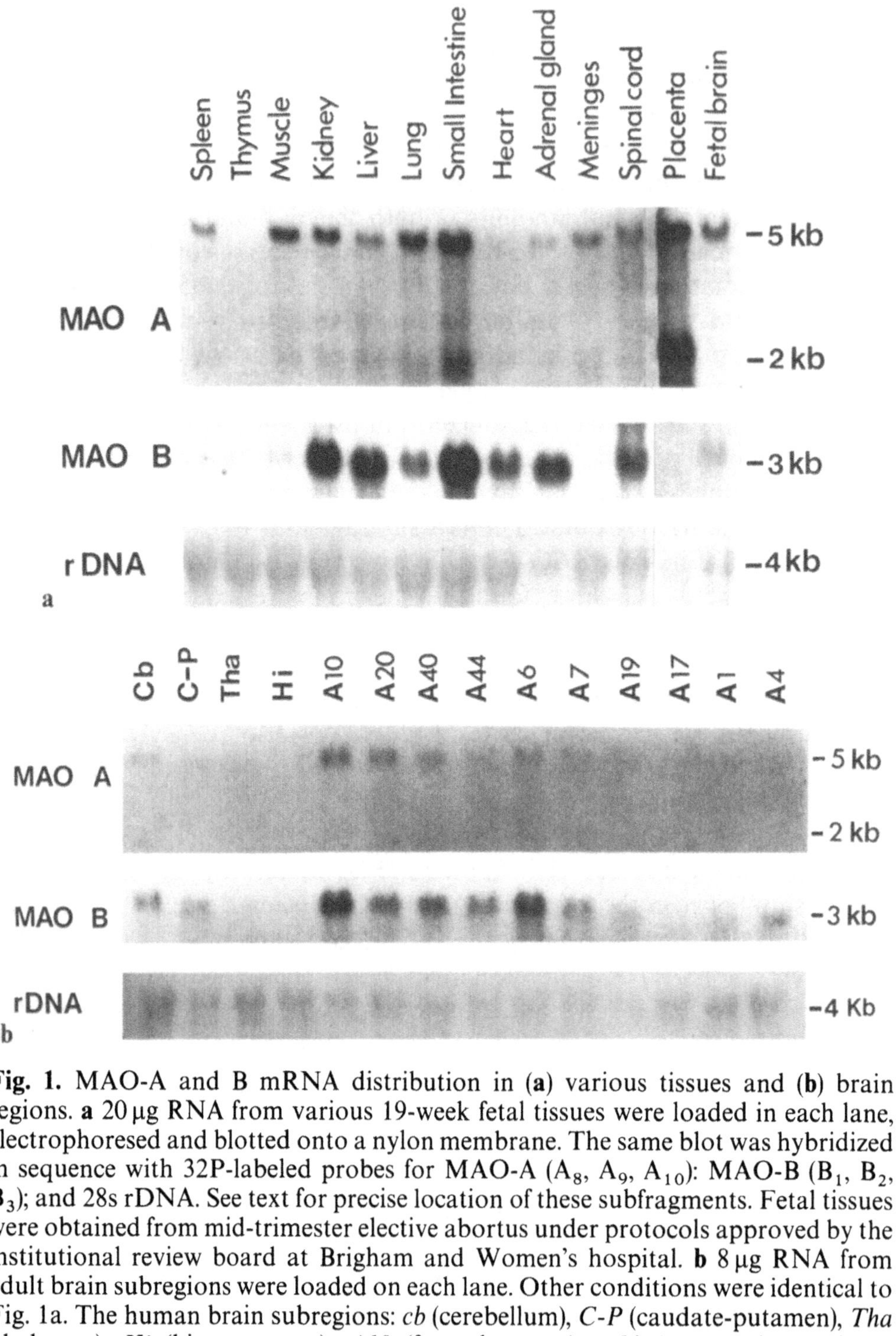

Fig. 1. MAO-A and B mRNA distribution in (**a**) various tissues and (**b**) brain regions. **a** 20 µg RNA from various 19-week fetal tissues were loaded in each lane, electrophoresed and blotted onto a nylon membrane. The same blot was hybridized in sequence with 32P-labeled probes for MAO-A (A_8, A_9, A_{10}): MAO-B (B_1, B_2, B_3); and 28s rDNA. See text for precise location of these subfragments. Fetal tissues were obtained from mid-trimester elective abortus under protocols approved by the institutional review board at Brigham and Women's hospital. **b** 8 µg RNA from adult brain subregions were loaded on each lane. Other conditions were identical to Fig. 1a. The human brain subregions: *cb* (cerebellum), *C-P* (caudate-putamen), *Tha* (thalamus), *Hi* (hippocampus), *A10* (frontal cortex), *A20* (temporal associated cortex), *A40* (posterior pensylvian cortex-supramarginal gyri), *A44* (anterior pensylvian cortex-opercular gyri), *A6* (nucleus locus coeruleus), *A7* (lateral pontine tegmention), *A19* (secondary visual cortex), *A17* (striatum), *A1* (ventrolateral medulla) and *A4* (motor cortex). The hybridization was carried out in 50% formamide/5 × SSC at 42 °C and washed with 0.1 × SSC at 65 °C. The blots were exposed for two days

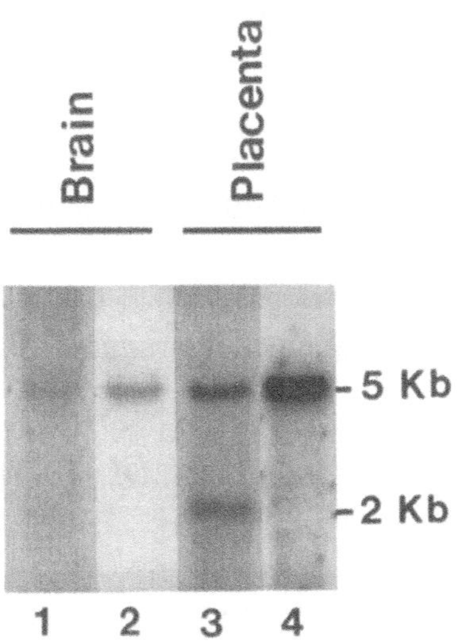

Fig. 2. Northern analysis of MAO-A transcripts. Total placenta RNA (10 μg, lane *3, 4*) and adult brain RNA (10 μg, lane *1, 2*) were electrophoresed on a 1% agarose/formaldehyde gel and blotted onto a nylon membrane. The same blot was hybridized in sequence with ^{32}P-labeled 1Kb Hind III-Sal 1 genomic fragment (lane *2, 4*) or MAO-A_8 cDNA fragment (lane *1, 3*), respectively. The hybridization was carried out in 50% formamide/5 × SSC at 37 °C and washed with 0.1 × SSC at 65 °C. The blots were exposed for two days

the 5Kb and 2Kb MAO-A transcripts is consistent with earlier reports (Bach et al., 1988; Hsu et al., 1988). Earlier data had demonstrated that MAO-A was expressed prior to MAO-B in the fetal brain (Lewinsohn, 1980). Our findings appear to corroborate this report, in that, at 19 weeks, a significant level of 5Kb transcript of MAO-A was found, while the 3Kb transcript for MAO-B was minimally detectable (Fig. 1a). Additionally, our findings are consistent with earlier reports which found that levels of MAO-B were greater in the adult brain and increased with aging (Shih, 1979; Fowler et al., 1980). The 3Kb MAO-B transcript was not detectable in the placenta under our experimental conditions which confirmed findings by Egashira and Yamanaka (1981). It is interesting to note that the small intestine had the highest level of both MAO-A and B transcript. This enzyme may play an important role in peripheral tissues. The presence of the 2Kb transcript in these two tissues suggests that MAO-A may be rapidly synthesized to meet the needs of the body.

The hybridization of the Hind III-Sal 1 fragment to placental and adult brain RNA was used to elucidate the difference between the 5Kb and 2Kb MAO-A transcript. This fragment lies 2.1Kb from the coding region of the 3′ terminal end of the MAO-A cDNA clone and hybridized with only the 5Kb transcript. This is consistent with the fact that the 5Kb transcript contains a longer 3′ untranslated region than does the 2Kb transcript. The 2Kb transcript seen in placenta may be a result of alternative polyadenylation. The fact that, when the 3′ terminal genomic clone was used as a probe, the intensity of the 5Kb transcript in placenta (Fig. 2, lane 4) was higher than that in the brain (Fig. 2, lane 2) may indicate that the 3′ ends may be different in the brain and placenta.

The tissue specificity of MAO-A and B gene expression has been demonstrated and may provide insight into their physiological roles.

Acknowledgements

This work was supported by grants R37 MH 39085 (Merit Award) and Research Scientist Award K05 MH 00796 from NIMH, DHHS. The support from Boyd and Elsie Weline professorship are also greatly appreciated.

References

Bach AWJ, Lan NC, Johnson DL, Abell CW, Bembenek ME, Kwan S-W, Seeburg PH, Shih JC (1988) cDNA cloning of human monoamine oxidase A and B: molecular basis of differences in enzymatic properties. Proc Natl Acad Sci USA 85:4934–4938

Cawthon RM, Pinter JE, Haseltine FP, Breakefield XO (1981) Differences in the structure of A and B forms of human monoamine oxidase. J Neurochem 37:363–372

Chirgwin JM, Przybyla AE, MacDonald RJ, Rutter WJ (1979) Isolation of biologically active ribonucleic acid from sources enriched in ribonuclease. Biochemistry 18:5294–5299

Denney RM, Fritz RR, Patel NT, Abell CW (1982) Human liver MAO A and MAO B separated by immunoaffinity chromatography with MAO B specific monoclonal antibody. Science 215:1400–1403

Egashira T, Yamanaka (1981) Further studies on the synthesis of A-form of MAO. Jpn J Pharmacol 31:763–770

Fowler CJ, Wiberg A, Oreland L, Marcusson J, Winblad B (1980) The effect of age on the activity and molecular properties of human brain monoamine oxidase. J Neural Transm 49:1–20

Fuller RW (1968) Kinetic studies and effects in vivo of a new monoamine oxidase inhibitor, N-[2-(o-chlorophenoxy)-ethyl]-cyclopropylamine. Biochem Pharmacol 17:2097–2107

Hsu Y-PP, Weyeler W, Chen S, Sims KB, Rinehart WB, Utterback M, Powell JF, Breakefield XO (1988) Structural features of human monoamine oxidase A elucidated from cDNA and peptide sequences. J Neurochem 51:1321–1324

Johnston JP (1968) Some observations upon a new inhibitor of monoamine oxidase in human brain. Biochem Pharmacol 17:1285–1297

Knoll J, Magyar K (1972) Some puzzling pharmacological effects of monoamine oxidase inhibitors. Adv Biochem Psychopharmacol 5:393–408

Lan NC, Chan C, Shih J (1989a) Expression of functional human monoamine oxidase A and B cDNAs in mammalian cells. J Neurochem 52:1652–1654

Lan NC, Heinzmann C, Gal A, Klisak I, Orth U, Lai E, Grimsby J, Sparkes RS, Mohandas T, Shih JC (1989b) Human monoamine oxidase A and B genes map to Xp11.23 and are deleted in a patient with Norrie disease. Genomics 4:552–559

Lewinsohn R, Glover W, Sandler M (1980) Development of benzylamine oxidase and monoamine oxidase A and B in man. Biochem Pharmacol 29:1221–1230
Shih JC (1979) Monoamine oxidase in aging human brain. In: Singer TP, Von Korff RW, Murphy DL (eds) Monoamine oxidase, structure, function and altered function. Academic Press, New York, pp 413–421
Shih JC, Bach AWJ, Seeburg PH, Lan NC, Bemberek M, Abell CW (1988) Molecular cloning of human liver monoamine oxidase B. In: Dahlstrom A, Belmaker H, Sandler M (eds) Progress in catecholamine research, part A. Basic aspects and peripheral mechanisms. Liss, New York, p 97
Singer TP, Von Korff RW, Murphy DL (1979) Monoamine oxidase structure, function and altered function. Academic Press, New York
Weyeler W, Salach JI (1985) Purification and properties of mitochondrial monoamine oxidase type A from human placenta. J Biol Chem 260:13199–13207
Wyatt RJ, Murphy DL (1975) Low platelet monoamine oxidase and vulnerability to schizophrenia. Probl Pharmacopyschiatry 10:38–56

Authors' address: Dr. J. C. Shih, Division of Biological Sciences, School of Pharmacy, University of Southern California, Los Angeles, CA 90033, U.S.A.

J Neural Transm (1990) [Suppl] 32: 49–53

Molecular neuroanatomy of MAO-A and MAO-B

J. Saura Marti, R. Kettler, M. Da Prada, and **J. G. Richards**

Pharma Research CNS, F. Hoffmann-La Roche Ltd, Basel, Switzerland

Summary. A selective, quantitative and high resolution technique (in vitro and in vivo enzyme radioautography) has been used to reveal the tissue distribution and abundance of MAO-A and MAO-B in the central nervous system and peripheral organs in the rat. The in vitro approach was also used to map the enzymes in human post-mortem brain. Furthermore, using in situ hybridization histochemistry, locus coeruleus and raphé neurons in the human brain were found to code for MAO-A and MAO-B respectively and not vice versa.

Introduction

Monoamine oxidase (MAO), a flavin-containing enzyme located in the outer membranes of neuronal, glial and other cell mitochondria, exists in two forms, MAO-A and MAO-B, which are identified by their inhibitor sensitivity and by their substrate selectivity. Since both forms are important for neurotransmitter regulation, fluctuations in functional MAO activity may be associated with neurological disorders such as depression, Parkinson's disease and perhaps Alzheimer's disease.

Novel, reversible and selective inhibitors of MAO-A and MAO-B (Ro 41-1049 [N-(2-aminoethyl)-5-(m-fluorophenyl)-4-thiazole-carboxamide. HCl] and Ro 19-6327 [N-(2-aminoethyl)-5-chloro-2-pyridine-carboxamide. HCl], respectively) (Da Prada et al., 1990) have been used as tritiated ligands, to map the distribution and abundance of the enzymes in microscopic regions of the rat CNS, peripheral organs and human brain by quantitative enzyme radioautography (Saura Marti et al., 1991). Extremely good correlations between binding and enzyme activity were obtained for various parameters, e.g. tissue selectivity (Fig. 1), IC_{50} values for inhibitors (Fig. 2), age-related changes and species-related differences [e.g. greater abundance of MAO-B than MAO-A in human brain (Fig. 3) and vice versa in rat brain]. Furthermore, the distribution, abundance and pharmacological selectivity of the binding to rat tissues in vitro and in vivo were very similar, MAO-A being concentrated i.a. in the locus coeruleus, paraventricular thalamus, raphé nuclei, solitary tract nucleus, inferior olives, interpeduncular nucleus, claustrum and numerous peripheral tissues including liver, vas deferens, heart, superior cervical ganglion, exocrine and endocrine pancreas. In contrast, MAO-B was concentrated i.a. in the ependyma, all circumventricular organs, raphé nuclei, paraventricular thalamus, posterior pituitary, liver and endocrine pancreas.

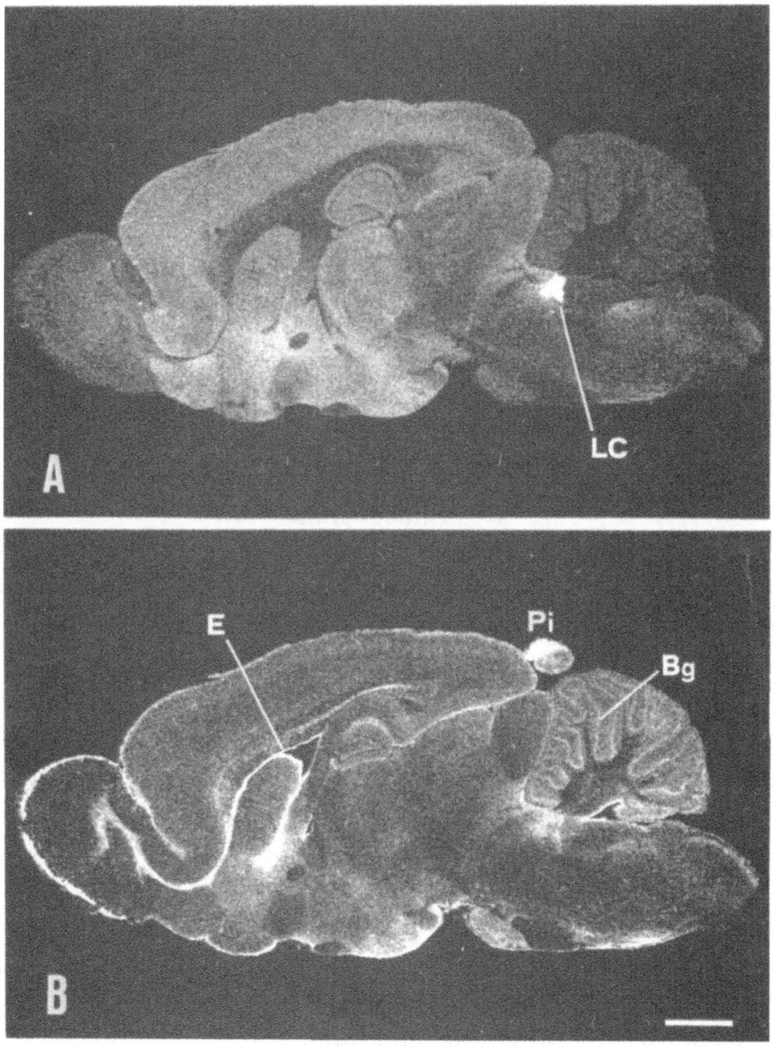

Fig. 1. Localization of MAO-A and MAO-B in adjacent parasagittal sections of rat brain revealed by enzyme radioautography. **A** [^{3}H]Ro 41-1049 and **B** [^{3}H]Ro 19-6327 binding in vitro. Note the high abundance (white areas) of MAO-A in the locus coeruleus (LC) and of MAO-B in the ependyma (E), pineal (Pi) and cerebellar Bergmann glia cells (Bg). Bar = 2 mm

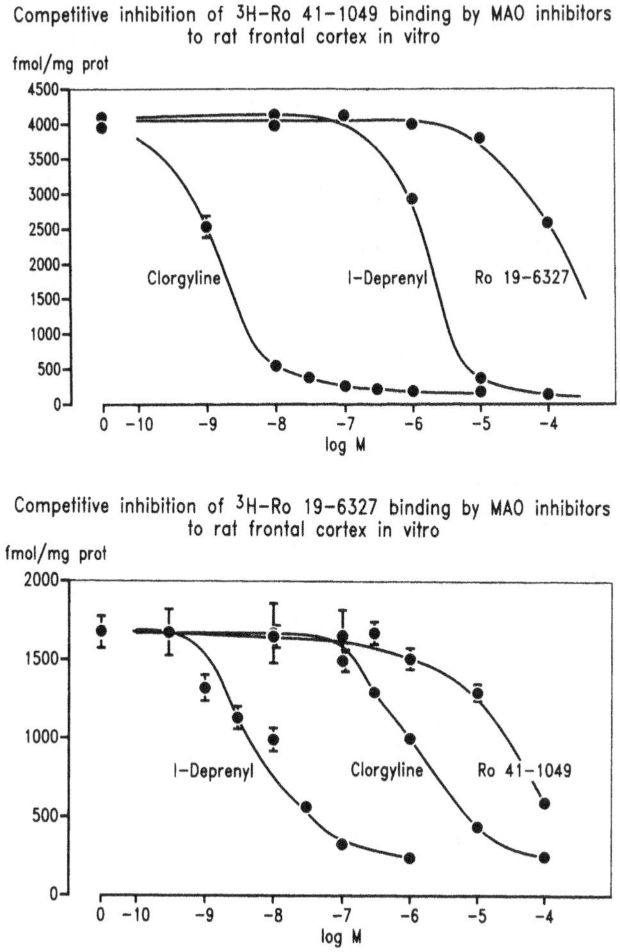

Fig. 2. IC_{50} values for various MAO inhibitors determined in competition binding assays by quantitative enzyme radioautography of radiolabelled sections from rat brain. The rank order of binding affinities correlates well with their potencies as MAO inhibitors

The kinetics of Ro 19-6327 binding to rat brain in vivo (Fig. 4) are such that it could be used to estimate the degree and time course of enzyme (i.e. MAO-B) inhibition in brains of healthy volunteers by PET analysis of [^{11}C]l-deprenyl binding.

Recently, using ^{35}S-labelled oligonucleotide probes, e.g. for human MAO-A and MAO-B, it was possible to identify, by in situ hybridization histochemistry, the cellular localization of enzyme synthesis in locus coeruleus and raphé neurons respectively but not vice versa.

The possible roles of MAO-A and MAO-B in the aetiology of depression, Parkinson's disease and Alzheimer's disease might be elucidated by studying their distribution and expression in diseased, post-mortem human brain using these selective, quantitative and high resolution techniques.

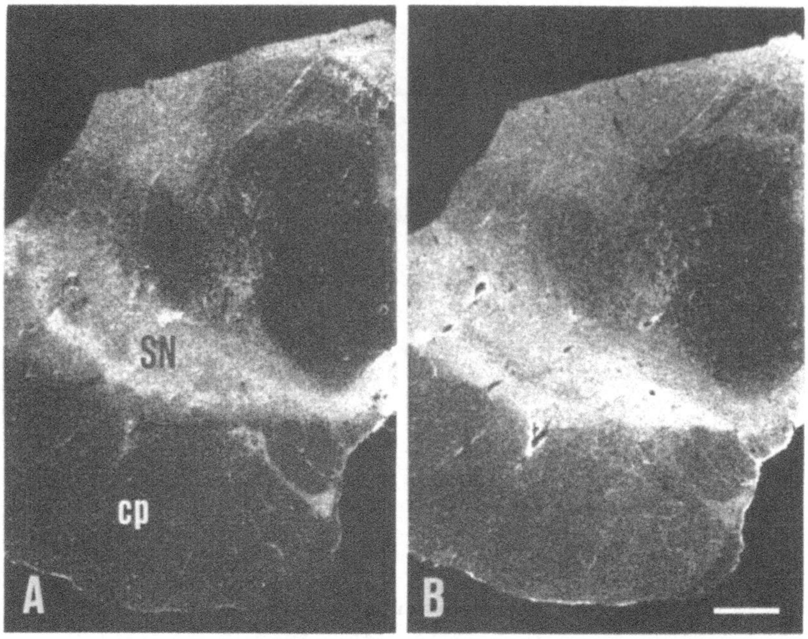

Fig. 3. Localization of MAO-A and MAO-B in adjacent sections of human substantia nigra revealed by enzyme radioautography. **A** [^{3}H]Ro 41-1049 and **B** [^{3}H]Ro 19-6327 binding in vitro. Note the higher abundance (white areas) of MAO-B, compared to MAO-A, in the substantia nigra (SN). *cp* cerebral peduncle. Bar = 2 mm

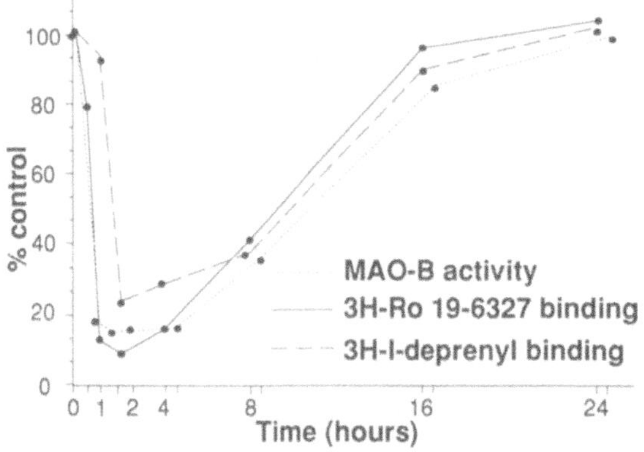

Fig. 4. Correlation between the time course of binding inhibition and enzyme inhibition by Ro 19-6327 in vivo. Rats were administered Ro 19-6327 (0.71 mg/kg p.o.) at various times prior to a tail vein injection of either [^{3}H]Ro 19-6327 or [^{3}H]l-deprenyl (s.a. 18.9 and 26.3 Ci/mmol, respectively; 1 mCi/kg) and sacrificed after 30 min or 1 h respectively. Some animals received non-radioactive Ro 19-6327 (12.5 μg/kg) i.v.; these were used to measure MAO-B inhibition by the orally administered Ro 19-6327 (radiochemical determination using PEA [20 μmol/l] as substrate)

References

Da Prada M, Kettler R, Keller HH, Cesura AM, Richards JG, Saura Marti J, Muggli-Maniglio D, Wyss P-C, Kyburz E, Imhof R (1990) From moclobemide to Ro 19-6327 and Ro 41-1049: the development of a new class of reversible, selective MAO-A and MAO-B inhibitors. J Neural Transm [Suppl] 29:179–192

Saura Marti J, Kettler R, Da Prada M, Richards JG (1991) Localization of MAO-A and MAO-B in rat CNS, peripheral organs and human brain: quantitative enzyme radioautography of [^{3}H]Ro 41-1049 and [^{3}H]Ro 19-6327. (Submitted)

Authors' addres: Dr. J. G. Richards, Pharma Research CNS, F. Hoffmann-La Roche Ltd, CH-4002 Basel, Switzerland

J Neural Transm (1990) [Suppl] 32: 55–59

Turnover of monoamine oxidase B (MAO-B) in pig brain by positron emission tomography using ^{11}C-L-deprenyl

L. Oreland[1], **S.S. Jossan**[1], **P. Hartvig**[2], **S.M. Aquilonius**[2], and **B. Långström**[3]

[1] Department of Medical Pharmacology, [2] University Hospital, and
[3] Department of Organic Chemistry, University of Uppsala, Uppsala, Sweden

Summary. Positron emission tomography (PET) was applied to investigate the rate of turnover of pig brain MAO-B. For this purpose ^{11}C-L-deprenyl was used as an irreversible ligand, which bind stoichiometrically to the enzyme. A tracer dose of ^{11}C-L-deprenyl was injected and PET scans performed to obtain baseline deprenyl binding. A high dose of unlabelled deprenyl was then administered to inhibit the enzyme and tracer doses of ^{11}C-L-deprenyl, with subsequent PET analyses, were given at 0, 2, 7, 21 and 42 days. The half-life for the turnover rate calculated was found to be 6.5 days.

Introduction

In mammalian brain, MAO exists in two different forms, an -A form, which is inhibited by low concentrations of clorgyline, and a -B form which is inhibited by low concentrations of deprenyl. Recently, it has been shown that these two forms represent separate proteins and are the product of different genes (Bach et al., 1988). Different approaches have been used to investigate the turnover rates of the two forms of MAO, e.g. estimation of the rate of return of the enzyme activity after complete inhibition of the enzyme, or estimation of the rate of disappearance of radiolabelled irreversible inhibitor from the tissue. The most recent development for estimation of the turnover rate of MAO-B is the use of positron emission tomography (PET). This technique gives the possibility for estimation of the turnover rate in vivo. Recently, Arnett et al. (1987) using PET with ^{11}C-L-deprenyl as ligand, found a half life for the MAO turnover of 30 days in the baboon, which is at variance with previous investigations using biochemical techniques. L-deprenyl has previously been found to bind irreversibly and stoichiometrically with rather low non-specific binding, both in studies using autoradiography and PET (Fowler et al., 1987; Jossan et al., 1989). In the present study, we have used ^{11}C-L-deprenyl as ligand in PET scans for estimation of the turnover rate of the MAO-B enzyme protein in pig brain. The aim of the study was to investigate whether there are pitfalls in using PET for this purpose, by comparing results obtained with PET with those obtained with conventional biochemical techniques. One reason for choosing the pig, without access of primates, was the similarities between the pig and the human brain with regard to the dominance of MAO-B over MAO-A.

Materials

The radionuclide ^{11}C-L-deprenyl was synthesised by alkylation of the N-desmethyl-L-deprenyl using ^{11}C methyl iodide at the Tandem Accelerator Laboratory, Uppsala. The doses administered ranged from 1.3 to 6.0 mCi. The specific activity of the radionuclide varied from 20 to 50 mCi/μmol. Non-labelled L-deprenyl hydrochloride was supplied by Farmos Group AB, Täby, Sweden.

Experimental procedure

The pig was anaesthetised with 4 mg/kg hypnodil (Stresnil, Janssen Pharmaceutica, Belgium) given intravenously and the anaesthesia maintained during experiment by continuous infusion of the drug in glucose. At the first PET session, a tracer dose of ^{11}C-L-deprenyl was given in an ear vein and PET scans recorded during 40 min. in order to obtain baseline deprenyl binding. One hour later, a high dose of non-labelled L-deprenyl (1 mg/kg) was injected to completely inhibit MAO-B. A second dose of ^{11}C-L-deprenyl was given 30 min. later and PET scans recorded. Arterial blood samples were initially drawn at intervals of 10 seconds with longer intervals later on up to 40 min. and radioactivity estimated. Tracer doses of ^{11}C-L-deprenyl with subsequent PET analyses were then given at 2, 7, 21 and 42 days after the administration of the high dose of non-labelled deprenyl.

Results

The total arterial blood radioactivities versus time is plotted in Fig. 1. It can be seen that the radioactivity was rapidly cleared from the blood. In Fig. 2 is shown the uptake curves for ^{11}C-L-deprenyl before and 0, 2 and 21 days after administration of the high dose (1 mg/kg body weight) of non-labelled deprenyl. For the first ^{11}C-L-deprenyl injection (baseline), the radioactivity reached the brain quickly and achieved a constant level after a few minutes. Immediately after the blockade of the MAO-B active sites with the high dose of L-deprenyl, the tracer dose of ^{11}C-L-deprenyl produced a similar initial uptake curve as before the inhibition, however, with radioactivity being rapidly cleared from the brain. Four additional PET scans were performed

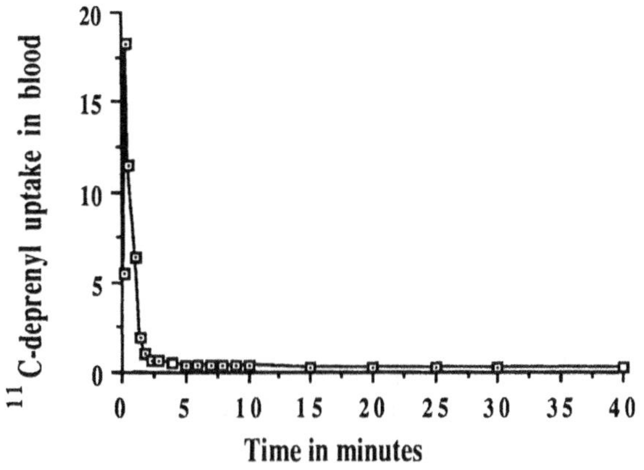

Fig. 1. Arterial blood clearance curve for ^{11}C deprenyl. Uptake values were calculated from the nanocuries per cubic centimeter, corrected for decay for blood, divided by the radioactive dose given per gram of body weight

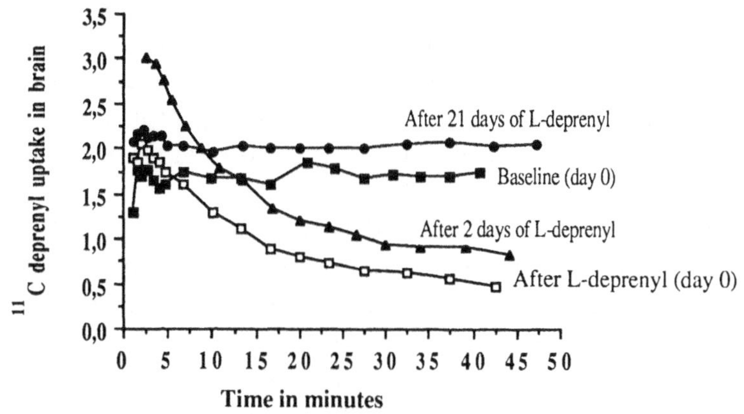

Fig. 2. Uptake of ^{11}C deprenyl in the brain. The uptake was obtained before (baseline) and after (0, 2 and 21 days respectively) a high dose of L-deprenyl. The plotted values were calculated from the nanocuries per cubic centimeter corrected for decay for brain divided by the radioactive dose given per gram of body weight

over a period of 42 days after the high dose of L-deprenyl, with gradually lower clearance until baseline conditions were achieved (Fig. 3).

Influx constants at the various times after the inhibition of MAO-B were calculated according to the method of Patlak et al. (1983). These constants were plotted versus time and the half life of MAO-B recovery calculated to be in the order of 6.5 days.

Discussion

The adaption of PET data for calculations of influx constants according to the graphical method of Patlak et al. (1983) is presupposing that a number of

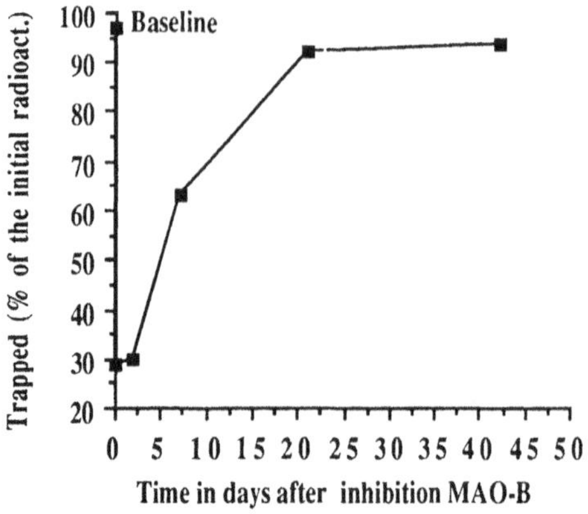

Fig. 3. Per cent radioactivity retained in brain after different times of MAO-B blockade with L-deprenyl

assumptions are fulfilled. Some of these assumptions, e.g. a single source of radioactivity, variation of concentration of the injected radioligand in the plasma with time, first order kinetics of transfer for the system and the absence of the radioligand at the start of the investigation, are easily fulfilled by injecting a single intravenous dose of a radioligand with high specific activity.

Another assumption to be fulfilled is that there is a rapid exchange of radioligand between blood and a reversible compartment in the brain. Analysis of Fig. 1 and bottom part of Fig. 2 (day 0) reveals that this was the case in the present investigation. The next assumption to be met with, is that the radioligand, from the blood or the reversible brain compartment, can enter a brain region where it is irreversibly trapped. Figure 2 shows that the concentration of ^{11}C-L-deprenyl in the brain is rather constant after about 20 minutes, indicating that also this assumption is fulfilled.

The present study was aimed at comparing the rate of recovery of MAO-B by using PET and by using biochemical ex vivo studies and the reason for this was the surprisingly slow rate of recovery found in one baboon by means of PET (Arnett et al., 1987). The present result, using PET, with a half life for the pig enzyme recovery of 6.5 days are in good agreement with that of Felner and Waldmeier (1979). They studied the rate of recovery of rat brain MAO-B, following a single injection of deprenyl, and found a half-life for the recovery of 7.9 days. Arnett et al. (1987), using PET, reported a half-life of 30 days for the turnover of baboon brain MAO-B. The great difference between their result on one hand, and those of the present study and of Felner and Waldmeier on the other, may be explained in various ways. Thus, one explanation might be that e.g. primates indeed have a slower rate of recovery for MAO-B than other species. Another explanation would be that the complicated calculations using PET might have induced a false value in the study on the baboon. Further studies, using PET on human brain MAO-B are in progress and will hopefully solve this problem.

References

Arnett CD, Fowler JS, MacGregor RR, Schyler DJ, Wolf AP, Långström B, Halldin CJ (1987) Turnover of brain monoamine oxidase measured in vivo by positron emission tomography using L-[^{11}C]deprenyl. J Neurochem 49: 522–527

Bach AW, Lan NC, Johnson DL, Abell CW, Bembeneck ME, Kwan SW, Seeburg PH, Shih JC (1988) cDNA cloning of human liver monoamine oxidase A and B: molecular basis of difference in enzymatic properties. Proc Natl Acad Sci USA 85(13):4934–4938

Felner AE, Waldmeier PC (1979) Cumulative effects of irreversible MAO inhibitors in vivo. Biochem Pharmacol 28:995–1002

Fowler JS, MacGregor RR, Wolf AP, Arnett CD, Dewey SL, Schyler D, Christman D, Logan J, Smith M, Sachs H, Aquilonius SM, Bjurling P, Halldin C, Harvig P, Leenders KL, Lundqvist H, Oreland L, Stålnacke CG, Långström B (1987) Mapping human brain monoamine oxidase A and B with ^{11}C-labelled suicide inactivators and PET. Science 235:481–485

Jossan SS, d'Argy R, Gillberg PG, Aquilonius SM, Långström B, Halldin C, Bjurling P, Stålnacke CG, Fowler J, MacGregor R, Oreland L (1989) Localization of monoamine oxidase B in human brain by autoradiographical use of ^{11}C-labelled L-deprenyl. J Neural Transm 77:55–64

Patlak CS, Balsberg RG, Fenstermacher JD (1983) Graphical evaluation of blood-to-brain transfer constants from multiple-time uptake data. J Cereb Blood Flow Metab 3:1–7

Authors' address: Prof. L. Oreland, Department of Medical Pharmacology, University of Uppsala, P.O.B. 593, S-751 24 Uppsala, Sweden.

J Neural Transm (1990) [Suppl] 32: 61–65

Visualization of brain monoamine oxidase B (MAO-B) in dementia of Alzheimer's type by means of large cryosection autoradiography: a pilot study

S. S. Jossan[1], **P.-G. Gillberg**[2], **I. Karlsson**[3], **C. G. Gottfries**[3], and **L. Oreland**[1]

[1] Department of Medical Pharmacology, and [2] Department of Neurology, University of Uppsala,
[3] Department of Psychiatry, St. Jörgens Hospital, University of Gothenburg, Sweden

Summary. Quantitative autoradiography, using ^{3}H-deprenyl, was applied to study the regional localization of MAO-B in large cryosections from 3 control brains and 3 brains from patients with Alzheimer's disease (AD). High ^{3}H-L-deprenyl binding was found in e.g. basal ganglia, cingulate gyrus, and insula cortex. Temporal cortex, parietal cortex, occipital cortex and various nuclei of the thalamus, showed moderate density, while low binding was observed in e.g. white matter. When the brains from the controls were compared with those with AD, the mean values for ^{3}H-deprenyl binding was higher in the latter group, however, more generalised than previously reported.

Introduction

Monoamine oxidase (MAO, EC: 1.4.3.4) exists in two different forms in human brain, an A-form which is sensitive to inhibition by low concentrations of clorgyline, and a B-form which is selectively inhibited by deprenyl. Recently it has been shown that these two catalytic forms are the products of two different genes (Bach et al., 1988). The activity of MAO-B increases with age and this increase is due to an increase in the number of enzyme molecules (Fowler et al., 1980). In neurodegenerative disorders such as AD (Oreland and Gottfries, 1986) Huntington's chorea (Mann et al., 1980; Riederer et al., 1983) and in Parkinson's disease (PD) this increase is further accelerated (Riederer et al., 1983). A selective increase in brain MAO-B activity can be induced by experimental neurodegenerative lesions, both by means of neurotoxins and by surgical lesions, and furthermore, the increased

MAO-B activity has been demonstrated to be localized extra-neuronally (Stenström et al., 1985). Those results have been interpreted that the increased MAO-B activity in aging and in various neurodegenerative disorders is a result of a reactive gliosis, most likely with astrocytes, which are rich in MAO-B, playing an important role.

In previous investigations on regional distribution of brain MAO, activities were measured in homogenates from dissected brain regions, which gives rather poor resolution. Recently, also immunological techniques, using monoclonal antibodies, have been used to study the cellular localization of both forms of enzyme in human brain (Konradi et al., 1988; Westlund et al., 1988). This technique, however, does not provide quantitative measurement of the enzyme. Another possibility, to study in detail the localization of MAO-B, is quantitative autoradiography using ^{3}H labelled irreversible MAO-B inhibitors. In the present investigation an in vitro quantitative autoradiographic method, using ^{3}H-L-deprenyl, has been applied to map the regional distribution of MAO-B in control brains and brains from patients with the diagnosis of Alzheimer's disease.

Materials and methods

Brains from 3 male patients, aged 65, 77 and 80 years with clinical symptoms of AD, were obtained at autopsy. As controls were used brains from 3 male patients aged, 62, 70 and 71 years with no clinical history of neurological disorder. Macroscopical *post mortem* examination of the brains confirmed the clinical diagnosis. The right hemispheres were cut in horizontal planes through the basal ganglia and thalamus into 5 mm thick slices. The slices were frozen between copper plates previously cooled with liquid nitrogen and stored at $-70\,^{\circ}C$ until used. ^{3}H-L-deprenyl was synthesized by alkylation of N-desmethyl-L-deprenyl using ^{3}H-methyl iodide and the product purified by HPLC. Non-labelled L-deprenyl hydrochloride was supplied by Farmos Group AB, Täby, Sweden.

Autoradiography

Large-section cryomicrotomy was performed according to the method described by Gillberg et al. (1986). Briefly the frozen slices were embedded into carboxymethyl cellulose. 100 µm thick sections were collected on tape/paper, mounted on gelatinized glass plates. The sections were dried and stored over desiccant for at least 1 week at $20\,^{\circ}C$. The sections were then incubated in 0.05 Na-K phosphate buffer (pH 7.4), containing 10 nM of ^{3}H-L-deprenyl, for 1 h and then washed twice (5 min. each) with the buffer and finally with distilled water. To determine non-specific binding, consecutive sections were incubated, in parallel, in the presence of unlabelled 10^{-6} M L-deprenyl. After drying, the sections and calibrated radioactive 'standards' (Amersham) were placed against tritium sensitive film (^{3}H-Ultrafilm, LKB, Sweden), exposed for 4 weeks and the film developed.

The quantitative analysis of the autoradiograms were performed using computer-assisted image analysis (Imteck Vision, Uppsala, Sweden). Mean grey values were obtained in areas of interest. To verify anatomical structures, sections were stained with Thionine.

Results

Regional differences in ^{3}H-deprenyl binding could easily be recognized. High ^{3}H-L-deprenyl binding was found in e.g. basal ganglia (caudate nucleus, globus pallidus, putamen), cingulate gyrus, and insula cortex. Other regions, including temporal cortex, parietal cortex, occipital cortex and various nuclei of the thalamus, showed moderate density of deprenyl binding. Low ^{3}H-deprenyl binding was observed in e.g. white matter. A lamination of ^{3}H-deprenyl binding was observed in the cortex, with high binding in superficial and low in deep layers. When the brains from the controls were compared with those with Alzheimer's disease, the mean values for ^{3}H-deprenyl binding was higher in the latter group in all regions investigated. The order of magnitude of the increase was, however, not similar for all regions. Thus, the relative increase in e.g. the white matter was considerably higher than that in other regions (Table 1).

Discussion

L-deprenyl is a suicide MAO-B inhibitor, which binds stoechiometrically to the enzyme. The nonspecific binding of L-deprenyl to human brain tissue under the conditions used in the present study was low and the reaction between the inhibitor and the enzyme reached completion. Thus, a high correlation between the concentration of MAO-B estimated by autoradiography and the MAO-B activity could be expected.

High density of L-deprenyl binding was found in caudate nucleus, cingulate gyrus, putamen, sulcus calcarinus, temporal cortex, globus pallidus, pulvinar and frontal cortex (Table 1). It can also be seen in Table 1 that cingulate gyrus and insula cortex showed the highest binding, while occipital and parietal cortex showed the lowest binding among the cortex regions studied. Low binding was, on the other hand, observed in white matter regions. These results are in good agreement with previously published data about MAO-B activities in various human brain regions (Fowler et al., 1980). The method also allowed for an easy recognition of differences in ^{3}H deprenyl binding between the laminae of the cerebral cortex. Thus, in agreement with a recent report (Jossan et al., 1987), a clear lamination of

Table 1. Quantitative autoradiography of ^{3}H-L-deprenyl (10 nM) binding in large cryosections from control brains and brains from patients with AD

Brain region	Control	AD/SDAT
Cerebral cortex		
Frontal cortex	340 ± 20	455 ± 20
Temporal cortex	345 ± 18	445 ± 18
Parietal cortex	292 ± 7	338 ± 10
Superficial layers	311 ± 8	382 ± 8
Deep layers	227 ± 12	298 ± 7
Insula cortex	375 ± 18	538 ± 51
Occipital cortex	212 ± 8	280 ± 19
Gyrus cinguli	432 ± 22	574 ± 24
Basal ganglia		
Caudate nucleus	445 ± 24	627 ± 9
Putamen	415 ± 8	578 ± 21
Globus pallidus	350 ± 12	512 ± 19
Thalamus		
Anterior nucleus	244 ± 25	348 ± 10
Ventrolateral nucleus	308 ± 13	370 ± 10
White matter		
Corpus callosum	94 ± 5	158 ± 12

Results were obtained from three brains in each group. Values are means (fmol/mg wet weight tissue) ± S.E.M.

L-deprenyl binding in cortex was observed with superficial layers having high and deep layers low binding.

When the binding of L-deprenyl in the brains from the patients with Alzheimer's disease were compared with that of the controls, an increase was observed in all regions investigated. The patients with Alzheimer's disease were older than the controls, but the differences persisted also when corrections for age were made according to the approximate increase in activity per decade (Oreland and Gottfries, 1986). A generalised increase in MAO-B binding is at variance with previous biochemical results on MAO-B (Oreland and Gottfries, 1986), and would indicate that pathological changes in the brain in Alzheimer's disease might be more generalized than indicated by a variety of other measures (Gottfries, 1988). In line with previous observations a relatively high increase was observed in the white matter, indicating that the pathological process might be relatively intense or even might have started in the white matter.

References

Bach AW, Lan NC, Johnson DL, Abell CW, Bembenek ME, Kwan SW, Seeburg PH, Shih J (1988) cDNA cloning of human liver monoamine oxidase A and B:

molecular basis of differences in enzymatic properties. Proc Natl Acad Sci USA 85(13):4934–4938

Fowler CJ, Wilberg Å, Oreland L, Marcusson J, Winblad B (1980) The effect of age on the activity and molecular properties of human brain monoamine oxidase. J Neural Transm 49:1–20

Gillberg PG, Jossan SS, Askmark H, Aquilonius SM (1986) Large-section cryomicrotomy for in vitro receptor autoradiography. J Pharmacol Methods 15: 169–180

Gottfries CG (1988) Alzheimer's disease. Compr Gerontol 2:47–62

Jossan SS, Aquilonius SM, d'Argy R, Bjurling P, Gillberg PG, Halldin C, Långström B, Stålnacke CG, Oreland L (1987) Autoradiographic studies on monoamineoxidase (MAO). Sixth International catecholamine symposium, Jerusalem, June 14–19, p 58 (Abstract)

Konradi C, Svoma E, Jellinger E, Riederer P, Denny R, Thiabult J (1988) Topographic immunocytochemical mapping of monoamine oxidase-A, monoamine oxidase-B and tryosine hydroxylase in human post mortem brain stem. Neuroscience 26(3):791–802

Mann JJ, Kaplan RD, Bird ED (1986) Elevated postmortem monoamine oxidase B activity in the caudate nucleus in Huntington's disease compared to schizophrenics and controls. J Neural Transm 65:277–283

Oreland L, Gottfries CG (1986) Brain and brain monoamine oxidase in aging and dementia of Alzheimer's type. Prog Neuro Psychopharmacol Biol Psychiatry 10:533–540

Riederer P, Jellinger K (1983) Neurochemical insights into monoamine oxidase inhibitors, with special reference to deprenyl (selegiline). Acta Neurol Scand [Suppl] 68:43–55

Stenström A, Arai Y, Oreland L (1985) Intra and extraneuronal monoamineoxidase-A and -B activities after central axotomy (hemisection) on rats. J Neural Transm 61:105–113

Westlund KN, Denny RM, Rose RM, Abell CW (1988) Localisation of distinct monoamine oxidase A and monoamine oxidase B cell populations in human brain stem. Neuroscience 25(2):439–456

Authors' address: Dr. S. S. Jossan, Department of Medical Pharmacology, University of Uppsala, P.O.B. 593, S-751 24 Uppsala, Sweden

J Neural Transm (1990) [Suppl] 32: 67–77

Immunofluorescence cytochemistry on thin frozen sections of human substantia nigra for staining of monoamine oxidase A and monoamine oxidase B: a pilot study

G. Moll[1], **R. Moll**[2], **P. Riederer**[1], **W. Gsell**[1], **H. Heinsen**[3], and **R. M. Denney**[4]

[1] Clinical Neurochemistry, Department of Psychiatry, University of Würzburg,
[2] Institute of Pathology, University of Mainz, and
[3] Morphological Brain Research, Department of Psychiatry, University of Würzburg, Federal Republic of Germany
[4] Department of Human Biology and Chemistry, University of Texas, Galveston, U.S.A.

Summary. Immunofluorescence cytochemistry on thin frozen sections of human substantia nigra for staining of MAO-A and MAO-B revealed that only about 10% of the melanin-containing neurons are positive for MAO-A, while they are substantially free of MAO-B.

Introduction

It has been reported that neurons of the substantia nigra neither contain monoamine oxidase A (MAO-A) and monoamine oxidase B (MAO-B) at all (Konradi et al., 1988) or only a very limited amount of MAO-A in a few melanin-containing neurons (Westlund et al., 1988). These authors used a free floating peroxidase based method on 30 to 50 μm sections. Since there might be problems with regard to the penetration of the antibody through the tissue, we have approached to the same questions using an alternative method based on immunofluorescence on thin sections (6 μm).

Materials and methods

Substantia nigra was taken at autopsy with a postmortem delay of 18 hours from a 24 year old male person, dying from a violent accident. The tissue was immediately frozen in liquid nitrogen. 6 μm thin serial sections were made using a cryostat and were fixed after air-drying either with 10% formalin solution for 20 minutes or with cold acetone of −20 °C for 15 minutes prior to formalin fixation or with acetone alone.

After washing in PBS, indirect immunofluorescence microscopy was performed using monoclonal antibodies against MAO-A (7B10) (Kochersberger et al., 1985), MAO-B (1C2) (Denney et al., 1982) and against neurofilaments (2F11, Viramed, Martinsried, FRG). Texas red coupled anti-mouse immunoglobulins were used as secondary antibodies (Dianova, Hamburg, FRG). Melanin-containing neurons were easily identified by phase-contrast microscopy. As control, sections through the medulla oblongata were taken, in which MAO-containing neurons are present.

Results

As a control and for identification of the melanin-containing cells as neurons neurofilament antibodies were used (Fig. 1). Note the positive reaction in the proximal portion of the melanin-containing neuron and the weaker reaction in the cytoplasma.

Melanin-free perikarya in the substantia nigra are strongly positive for MAO-A (Fig. 2) or MAO-B (Fig. 3).

In contrast, only about 10% of the melanin containing neurons are positive for MAO-A antibody, with staining reaching from weak but significant positive in the two neurons of Fig. 4a and b, by positive with partly fibrillar reaction in the neuron of Fig. 5a and b, to strongly positive within the group of three neurons of Fig. 6a and b. Nevertheless, the majority of the melanin-containing neurons in the substantia nigra (about 90%) was negative for MAO-A, with only coarsely granular staining in the surrounding

Fig. 1. Identification of melanin-containing cells as neurons by staining with neurofilament antibodies

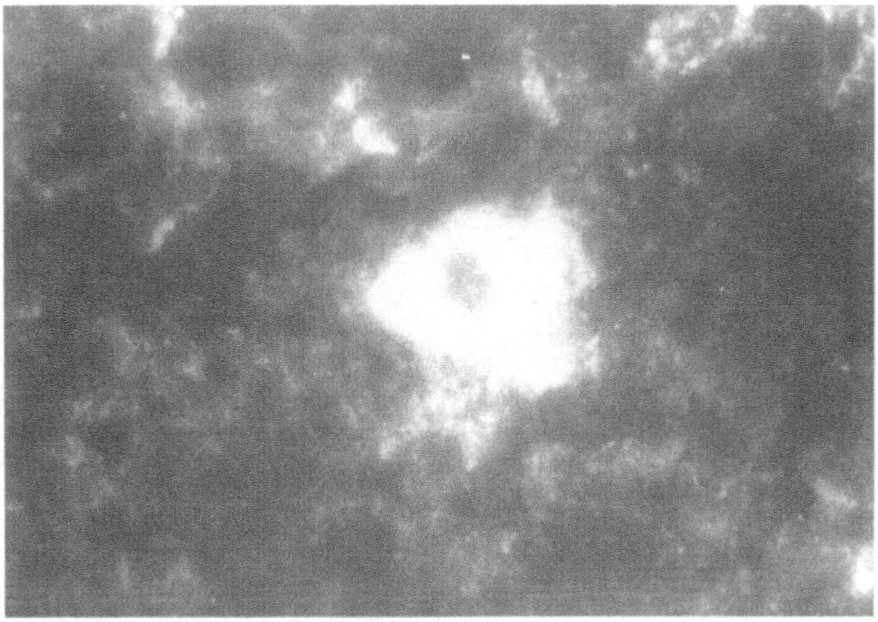

Fig. 2. Melanin-free perikaryon in the substantia nigra strongly positive for MAO-A

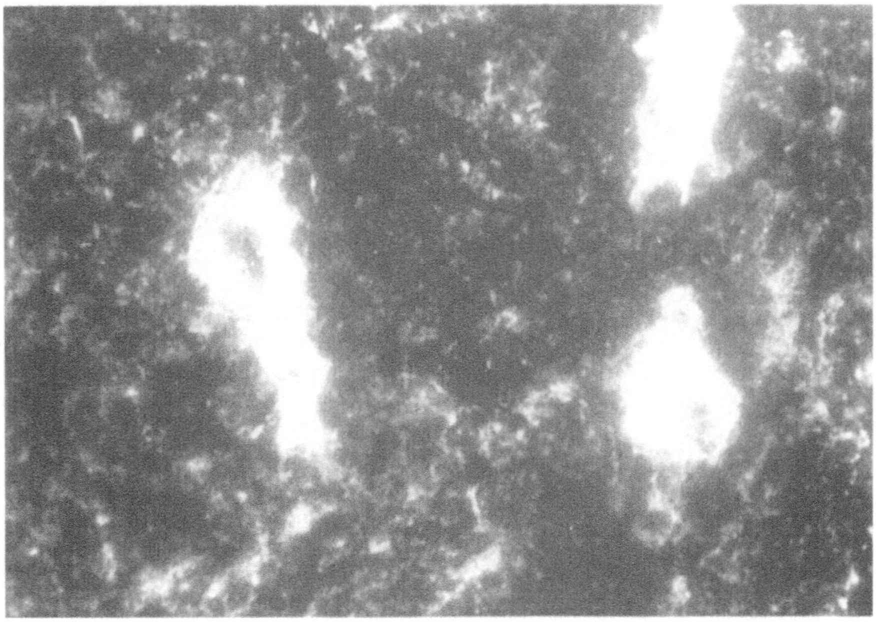

Fig. 3. Three melanin-free perikarya in the substantia nigra strongly positive for MAO-B

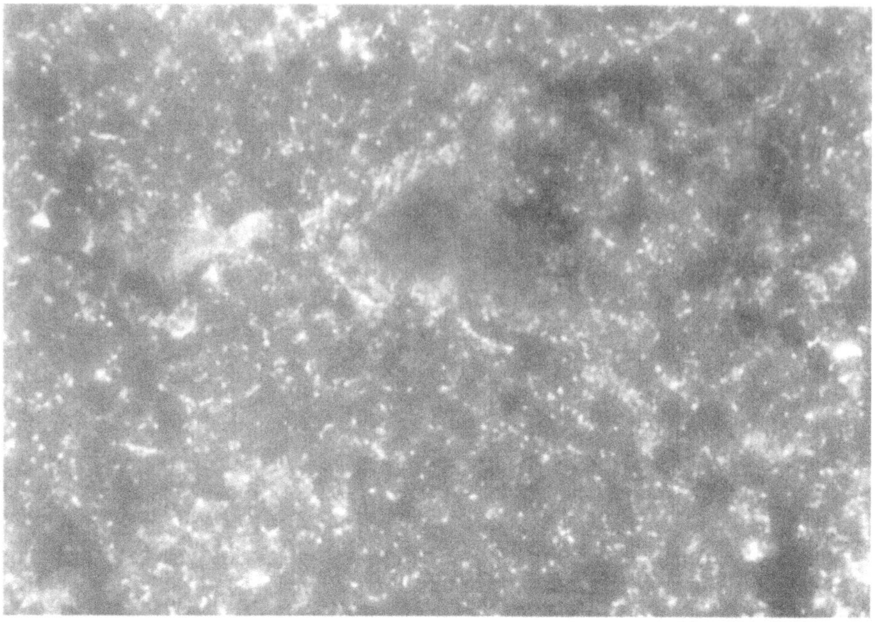

a

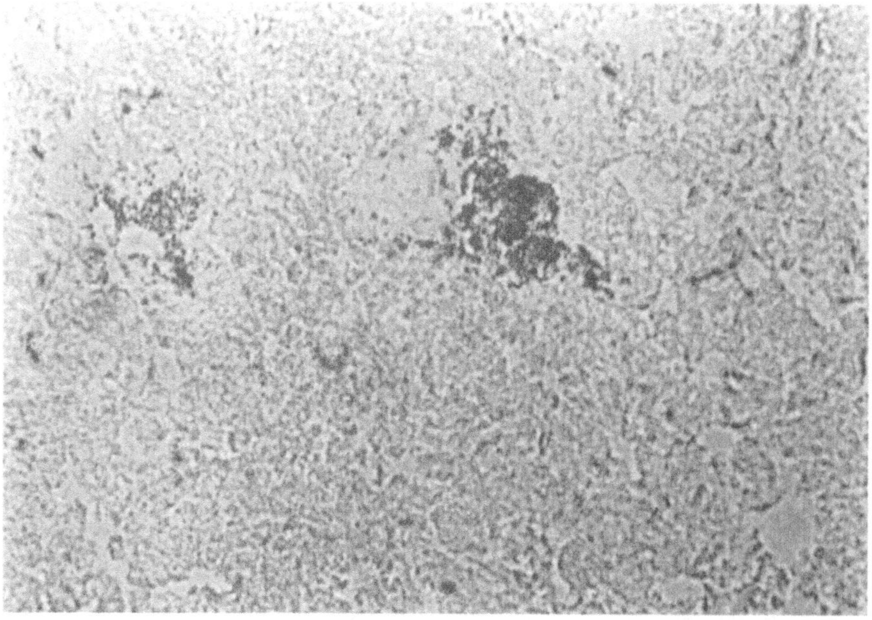

b

Fig. 4. Fluorescence (**a**, upper picture) and light (**b**, lower picture) microscopy of a group of two neurons, being weakly, but significantly positive for MAO-A antibody

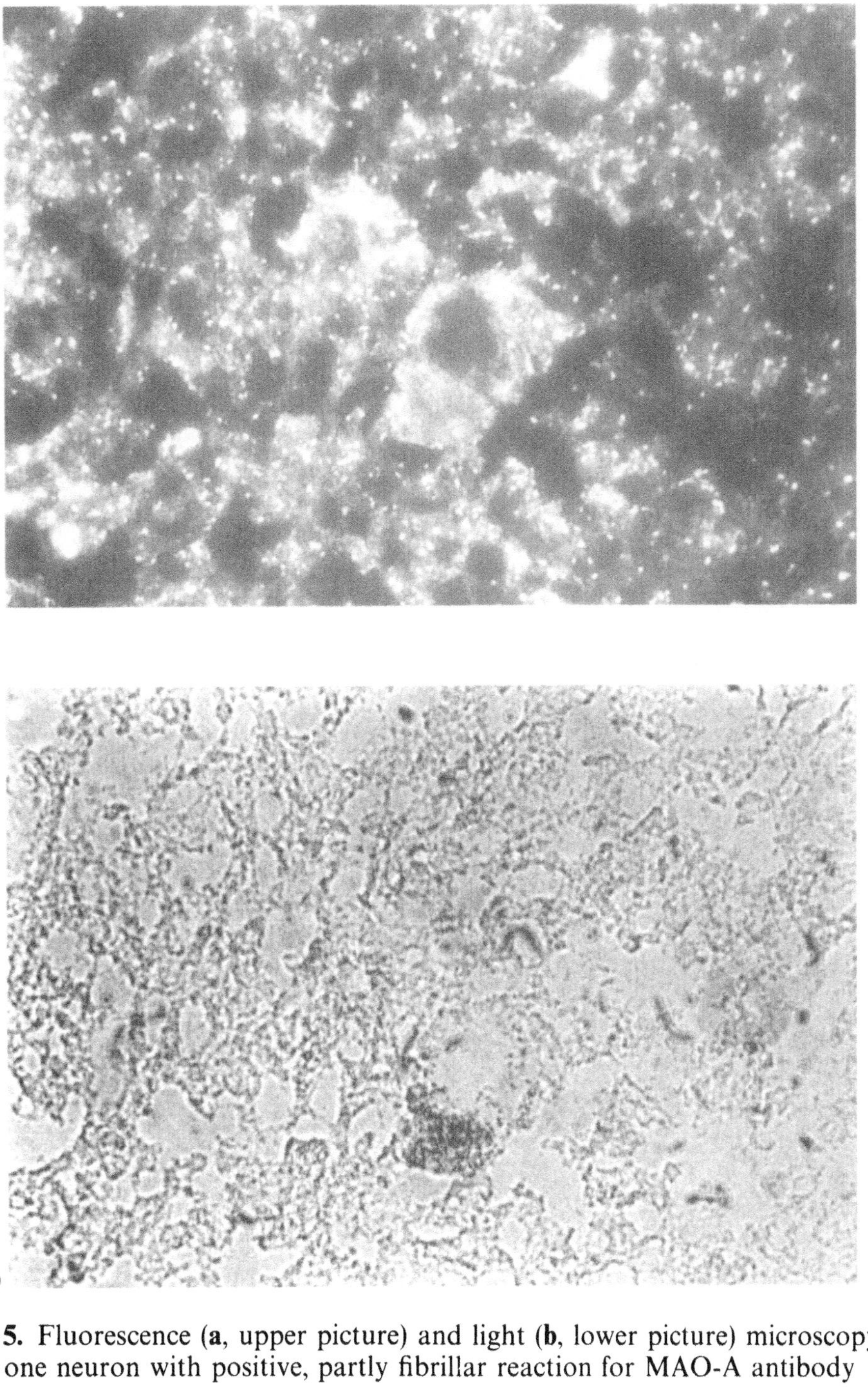

Fig. 5. Fluorescence (**a**, upper picture) and light (**b**, lower picture) microscopy of one neuron with positive, partly fibrillar reaction for MAO-A antibody

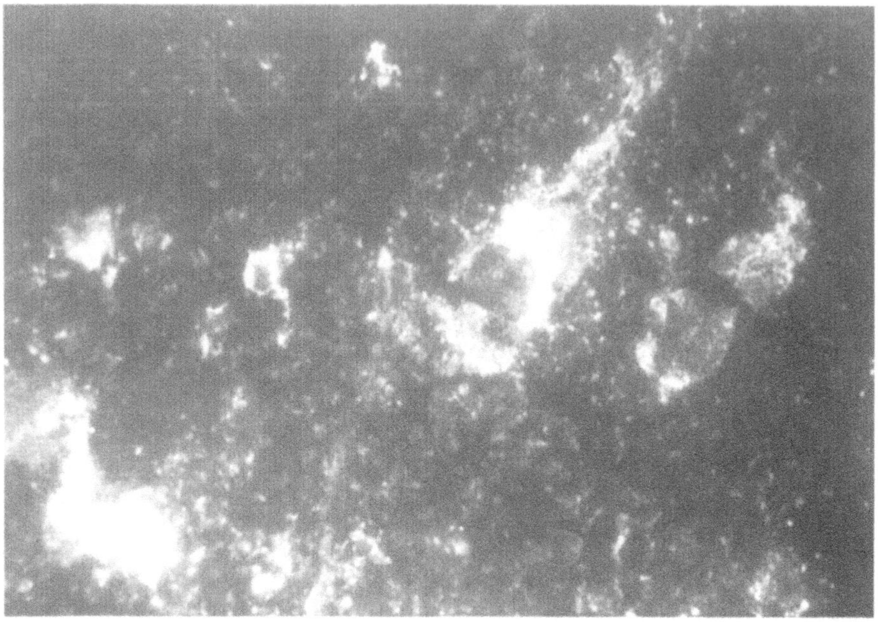
a

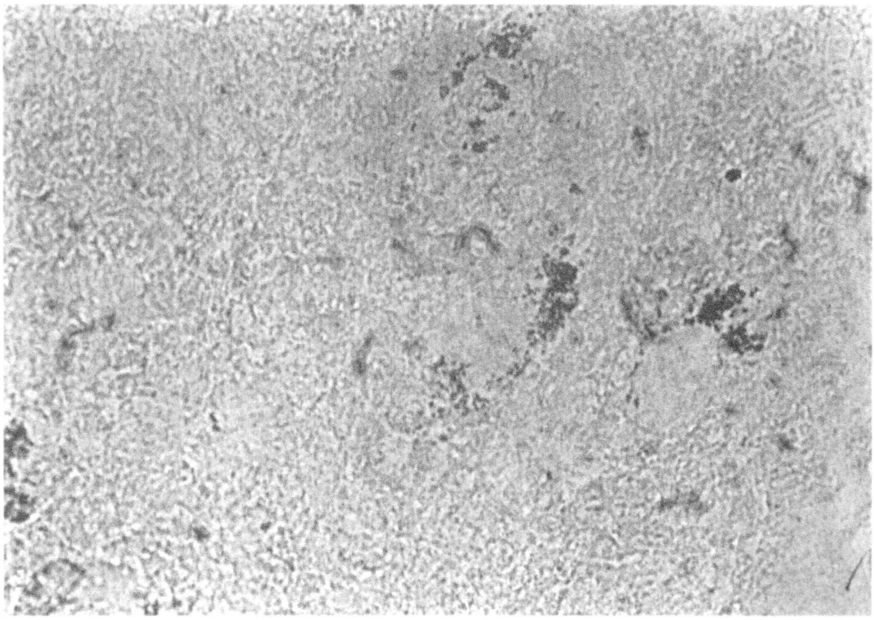
b

Fig. 6. Fluorescence (**a**, upper picture) and light (**b**, lower picture) microscopy of a group of three neurons strongly positive for MAO-A antibody

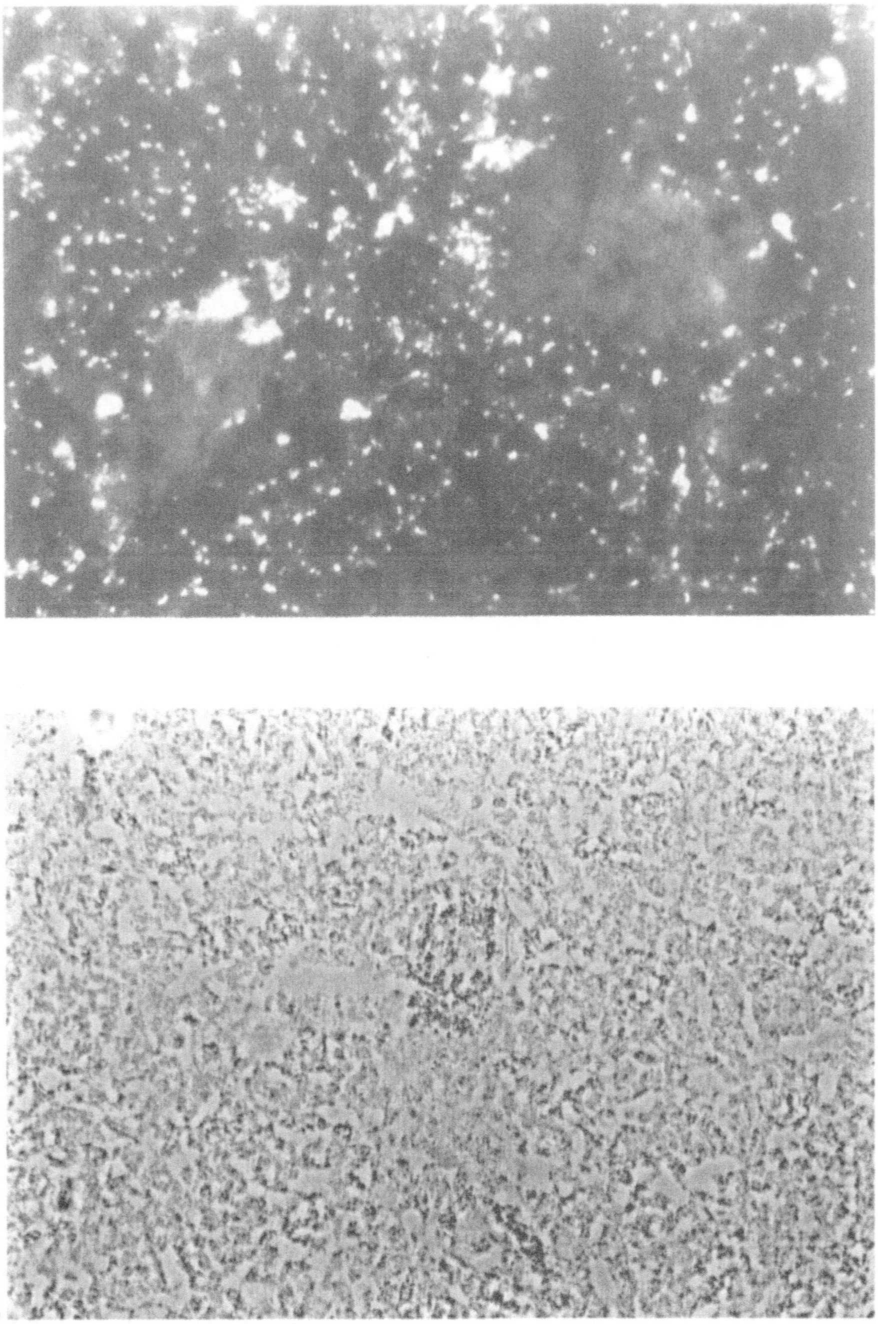

a

b

Fig. 7. Fluorescence (**a**, upper picture) and light (**b**, lower picture) microscopy of a group of two neurons, which were negative for MAO-A antibody

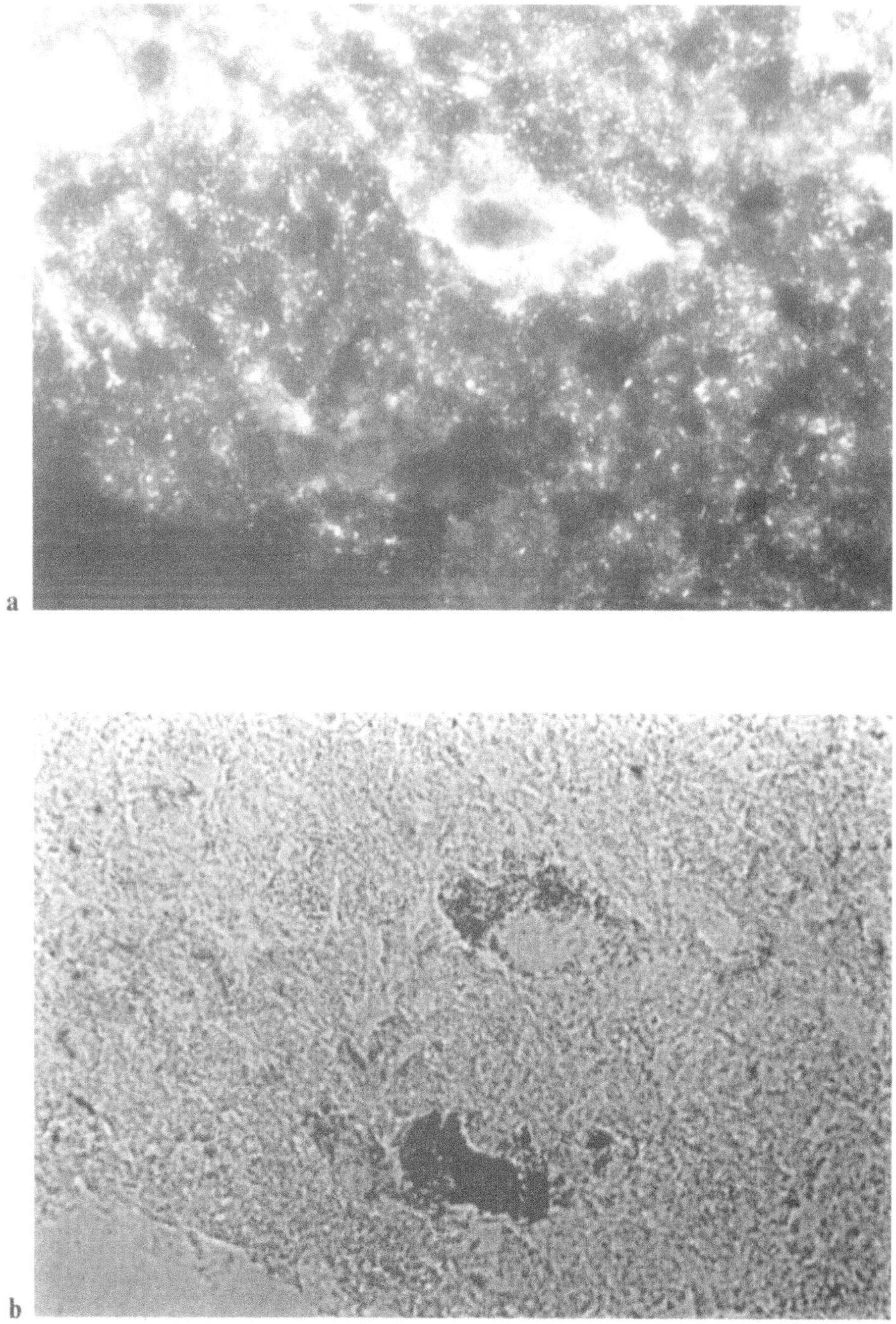

Fig. 8. Fluorescence (**a**, upper picture) and light (**b**, lower picture) microscopy of a group of two neurons, where the neuron containing the lower amount of melanin exhibits the stronger reaction for MAO-A (upper neuron) and the neuron with high melanin density exhibits a very reduced reaction for MAO-A

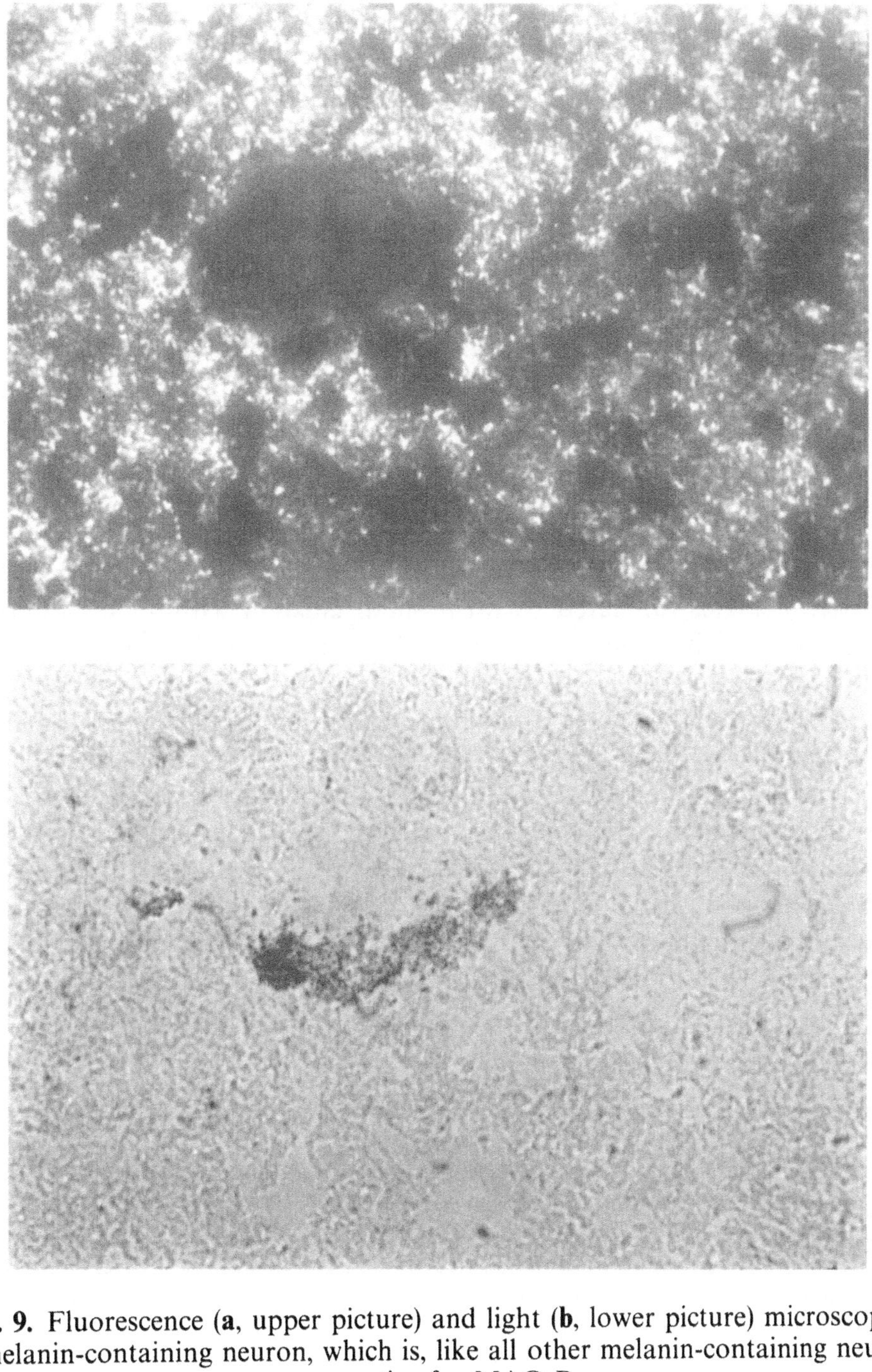

a

b

Fig. 9. Fluorescence (**a**, upper picture) and light (**b**, lower picture) microscopy of a melanin-containing neuron, which is, like all other melanin-containing neurons negative for MAO-B

fiber networks (Fig. 7a and b). In general, we assume that only weakly pigmented neurons exhibit a strong reaction for MAO-A (upper neuron) whereas heavily pigmented neurons are negative for MAO-A (lower neuron; Fig. 8a and b).

All studied melanin-containing neurons in the substantia nigra, pars compacta, are negative for MAO-B (Fig. 9a and b).

Conclusions

Our results confirm and extent data from Westlund et al. (1988) and are essentially in agreement with Konradi et al. (1988). The perikarya of all pigmented neurons in the substantia nigra were negative for MAO-B and the great majority was negative for MAO-A. We were able to demonstrate a positive reaction for MAO-A in approximately 10% of the melanin-containing neurons suggesting the presence of this enzyme in the perikaryon. Statements about the presence of MAO-A in the processes are not possible with the method used. The staining intensity was variable ranging from weak to strong and seems dependent on the melanin content. The MAO-A-positive neurons appeared to form small clusters within the subtantia nigra but till yet the exact topography of these clusters is still unknown.

This findings are in contrast to findings on substrate specific MAO activity in other regions of the human brain stem and basal ganglia, reported by Riederer et al. (1986). While locus coeruleus, raphe nuclei and reticular formation exhibit about the same MAO-A than MAO-B activity, MAO-B is the dominating form in caudate nucleus. On the other hand, our data confirm earlier suggestions that treatment of Parkinson's disease with MAO-B-inhibitors like l-deprenyl act exclusively at the extraneuronal site of the substantia nigra. The missing MAO-B within melanin-containing neurons may be of importance in the etiology of Parkinson's disease.

References

Denney RM, Patel NT, Fritz RR, Abell CW (1982) A monclonal antibody elicited to human platelet monoamine oxidase. Isolation and specifity for human monoamine oxidase-B but not for A. Mol Pharmacol 22:500–508

Kochersperger LN, Waguespack A, Patterson JC, Hseih CCW, Weyler W, Salach JI, Denney RM (1985) Immunological uniqueness of human monoamine oxidase A and B. New evidence from studies with monoclonal antibodies to human MAO A. J Neurosci 11:2874–2881

Konradi C, Svoma E, Jellinger K, Riederer P, Denney R, Thibault J (1988) Topographic immunocytochemical mapping of monoamine oxidase-A, monoamine oxidase-B and tyrosine hydroxylase in human post mortem brain stem. Neuroscience 26:791–802

Riederer P, Konradi C, Schay V, Kienzl E, Birkmayer G, Danielczyk W, Sofic E, Youdim MBH (1986) Localization of MAO-A and MAO-B in human brain: a step in understanding the therapeutic action of 1-deprenyl. Adv Neurol 46:111–118

Westlund KN, Denney RM, Rose RM, Abell CW (1988) Localization of distinct monoamine oxidase-A and monoamine oxidase-B cell populations in human brain stem. Neuroscience 25:439–456

Authors' address: Dr. G. Moll, Clinical Neurochemistry, Department of Psychiatry, University of Würzburg, Füchsleinstrasse 15, D-8700 Würzburg, Federal Republic of Germany.

J Neural Transm (1990) [Suppl] 32: 79–84

Effect of selective and reversible MAO inhibitors on dopamine outflow in rat striatum: a microdialysis study

A. Colzi, F. d'Agostini, R. Kettler, E. Borroni, and **M. Da Prada**

Pharmaceutical Research Department, F. Hoffmann-La Roche Ltd, Basel, Switzerland

Summary. The effect of reversible inhibitors of the monoamine oxidase type A (MAO-A), moclobemide (Aurorix®) and Ro 41-1049 (20 mg/kg i.p. each), as well as of reversible inhibitors of the MAO type B (MAO-B), Ro 19-6327 (1 mg/kg i.p.), on the outflow of dopamine (DA), 3,4-dihydroxyphenylacetic acid (DOPAC) and homovanillic acid (HVA) was studied in the rat by trans-striatal microdialysis. Reversible MAO-A inhibitors markedly increased the output of DA and concomitantly decreased the output of DOPAC and HVA. These effects were absent with the highly selective MAO-B inhibitor Ro 19-6327.

Introduction

Microdialysis is a recently developed method for studying the effects of drugs on the dynamics of neurotransmitters and their main metabolites at specific brain areas of freely moving animals (Di Chiara, 1990). Transstriatal microdialysis in the rat showed that pargyline, an irreversible inhibitor of both monoamine oxidase type A (MAO-A) and type B (MAO-B), induced a significant increase of dopamine (DA) and a parallel decrease of its main metabolites 3,4-dihydroxyphenylacetic acid (DOPAC) and homovanillic acid (HVA) (Imperato and Di Chiara, 1984). Although it is generally accepted that DA in human brain is largely deaminated by MAO-B (Glover et al., 1977), DA in the rat striatum seems to be almost exclusively metabolized by MAO-A (Waldmeier and Maître, 1975). In this study we used the microdialysis technique to investigate the effect of reversible and selective inhibitors of MAO-A (moclobemide and Ro 41-1049) and MAO-B (Ro 19-6327) (Da Prada et al., 1988a, 1990) on the DA deamination in rat striatum.

Materials and methods

Male albino rats (Fü-albino, 250–300 g body weight) were anaesthetized with Vetanarcol (0.4 ml/kg, Veterinaria AG, Zürich). Trans-cerebral dialysis probes (Filtral 12, AN 69 HF, HOSPAL-Industrie, France) were transversally inserted through the dorsal caudate nucleus (coordinates A + 1 mm, V − 5.5 mm from bregma) and secured to the skull using dental cement, according to Imperato and Di Chiara (1984).

After surgery, the rats were allowed to recover for 24 h in their home cages with free access to food and water. Dialysis probes were perfused with Ringer solution (2 µl/min), using polyethylene tubing, connected to a microinfusion pump (CMA 100, Carnegie Medicin, Sweden).

The perfusates were collected every 15 min and 20 µl portions of the Ringer were put into Eppendorf tubes (500 µl) containing 10 µl of internal standard solution. The samples were then assayed for DA, DOPAC and HVA in a single run high pressure liquid chromatography (HPLC) system with a reverse phase column (Nucleosyl 7 µm, C18) and coulometric detection (glassy-carbon electrodes, Analytical cell ESA, mod. 5011). The potential of the two electrodes was maintained at + 30 and − 20 V.

The optimized mobile phase consisted of 0.04 M citric acid, 0.06 M Na_2HPO_4 $2H_2O$, 0.10 mM EDTA, 0.5 mM 1-octanesulphonic acid sodium salt monohydrate and 15% methanol; the pump flow rate was 1.3 ml/min. The retention time for DOPAC, DA and HVA was 7, 8.2 and 15 min, respectively. The detection limit for DA was 0.04 pmoles/sample (20 µl).

Moclobemide (p-chloro-N-[2-morpholinethyl]benzamide), Ro 41-1049 (N-[2-aminoethyl]-5-[m-fluorophenyl]-4-thiazole carboxamide HCl), Ro 19-6327 (N-[2-aminoethyl]-5-chloro-2-pyridine carboxamide HCl) were suspended in all glass Potter-Elvhenjen homogenizer with 0.9 NaCl and 0.3% Tween-80. The volumes for i.p. administration were 2 ml/kg body weight.

The absolute values were expressed in pmoles (means ± SEM); at a time at which the spontaneous release had stabilized (about 2 h after Ringer perfusion) the mean ± SEM of three values were taken as 100% and absolute values transformed in percent values.

Results

As already shown for whole rat brain (Da Prada et al., 1990), moclobemide (20 mg/kg i.p.) and Ro 41-1049 (20 mg/kg i.p.) produced virtually the same inhibition of MAO-A activity in the striatum (Table 1). In contrast to moclobemide, however, Ro 41-1049 induced an almost selective inhibition of MAO-A. The potent and reversible MAO-B inhibitor, Ro 19-6327 (1 mg/kg i.p.), induced a marked and highly selective blockade of MAO-B (Table 1). In parallel experiments the rat striatum was also prepared from animals sacrificed by decapitation and the levels of DA, DOPAC and HVA were measured in striatal homogenates by HPLC with electrochemical detection (Da Prada et al., 1989). These ex vivo experiments showed that both moclobemide and Ro 41-1049 at the doses used, increased DA levels in striatal tissue 90 min after dosing (135 ± 3.2 and 142 ± 8.1% increase, respectively). In parallel, DOPAC and HVA were markedly decreased to 16.0 ± 2 and 30.0 ± 4.4 after moclobemide and to 11.0 ± 0.9 and 17.0 ± 2.3 after Ro 41-1049. As previously observed for rat whole brain after very high oral doses (Da Prada et al., 1988b), Ro 19-6327 injected i.p. failed to modify the levels of DA and of its main metabolites in rat striatum. In the microdialysis experiments moclobemide (20 mg/kg i.p.) increased the outflow of striatal DA with a maximum increase (199 ± 44% increase) 45 min post-dosing (Fig. 1). DOPAC and HVA showed a marked parallel decrement (> 80% decrement) which, however, lasted for more than 120 min

Table 1. Ex vivo MAO-A (5-HT) and MAO-B (PEA) activity in rat striatum

Inhibitors (mg/kg i.p., 90 min)	Activity in % of control (= 100) (Means ± SEM, n = 4) 5-HT	PEA
Moclobemide (20)	12.4 ± 0.9	34.1 ± 4.2
Ro 41-1049 (20)	6.0 ± 0.3	76.0 ± 1.7
Ro 19-6327 (1)	104.6 ± 3.4	12.1 ± 0.7

MAO-A activity was measured radiochemically using [^{14}C]5-HT 200 µM and MAO-B using [^{14}C]PEA 20 µM as substrates (Da Prada et al., 1989).
Control values (nmoles/h/mg wet tissue, means ± SEM) were: 5-HT = 10.3 ± 0.3; PEA = 2.9 ± 0.1

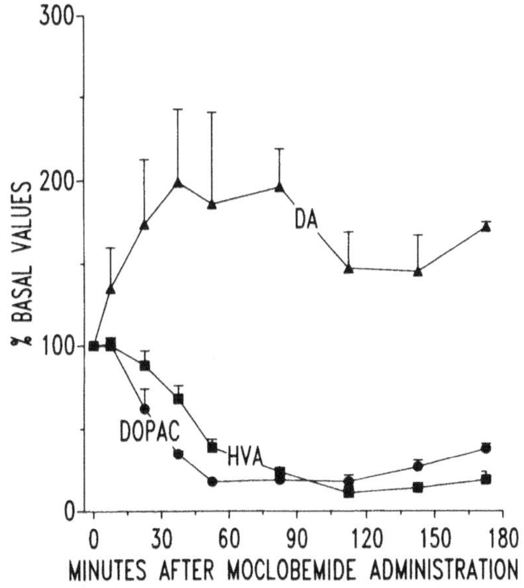

Fig. 1. Effect of moclobemide (20 mg/kg i.p.) on DA (▲), DOPAC (●) and HVA (■) outputs. Results are means ± SEM, n = 3 rats. Basal outputs (pmoles/15 min, means ± SEM) were: DA, 0.227 ± 0.027; DOPAC, 23.9 ± 1.8 and HVA, 12.9 ± 0.6

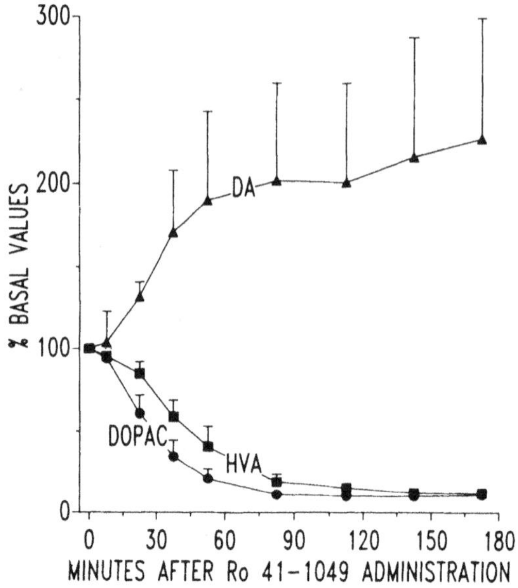

Fig. 2. Effect of Ro 41-1049 (20 mg/kg i.p.) on DA (▲), DOPAC (●) and HVA (■) outputs. Results are means ± SEM, n = 3 rats. Basal outputs (pmoles/15 min, means ± SEM) were: DA, 0.255 ± 0.048; DOPAC, 37.6 ± 5.0 and HVA 22.8 ± 3.4

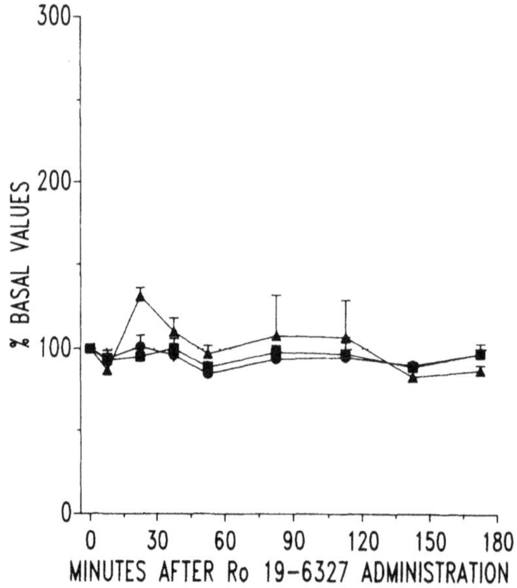

Fig. 3. Effect of Ro 19-6327 (1 mg/kg i.p.) on DA (▲), DOPAC (●) and HVA (■) outputs. Results are means ± SEM, n = 3 rats. Basal outputs (pmoles/15 min, means ± SEM) were: DA, 0.235 ± 0.029; DOPAC, 28.2 ± 1.7 and HVA, 20.9 ± 0.4

after dosing. Ro 41-1049 (20 mg/kg i.p.) induced a marked increase of the DA outflow which was longer-lasting than after moclobemide; concomitantly, DOPAC and HVA were decreased more consistently than after moclobemide (Fig. 2). In contrast, the MAO-B inhibitor Ro 19-6327 did not modify the outflow of DA or of DOPAC and HVA (Fig. 3).

Discussion

This trans-striatal microdialysis study in the rat shows that reversible inhibition of striatal MAO-A induced by moclobemide and Ro 41-1049, is accompanied by a marked increase of the overflow of DA and by a parallel decrement of its metabolites (DOPAC, HVA).

The time-course of the DA outflow indicates that moclobemide produced quantitatively similar changes as Ro 41-1049 although the effects of the latter were longer-lasting. Previous ex vivo experiments with rat whole brain had shown that these inhibitors induced similar MAO-A inhibitory effects after oral administration (Da Prada et al., 1990). As expected, the results in Table 1 indicate that moclobemide and Ro 41-1049, 90 min after dosing produce a similar MAO-A inhibition also in the rat striatum.

Since the highly selective MAO-B inhibitor Ro 19-6327 does not change the outflow of DA and of its metabolites, our results fully support the view that DA in rat striatum and in physiological conditions is virtually not metabolized by MAO-B (Waldmeier and Maître, 1975). These results also suggest that microdialysis in vivo provides an additional technique to study in a more refined way neurotransmitter changes occurring in restricted brain areas after administration of RIMA (reversible inhibitor of the MAO-A) antidepressants.

References

Da Prada M, Kettler R, Cesura AM, Richards JG (1988a) Reversible, enzyme-activated monoamine oxidase inhibitors: new advances. Pharmacol Res Commun [Suppl] 4:21–33

Da Prada M, Kettler R, Zürcher G, Keller HH (1988b) Hemmer der MAO-B und COMT: Möglichkeiten ihrer Anwendung bei der Parkinson-Therapie aus heutiger Sicht. In: Fischer PA (Hrsg) Modifizierende Faktoren bei der Parkinson-Therapie. Editiones Roche, S 309–322

Da Prada M, Kettler R, Keller HH, Burkard WP, Muggli-Maniglio D, Haefely WE (1989) Neurochemical profile of moclobemide, a short-acting and reversible inhibitor of monoamine oxidase type A. J Pharmacol Exp Ther 248:400–413

Da Prada M, Kettler R, Keller HH, Cesura AM, Richards JG, Saura Marti J, Muggli-Maniglio D, Wyss P-C, Kyburz E, Imhof R (1990) From moclobemide to Ro 19-6327 and Ro 41-1049: the development of a new class of reversible, selective MAO-A and MAO-B inhibitors. J Neural Transm [Suppl] 29:279–292

Di Chiara G (1990) In vivo brain dialysis of neurotransmitters. TIPS 11:116–121
Glover V, Sandler M, Owen F, Riley GJ (1977) Dopamine is a monoamine oxidase B substrate in man. Nature 265:80–81
Imperato A, Di Chiara G (1984) Trans-striatal dialysis coupled to reverse phase high pressure liquid chromatography with electrochemical detection: a new method for the study of the in vivo release of endogenous dopamine and metabolites. J Neurosci 4:966–977
Waldmeier PC, Maître L (1975) Lack of significance of MAO-B for the in vivo deamination of dopamine. Naunyn Schmiedebergs Arch Pharmacol 287:R2–R3

Authors' address: Prof. Dr. M. Da Prada, c/o F. Hoffmann-La Roche Ltd, Pharmaceutical Research Department, CH-4002 Basel, Switzerland.

J Neural Transm (1990) [Suppl] 32: 85–88

In vivo studies on the effect of monoamine oxidase inhibitors on dopamine and serotonin metabolism in rat brain areas

H.-W. Clement, Chr. Grote, and **W. Wesemann**

Department of Neurochemistry, Institute of Physiological Chemistry, Philipps University, Marburg, Federal Republic of Germany

Summary. In vivo voltammetric analyses on the action of the monoamine oxidase inhibitors pargyline, deprenyl, and clorgyline on dopamine and serotonin metabolism in rat brain showed strain and regional differences depending on the compound under study.

Introduction

In vivo voltammetry with carbon-fibre electrodes according to Ponchon et al. (1979) and Cespuglio et al. (1981) is a suitable tool to study metabolism of the neurotransmitters serotonin (5-HT), dopamine (DA), and norepinephrine in small discrete areas of the rat brain. The aim of our study was to analyze the effects of three types of monoamine oxidase (MAO) inhibitors in the striatum, the N. raphe dorsalis (NRD) and the locus coeruleus (LC) of the rat: pargyline as a MAO-A and B inhibitor; clorgyline and deprenyl which show preferentially MAO-A and MAO-B inhibition, respectively.

Materials and methods

The effect of MAO inhibitors on catecholamines and 5-HT metabolism in striatum, LC, and NRD of Wistar WU and OFA rats was measured by in vivo voltammetric analyses of 3,4-dihydroxyphenyl-acetic acid, DOPAC, and 5-hydroxyindole-acetic acid, 5-HIAA, according to Ponchon et al. (1979) and Cespuglio et al. (1981). The assays were performed under acute and chronic conditions.

Acute conditions

1 h before anesthesia (chloralhydrate 500 mg/kg i.p.) rats were pretreated with allopurinol (40 mg/kg i.p.), an inhibitor of xanthinoxidase. Thereafter the electrodes

were stereotaxically implanted with coordinates according to Paxinos and Watson (1982). Measurements started directly after implantation of the electrodes and were performed for up to 4 h.

Chronic conditions

Electrodes were implanted stereotaxically under chloralhydrate anesthesia (500 mg/kg i.p.) and connected with the polarograph by use of a flexible wire and a sliding contact. Measurements in the freely moving rat started 5 to 10 days after implantation.

Results

Pargyline

The experiments with pargyline (75 mg/kg i.p.) in the striatum, were performed in Wistar WU and OFA rats under acute conditions. The DOPAC signal in both strains disappeared completely after 1 h in contrast to the 5-HIAA signal which was different in the two strains. In OFA rats the signal disappeared after 3 h, whereas only a small decrease was observed in the Wistar WU rats (Fig. 1). In the LC of Wistar WU rats the DOPAC signal, as a measure for neuronal activity of NE containing cells, declined as rapidly as in the striatum. The 5-HIAA signal showed a slow decrease, and 20–30% of the signal remained even 3 h after injection. The experiments in the NRD of Wistar WU rats were performed under chronic conditions 5 and 10 days after implantation. After 5 days pargyline lead to a 60–70% decrease of the signal in the NRD. The remaining peak seems to be due to uric acid, as was shown by Cespuglio et al. (1986). 10 days after implantation the signal in the NRD disappeared completely after 1 h. The same results were obtained by

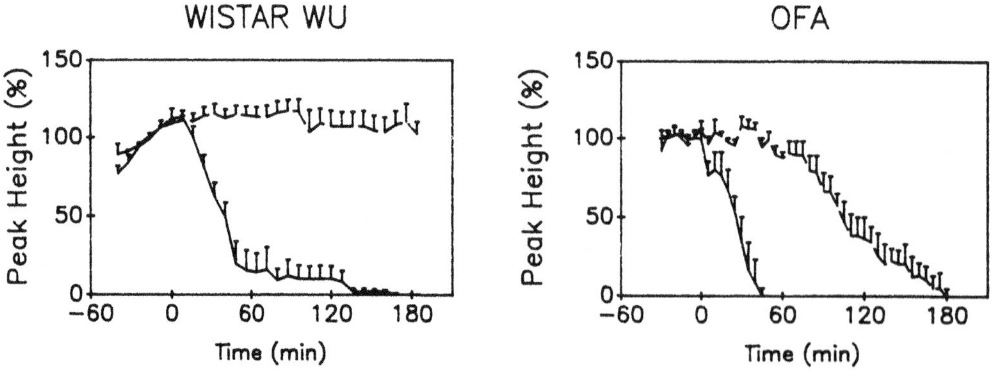

Fig. 1. Effect of pargyline (75 mg/kg i.p., n = 5) on extracellular DOPAC (solid line) and 5-HIAA (dashed line) concentrations in the striatum of Wistar WU and OFA rats. Concentrations are expressed as % peak height (± SEM) of control (100%)

Cespuglio et al. (1986) in OFA rats. According to the measurements in the NRD there is no difference between the two strains.

Deprenyl

The effect of deprenyl (3 mg/kg i.p.) was studied in the striatum of Wistar WU rats under acute conditions. The DOPAC peak disappeared completely while the 5-HIAA signal was not affected. This is in accordance with the well-known MAO type B inhibitor activity of this substance. Under chronic conditions the 5-HIAA signal was also unchanged in the NRD.

Clorgyline

The MAO type A inhibitor clorgyline (10 mg/kg i.p.) lead to a disappearance of the DOPAC signal in the striatum after nearly 3 h. The 5-HIAA signal in the striatum was unaffected while the signal in the NRD was reduced.

Discussion

By using differential pulse voltammetry with carbon fibre electrodes we observed strain differences regarding the action of pargyline on the 5-HIAA peak in the striatum of Wistar WU and OFA rats. The identity of this peak with the 5-HIAA signal was confirmed in both strains by administration of tryptophan, 5-HTP, and the decarboxylase inhibitor NSD 1015 (Cespuglio et al., 1981; Clement et al., 1989). Pretreatment of the rats with allopurinol prevents contamination of the signal with uric acid. The latter compound is important for experiments under acute conditions. This contamination is due to the destruction of the blood brain barrier, which seems to be renewed 10 days after implantation (Cespuglio et al., 1986). This may explain our findings with pargyline in the NRD under chronic conditions. The observation that deprenyl has no effect on the 5-HIAA signal in the NRD is in accordance with findings of Cespuglio et al. (1990). Also in the striatum we found in accordance with published data on ex vivo studies no deprenyl effect on the 5-HIAA peak, an indication that 5-HT is not a substrate for MAO B. The rapid and complete disappearance of the striatal DOPAC signal is in agreement with the activity of deprenyl as a pure MAO B inhibitor. The decrease of 5-HIAA levels in the NRD after clorgyline application meets the requirements of a MAO A inhibitor and is in agreement with results published by Cespuglio et al. (1990). However, no explanation can be given so far why clorgyline and pargyline have no effect on 5-HIAA in the striatum. Since dopamine is a substrate for MAO A and B, the reduced striatal DOPAC signal is due to MAO A inhibition by clorgyline. The strain and regional differences of monoamine metabolites found after MAO inhibition underline the importance of in vivo measurements in discrete brain areas.

References

Cespuglio R, Faradji H, Ponchon JL, Buda M, Riou F, Gonon F, Pujol JF, Jouvet M (1981) Differential pulse voltammetry in brain tissue. I. Detection of 5-hydroxyindoles in the rat striatum. Brain Res 223:287–298

Cespuglio R, Sarda N, Gharib H, Faradji H, Jouvet M (1986) Differential pulse voltammetry in vivo with working carbon fiber electrodes: 5-hydroxyindole compounds or uric acid detection? Exp Brain Res 64:588–595

Cespuglio R, Sarda N, Gharib H, Chastrette N, Houdouin F, Rampin C, Jouvet M (1990) Voltammetric detection of the release of 5-hydroxyindole compounds throughout the sleep-waking cycle of the rat. Exp Brain Res 80:121–128

Clement HW, Grote C, Wesemann W (1989) Differential pulse voltammetry as a tool to study Parkinson's disease – effect of drugs on the dopaminergic and serotonergic system. The fourth meeting on electrochemical detection, HPLC and in vivo monitoring in the bioscience, Nottingham, September 19–21. J Neurosci Methods 29:298

Paxinos G, Watson C (1982) The rat brain in stereotaxic coordinates. Academic Press, Sydney

Ponchon JL, Cespuglio, R, Gonon F, Jouvet M, Pujol JF (1979) Normal pulse polarography with carbon fiber electrodes for in vitro and in vivo determination of catecholamines. Anal Chem 51:1483–1486

Authors' address: Prof. Dr. W. Wesemann, Department of Neurochemistry, Institute of Physiological Chemistry, Philipps University, Hans-Meerwein-Strasse, D-3550 Marburg/Lahn, Federal Republic of Germany.

J Neural Transm (1990) [Suppl] 32: 89–92

Some pharmacological implications of MAO-mediated deamination of branched aliphatic amines: 2-Propyl-1-aminopentane and N-(2-propylpentyl)glycinamide as valproic acid precursors

P. H. Yu and **B. A. Davis**

Neuropsychiatric Research Unit, Department of Psychiatry, University of Saskatchewan, Saskatoon, Saskatchewan, Canada

Summary. 2-Propyl-1-aminopentane (2-PAP) and N-(2-propylpentyl)glycinamide (PPG) were readily deaminated by rat liver monoamine oxidase B and rat aorta semicarbazide-sensitive amine oxidase. The deaminated product, valproic acid (VPA), was identified by HPLC-fluorometric assessment. Absorption and biotransformation of these compounds and their VPA metabolite into the brain were rapid processes. An investigation was conducted to examine whether these compounds can be used as VPA prodrugs. Both compounds, however, at relatively low doses exhibited distinct tremor effects in mice and rats. They also potentiate the convulsant effect induced by mercaptopropionic acid (MPA).

Introduction

Monoamine oxidase (MAO, EC 1.4.3.4.) is well known for its catalytic activities on endogenous arylalkylamine substrates, such as neuronal catecholamines, indolylethylamines and trace amines (Yu, 1986). Interestingly, straight chain aliphatic amines in the range 3 up to 18 carbon atoms are also found to be deaminated by MAO-B with high affinity, and by semicarbazide-sensitive amine oxidase (SSAO, EC 1.4.3.6.) (Blascho, 1952; Yu, 1989, 1990).

Recently, the deamination of an antiepileptic prodrug 2-n-pentyl-aminoacetamide (Milacemide) (de Varebeke et al., 1988), which can cross the blood brain barrier and then be oxidized by MAO to form glycinamide and glycine, has been shown to contribute to the delivery of glycine to the CNS. MAO-mediated deamination of aliphatic amines may prove useful in the target-directed delivery of drugs.

Valproic acid (VPA), a branched aliphatic acid, has been widely used as an anticonvulsant. We became interested in investigating whether 2-PAP or 2-PPG could be deaminated by MAO or SSAO to yield VPA. In addition we wished to assess whether the VPA precursors also possessed anticonvulsant activity and could thus be used as a VPA prodrug.

Experimentals

Both 2-PAP and 2-PPG were chemically synthesized in our laboratory. The identity and purity of the products were assured by mass spectroscopic analysis.

The rat liver mitochondrial MAO was prepared by differential centrifugation as previously described and SSAO was prepared from the rat aorta according to a previously described method.

A fluorometric method based on the measurement of hydrogen peroxide released during the oxidation of the amines (Yu and Boulton, 1980) was used for MAO assays. The conversion of aldehyde products to their corresponding acids was carried out when necessary by including yeast aldehyde dehydrogenase and β-NAD in the enzyme reaction mixture (Yu and Davis, 1988).

For detection of VPA, a slightly modified procedure was adopted, using 4-bromomethyl-6,7-dimethoxycoumarin (BMDC) for derivatization and the products were separated by high-performance liquid chromatography with caproic acid as internal standard and detected fluorometrically (Dunges, 1977).

Tremors induced by 2-PAP and 2-PPG in mice and rats were measured with different doses and routes of administration. In the study of the potentiation of convulsion by 2-PAP and 2-PPG, the occurrence and intensity of clonic and tonic seizure episodes, tremor and mortality of the animals after administration of MPA were recorded.

Results and Discussion

When 2-PAP and 2-PPG were oxidized, i.e. with formation of H_2O_2, by rat mitochondrial MAO the reactions were linear with time (for at least 10 min) and with respect to increasing enzyme concentration. The deamination of 2-PAP and 2-PPG by the enzymes in vitro has also been confirmed by direct analysis of VPA after enzymatic conversion of the aldehyde intermediate metabolite.

The IC_{50} for 1-deprenyl and clorgyline are about 1×10^{-7} M and 1×10^{-5} M respectively. Consistent with the deamination of straight chain aliphatic amines, 2-PAP and 2-PPG are also predominantly deaminated by MAO-B. Both 2-PAP and 2-PPG can also be deaminated by SSAO prepared from rat aorta tissues.

The apparent K_m and V_{max} values of rat liver mitochondrial MAO-B (enzyme pretreated with 5×10^{-7} M clorgyline) and rat aorta SSAO (enzyme pretreated with 1×10^{-4} M clorgyline) towards 2-PAP and 2-PPG are in the micromole range, indicating that both compounds are high affinity substrates for the enzymes.

VPA, the deaminated product of 2-PAP, was identified in the blood and brain following administration of a single i.p. dose of the drugs. Absorption and biotransformation of these compounds into blood and transfer of the drug and its metabolite into the brain were found to be very rapid. The half-life of VPA in the blood was about 4 hours, which is longer than following administration of VPA. However, the maximal levels of VPA after injection of 2-PAP or 2-PPG were much lower. It is interesting that the VPA levels were similar after 5 hours of administration of both prodrugs and VPA.

Since 2-PAP can be oxidized in vivo to produce VPA we tried therefore to evaluate whether 2-PAP also exhibits anticonvulsant activity. Unfortunately, any such activity proved to be difficult to measure, since both 2-PAP and 2-PPG induced an intense tremor effect in mice. On administration of lower doses of the drugs, animals exhibit a sporadic shivering effect. At higher doses both drugs induced a continuous, intense, tremor of the whole body for at least 60 min. Straight chain aliphatic amines such as n-pentylamine and n-octylamine, at relatively high doses do not cause similar tremor effects.

Furthermore we noticed that 2-PAP and 2-PPG also potentiate the convulsions induced by convulsant MPA. This includes an increase of clonic seizure (episodes of spasm), tonic seizure (i.e. hindlimb extension), as well as mortality. The time required for the on-set of first clonic seizure, however, was not affected.

Our findings that 2-PAP is readily deaminated by rat liver MAO-B and aorta SSAO, further confirms that both MAO and SSAO are not only capable of metabolizing straight chain aliphatic amines, but are also able to deaminate branched aliphatic amines in vitro and in vivo. They are high-affinity substrates in comparison to many catecholamines and indolylamines for these amine oxidases.

Although VPA can be detected in the brain following peripheral administration of these VPA precursors, the latter can not be considered to a VPA prodrug as we had previously hoped. Therapeutic levels of VPA in the brain and serum following administration could not be achieved because of the severe behavioral effects.

The mechanism of such behavioral effects is not yet clear. The branched chain structure of these compounds appears to be essential for the induction of the tremor, since straight chain aliphatic amines, such as n-pentylamine and n-octylamine at the same concentrations are without effect. It is possible that the observed tremor induced by 2-PAP and 2-PPG is related to a toxic effect of VPA; it is known that tremor can be a side effect of chronic administration of VPA.

Since the discovery of the therapeutic properties of VPA many chemical analogues have been tested for their anticonvulsant activity. Among them several VPA analogues, such as 2-butylhexanoic acid, 2-propylheptanoic acid and 2-propylhexanoic acid, exhibit enhanced anticonvulsant activity. In addition the amide and various esters of VPA have also been tested. Dipropylacetamide also possess anticonvulsant activity in man, apparently as a consequence of its conversion to VPA in the gastrointestinal tract. The amide itself also seems to possess anticonvulsant activity. It was surprising therefore to find, that 2-PAP and 2-PPG, VPA precursors, enhanced the convulsions induced by MPA.

Acknowledgements

We thank D. Young and L. Schira for their technical assistance and Canadian Medical Research Council and Saskatchewan Health for their continuing financial support.

References

Blaschko H (1981) Amine oxidase and amine metabolism. Pharmacol Rev 4: 414–458

De Varebeke PJ, Cavalier R, David-Remacle M, Youdim MBH (1988) Formation of the neurotransmitter glycine from the anticonvulsant milacemide is mediated by brain monoamine oxidase-B. J Neurochem 50:1011–1016

Dunges W (1977) 4-Bromomethyl-7-methoxycoumarin as a new fluorescence label for fatty acids. Anal Chem 49:442–444

Yu PH (1986) Monoamine oxidase. In: Boulton AA, Baker GB, Yu PH (eds) Neuromethods, vol 5. Humana, Clifton NJ, pp 235–272

Yu PH (1989) Deamination of aliphatic amines of different chain lengths by rat liver monoamine oxidase A and B. J Pharm Pharmacol 41:205–208

Yu PH (1990) Oxidative deamination of aliphatic amines by rat aorta semi-carbazide-sensitive amine oxidase. J Pharm Pharmacol (in press)

Yu PH, Boulton AA (1980) Regional oxidation of tyramine isomers in rat brain. J Neurochem 35:255–257

Yu PH, Hertz L (1982) Differential development of type A and type B monoamine oxidase of mouse astrocytes in primary cultures. J Neurochem 39:1492–1495

Yu PH, Davis BA (1988) Stereospecific deamination of benzylamine catalyzed by different amine oxidases. Int J Biochem 20:1197–1201

Authors' address: Dr. P. H. Yu, Neuropsychiatric Research Unit, Department of Psychiatry, University of Saskatchewan, Saskatoon, Saskatchewan, Canada, S7N 0W0.

J Neural Transm (1990) [Suppl] 32: 93–102

[^{3}H]Harman labels selectively and with high affinity the active site of monoamine oxidase (EC 1.4.3.4) subtype A (MAO-A) in rat, marmoset, and pig

T. May, S. Strauss, and **H. Rommelspacher**

Department of Neuropsychopharmacology, Free University, Berlin, Federal Republic of Germany

Summary. [^{3}H]Harman was used in binding studies with CNS tissue of rat, pig, and marmoset and with visceral organs of the rat.

In the mitochondrial fractions of the CNS of the 3 species [^{3}H]harman binding exhibits the same pharmacological profile in displacement studies. A detailed analysis reveals a high specificity for MAO-A.

Furthermore, we applied [^{3}H]harman binding to quantify the MAO-A content in 6 CNS regions of each species as well as in 6 visceral organs of the rat.

Introduction

The amine harman (1-methyl-β-carboline, Fig. 1) has been found in plants and mammals (e.g. Rommelspacher and Susilo, 1985; Bosin et al., 1989) and elicits neurotropic effects (e.g. tremor and hallucinations; reviewed by Airaksinen and Kari, 1981).

We have determined by a novel binding assay with [^{3}H]harman, that harman binds selectively and with high affinity to monoamine oxidase (EC 1.4.3.4) subtype A (MAO-A) in the cerebral cortex of the rat. The K_i-values of [^{3}H]harman displacement correlate highly significant with the corresponding K_i-values of inhibition of MAO-A but not MAO-B (May et al., 1990a).

Furthermore, the B_{max}-values from [^{3}H]harman binding at different CNS tissues of rat and marmoset determined by Scatchard analyses correlate highly significant with the respective V_{max}-values of MAO-A but not MAO-B, determined by Lineweaver-Burk plots with kynuramine as substrate (May et al., 1990b).

In this paper, we have compared the binding characteristics of [^{3}H]harman in 6 different CNS tissues of the 3 mammalian species rat,

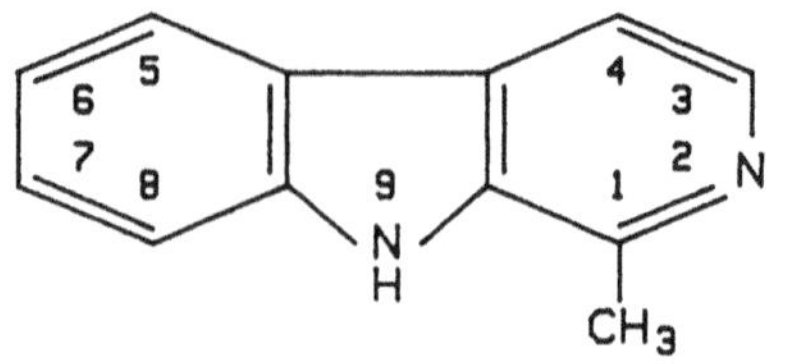

Fig. 1. Chemical structure of harman or 1-methyl-β-carboline

marmoset, and pig and determined the maximum binding densities of [^{3}H]harman in 6 visceral organs of the rat in relation to the corresponding maximum MAO-A and MAO-B activities.

Materials and methods

Animals

3–6 months and 26 days old male Wistar rats (animal farm Hagemann, Boesingfeld, F.R.G.) and 18 months old female marmosets (Callithrix jacchus; provided by Dr. Neubert, Dept. Embryonal Pharmacology and Toxicology, Free University, Berlin, F.R.G.) were used. Pig brains were obtained within 2 hours after death from a local slaughterhouse.

Chemicals

[6-^{3}H]Harman (specific activity 26.6 Ci/mmol) was custom-labelled by Amersham, G.B.

Harmaline, harmine, norharman, kynuramine, 5-F-α-methyltryptamine, 4-hydroxyquinoline, MOPS, Tris, bovine serum albumine were purchased from Sigma, Munich, F.R.G.; pargyline from Serva, Heidelberg, F.R.G.; clorgyline and L-deprenyl from RBI, Biotrend, Koeln, F.R.G.

Lilly 51641 (Eli Lilly and Corporate Center, Indianapolis, U.S.A.) and brofaromine (Ciba Geigy, Basel; Switzerland) were gifts.

Preparation of crude mitochondria

The crude mitochondrial or P_2-fraction of the different tissues were prepared according to Whittaker (1969) in a slightly modified fashion as described previously (May et al., 1990b).

MAO assay

Specific MAO activity was measured according to Kraml (1965) using crude mitochondria of different visceral organs as enzyme sources. The mitochondria homogenate was pretreated (30 minutes at 37 °C) with 1.0 µM L-deprenyl and 1.0 µM clorgyline, respectively, for determining maximum MAO-A or MAO-B activity (30 minutes at 37 °C) by Lineweaver-Burk analyses.

[^{3}H]Harman binding assay

The assay was performed in 1 ml 25 mM sodium-Mops, pH 7.4, 90 minutes at 0°C, containing 100 or 200 µl homogenate of the various tissues as described previously (May et al., 1990a).

Protein determination

Protein was determined according to Lowry et al. (1951) with bovine serum albumine as the standard.

Data analyses

All data were analyzed by Hewlett Packard table computers (series 200/300) as described previously (May et al., 1990a, b).

The average values are presented from at least 2 independent experiments. The SEM-values are between 5 and 30%.

Results and discussion

The potent and selective inhibitors of MAO-A, Lilly 51641 (Fuller, 1968; Fig. 2) and clorgyline (Johnston, 1968; Fig. 2), as well as harmaline (Buckholtz and Boggan, 1977), harmine (Gorkin and Tatyanenko, 1967), 5-F-α-methyltryptamine (Kinemuchi and Arai, 1986), and brofaromine (Waldmeier et al., 1983) displace the [^{3}H]harman binding at pig mitochondria of cerebral cortex in the lower and middle nM concentration range (Table 1).

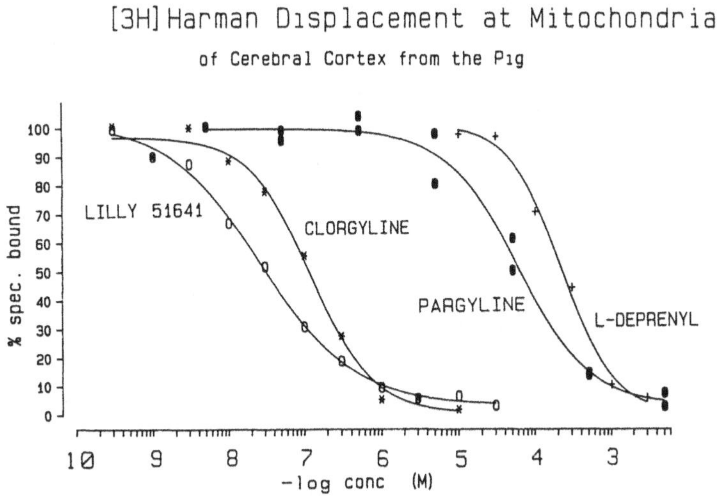

Fig. 2. 300 µg protein of crude mitochondrial fraction and 5.5 nM [^{3}H]harman were used for this representative displacement studies. Nonspecific binding was defined as residue binding in the presence of 5 µM harmaline

Table 1. [^{3}H]Harman displacement at crude mitochondria of cerebral cortex from 3 mammalian species

Compound	Pig K_i(nM)	n	Marmoset K_i(nM)	n	Rat K_i(nM)	n
Harmine	6.4	2	1.6	2	1.6	3
Harmaline	10.2	2	4.2	2	2.2	4
Clorgyline	35.8	2	7.3	2	8.0	4
5-F-α-methyltryptamine	29.7	2	22.3	2	15.5	2
Lilly 51641	8.4	2	16.7	2	17.2	2
Brofaromine	153.	2	94.	2	65.	3
Norharman	34.4	2	n.d.		100.	9
Pargyline	19,000.	2	n.d.		2,400.	4
L-deprenyl	71,000.	3	n.d.		5,500.	4

However, the potent and selective MAO-B inhibitors L-deprenyl (Knoll and Magyar, 1972) and pargyline (Squires, 1972), displace the binding of [^{3}H]harman at the same tissue fraction only in a 3 order higher concentration (Fig. 2 and Table 1), where both compounds inhibit MAO-A activity, too. These data suggest that [^{3}H]harman binds selectively to MAO-A in the pig, as described previously for the rat (May et al., 1990a) and the marmoset (May et al., 1990b). Analyses of the logarithmic K_i-values of the inhibitors in Table 1 shows highly significant correlations between rat and pig ($r = 0.94$, $p < 0.001$, $n = 9$) as well as between rat and marmoset ($r = 0.98$, $p < 0.001$, $n = 6$), and a significant correlation between marmoset and pig ($r = 0.82$, $p < 0.05$, $n = 6$). In conclusion, these displacement experiments reveal the same pharmacological profile of [^{3}H]harman binding in all 3 mammalian species, characteristic for the active site of MAO-A.

We determined the [^{3}H]harman binding characteristics further in several CNS regions of the 3 species (Table 2): Only a single high affinity binding site has been detected in all tissues and species with a very similar K_D about 2 nM (Scatchard plots with r between -0.91 and -0.99). However, the B_{max}-values vary among the species and tissues. The rat shows the highest binding densities in all regions: between 2-fold (cerebellum) and 4-fold (hippocampus) higher relative to the pig, and between 60% (hypothalamus) and 4-fold (hippocampus) higher relative to the marmoset. By comparing the B_{max}-values of pig and marmoset, a similar density is in all regions but not in the hypothalamus, where the marmoset has a 2-fold higher B_{max} (Table 2).

The finding of solely MAO-B in pig brain by former investigators (Tipton and Spires, 1968; Squires, 1972) can be explained by experimental restrictions. They determined the relative part of the MAO subtypes by selective inhibition of deamination with different substrates. With this

Table 2. Regional distribution of specific [^{3}H]harman binding at crude mitochondria of the CNS determined by Scatchard analyses from 2–5 experiments

Region	Pig K_D	B_{max}	Marmoset K_D	B_{max}	Rat K_D	B_{max}
Hypothalamus	2.1	1.59	1.6	3.44	2.3	5.54
Thalamus	2.9	1.45	n.d.		n.d.	
Hippocampus	2.9	1.02	1.4	1.10	1.9	4.03
Cerebral cortex	2.4	0.89	1.6	1.35	2.4	3.64
Striatum	1.9	1.09	1.4	1.41	2.1	3.60
Cerebellum	1.7	1.42	2.0	1.23	2.2	3.19
Spinal cord	n.d.		2.6	0.90	2.6	2.06

K_D-values: nM; B_{max}-values: pmol/mg protein

procedure, they found only sigmoid curves indicating a pharmacological profile of the single enzyme type B in pig brain. In brains of rats is a MAO-A: MAO-B ratio of 50:50 (Johnston, 1968; Fig. 4), whereas in primates the MAO-B part displays ~ 70–95% (Garrick and Murphy, 1980). The pig brain contains possibly at least 95% MAO activity of the B type: Thus, the MAO-A part cannot accurately be detected by the used enzymatical methods.

Finally, for investigation of extraneuronal tissues with the novel binding assay, we applied the method to visceral organs of the rat (Figs. 3, 4). In all 6 tissues (liver, thymus, kidney, lung, spleen, heart), the Scatchard plots are linear with calculated K_D-values between 2 and 3 nM (Fig. 3A–C). The highest binding densities (B_{max}, Figs. 3A–C, 4) and maximum MAO-A activities (V_{max}, Figs. 3G–F, 4) were determined in hearts of 6 months old rats, whereas in hearts of 26 days old rats both values are ~ 20-fold lower (Figs. 3C, F, 4A). A strong increase of cardiac MAO-A activity during ontogeny in rats (especially in male rats) has been described previously (Horita and Lowe, 1972; Edwards et al., 1979). The MAO-B activity, however, remains low and unchanged during ontogeny (Figs. 3I, 4B). Liver and lung of 6 months old rats possess ~ 1:4 of the heart B_{max} and MAO-A V_{max} (Figs. 3A, B, D, E, 4A) but more than one range order higher MAO-B activity (Figs. 3G, H, 4B). Plotting of the relationship between B_{max}-values of [^{3}H]harman binding versus V_{max}-values of the MAO assays reveals a calculated highly significant positive correlation with MAO-A (Fig. 4A) but not MAO-B (Fig. 4B).

All presented results are consistent and directs to a single conclusion: [^{3}H]Harman can be used as a selective label at different tissues and mammalian species for determination of concentration, distribution pattern and inhibitor specificity of the MAO-A site.

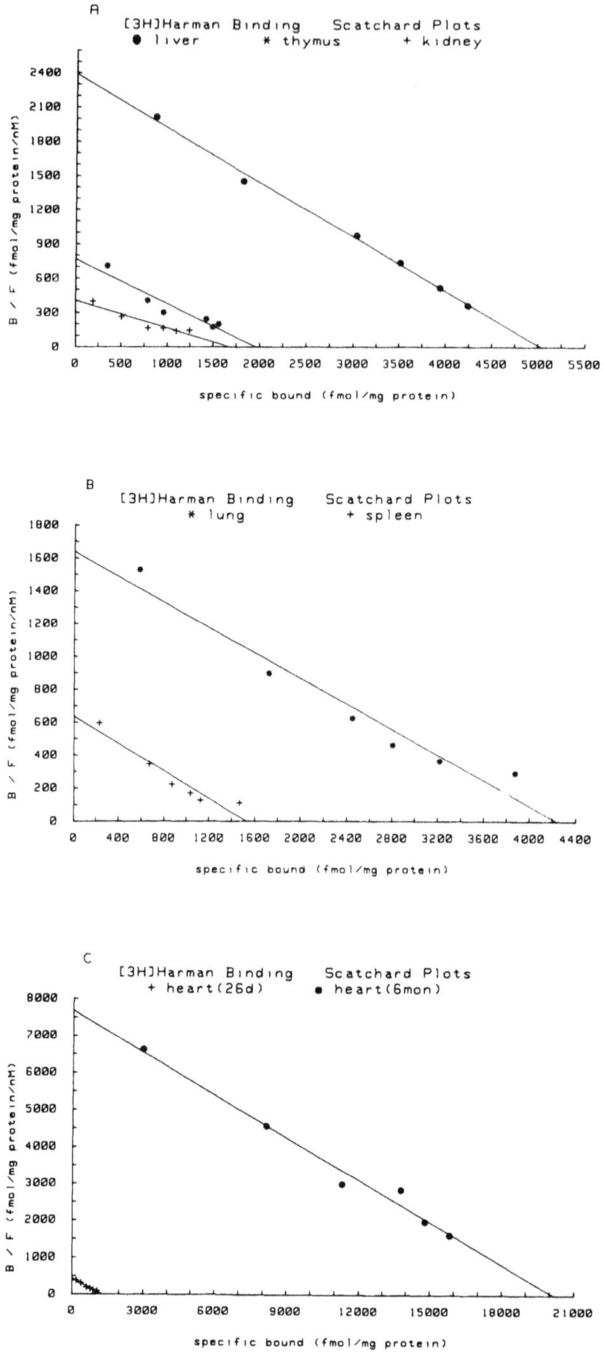
A
[3H]Harman Binding Scatchard Plots
● liver * thymus + kidney
B / F (fmol/mg protein/nM)
specific bound (fmol/mg protein)
B
[3H]Harman Binding Scatchard Plots
* lung + spleen
B / F (fmol/mg protein/nM)
specific bound (fmol/mg protein)
C
[3H]Harman Binding Scatchard Plots
+ heart(26d) ● heart(6mon)
B / F (fmol/mg protein/nM)
specific bound (fmol/mg protein)

Fig. 3A–C

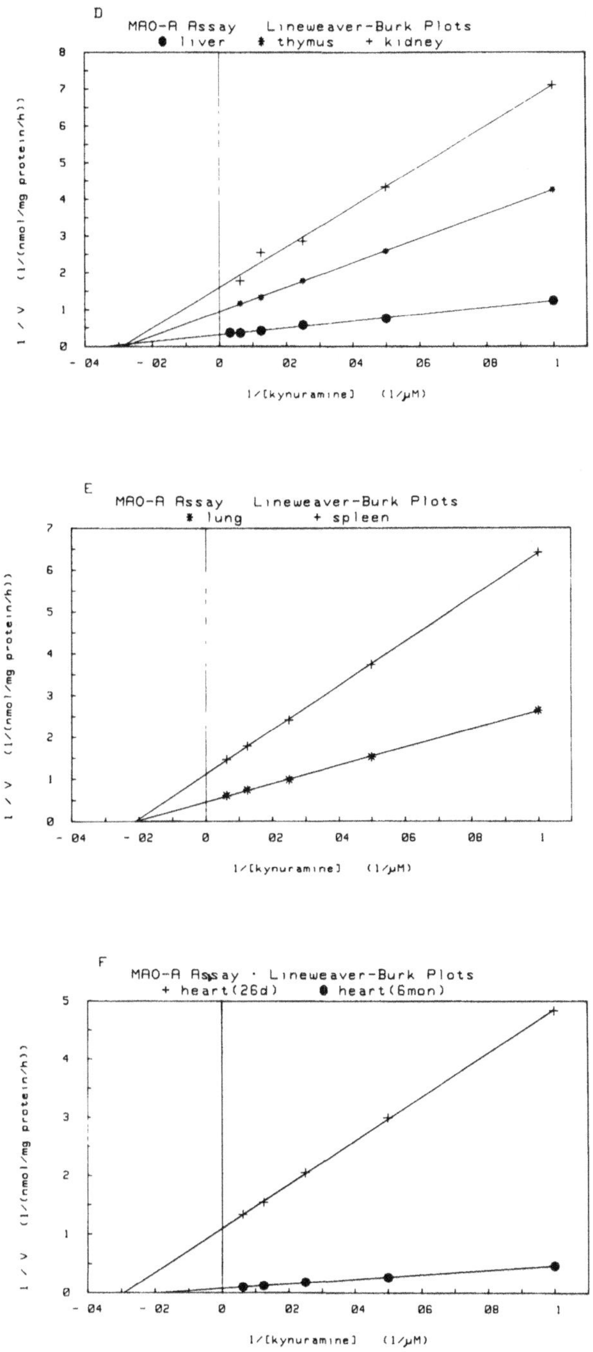
D
MAO-A Assay Lineweaver-Burk Plots
● liver * thymus + kidney
1 / V (1/(nmol/mg protein/h))
1/[kynuramine] (1/μM)
E
MAO-A Assay Lineweaver-Burk Plots
* lung + spleen
1 / V (1/(nmol/mg protein/h))
1/[kynuramine] (1/μM)
F
MAO-A Assay · Lineweaver-Burk Plots
+ heart(26d) ● heart(6mon)
1 / V (1/(nmol/mg protein/h))
1/[kynuramine] (1/μM)

Fig. 3D–F

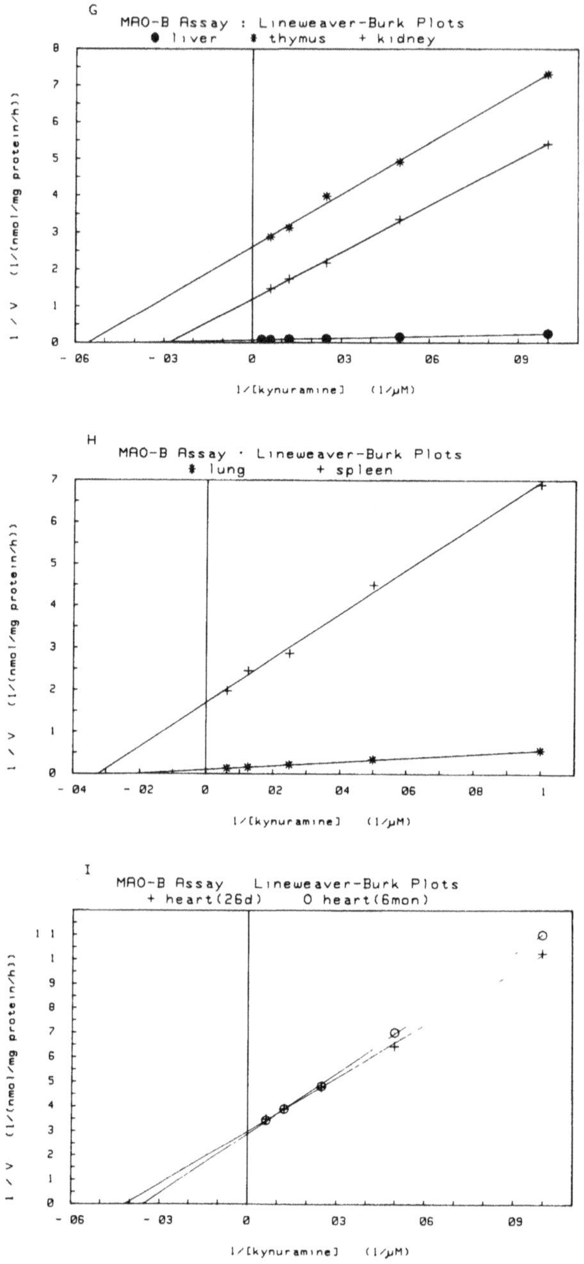

Fig. 3G–I

Fig. 3. [^{3}H]Harman binding and MAO assays were carried out with the same crude mitochondrial fraction of visceral organs from 6 months old rats and of hearts from 26 days old rats, too

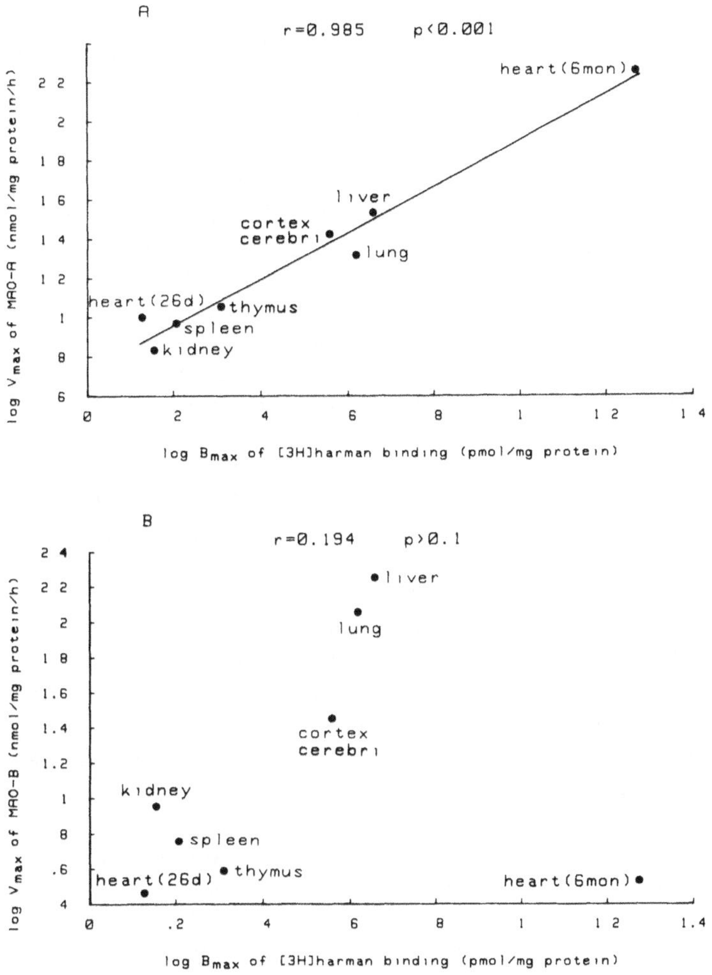

Fig. 4. The logarithmic B_{max}-values of Scatchard analyses were plotted versus the respective V_{max}-values of Lineweaver-Burk analyses determined with crude mitochondria from visceral organs and cerebral cortex of the rat

References

Airaksinen MM, Kari I (1981b) β-carbolines, psychoactive compounds in the mammalian body, part II. Effects. Med Biol 59:190–211

Bosin TR, Borg S, Faull KF (1989) Harman in rat brain, lung and human CSF: effects of alcohol consumption. Alcohol 5:505–511

Buckholtz NS, Boggan WO (1977) Monoamine oxidase inhibition in brain and liver produced by β-carbolines: structure-activity relationships and substrate specificity. Biochem Pharmacol 26:1991–1996

Edwards DJ, Pak KY, Venetti MC (1979) Developmental aspects of rat heart monoamine oxidase. Biochem Pharmacol 28:2337–2343

Fuller RW (1968) Kinetic studies and effects in vivo of a new monoamine oxidase inhibitor, N[2-(o-chlorophenoxy)-ethyl]-cyclopropylamine. Biochem Pharmacol 17:2097–2106

Garrick NA, Murphy DL (1980) Species differences in the deamination of dopamine and other substrates for monoamine oxidase in brain. Psychopharmacology 72:27–33

Gorkin VZ, Tatyanenko LV (1967) On the inhibition by harmine of the oxidative deamination of biogenic monoamines. Life Sci 6:791–795

Horita A, Lowe MC (1972) On the extraneuronal nature of cardiac monoamine oxidase in the rat. In: Costa E, Sandler M (eds) Advances in biochemical psychopharmacology, vol 5. Raven Press, New York, pp 227–242

Johnston JP (1968) Some observations upon a new inhibitor of monoamine oxidase in brain tissue. Biochem Pharmacol 17:1285–1297

Kinemuchi H, Arai Y (1986) Selective inhibition of monoamine oxidase A and B by two substrate analogues, 5-F-α-methyltryptamine and p-chloro-β-methylphenethylamine. Res Commun Chem Pathol Pharmacol 54:125–128

Knoll J, Magyar K (1972) Some puzzling pharmacological effects of monoamine oxidase inhibitors. In: Costa E, Sander M (eds) Advances in biochemical psychopharmacology, vol 5. Raven Press, New York, pp 393–408

Kraml M (1965) A rapid microfluorometric determination of monoamine oxidase. Biochem Pharmacol 14:1683–1685

Lowry OH, Rosebrough NJ, Farr AL, Randall RJ (1951) Protein measurement with the Folin phenol reagent. J Biol Chem 193:265–275

May T, Rommelspacher H, Pawlik M (1990a) [^{3}H]Harman binding experiments. I. A reversible and selective radioligand for monoamine oxidase subtype A in the CNS of the rat. J Neurochem (in press)

May T, Pawlik M, Rommelspacher H (1990b) [^{3}H]Harman binding experiments. II. Regional and subcellular distribution of specific [^{3}H]harman binding and MAO-A and -B activity in marmoset and rat. J Neurochem (in press)

Rommelspacher H, Susilo R (1985) Tetrahydroisoquinolines and β-carbolines: putative natural substances in plants and mammals. In: Jucker E (ed) Progress in drug research, vol 29. Birkhäuser, Basel, pp 415–459

Squires RF (1972) Multiple forms of monoamine oxidase in intact mitochondria as characterized by selective inhibitors and thermal stability: a comparison of eight mammalian species. In: Costa E, Sandler M (eds) Advances in biochemical psychopharmacology, vol 5. Raven Press, New York, pp 355–370

Tipton KF, Spires IPC (1968) The homogeneity of pig brain monoamine oxidase. Biochem Pharmacol 17:2137–2141

Waldmeier PC, Felner AF, Tipton KF (1983) The monoamine oxidase inhibiting properties of CGP 11305 A. Eur J Pharmacol 94:73–83

Whittaker VP (1969) The synaptosome. In: Lajtha A (ed) Handbook of neurochemistry, vol 2. Plenum Press, London New York, pp 327–364

Authors' address: Dr. T. May, Department of Neuropsychopharmacology, Free University, Ulmenalle 30, D-1000 Berlin 19, Federal Republic of Germany.

J Neural Transm (1990) [Suppl] 32: 103–105

Inhibition of MAO by substituted tryptamine analogues

D. Balsa[1], **E. Fernandez-Alvarez**[2], **K. F. Tipton**[1], and **M. Unzeta**[1]

[1] Departament de Bioquímica i Biología Molecular, Facultat de Medicina, Universitát Autónoma de Barcelona, and
[2] Departmento de Química Orgánica, Consejo Superior de Investigaciones Científicas, Madrid, Spain

Summary. Three different acetylenic analogues of tryptamine, in which the side chain is attached at the 2 position of the heterocyclic ring, were studied as inhibitors of MAO-A and MAO-B.

IC_{50} values were determined after 30 min preincubation of the enzyme and inhibitor, at 37 °C before assay. Irreversibility and time-dependence of the inhibiton were also established in each case. The kinetic parameters defining non-covalent complex formation and covalent adduct formation were calculated for the mechanism-based inhibition of both MAO-A and MAO-B by these compounds.

Introduction

Cruces et al. (1988) have shown that N-acetylenic analogues of tryptamine in which the side-chain is attached at the 2, rather than the 3 position of the heterocyclic ring, can be potent and selective inhibitors of MAO-A. In the present study, we have investigated the effects of different substitutions in the molecule on the potency of inhibition.

Materials and methods

Rat liver mitocondrial MAO activity was assayed towards 2-phenylethylamine (20 μM) and 5-hydroxytryptamine (100 μM), using the radiochemical procedure previously described (Fowler and Tipton, 1981). IC_{50} values were determined after 30 min preincubation of inhibitor (in the concentration range 10^{-2}–10^{-12} M) and enzyme at 37 °C before assay. The time-dependence of the process was determined at the inhibitor concentration that rendered total loss of activity in the above studies, after different times of enzyme-inhibitor preincubation. Reversibility of inhibition was determined by repeated washing, by centrifugation and resuspension (see Waldmeier et al., 1983) of samples which had been preincubated with appropriate inhibitor concentrations for 30 min at 37 °C.

The kinetic parameters for inhibition according to the mechanism (see Fowler et al., 1982):

$$E + I \xrightarrow{K_I} EI \xrightarrow{k} E\text{-}I$$

were determined spectrophotometrically by direct analysis of reaction progress curves (Walker and Elmore, 1984), in the presence of varying inhibitor concentrations. Benzylamine (333 μM) was used as a substrate for MAO-B in these studies, and kynuramine (40 μM) was used as the substrate for MAO-A. In the latter case, the enzyme preparations were preincubated with (–)-deprenyl (10^{-7} M) to inhibit the MAO-B activity (see Fowler and Tipton, 1982).

Results and discussion

Table 1 shows the effect of substituting methyl groups at the side-chain nitrogen and the terminal of the acetylenic group, on the potency and selectivities of MAO inhibitors. The compound lacking the methyl group on the side-chain N showed little selectivity, whereas the other two compounds were selective towards MAO-A, but their relative potencies and selectivities were significantly affected by the substitutions. In all cases inhibition was time-dependent and irreversible (results not shown).

It can be seen that substitution of a methyl group for the H on the side-chain amino group, significantly increases the potency of inhibition of

Table 1. Inhibition of MAO by acetylenic tryptamine derivatives

Compound	MAO	K_I (μM)	k(min^{-1})	IC_{50} (μM)
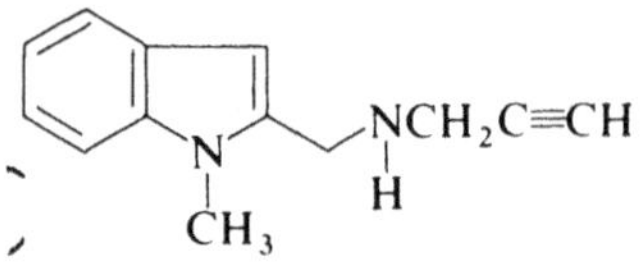	A	0.91	0.34	0.2
	B	3.7	3.0	0.3
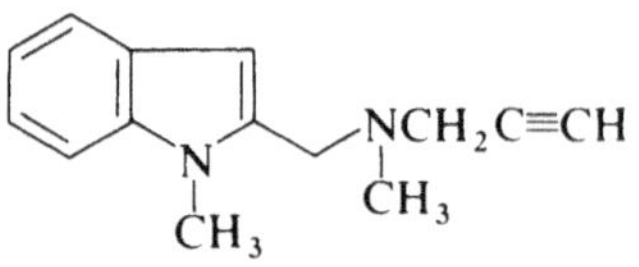	A	0.0027	1.46	0.003
	B	0.027	1.23	0.1
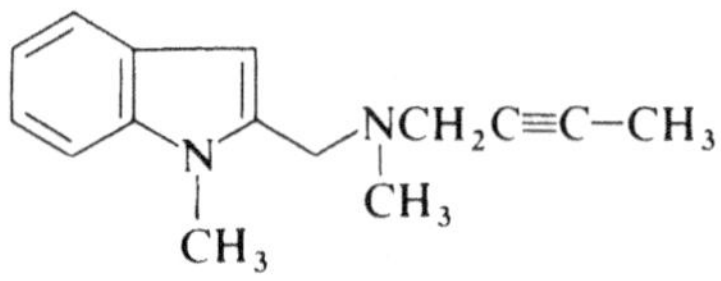	A	0.04	9.35	0.003
	B	0.047	0.89	0.1

MAO-A whilst having little effect on that of MAO-B. The enhanced selectivity for MAO-A can be seen to be essentially due to an increased affinity for non-covalent binding to that form of the enzyme. Although the further substitution of a methyl group at the end of the acetylenic group retained the MAO-A selectivity, the affinity for non-covalent binding was, in fact, significantly reduced, whereas the rate constant for the formation of the covalent adduct was greatly increased. These results are consistent with previous observations (see Tipton et al., 1983) that small changes in inhibitor structure can have profound and, sometimes, opposite effects on the two processes that combine to give mechanism-based inhibition.

It is hoped that extending this approach to a wider series of derivatives of this class will enable the structural features necessary for optimal inhibition to be elucidated. Such a process will involve achieving an appropriate combination of effects on the enzyme-inhibitor affinity, as measured by K_i, and the interactions within the noncovalent complex, that lead to irreversible inhibition, as measured by k (see Fowler et al., 1982). The possibility that the behaviour of these compounds in vivo may be enhanced by their acting as specific substrates for pre-synaptic amine transport systems remains to be assessed.

References

Cruces MA, Elorriaga C, Fernandez-Alvarez E, Nieto O (1988) Synthesis and biochemical properties of N-acetylenic and N-allenic derivatives of 2-amine-methylindoles as selective inhibitors of monoamine oxidase. Pharmacol Res Commun 20:105–107

Fowler CJ, Tipton KF (1981) Concentration dependence of the oxidation of tyramine by the two forms of rat liver mitochondrial monoamine oxidase. Biochem Pharmacol 30:3329–3332

Fowler CJ, Tipton KF (1982) Deamination of 5-hydroxytryptamine by both forms of monoamine oxidase in rat brain. J Neurochem 38:733–736

Fowler CJ, Mantle JT, Tipton KF (1982) The nature of inhibition of rat liver monoamine oxidase types A and B by the acetylenic inhibitors clorgyline, l-deprenyl and pargyline. Biochem Pharmacol 31:3555–3561

Tipton KF, O'Carroll A-M, Mantle TJ, Fowler CJ (1983) Factors involved in the selective inhibition of monoamine oxidase. Mod Probl Pharmacopsychiatry 19:15–30

Waldmeier PC, Felner AE, Tipton KF (1983) The monoamine oxidase inhibiting properties of CGP 1135A. Eur J Pharmacol 94:73–83

Walker B, Elmore DT (1984) The irreversible inhibition of miokinase, kidney-cell plasminogen activator, plasmin and β-trypsin by 1-[N-6 amine-n-hexyl]-carbamoyl imidazole. Biochem J 221:277–280

Authors' address: Dr. Dolors Balsa, Department de Bioquímica i Biología Molecular, Facultat de Medicina, Universitat Autónoma de Barcelona, 08193-Bellaterra (Barcelona), Spain.

J Neural Transm (1990) [Suppl] 32: 107–112

Ring-substituted analogues of tranylcypromine as monoamine oxidase inhibitors

R. L. Sherry, G. B. Baker, R. T. Coutts, and **D. D. Mousseau**

Neurochemical Research Unit and PMHAC Research Unit, Department of Psychiatry, University of Alberta, Edmonton, Alberta, Canada

Summary. 4-Fluorotranylcypromine and 4-methoxytranylcypromine, in which the 4-position of the phenyl ring is protected from metabolic ring hydroxylation, were tested for their ability to inhibit, relative to tranylcypromine, monoamine oxidase (MAO) -A and -B in rat brain after administration of low doses (1.2 and 3.7 µmol/kg) of the drugs. One hour after intraperitoneal injection of the lower dose, tranylcypromine was weaker than 4-fluorotranylcypromine and 4-methoxytranylcypromine at inhibiting MAO-A. After long-term (28 day) administration of a dose of 3.7 µmol/kg/day (administered via osmotic minipumps), 4-fluorotranylcypromine had a slightly stronger inhibitory effect on MAO-B than did the other two drugs. At the same time and dose both 4-substituted analogues were slightly more potent inhibitors of MAO-A than was tranylcypromine. After 28 days of administration at a daily dose of 1.2 µmol/kg/day, both analogues produced greater inhibition of MAO-A and -B than did tranylcypromine. 4-Methoxytranylcypromine and 4-fluorotranylcypromine were similar in their extent of inhibition of MAO-B but the former was more potent than the latter at inhibition MAO-A.

Introduction

Tranylcypromine (TCP) [(±)-2-phenylcyclopropylamine] is a commercially available nonhydrazine MAO inhibitor with a close structural similarity to amphetamine. Although in recent years there has been an increased recognition of the efficacy of and increased use of MAO inhibitors (Murphy et al., 1985), the prescribing of TCP by physicians was greatly reduced for several years because of reports of side effects, including cardiovascular ones. Ironically, both hypotension and hypertension can be problems with MAO inhibitors, the former being one of the most frequent side effects of this class of drugs (Murphy et al., 1985; Rabkin et al., 1985) and the latter usually being the result of concomitant consumption of pressor amines in the diet or of certain classes of interacting drugs (Murphy et al., 1985). Mallinger et al. (1986) found that in human subjects mean plasma TCP concentrations were correlated with mean orthostatic drop of systolic blood pressure and rise of pulse rate. The authors suggested that patients who have hypotensive reactions to this drug may benefit from dose regimen changes aimed at minimizing peak TCP levels. However, such adjustments could be problematic considering the relatively short half-life of TCP (Edwards et al., 1985; Mallinger et al., 1986).

In addition to being an inhibitor of MAO, TCP (which is structurally very similar to amphetamine) also alters uptake and release of the catecholamines and 5-hydroxytryptamine (5-HT) (reviews: Baker and Coutts, 1989; Coutts et al., 1987). Presumably the release and/or inhibition of reuptake of these putative neurotransmitter amines, particularly noradrenaline, contributes to the cardiovascular side effects of this drug.

It is well known that the tricyclic antidepressants undergo metabolic ring hydroxylation (Rudorfer and Potter, 1985 for review). It has been suggested that these hydroxylated metabolites contribute to cardiotoxic side effects of these drugs (Kutcher et al., 1985), but less work has been done with the MAO inhibitors in this regard (Baker and Coutts, 1989 for review). The 4-hydroxy metabolite of TCP has now been identified unequivocally in our laboratories using combined gas chromatography-mass spectrometry (Baker et al., 1986) and has been demonstrated to be present in both the brain and heart of the rat (Nazarali et al., 1987) following intraperitoneal injection of TCP. Kang and Chung (1984) also reported the presence of 4-hydroxy-N-acetyl-TCP in urine of rats treated with TCP. This ring hydroxylation may also contribute to the observation, made by several research groups, that TCP reaches high brain and plasma levels after administration but is cleared relatively quickly, i.e. it has a short elimination half-life (reviewed in Coutts et al., 1987). The degree of hydroxylation of tricyclic antidepressants is known to vary widely among individuals (Bock et al., 1983). If the same situation exists with TCP, it is conceivable that the TCP/4-hydroxy-TCP ratio may be important in determining the susceptibility of patients to cardiovascular side effects when being administered TCP.

In an effort to develop analogues of TCP that might produce more consistent brain and plasma concentrations than the parent drug, we synthesized in our laboratories some analogues in which the 4-position of the ring was blocked from hydroxylation by replacing the hydrogen at that position with another group. We found that four of these analogues, namely 4-fluoro-TCP (FTCP), 4-methoxy-TCP (MTCP), 2-naphthylcyclopropylamine and 3,4-methylenedioxy-TCP (all drugs are racemates) were equipotent to or more potent that TCP itself at inhibiting MAO in vitro in brain homogenates (Rao et al., 1986). One of them, FTCP, which was ten times more potent than TCP at inhibiting MAO, was tested further and found, after injection at equimolar doses, to attain higher, more sustained brain levels than TCP itself (Coutts et al., 1987). Thus, this drug seems to fulfill the criteria for a safe version of TCP: it is a stronger inhibitor of MAO than is TCP, so, presumably, lower doses will be required to produce the same degree of inhibition. In addition, it appears that protection from ring hydroxylation provides higher, more consistent brain levels than for TCP (Coutts et al., 1987), thus perhaps reducing even further the dosage and frequency of dosage required.

We have now conducted additional short- and long-term experiments on the effects of FTCP and MTCP on inhibition of MAO and on brain levels of putative neurotransmitter amines and their acid metabolites. Some of the findings have been published recently (Sherry et al., 1990a) and some are presented here.

Methods

In the short-term experiments, equimolar amounts (3.7 or 1.2 μmol/kg) of the drugs were administered i.p. and the rats (male Sprague-Dawley) were killed 1 h later. In the long-term studies, the rats were administered equimolar (3.7 or 1.2 μmol/kg/day) doses of the drugs for 28 days via osmotic minipumps implanted in the dorsal thoracic region. At the time of sacrifice, all rats were killed by cervical dislocation and decapitation. The brains were removed rapidly and frozen immediately in isopentane on solid carbon dioxide. Following freezing, the brains were removed to plastic receptacles and stored at $-80\,^{\circ}C$ until the time of analysis.

Monoamine oxidase activity was measured by the radiochemical procedure of Wurtman and Axelrod (1963), using ^{14}C-5-hydroxytryptamine and ^{14}C-β-phenylethylamine as substrates for MAO-A and MAO-B respectively. In the short-term study, whole brains were used for analysis. In the long-term experiments, the hippocampus, hypothalamus and pons-medulla were removed (for analysis of levels of neurotransmitter amines) and the remainder of the brain used for measurement of MAO activity.

Data were analyzed by analysis of variance followed by the Newman-Keuls test.

Results and discussion

The results of the short- and long-term investigations are shown in Tables 1 and 2, respectively. A dose of 3.7 μmol/kg was chosen for initial studies since this corresponds to the dose, on a mg/kg basis, of TCP used clinically. At this dose the drugs were similar in their effects on inhibition of MAO. Additional studies at this time interval of 1 h indicated that a dose 1/3 of this value, TCP was considerably weaker than FTCP and MTCP at inhibiting MAO-A. Long-term studies (28 days) at doses of 3.7 and 1.2 μmol/kg/day

Table 1. Effects of TCP, FTCP and MTCP on rat whole brain MAO activity 1 h after injection

% Inhibition[a]			
Drug	Dose[b]	MAO-A	MAO-B
TCP	3.7	80.1 ± 5.1	77.8 ± 4.9
FTCP	3.7	86.5 ± 4.6	86.6 ± 9.2
MTCP	3.7	89.9 ± 1.6	89.8 ± 2.0
TCP	1.2	33.7 ± 2.7	43.0 ± 11.5
FTCP	1.2	59.0 ± 3.0*	75.7 ± 20.2
MTCP	1.2	62.3 ± 8.2*	59.0 ± 15.8

[a] mean ± S.E.M. (N = 5–14)
[b] μmol/kg i.p.
* $p < 0.05$, compared to values in TCP-treated rats

Table 2. Effects of TCP, FTCP and MTCP on rat brain MAO activity following chronic (28 day) drug administration

% Inhibition[a]			
Drug	Dose[b]	MAO-A	MAO-B
TCP	3.7	85.8 ± 0.7	91.0 ± 0.5
FTCP	3.7	90.6 ± 0.6*	94.1 ± 0.2*
MTCP	3.7	92.0 ± 0.6*	89.9 ± 1.1
TCP	1.2	44.0 ± 3.8	67.0 ± 1.5
FTCP	1.2	64.1 ± 2.5*	81.2 ± 1.2*
MTCP	1.2	82.0 ± 2.7*	82.9 ± 1.0*

[a] mean ± S.E.M. (N = 7–16)
[b] μmol/kg i.p.
* $p < 0.05$, compared to values in TCP-treated rats

demonstrated that at the former dose FTCP had a stronger inhibitory effect on MAO-B than did TCP and MTCP; at this same dose both analogues produced a slightly greater effect than TCP on inhibition of MAO-A.

After 28 days of administration of the three drugs at a daily dose of 1.2 μmol/kg, the analogues produced a markedly greater inhibition of MAO-A and MAO-B than did TCP. MTCP and FTCP were similar in their inhibition of MAO-B (81.2 and 82.9 % inhibition, respectively) while the former was more potent than the latter at inhibition MAO-A (82.0 and 64.1% inhibition, respectively).

One concern with MTCP was that it might be neurotoxic to 5-HT neurones because of the reported neurotoxicity of the structurally related methylenedioxyamphetamines (Schmidt et al., 1987). However, the 28-day study with this drug indicated that there was an elevation of 5-HT over control values in hippocampus, pons medulla and hypothalamus and in fact the elevation with MTCP was greater than with TCP (Sherry et al., 1990b).

The results reported here indicate that FTCP and MTCP are potent MAO inhibitors ex vivo and should be studied further with regard to possible antidepressant effects.

Acknowledgements

Funded by the Alberta Provincial Mental Health Advisory Council (PMHAC) and the MRC of Canada. The authors are grateful to Dr. A. J. Greenshaw for discussion of statistical analysis and to Ms. S. Omura for typing this manuscript.

References

Baker GB, Hampson DR, Coutts RT, Micetich RG, Hall TW, Rao TS (1986) Detection and quantitation of a ring-hydroxylated metabolite of the antidepressant drug tranylcypromine. J Neural Transm 65:233–244

Baker GB, Coutts RT (1989) Metabolism of monoamine oxidase inhibitors. Prog Neuro Psychopharmacol Biol Psychiatry 13:405–417

Bock JL, Nelson JC, Gray S, Jatlow PI (1983) Desipramine hydroxylation: variability and effect of antipsychotic drugs. Clin Pharmacol Ther 33:322–328

Coutts RT, Rao TS, Baker GB, Micetich RG, Hall TW (1987) Neurochemical and neuropharmacological properties of 4-fluorotranylcypromine. Cell Mol Neurobiol 7:271–290

Edwards DJ, Mallinger AG, Knopf S, Himmelhoch J (1985) Determination of tranylcypromine in plasma using gas chromatography-chemical ionization mass spectrometry. J Chromatogr Biomed Appl 344:356–361

Kang GI, Chung SY (1984) Identification of N-acetyl and hydroxylated N-acetyltranylcypromine from tranylcypromine-dosed rat urine. Arch Pharmacol Res 7:65–68

Kutcher SP, Reid K, Dubbin JD, Shulman KI (1985) Electrocardiogram changes and therapeutic desipramine and 2-hydroxydesipramine concentrations in elderly depressives. Br J Psychiatry 148:676–679

Mallinger AG, Edwards DJ, Himmelhoch JM, Knopf S, Elher J (1986) Pharmacokinetics of tranylcypromine in patients who are depressed: relationship to cardiovascular effects. Clin Pharmacol Ther 40:444–450

Murphy DI, Sunderland T, Campbell I, Cohen RM (1985) Monoamine oxidase inhibitors as antidepressants. In: Dewhurst WG, Baker GB (eds) Pharmacotherapy of affective disorders: theory and practice. Croom Helm, London, pp 238–262

Nazarali AJ, Baker GB, Coutts RT, Wong JTF (1987) N-(2-cyanoethyl)-tranylcypromine, a potential prodrug of tranylcypromine: its disposition and interaction with catecholamine neurotransmitters in brain. Pharmaceut Res 4:16–20

Rabkin J, Quitkin F, McGrath P, Harrison W, Tricamo EP (1985) Adverse reactions to monoamine oxidase inhibitors. Part II. Treatment correlates and clinical management. J Clin Psychopharmacol 5:2–9

Rudorfer MV, Potter WZ (1985) Metabolism of drugs used in affective disorders. In: Dewhurst WG, Baker GB (eds) Pharmacotherapy of affective disorders: theory and practice. Croom Helm, London, pp 382–448

Schmidt CJ, Levin JA, Lovenberg W (1987) In vitro and in vivo neurochemical effects of methylenedioxymethamphetamine on striatal monoaminergic systems in the rat brain. Biochem Pharmacol 36:747–755

Sherry RL, Baker GB, Coutts RT (1990a) Effects of low dose 4-fluorotranylcypromine on rat brain monoamine oxidase and neurotransmitter amines. Biol Psychiatry (in press)

Sherry RL, Baker GB, Coutts RT, Mousseau DD, Greenshaw AJ (1990b) Effects of chronic administration of 4-methoxytranylcypromine, a novel analogue of tranylcypromine, in the rat. Proc 13th Ann Meet Can Coll Neuropsychopharmacol, Banff, Canada (in press)

Wurtman RJ, Axelrod J (1963) A sensitive and specific assay for the estimation of monoamine oxidase. Biochem Pharmacol 12:1439–1440

Authors' address; R. L. Sherry, Neurochemical Research Unit, Department of Psychiatry, University of Alberta, Edmonton, Alberta, Canada, T6G 2B7.

J Neural Transm (1990) [Suppl] 32: 113–118

Recent studies on the MAO inhibitor phenelzine and its possible metabolites

K. F. McKenna, G. B. Baker, R. T. Coutts, G. Rauw, A. Mozayani, and **T. J. Danielson**

Neurochemical Research Unit and PMHAC Research Unit, Department of Psychiatry and Faculty of Pharmacy and Pharmaceutical Sciences, University of Alberta, Edmonton, Canada

Summary. Although N^2-acetylphenelzine (N^2AcPLZ) appears to be only a minor metabolite of phenelzine (PLZ), other investigations have demonstrated that it may be worthy of study as an antidepressant in its own right. In the present report, the possibility of ring hydroxylation as a metabolic route for PLZ was investigated in the rat. Indirect evidence for such a route was obtained using iprindole, a drug known to block ring hydroxylation. Treatment of rats with iprindole followed by PLZ was demonstrated to result in increased brain levels of PLZ and β-phenylethylamine (control rats were treated with vehicle and then PLZ). The possibility that hydroxylation in the *para*-position might be a metabolic route for PLZ has led to interest in the possible use of analogues in which this position is blocked with a substituent. In preliminary acute studies at a dose of 0.1 mmol/kg *p*-chloro-PLZ was found to have a similar effect to PLZ on the inhibition of MAO and to lead to an elevation of catecholamines and 5-hydroxytryptamine (5-HT) in rat whole brain.

Introduction

Phenelzine [2-phenylethylhydrazine (PLZ)] is a potent irreversible non-selective inhibitor of monoamine oxidase (MAO) which is used in the treatment of depression and panic disorder. Despite years of clinical use the metabolism of PLZ in humans is poorly understood (Baker and Coutts, 1989). Previously it was assumed that acetylation occurred because of the similarity in structure of PLZ to the hydrazine derivatives isoniazid and hydralazine, which are known to be acetylated in vivo (Testa and Jenner, 1976) and because PLZ is a substrate for human N-acetyltransferases in vitro (Tilstone et al., 1979). Acetylated metabolities of PLZ have not been conclusively identified in man and at present the evidence suggests that oxidation rather then acetylation is the main metabolic pathway in humans (Robinson et al., 1985). When 6 human subjects were administered single doses of $^{13}C_6$-PLZ (30 mg), 66–79% of the administered dose was excreted as either phenylacetic acid (PAA) or *para*-hydroxyphenylacetic acid (*p*-OHPAA). Labelled PLZ, *para*-hydroxy PLZ (*p*-OHPLZ), and mono- or di-acetylated forms of PLZ were not detected in urine (Robinson et al., 1985). In this study 21–34% of the labelled drug was not accounted for in the first 96 h after administration. As the label was incorporated on the phenyl ring rather than the side chain, the possibility of acetylhydrazine as a metabolite cannot be ruled out. Any metabolites that were present in small amounts or that had a short half-life would be more difficult to identify. Using a deuterated analogue of PLZ ($\alpha,\alpha,\beta,\beta$,-^{2}H-PLZ) it has been shown that β-phenylethylamine (PEA) is a metabolite of PLZ in vivo using a rat model (Dyck et al., 1985).

Several possible metabolites have not been directly identified to date in humans. In our laboratories we have identified, using gas chromatographic/mass spectrometric methods, 1-acetyl-2-(2-phenylethyl)hydrazine [N^2-acetylphenelzine (N^2AcPLZ)] in rat brain and plasma after administration of PLZ (Mozayani et al., 1988). Although this metabolite is present in only small amounts, other investigations have demonstrated that it may be worthy of study as a potential antidepressant drug in its own right. N^2AcPLZ has been shown in acute and chronic studies to be a good inhibitor of MAO and to cause significant elevations of rat brain levels of biogenic amines (Danielson et al., 1988; McKenna et al., submitted). Unlike PLZ, N^2AcPLZ does not cause an elevation of brain levels of γ-aminobutyric acid (GABA) at doses up to 0.2 mmol/kg i.p. (McKenna et al., submitted).

In more recent experiments, we have investigated the possibility of ring hydroxylation as a metabolic route for PLZ in the rat. Indirect evidence for such a route has been obtained using iprindole, a drug known to block ring hydroxylation (Freeman and Sulser, 1972; Mosnaim et al., 1989). We have demonstrated increased brain levels of both PLZ and PEA in rats treated with iprindole prior to PLZ administration. The possibility that hydroxylation in the *para*-position of the phenyl ring might be a metabolic route for PLZ has led to interest in the possible use of analogues in which this position is blocked by a substituent. In preliminary acute studies we have found that *p*-chloro-PLZ (ClPLZ) has a similar effect to PLZ on inhibition of MAO and elevation of catecholamines and 5-hydroxytryptamine (5-HT) in rat whole brain, and those data are presented in this report.

Materials and methods

Animals and drugs

Phenelzine sulfate (PLZ) was obtained from Sigma Chemical Co. (St. Louis, MO, USA), iprindole from Wyeth (UK) and ClPLZ from R. T. Coutts. Sprague-Dawley rats (male) were administrated intraperitoneal (i.p.) injections of vehicle or of iprindole HCl (11.2 mg/kg) dissolved in 0.9% NaCl. One h later rats were injected i.p. with vehicle or with PLZ (sulfate salt) (0.2 mmol/kg) dissolved in 0.01 M sodium phosphate buffer (pH = 7.4). Animals were sacrificed by decapitation at 1 and 3 h after this second injection and whole brains were immediately removed and frozen solid in isopentane on solid CO_2. Brains were stored at $-80°$ until the time of analysis. In the analogue study, ClPLZ (0.1 mmol/kg) dissolved in 0.9% NaCl was administered i.p. and animals were sacrificed at 1, 2 and 4 h.

Monoamine oxidase inhibition

Frozen brains were homogenized in 5 vol of ice-cold isotonic KCl. Brain MAO activity was measured ex vivo by a modification of the radiochemical method of Wurtman and Axelrod (1963) using ^{14}C-5-HT as the substrate for MAO-A and ^{14}C-β-PEA as the substrate for MAO-B.

Neurotransmitter amines and acid metabolites

Frozen brains were homogenized in 5 vol of ice-cold 0.1 N perchloric acid containing 10 mg% EDTA and 0.05 mM ascorbic acid, and the homogenates were centrifuged at 10,000 rpm for 15 min at 4 °C. High-pressure liquid chromatography with electrochemical detection (HPLC-EC) was used to determine brain levels of noradrenaline (NA), dopamine (DA), 5-HT, homovanillic acid (HVA), 3,4-dihydroxyphenylacetic acid (DOPAC) and 5-hydroxyindole-3-acetic acid (5-HIAA). The HPLC-EC method and apparatus used are described in detail in Baker et al. (1987).

Brain levels of PLZ and PEA

An electron-capture gas chromatographic method following extraction and derivatization with pentafluorbenzoyl chloride was used (Rao et al., 1987).

Statistical analysis

Statistics consisted of a one-way analysis of variance followed by Newman-Keuls tests in the case of multiple pairwise comparisons ($\alpha = 0.05$).

Results and discussion

The results of the iprindole study are presented in Table 1. Whole brain levels of PLZ are significantly higher at 1 h after administration than at 3 h in the vehicle/PLZ treated groups. Pretreatment with iprindole results in significantly higher PLZ and PEA levels compared to controls (vehicle/PLZ) at 3 h after PLZ administration. In the presence of iprindole, PLZ levels at 3 h were similar to the levels normally seen at 1 h after administration. PEA levels also rose at 3 h, presumably due in part to metabolism of PLZ to PEA. These results show that PLZ levels in rat brain are raised by pretreatment with a drug that blocks ring hydroxylation. This provides indirect evidence that PLZ is normally ring-hydroxylated to *p*-OHPLZ. However it must be remembered that iprindole may block other metabolic reactions; for example recent experiments in our laboratories have shown that iprindole can block N-dealkylation (Hegadoren, Baker and Coutts, unpublished). We are currently developing a gas chromatographic assay that will allow us to directly assess the presence of *p*-OHPLZ in rat brain and human urine in order to further evaluate this possible pathway.

Table 1. Effects of pretreatment with iprindole (1 h previous) on rat whole brain levels of phenelzine (PLZ) and phenylethylamine (PEA)

Drug treatment	PLZ	PEA
A. Vehicle/PLZ, 1 h	183 ± 11	29 ± 5
B. Iprindole/PLZ, 1 h	209 ± 31	27 ± 4
C. Vehicle/PLZ, 3 h	109 ± 11	28 ± 4
D. Iprindole/PLZ, 3 h	205 ± 34*	46 ± 6*

Concentrations of PLZ and PEA are expressed as ng/g (mean ± S.E.M., n = 8). * $p < 0.05$ compared to values in treatment group C

Table 2. Effects of ClPLZ (0.10 mmol/kg i.p.) on MAO activity and whole brain levels of neurotransmitter amines and acid metabolites

	Vehicle	1 h	2 h	4 h
Inhibition of MAO-A		91 ± 4*	95 ± 1*	89 ± 8*
Inhibition of MAO-B		65 ± 6*	66 ± 2*	71 ± 9*
NA	317 ± 11	397 ± 15*	411 ± 21*	437 ± 20*
DA	667 ± 15	817 ± 16*	799 ± 25*	852 ± 12*
5-HT	446 ± 18	710 ± 67*	781 ± 35*	832 ± 73*
DOPAC	75 ± 4	15 ± 10*	N.D.	23 ± 12*
HVA	91 ± 4	35 ± 13*	25 ± 4*	13 ± 10*
5-HIAA	346 ± 10	250 ± 15*	230 ± 17*	231 ± 29*

Results for MAO expressed as % inhibition compared to controls (mean ± S.E.M.). In the case of neurotransmiters and metabolites, results are expressed as ng/g (mean ± S.E.M.). Control n = 18, ClPLZ n = 6, except 4 h where n = 4. * $p < 0.05$ compared to controls. *N.D.* not detectable

In order to examine the effects of blocking ring hydroxylation, we studied a PLZ analogue with a chlorine substituent on the *para* position of the phenyl ring. The results of the ClPLZ experiments are shown in Table 2. This drug inhibits MAO-A and MAO-B ex vivo, which results in a significant elevation of NA, DA, 5-HT and a significant decrease in the metabolites HVA, DOPAC and 5-HIAA. There were no significant differences between the different times studied. The only previous report on this drug that we are aware of demonstrated that ClPLZ raised whole brain levels of the trace amines *m*- and *p*-tyramine and *m*- and *p*-octopamine (Juorio et al., 1980). We now intend to extend these findings by comparing ClPLZ to PLZ at lower doses in order to determine if blocking ring-hydroxylation by halogen substitution produces a more potent MAO inhibitor which attains higher levels in brain than does the parent drug.

Acknowledgements

Funded by Alberta Heritage Foundation for Medical Research, the Alberta Provincial Mental Health Advisory Council (PMHAC) and the MRC of Canada.

References

Baker GB, Coutts RT, Rao TS (1987) Neuropharmacological and neurochemical properties of N-(2-cyanoethyl)-2-phenylethylamine, a prodrug of 2-phenylethylamine. Br J Pharmacol 92:243–255

Baker GB, Coutts RT (1989) Metabolism of monoamine oxidase inhibitors. Proc West Pharmacol Soc 25:417–420

Danielson TJ, Coutts RT, Baker GB, Chan MC (1988) Potential prodrugs of phenelzine: N^2-ethoxycarbonylphenelzine. J Pharmacol Sci 77:498–499

Dyck LE, Durden DA, Boulton AA (1985) Formation of β-phenylethylamine from the antidepressant, β-phenylethylhydrazine. Biochem Pharm 34:1925–1929

Freeman JJ, Sulser F (1972) Iprindole-amphetamine interactions in the rat: the role of aromatic hydroxylation of amphetamine in its mode of action. J Pharmacol Exp Ther 183:307–315

Juorio AV, Davis BA, Boulton AA (1980) The effect of phenelzine and some if its *para*-halogenated derivatives on the levels of brain tyramine and octopamine in the mouse. Res Commun Psychol Psychiat Behav 5:255–264

Mosnaim AD, Callaghan OH, Wof MR (1989) Hydroxylation of phenylethylamine by rat liver preparations: inhibition studies. Gen Pharmacol 20:463–467

Mozayani A, Coutts RT, Danielson TJ, Baker GB (1988) Metabolic acetylation of phenelzine in rats. Res Commun Chem Pathol Pharmacol 62:397–406

Rao TS, Baker GB, Coutts RT, Yeung JM, McIntosh GJA, Torok-Both GA (1987) Analysis of the antidepressant phenelzine in brain tissue and urine using electron-capture gas chromatography. J Pharmacol Methods 17:297–304

Robinson DS, Cooper TB, Satya PJ, Corcella J, Lutz T (1985) Metabolism and pharmacokinetics of phenelzine: lack of evidence for acetylation pathway in humans. J Clin Psychopharmacol 5:333–337

Testa B, Jenner P (1976) Drug metabolism: chemical and biochemical aspects. Marcel Dekker, New York, pp 180–186

Tilstone WJ, Margot P, Johnstone C (1979) Acetylation of phenelzine. Psychopharmacology 60:261–263

Wurtman RJ, Axelrod J (1963) A sensitive and specific assay for the estimation of monoamine oxidase. Biochem Pharmacol 12:1439–1440

Authors' address: Dr. K. F. McKenna, Neurochemical Research Unit, Department of Psychiatry, University of Alberta, Edmonton, Canada, T6G 2B7

J Neural Transm (1990) [Suppl] 32: 119–122

Stylbasole analogues of MPTP as monoamine oxidase (MAO) substrates

S. O. Sablin, S. O. Bachurin, and **S. E. Tkachenko**

Institute of Physiologically Active Substances, Academy of Sciences of USSR, Moscow, USSR

Summary. Stylbasole analogs of 1-methyl-4-phenyl-1,2,3,6-tetrahydropyridine (MPTP) were studied as monoamine oxidase (MAO) substrates. Dehydrogenation of these compounds was shown to be catalyzed by both serotonine specifical and benzylamine specifical MAO activities. Markedly high affinity of stylbasoles to B type of MAO was found. Influence of substrate structure on its biotransformation effectiveness is realized by the principle – "better binding-worse catalysis". MAO inactivation during the reaction is appeared to be realized as result of product inhibition and perhaps of substrate inhibition.

Introduction

Last years it's established that MAO's are capable to catalyze dehydrogenation of some tetrahydropyridines besides the reactions of oxidative desamination of arylalkylamines. As for MPTP such transformation leads to the formation of neuroactive metabolites 1-methyl-4-phenyl-2,3-dihydropyridinium and 1-methyl-4-phenylpyridinium species causing degradation of dopaminergic neurons close to the picture of Parkinson's disease (Heikkilla, 1984). In previous study with tolyl MPTP homologs we have showed structural kinetic regularities of dehydrogenation process can determine the efficacy of whole neurotropic action of substances (Bachurin, 1989). In present paper we report results of kinetic studies of MAO catalyzed dehydrogenation of stylbasole analogs of MPTP which metabolites are also of interest as protectors of anticholinergic agents (Gray, 1988).

Materials and methods

MAO was isolated as in our previous work (Bachurin, 1989). Rates of substrates oxidation were measured spectrophotometrically by dihydropyridinium product accumulation. Other materials and methods were, as before (Bachurin, 1989).

Results

The series of stylbasole derivatives of MPTP as MAO substrates were studied. Their enzymatic transformation leads to consecutive formation of 2,3-dihydropyridinium and pyridinium products which were identified spectrophotometrically. Oxidation of compounds (I)–(IX) can be characterized by two pairs of parameters K_m and V_{max} (Table 1). Oxidation on the high affinity center was inhibited competitively reversibly by specific MAO-B substrate benzylamine while oxidation on the low affinity center was inhibited by specific MAO-A substrate serotonine. Compound (X) was also oxidized by MAO but rate of the reaction was very low and hard detectable. For all of compounds studied enzyme inactivation during the reaction was shown.

Table 1. Kinetic parameters of stylbasol derivatives

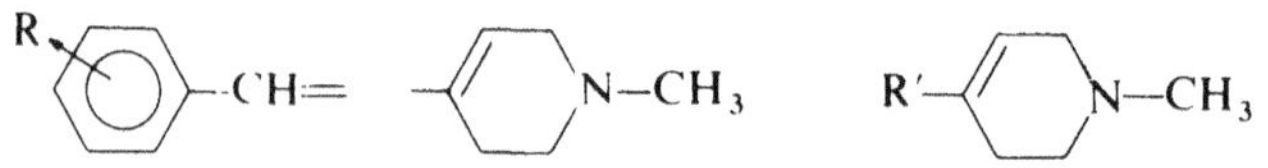

No	R	MAO A		MAO B	
		K_m(mM)	V^*_{max}	K_m(μM)	V^*_{max}
I	H	1.6	22.5	3.4	0.6
II	o-CH_3	8.0	32	9	1.2
III	o-OCH_3	11	70	10	2.9
IV	p-CH_3	1.25	1.47	0.47	0.18
V	p-OCH_3	5.3	4.67	1.2	0.81
VI	p-$N(CH_3)_2$	1.4	0.38	0.5	0.07
VII	p-F	1.3	5.1	2	2.1
	R′				
VIII	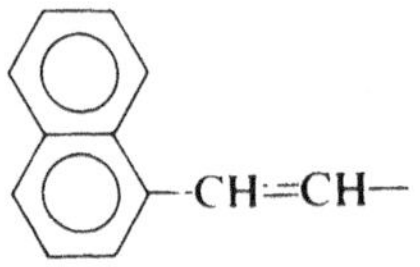	5.0	620	< 0.1	1.6
IX	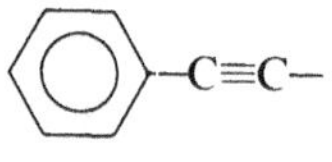	> 10	> 100	< 0.1	2.8
X	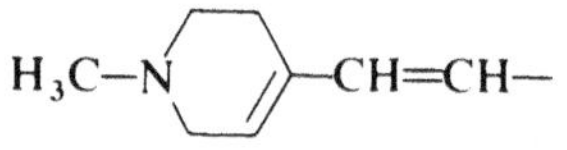	–	–	–	
XI	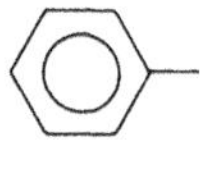	–	–	330	100

* † maximum velocity (%) relative to MPTP

Discussion

Most of the known substrates of dehydrogenation function of MAO are direct MPTP analogs. They are oxidized by both types of MAO and have values of Michaelis constants as a rule higher than 50 μM. Compounds of stylbasole group are also substrates of both MAO types. Their oxidation kinetics are well described by equation

$$v = \frac{V^a_{max}[S]_0}{K^a_m + [S]_0} + \frac{V^b_{max}[S]_0}{K^b_m + [S]_0}$$

Relative velocity of catalysis by both MAO types is mainly lower compared to that of MPTP but extraordinary high affinity of stylbasoles to MAO B is observed. Regularity of changes of substrate properties from para to ortho derivatives is the same as in the case of MPTP homologs (Bachurin, 1989). Para derivatives have higher affinity to the both types of the enzyme but lower values of maximum velocity compared to that of ortho ones. High correlation between steric properties of substrate molecules (among stylbasoles and MPTP analogs studied before) and their affinity towards MAO-B was revealed (Fig. 2). Enzyme inactivation type during the stylbasole oxidation is appeared to differ considerably depending on initial substrate concentration (Fig. 1). At low substrate concentrations (lower than 0.1 mM) plot of inactivation versus time has S-mode type like in the case of tolyl homologs of MPTP when inactivation mechanism by dihydropyridinium product of reaction takes place (kinetic analysis of inactivation mechanisms discrimination has carried out by authors earlier in (Bachurin, 1989). In the range of higher substrate concentration (higher than 1 mM) an exponential enzyme inactivation is observed. This fact can be evidence of prevalent contribution of enzyme inhibition by substrate. Thus enzyme inactivation

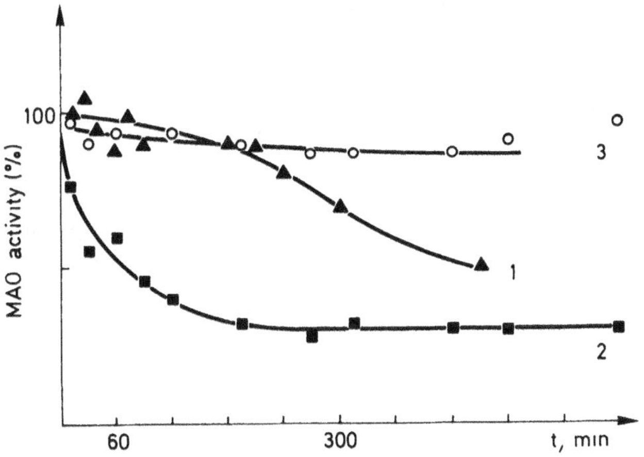

Fig. 1. Enzyme inactivation during compound (V) oxidation. Substrate concentration, mM: *1* 0.1; *2* 1.0; *3* 0 (control)

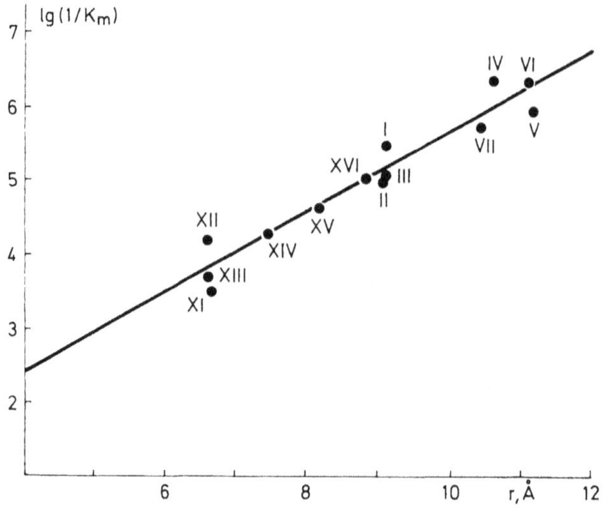

Fig. 2. Correlation between linear lengths of substrate molecules (r) and values of Michaelis constants in MAO-B catalysed reactions. **XII** o-Me-MPTP: **XIII** o-MeO-MPTP; **XIV** m-Me-MPTP; **XV** p-Me-MPTP; **XVI** p-MeO-MPTP. Values of K_m from Bachurin (1989). $\lg(1/K_m) = 0.30\ (\pm 0.13) + 0.53\ (\pm 0.002)\ .r.$ Corr.coef. 0.969

can be realized by different mechanisms depending on the condition of reaction.

References

Bachurin SO, Sablin SO, Grishina GV, Gaydarova EL, Dubova LG, Zubov ND (1989) Monoamine oxidase catalysis of bioconversion of physiologically active 1-methyl-4-aryl-1,2,3,6-tetrahydropyridines. Bioorganicheskaya khimia 15: 620–626 (russ)

Gray AP, Platz RD, Henderson TR, Chang TCP, Takahashi K, Drecthen KL (1988) Approaches to protection against nerve agent poisoning. (Naphthyl-vinil)pyridine derivatives as potential antidotes. J Med Chem 31:807–814

Heikkilla RE, Manzino L, Cabbat FS, Duvoisin RC (1984) Protection against the dopaminergic neurotoxicity of 1-methyl-4-phenyl-1,2,3,6-tetrahydropyridine by monoamine oxidase inhibitors. Nature 311:467–469

Authors' address: Dr. S. O. Bachurin, Institute of Physiologically Active Substances, Academy of Sciences of USSR, Chernogolovka, Moscow Region, 142432, USSR

J Neural Transm (1990) [Suppl] 32: 123–129

MAO activity, metabolism and anticonvulsant activity of milacemide in rats and mice

M. Colombo, M. Strolin Benedetti, A. Bonsignori, G. Cocchiara, R. Roncucci, and P. Dostert

Research and Development – Erbamont Group, Farmitalia Carlo Erba, Milan, Italy

Summary. Milacemide was found to protect Swiss albino CD1 mice but not Sprague Dawley rats against bicuculline-induced lethality. Since it had been previously suggested that the anticonvulsant activity of milacemide might be related to MAO-B- mediated glycine formation, brain and liver MAO-A and-B activities and the urinary metabolic pattern of milacemide were determined in the same mice and rat strains. Similar brain and liver MAO activities were found in the two species, except for liver MAO-A activity which was higher in rats. After the same oral dose of milacemide, the percent of the dose excreted as glycinamide was significantly higher in mice than in rats, whereas that excreted as metabolite UK_1 was significantly higher in rats. These results support the hypothesis of a glycine-mediated anticonvulsant activity for milacemide and suggest that the increased formation of UK_1 to the detriment of glycinamide might account for the lack of protection against bicuculline-induced lethality by milacemide in rats.

Introduction

Milacemide [2-(pentylamino)acetamide] (Roncucci et al., 1980) has been reported to have anticonvulsant activity in some animal models of epilepsy, with antagonism of bicuculline-induced convulsions and lethality representing the most sensitive test of anticonvulsant potency (van Dorsser et al., 1983).

The mechanism(s) of the anticonvulsant action of milacemide has not been elucidated. The acute administration of milacemide to rats was shown to significantly increase glycine concentrations in most brain structures (Christophe et al., 1983; Chapman and Hart, 1988) and GABA levels in substantia nigra (Janssens de Varebeke et al., 1983), but this last result was not confirmed (Chapman and Hart, 1988). Acute but not repeated administration of milacemide was found to result in significantly increased glutamate decarboxylase activity in synaptosomal fractions prepared from rat striatum (Cini et al., 1990). Milacemide was also reported to significantly decrease DOPAC levels in rat striatum and substantia nigra (Dostert et al., 1988) and to bind to the NMDA receptor (Reynolds and Miller, 1989).

Milacemide was shown to be a selective substrate and inhibitor of MAO-B (Janssens de Varebeke et al., 1988, 1989). However, selectivity of milacemide toward MAO-B has been questioned by Truong et al. (1989), who found potentiation of the 5-HTP-induced "wet dog shake" syndrome by milacemide in the rat; such a potentiation was observed after clorgyline, but not after 1-deprenyl (Tadano et al., 1989).

In the course of our studies on the mechanism(s) of action of milacemide, determination of its anticonvulsant potency was performed in mice and rats. Contrary to previous results (van Dorsser et al., 1983). oral milacemide was found to display weak antibicuculline activity in mice and to be ineffective against bicuculline-induced lethality in rats. In an attempt to investigate whether the anticonvulsant activity of milacemide could be related to the amount of unchanged drug or to the formation of active metabolite(s), its urinary metabolic profile, as well as liver and brain MAO activities, were determined in the rat and mice strains used in the antibicuculline activity test. The results of these experiments are reported below.

Materials and methods

Male Swiss albino mice (CD1, Charles River) and male Sprague Dawley rats (Charles River) were used. Milacemide HCl (M) was synthesized at Farmitalia Carlo Erba and ^{14}C-milacemide HCl (^{14}C-M), labelled in the CH_2 of the acetamide residue, specific activity 15.8 mCi/mmol, radiochemical purity $\geq 98\%$, was synthesized by Prof. C. Luu-Duc at the Laboratoire de Chimie Pharmaceutique, Grenoble, France. 5-Hydroxy[side chain-2-^{14}C]tryptamine creatinine sulphate (5-HT) and 2-phenyl[1-^{14}C]ethylamine hydrochloride (PEA) were purchased from Amersham, UK.

Antagonism of bicuculline-induced convulsions and lethality in mice and rats

Antibicuculline activity in mice and rats was essentially measured as described by van Dorsser et al. (1983). Bicuculline was injected intravenously at the dose of 0.6 mg/kg in mice (18–20 g body weight) and of 1 mg/kg in rats (180–200 g body weight) 3 h after oral administration of 3 graded doses of M. Ten animals per dose were used. The activity of M was expressed as the dose calculated by regression analysis which protected 50% of the animals from convulsions and death for 1 h after injection of bicuculline (ED_{50} and 95% confidence limits).

Urinary metabolism of milacemide in mice and rats

Four rats weighing 203 ± 3 g (mean ± SD) and 3 mice weighing 25 ± 2 g (mean ±SD) were used. Animals were individually housed in glass metabolism cages. ^{14}C-M (20 mg/kg) was given by gastric gavage to rats (final specific activity 6.54 μCi/mg) and mice (final specific activity 12.91 μCi/mg). Then, urine was collected in ice-cooled containers in the intervals 0–8, 8–24 h and at 24 h intervals for 3 additional days; rats faeces were collected at 24 h intervals for 96 h. Urines and faeces were stored at − 20 °C in the dark until radiometric analysis.

Urine radioactivity was measured in duplicate using liquid scintillation; the counting error was ≤ 3%. Rat faeces were lyophilized, homogenized and the radioactivity content was determined in triplicate by combustion. The urinary metabolic pattern was analyzed by TLC using silica gel 60 F_{254} and ethylacetate/methanol/ammonia (30%) 2:1:1, v:v:v) as eluent. The radioactive peaks were detected and quantified using a RAYTEST RITA TLC-analyzer (Nuclear Interface) and the R_f values compared with those of milacemide, glycinamide and glycine. The reference compounds were detected by spraying an alcoholic solution of ninhydrin.

Brain and liver MAO activities in mice and rats

Three mice (body weight 18–20 g) and three rats (body weight 190–230 g) were used. Animals were killed by decapitation; brains and livers were removed, rinsed in saline, blotted on filter paper, frozen in liquid nitrogen and stored at − 20 °C until use. MAO activity was assayed basically as described by Cao Danh et al. (1984). 5-HT (MAO-A) and PEA (MAO-B) were used at final concentrations of 400 μM and 50 μM, respectively, for both tissues. The deaminated products were extracted in 7 ml of toluene-ethylacetate (1/1, v/v) and the organic phase obtained after centrifugation (1800 g, 10 min) was counted in 10 ml of scintillation cocktail Insta-Fluor II (Canberra-Packard, USA). MAO activity was calculated after 2 min and 10 min incubation for PEA and 5-HT, respectively. Enzyme activity was expressed as nmol/mg of tissue/min; values were not corrected for the extraction coefficients of the deaminated metabolites.

Results

Given orally 3 h before bicuculline, milacemide was found to protect the mice against tonic convulsions and death with an ED_{50} value of 427 mg/kg (265–681: 95% confidence limits). In rats, the oral administration of milacemide did not prevent bicuculline-provoked death up to 1.5 g/kg, but an ED_{50} value of 286 mg/kg (124–661) was found for the prevention of tonic convulsions. Under the same experimental conditions, valproate, diphenylhydantoin and phenobarbital displayed a protective effect against bicuculline-induced death in rats with ED_{50} values of 312, 241 and 14.8 mg/kg, respectively.

Brain and liver MAO-A and MAO-B activities in rats and mice are reported in Table 1. Similar MAO-A and MAO-B activities were found in

Table 1. Brain and liver MAO-A and MAO-B activities in Swiss albino CD1 mice and Sprague Dawley rats

species	tissues	MAO activity (nmol/mg tissue/min) (mean $\pm$ SD)	
		MAO-A	MAO-B
mouse (3)*	brain	0.165 ± 0.015	0.079 ± 0.006
	liver	0.130 ± 0.016	0.870 ± 0.087
rat (3)	brain	0.209 ± 0.020	0.119 ± 0.005
	liver	0.591 ± 0.058	0.781 ± 0.075

* number of animals is given in parentheses

the two species, except for liver MAO-A activity which was more than 4 times higher in rats than in mice.

In rats, radioactivity was mainly excreted over 24 h and accounted for 47.8 ± 4.5 (mean $\pm$ SD) and $3.1 \pm 3.7\%$ of the dose administered in urine and faeces, respectively; after 96 h, the corresponding values reached 50.4 ± 4.5 and $4.3 \pm 3.8\%$ of the dose, respectively. In rats and mice, the percent amount of the dose eliminated in the 8 h-urines was 34.3 ± 7.2 and 38.6 ± 7.0, respectively, whereas the corresponding values were 47.8 ± 4.5 and $51.1 \pm 1.6\%$ after 24 h (Table 2).

Labelled free glycine was not detected in the urine of rats and mice. The five major components detected in rat and mice urines are reported in Table 2. Three non-identified radioactive peaks named UK_1, UK_2, and UK_3 were observed, whose homogeneity has not yet been established. While unchanged milacemide was excreted to a similar extent in the urine of rats and mice, the percent of milacemide excreted as urinary glycinamide was significantly higher in mice than in rats, whereas the unknown metabolite UK_1 accounted for a significantly higher percent of the dose in rats than in mice.

Discussion

Although we adopted experimental conditions very similar to those retained by van Dorsser et al. (1983), oral milacemide was found to have considerably lower antibicuculline potency than reported by these authors, who found ED_{50} values of 5.7 and 31.9 mg/kg in mice and rats, respectively. At oral doses below 100 mg/kg, less than 20% of the mice were found to be protected. For valproate, van Dorsser et al. (1983) reported an ED_{50} value of 29.6 mg/kg p.o. in mice, whereas we found an ED_{50} value of 546 mg/kg, in good agreement with the values (ED_{50} values from 250 to 600 mg/kg p.o.) obtained in similar conditions by other authors (Frey and Löscher, 1976; Meldrum et al., 1986; Chambon et al., 1986).

Table 2. Major radioactive components in the 24 h-urine after oral administration of ^{14}C-milacemide HCl (20 mg/kg) in mice and rats. Results are expressed as percent of the administered dose

	% of the dose (mean ± SD)					
Species	UK_1	Glycinamide	UK_2	UK_3	Milacemide	Total radioactivity
mouse (3)	6.6 ± 0.6	22.9 ± 3.5	11.1 ± 1.6	1.0 ± 0.2	9.5 ± 1.7	51.1 ± 1.6
rat (4)	9.6 ± 0.5**	14.3 ± 2.5*	10.5 ± 1.0	1.3 ± 0.3	12.1 ± 2.5	47.8 ± 4.5

UK_1, UK_2, UK_3 unknown metabolites 1, 2 and 3; * $p < 0.05$, ** $p < 0.01$; Student's t-test for unpaired data

The difference in antibicuculline potency of milacemide between mice and rats remains unclear. Milacemide was shown to be a quite selective substrate of MAO type B, with apparent K_m values of 30–90 μM and 1300 μM for MAO-B and MAO-A, respectively (Janssens de Varebeke et al., 1988). In the two tissues examined in this study, MAO-B activity was found to be similar in rats and mice. This and the fact that the percent of the dose of milacemide excreted in the 24 h-urine as UK_2, UK_3, unchanged milacemide and total radioactivity was similar in both species, seem to indicate that the higher percent of the dose excreted as glycinamide in mice does not result from a more extensive metabolism of milacemide in mice compared to rats. The increased excretion of metabolite UK_1 in rats suggests that the metabolic pathway leading to the formation of UK_1 might be detrimental to milacemide as an anticonvulsant in the rat. Characterization of the chemical structures of UK_1, UK_2 and UK_3 is in progress. The relation between increased glycinamide formation and antibicuculline activity after milacemide in mice compared to rats is in keeping with the hypothesis that the anticonvulsant activity of milacemide might be due, at least partially, to its ability to behave as a glycine-prodrug (Youdim et al., 1988; Janssens de Varebeke et al., 1988).

References

Cao Danh H, Strolin Benedetti M, Dostert P (1984) Differential changes in monoamine oxidase A and B activity in aging rat tissues. In: Tipton KF, Dostert P, Strolin Benedetti M (eds) Monoamine oxidase and disease. Prospects for therapy with reversible inhibitors. Academic Press, London, pp 301–317

Chambon JP, Hallot A, Bizière K (1986) CM 40907 and SR41378: two piperidino-pyridazine derivatives. Curr Probl Epilepsy 4:313–316

Chapman AG, Hart GP (1988) Anticonvulsant drug action and regional neurotransmitter amino acid changes. J Neural Transm 72:201–212

Christophe J, Kutzner R, Nguyen-Bui ND, Damien C, Chatelain P, Gillet L (1983) Conversion of orally administered 2-n. pentylaminoacetamide into glycinamide and glycine in the rat brain. Life Sci 33:533–541

Cini M, Varasi M, Roncucci R, Dostert P (1990) Effects of milacemide on glutamate decarboxylase activity in rat striatum synaptosomal fractions. Eur J Pharmacol 183:2340

Dostert P, Ricciardi S, Roncucci R (1988) Effect of milacemide on the levels of dopamine, serotonin and their metabolites in rat substantia nigra and striatum. Pharmacol Res Commun 20 [Suppl] 4:133–134

Frey H-H, Löscher W (1976) Di-n-propylacetic acid – Profile of anticonvulsant activity in mice. Arzneim-Forsch 26:299–301

Janssens de Varebeke P, Cavalier R, David-Remacle M, Youdim MBH (1988) Formation of the neurotransmitter glycine from the anticonvulsant milacemide is mediated by brain monoamine oxidase B. J Neurochem 50:1011–1016

Janssens de Varebeke P, Niebes P, Pauwels G, Roba J, Korf J (1983) Effect of milacemide, a glycinamide derivative, on the rat brain γ-aminobutyric acid system. Biochem Pharmacol 32:2751–2755

Janssens de Varebeke P, Pauwels G, Buyse C, David-Remacle M, De Mey J, Roba J, Youdim MBH (1989) The novel neuropsychotropic agent milacemide is a specific enzyme-activated inhibitor of brain monoamine oxidase B. J Neurochem 53:1109–1116

Meldrum BS, Kehr W, Stephens DN, Bhargava AS (1986) ZK 91296 and ZK 93423: β-carbolines with partial or full agonist activity at benzodiazepine receptors. Curr Probl Epilepsy 4:303–308

Reynolds IJ, Miller RJ (1989) Ifenprodil is a novel type of N-methyl-D-aspartate receptor antagonist: interaction with polyamines. Mol Pharmacol 36:758–765

Roncucci R, Gillet C, Cordi A, Martens M, Roba J, Niebes P, Lambelin G, van Dorsser W (1980) Glycinamide derivatives and their use. Ger Offen 3,010,599

Tadano T, Satoh S, Satoh N, Kisara K, Arai Y, Kim SK, Kinemuchi H (1989) Potentiation of para-hydroxyamphetamine-induced head-twich response by inhibition of monoamine oxidase type A in the brain. J Pharmacol Exp Ther 250:254–260

Truong DD, Diamond B, Pezzoli G, Mena MA, Fahn S (1989) Monoamine oxidase inhibitory properties of milacemide in rats. Life Sci 44:1059–1066

van Dorsser W, Barris D, Cordi A, Roba J (1983) Anticonvulsant activity of milacemide. Arch Int Pharmacodyn 266:239–249

Youdim MBH, Kerem D, Duvdevani Y (1988) The glycine-prodrug, milacemide, increases the seizure threshold due to hyperbaric oxygen; prevention by 1-deprenyl. Eur J Pharmacol 150:381–384

Authors' address: Dr. P. Dostert, R&D, Farmitalia Carlo Erba, Via C. Imbonati 24, I-20159 Milan, Italy.

J Neural Transm (1990) [Suppl] 32: 131–137

Kinetic evaluation of MAO-B-activity following oral administration of selegiline and desmethyl-selegiline in the rat

H. O. Borbe, G. Niebch, and **B. Nickel**

Department of Biochemistry, Asta Pharma AG, Frankfurt a.M., Federal Republic of Germany

Summary. The monoamine oxidase (MAO) B activity of rat brain was inhibited by selegiline and its desmethyl-metabolite in vitro with IC_{50}-values of 11.25 nmol/l and 625.00 nmol/l, respectively. When measured in an ex vivo experiment following oral treatment of rats, the large difference in potency was distinctly reduced, from factor 60 in vitro to factor 3 ex vivo. Restoration experiments of MAO-B-activity after cessation of treatment revealed a nearly identical time course for both compounds. It is concluded that desmethyl-selegiline is an irreversible blocker of MAO-B, nearly equipotent to selegiline after multiple oral administration. No pharmacologically relevant inhibition of MAO-A was found with both compounds.

Introduction

A decisive advance in the pharmaco-therapy of Morbus Parkinson was the introduction of levodopa by Birkmayer and Hornykiewicz and Barbeau and his coworkers in 1961. Parkinsonism was first described as paralysis agitans by James Parkinson in 1817 and can be paraphrased today as striatal dopamine-deficiency syndrome. As rational consequence of this recognition levodopa was used to replenish the depleted dopamine stores in the striatum of Parkinson patients. The later introduction of peripheral decarboxylase inhibitors, such as benserazide or carbidopa has further increased the clinical value of levodopa therapy.

However, there are several reasons to try to reduce the dose of levodopa, necessary for an effective therapy. The most common side effects early in levodopa-therapy are nausea, vomiting and anorexia. Cardiac arrhythmias occur in some patients, especially those with preexisting disturbances in cardiac conduction. The majority of patients on long term treatment develop abnormal involuntary movements, which often limit the tolerated dosage of levodopa. Psychiatric disturbances are produced by levodopa in a significant proportion of patients and frequently limit also the dose that can be tolerated. However, all undesired side effects are reversible and can generally be controlled by a reduction of the dosage.

Reduction of the levodopa dosis, without loss of the therapeutic efficacy, is possible by stabilizing the dopamine level in the brain by applying a selective inhibitor of monoamine oxidase, EC 1.4.3.4, type B (MAO-B) in order to prevent a too rapid degradation of dopamine (Birkmayer et al., 1977; Glover et al., 1977). Selegiline (R(–)-N-methyl-N-(1-phenyl-2-propyl)-2-propinylamine hydrochloride), (R(–)-deprenyl) was found to be a highly selective, irreversible inhibitor of MAO-B (Knoll, 1976; Mayrock et al., 1976; Salach et al., 1979; Finberg and Youdim, 1983). However, pharmacokinetic studies failed in detecting unchanged selegiline in plasma or urine. Selegiline is rapidly metabolized via N-demethylation and oxidative dealkylation (for review see Mutschler and Möhrke, 1983; Yoshida et al., 1986). In the present study the effects of desmethyl-selegiline (R(–)-nordeprenyl) on MAO-B and MAO-A of rat brain were investigated in vitro and ex vivo following oral administration of either compounds.

Materials and methods

Animals

Male Sprague-Dawley rats (Ivanovas, Kisslegg, FRG) of 220–240 g (6–8 weeks old) were used for all MAO-A and MAO-B assays.

Rats were decapitated and brains($\approx$ 2.0 g) rapidly removed and homogenized (glass homogenizer with teflon pestle) in 10 ml icecold 5 mmol/l potassium-phosphate-buffer pH 7.5 containing 10 mmol/l dithiothreitol, 2.5 mmol/l EDTA, 0.5 mg/ml bovine serum albumine. After centrifugation of the crude homogenate (10 min at 16,000 × g) the resulting pellet was resuspended in 10 ml of the above mentioned buffer. The brain homogenate was refrigerated and stored at – 50 °C until use.

Monoamine oxidase A and B assay

The MAO-A and MAO-B-assay in vitro was performed according to the method of Young et al. (1986).

For the MAO-A assay tissue samples were diluted 1:10 (vol/vol) with 50 mmol/l potassium-phosphate-buffer pH 7.5. Aliquots of 50 µl rat brain were preincubated together with 100 µl of the inhibitors for 30 min at 25 °C. The enzymatic reaction was started by addition of 300 µmol/l ^{14}C-5-hydroxy-tryptamine (111 MBq/mmol; Amersham Buchler, Braunschweig, Federal Republic of Germany). After incubation of the samples for 10 min at 37 °C the reaction was terminated by adding 250 µl 4 mol/1 HCl. The ^{14}C-labeled deaminated reaction product was extracted with 5 ml toluol/ethylacetate (1:1; 2 min) and the radioactivity quantified by liquid scintillation counting. "Blank" samples were prepared in analogy after a preincubation for 5 min at 95 °C.

For the MAO-B assay tissue samples were diluted 1:10 (vol/vol) with 50 mmol/l potassium-phosphate-buffer pH 7.5. Aliquots of 10 µl rat brain were preincubated together with 100 µl of the inhibitors for 30 min at 25 °C. The enzymatic reaction was started by addition of 10 µmol/l ^{14}C-β-phenyl-ethylamine diluted in 30 µl of the incubation buffer. After incubation of the samples for 6 min at 37 °C the reaction was terminated by adding 250 µl 4 mol/l HCl. The ^{14}C-labeled deaminated reaction product was extracted with 5 ml toluol (2 min) and the radioactivity quantified by liquid scintillation counting. "Blank" samples were prepared in analogy after a preincubation for 5 min at 95 °C.

Ex vivo experiments

In order to establish a *dose-response-relationship* increasing single oral doses of selegiline or desmethyl-selegiline were used (range: 0.1–5.0 mg/kg). After sacrificing the animals MAO-activity was measured according to the procedure described above.

For investigation of the *time-course* of MAO-B-inhibition following a single oral dose of selegiline or desmethyl-selegiline rats were treated orally with 5.0 mg/kg of either compound.

The *restoration* of the MAO-B-inhibition was studied following a three-days oral treatment (single daily doses; 5 mg/kg) with both compounds.

Results

MAO-inhibition in vitro

Selegiline was found to inhibit MAO-B-activity in rat brain in a concentration-dependent fashion with an IC_{50}-value ($\pm$ SEM) of 11.25 $\pm$ 1.14 nmol/l (Table 1). Desmethyl-selegiline was by factor 60 less active IC_{50}: 625.00 $\pm$ 80.12 nmol/l). Both values are in good agreement with data found by Knoll: IC_{50}-values in rat brain mitochondria for selegiline were calculated with 15 nmol/l, and for desmethyl-selegiline with 897 nmol/l (unpublished data).

The inhibition of MAO-A for selegiline and desmethyl-selegiline in vitro was clearly less. For selegiline an IC_{50}-value of 1.07 $\pm$ 0.05 µmol/l (selectivity for MAO-B: factor 95) and for desmethyl-selegiline of 124.5 $\pm$ 20.9 µmol/l (selectivity for MAO-B: factor 200) was determined.

MAO-B-inhibition ex vivo

In order to further characterize the inhibitory action of both compounds, selegiline (0.1–5.0 mg/kg) and desmethyl-selegiline (0.1–5.0 mg/kg) were administered orally to rats in different concentrations. The half-maximal inhibitory dosis (ID_{50}) for selegiline was calculated to be 0.43 $\pm$ 0.02 mg/kg. Surprisingly, desmethyl-selegiline was found to be nearly equipotent (only

Table 1. Inhibition of monoamine oxidase (EC 1.4.3.4, MAO) type B and A by selegiline, N-desmethyl-selegiline, clorgyline and pargyline in rat brain homogenates in vitro and in rat brain ex vivo

Compound	IC_{50}/ID_{50}	
	MAO-A	MAO-B
In vitro (IC_{50} *nmol/l*)		
Selegiline	1067.50 ± 50.42	11.25 ± 1.14
N-Desmethyl-selegiline	124500.00 ± 20861.15	625.00 ± 80.12
Clorgyline	1.66 ± 0.31	2562.50 ± 202.60
Pargyline	585.00 ± 43.23	20.75 ± 0.82
Ex vivo (ID_{50} *mg/kg*)		
Selegiline		0.43 ± 0.02
N-Desmethyl-selegiline		1.21 ± 0.06

by factor 3 less potent) with an ID_{50}-value of 1.21 ± 0.06 mg/kg. This result indicates that desmethyl-selegiline like the parent compound might be an irreversible blocker of MAO-B.

Time-course of MAO-B-inhibition ex vivo (single dose)

For investigation of the time-course of MAO-B-inhibition following single oral admininstration of selegiline or desmethyl-selegiline to rats (5.0 mg/kg), a slight inhibition of about 10% of the basal MAO-B-activity could be observed with either compounds at the first time points of measurement (1 and 2 min). After 5 min the inhibition of MAO-B-activity of the selegiline-treated animals reached 50% and after 10 min the maximal inhibition of about 85% was found. Following oral treatment with desmethyl-selegiline a somewhat delayed inhibition of MAO-B-activity was observed as compared with the parent compound (Fig. 1). After 120 min an inhibition of about 75% was found.

Half-maximal inhibition of the enzyme was estimated by interpolation and can be given for selegiline with about 4 min and for desmethyl-selegiline with 20–30 min.

Duration of MAO-B-inhibition ex vivo (multiple doses)

To prove the assumption that desmethyl-selegiline is also an effective irreversible inhibitor of MAO-B the activity of MAO-B after multiple doses (selegiline: 5.0 mg/kg or desmethyl-selegiline: 5.0 mg/kg; 3 days, once a day to groups of 5 animals) was measured over 28 days following the last dose. The

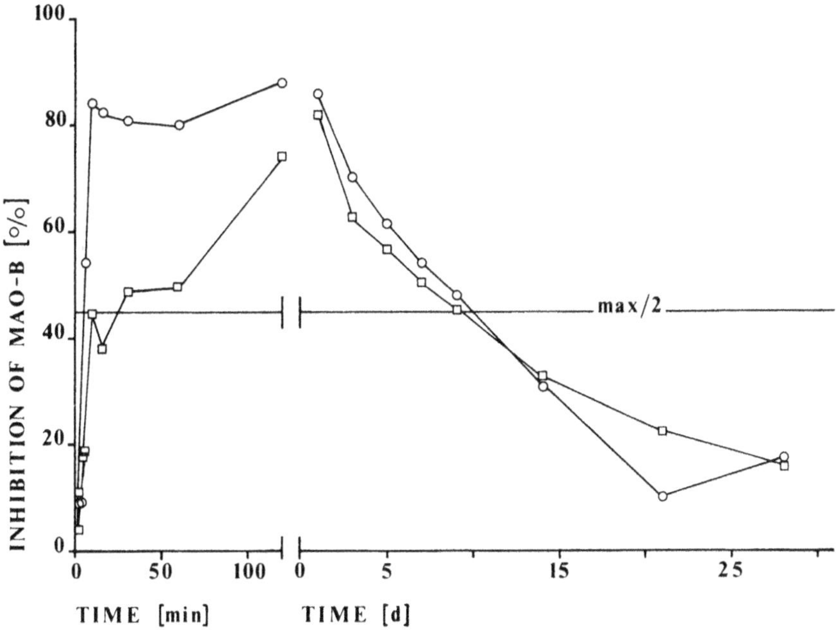

Fig. 1. Time dependence of the inhibition of monoamine oxidase (EC 1.4.3.4, MAO) type B activity by selegiline (○), N-desmethyl-selegiline (□) in rat brain following single (5.0 mg/kg, orally) or three-days-treatment (5.0 mg/kg once daily, orally)

time course of the regenerated MAO-B-activity in rat brain is demonstrated in Fig. 1. As obvious from the curves, multiple administration of either selegiline or desmethyl-selegiline leads to a nearly complete inhibition of MAO-B-activity in rat brain as measured 24 h following the last dose (selegiline: 85.83 ± 1.22%; desmethyl-selegiline: 82.6 ± 0.34%). On the following days after treatment MAO-B-activity increased constantly, reaching 80% (residual inhibition – selegiline: 18.07 ± 0.52%; desmethyl-selegiline: 17.40 ± 1.38%) of the activity of the control groups after 28 days. Half of the basic MAO-B-activity had been reached approximately 10 days after stopping the administration of either selegiline or desmethyl-selegiline. All data of this particular experiment were compared to a control animal group, treated with vehicle.

In the respective investigations on MAO-A-activity only a slight inhibition of about 10–20%, without any time-dependence could be observed for selegiline or desmethyl-selegiline.

Discussion

In vitro experiments showed that selegiline is a potent inhibitor of MAO-B and that the desmethyl-metabolite is by factor 60 less potent. It could be

Table 2. Inhibition of monoamine oxidase (EC 1.4.3.4, MAO) type B by selegiline, N-desmethyl-selegiline (means ± S.E.M.) in rat brain following single (5.0 mg/kg) or three-day-treatment (5.0 mg/kg once daily).

Time	Inhibition (%)	
	Selegiline	Desmethyl-selegiline
Rat brain ex vivo (single dose)		
1 min	9.43 ± 2.55	11.47 ± 2.57
2 min	9.54 ± 1.48	4.72 ± 1.91
5 min	54.70 ± 9.39	18.91 ± 4.38
10 min	84.30 ± 5.51	43.85 ± 8.04
15 min	82.66 ± 3.86	38.11 ± 11.33
30 min	81.53 ± 2.34	49.19 ± 14.66
60 min	80.37 ± 2.44	49.96 ± 6.66
120 min	88.20 ± 1.44	74.56 ± 3.12
Rat brain ex vivo (multiple doses)		
1 d	85.83 ± 1.22	82.60 ± 0.34
3 d	70.60 ± 0.95	61.90 ± 3.11
5 d	61.83 ± 1.59	56.80 ± 0.50
7 d	54.18 ± 1.03	50.80 ± 1.29
9 d	48.18 ± 3.67	45.90 ± 1.31
14 d	31.38 ± 2.30	33.30 ± 1.91
21 d	10.39 ± 1.44	22.80 ± 1.46
28 d	18.07 ± 0.52	17.40 ± 1.38

demonstrated ex vivo that the desmethyl-metabolite of selegiline possesses nearly the same inhibitory potency as it has been described for selegiline. The results indicate that in the case of irreversible enzyme inhibitors in vitro data or single dose data might be misleading because total inhibition is only a matter of time.

The results of investigations into the time-course of the inhibition of MAO-B-activity following single oral administration indicate a rapid availability of the active principle in the brain.

After stopping the administration of selegiline or desmethyl-selegiline, MAO-B-activity reappeared in a time-dependent fashion (Fig. 1), that might reflect the de-novo-synthesis of the enzyme.

Only a slight inhibition of about 10–20% was found for the inhibition of MAO-A by either selegiline or desmethyl-selegiline, that showed no time-dependence. Despite the results at day 1 and 3, where no inhibition could be observed with both compounds, the inhibitory values at all other time points (up to 14 days) lie in the same range. This inhibition might be due to a cross-reactivity of the MAO-A- and the MAO-B-assay that has also been described to reach about 15% (Young et al., 1986).

Taking into account that selegiline is metabolized to desmethyl-selegiline, strong evidence occurs that the MAO-B-inhibitory effect after orally administered selegiline might be in part due to its desmethyl-metabolite.

References

Barbeau A, Murphy CF, Sourkes TL (1961) Excretion of dopamine in diseases of basal ganglia. Science 133:1706

Birkmayer W, Hornijkiewicz O (1961) Der L-3,4 Dioxyphenylalanin (L-Dopa)-Effekt bei der Parkinson-Akinesie. Wien Klin Wochenschr 73:787

Birkmayer W, Riederer P, Ambrozi L, Youdim MBH (1977) Implications of combined treatment with 'Madopar' and L-deprenyl in Parkinson's disease. Lancet i:439–443

Finberg JPM, Youdim MBH (1983) Selective MAO-A and MAO-B inhibitors: their mechanism of action and pharmacology. Neuropharmacology 22:441–446

Glover V, Sandler M, Owen F, Riley GJ (1977) Dopamine is a monoamine oxidase B substrate in man. Nature 265:80–81

Knoll J (1976) Analysis of the pharmacological effects of selective monoamine oxidase inhibitors. In: Wolstenholme GEW, Knight J (eds) Monoamine oxidase and its inhibition. Elsevier, Amsterdam, pp 135–161

Maycock AL, Abeles RH, Salach JI, Singer TP (1976) The action of acetylenic inhibitors on mitochondrial monoamine oxidase: structure of the flavin site in the inhibited enzyme. In: Wolstenholme GEW, Knight J (eds) Monoamine oxidase and its inhibition. Elsevier, Amsterdam, pp 33–47

Mutschler E, Möhrke W (1983) Kinetics of MAO inhibitors. Mod Probl Pharmacopsychiatry 19:126–134

Salach JI, Detmer K, Youdim MBH (1979) The reaction of bovine and rat liver monoamine oxidase with ^{14}C-clorgyline and ^{14}C-deprenyl. Mol Pharmacol 16: 234–241

Yoshida T, Yamada Y, Yamamoto T, Kuroiwa Y (1986) Metabolism of deprenyl a selective monoamine oxidase (MAO) B inhibitor in rat: relationship to MAO-B inhibitory potency. Xenobiotica 16(2):129–136

Young WF, Laws ER, Sharbrough FW, Weinshilboum RM (1986) Human monoamine oxidase. Arch Gen Psychiatry 43:604–609

Authors' address: Dr. H. O. Borbe, Asta Pharma AG, Department of Biochemistry, Weismüllerstrasse 45, D-6000 Frankfurt/Main, Federal Republic of Germany.

J Neural Transm (1990) [Suppl] 32: 139–144

Effect of selegiline and desmethyl-selegiline on cortical electric activity in rats

B. Nickel, H. O. Borbe, and **I. Szelenyi**

Departments of Pharmacology and Biochemistry, ASTA Pharma AG, Frankfurt/Main, Federal Republic of Germany

Summary. The pharmaco-EEG changes caused by the monoamine oxidase (MAO) B inhibitor selegiline were compared with the frequency band alterations aroused by its desmethyl metabolite after oral administration in rats. After single administration (5 mg/kg) the EEG changes caused by selegiline or desmethyl-selegiline differed significantly. Distinct decreases in delta and clear increases in theta EEG frequency bands were obvious after administration of selegiline. The single oral dose of desmethyl-selegiline (5 mg/kg) caused only trendly the same EEG changes observed after giving the mother compound.

Following repeated administration on four consecutive days no significant differences in the frequency band changes could be seen after selegiline or desmethyl-selegiline. Based on present results it is likely that the mode of action of desmethyl-selegiline appears to be similar or identical with the mode of action of the parent compound, selegiline.

Introduction

Following the discovery of the nigrostriatal dopamine deficiency in Parkinson's disease, one obvious therapeutic strategy was to inhibit monoamine oxidase (MAO), the major enzyme metabolizing it both in human and animal brain, in order to conserve the attenuated supply. The next great milestone in the management of the Parkinson's disease was the discovery that MAO exists in two forms, MAO-A and MAO-B, each with different substrate specificities and inhibitory sensitivities. This classification was made by Johnston (1968) using clorgyline a selective MAO-A inhibitor. Some years later, Knoll and Magyar (1972) have developed a new class of MAO-inhibitors without the "cheese effect". The most active compound, selegiline selectively inhibits MAO-B both in vitro and in vivo. Three years following the experimental results with selegiline, the compound has successfully been introduced into the therapy of the Parkinson's disease by Birkmayer et al. (1975).

In the present study experiments were performed in order to establish differences or similarities in changes of EEG of freely moving rats induced by selegiline and its desmethyl metabolite.

Materials and methods

Animals

Female Sprague-Dawley rats (SIV 50, Fa. Savo, Kisslegg, FRG) weighing 160–220 g were used in EEG experiments. Animals received rat pellet food (Altromin®) purchased from Fa. Altrogge, Lage/Lippe, FRG. Both food and water were usually allowed ad libitum. The animals were kept under constant environmental conditions (temperature: 21–23 °C; relative humidity 55–60%; light-dark rhythm: 12/12 hours).

Registration of cortical electrical activity in rats

The method has been described in our previous papers in more details (Nickel and Zerrhan, 1987; Nickel and Szelenyi, 1989). Briefly, under general anesthesia, four cortical electrodes based on coordinates given by the atlas of Zilles (1985) were implanted. After surgery the animals were allowed to recover for two weeks. Rats were housed individually. They were kept under constant environmental conditions as the animals used in dependency studies. However, they had reversed light-dark rhythm, 8:00 a.m.–8:00 p.m. dark.

After beginning of the active phase (8:00 a.m.), the animals were transferred to the recording room in a sound-isolated box placed in a Faraday cage and allowed to adapt for 30 min. Thereafter, 2 periods of 15 min were recorded as reference periods (base line activity). The particular drug was given in a volume of 10 ml/kg by gavage followed by a lack period of 5 min before the electroencephalographic (EEG) recording was restarted. Drug effects were observed then for 165 min in 11 successive periods, each lasting 15 min. Following drugs were tested orally in 6 experiments in a total of 6 animals: selegiline (5 mg/kg) and desmethyl-selegiline (5 mg/kg). The compounds were given in a single dose or once daily over four days.

Analysis of EEG

The EEG was analyzed by Fourier analysis as originally suggested by Dietsch (1932). The EEG signals were digitized with a sample rate of 512 Hz and a resolution of 12 bits. Four successive samples were averaged to give an effective sampling rate of 128 Hz. After artefact rejection, elimination of DC-terms from the signals and windowing power density spectra were sampled online for periods of 4 sec by fast Fourier transformation on a PDP 11/23 (Digital Equipment, U.S.A.) and averaged on each channel over time blocks of 15 min. The power spectra were cut into 6 frequency bands in congruence with clinical standards. Delta activity was defined from 1.5 to 3.5 Hz, theta activity from 3.7 to 7.5 Hz, $alpha_1$ from 7.8 to 13.5 Hz and beta from 13.8 to 30.0 Hz.

Each frequency band was evaluated separately in percentage of the baseline activity, measured during the first 2 pre-drug recording periods of 15 min. For more details see Nickel and Zerrhan (1987).

Drugs

Selegiline and desmethyl-selegiline were generous gifts of Chinoin (Budapest, Hungary). All drugs suspended in 1% carboxymethylcellulose were prepared daily and given in a volume of 10 ml/kg orally.

Statistical evaluation

Significance was calculated by Student's t-test for paired samples (Tallarida and Murray, 1984).

Results

Changes of the EEG frequency bands recorded in freely moving rats after oral administration of test compounds are summarized in Table 1. Single oral doses of selegiline induced significant decreases of mean power values in delta and increases in theta frequency bands (Table 1 and Fig. 1).

Only small not significant EEG frequency changes could be observed after single dose (5 mg/kg) of desmethyl-selegiline. But the tendency of the power spectrum changes are similar to those seen after administration of selegiline (Table 1 and Fig. 1).

Table 1. Changes of the EEG frequency bands in freely moving rats ($n = 6$) after single or repeated oral administration of selegiline and desmethyl-selegiline in percentage, related to the control value for the same animal

Compounds	Dose mg/kg p.o.	Minutes after treatm.	Delta 1.5–3.5 Hz (%)	Theta 3.7–7.5 Hz (%)	Alpha 7.8–13.5 Hz (%)	Beta 13.8–30.0 Hz (%)
Selegiline	5	60	– 22**	+ 21**	– 4	+ 2
Desmethyl-selegiline	5	60	– 6	+ 8	+ 2	+ 4
Selegiline	4 × 5	60	– 25**	+ 20**	– 8	+ 7
Desmethyl-selegiline	4 × 5	60	– 21**	+ 24**	– 4	– 2

Asterisks indicate that EEG frequency changes after drug administration differed significantly from the baseline value obtained before drug application (* $p < 0.05$; ** $p < 0.01$)

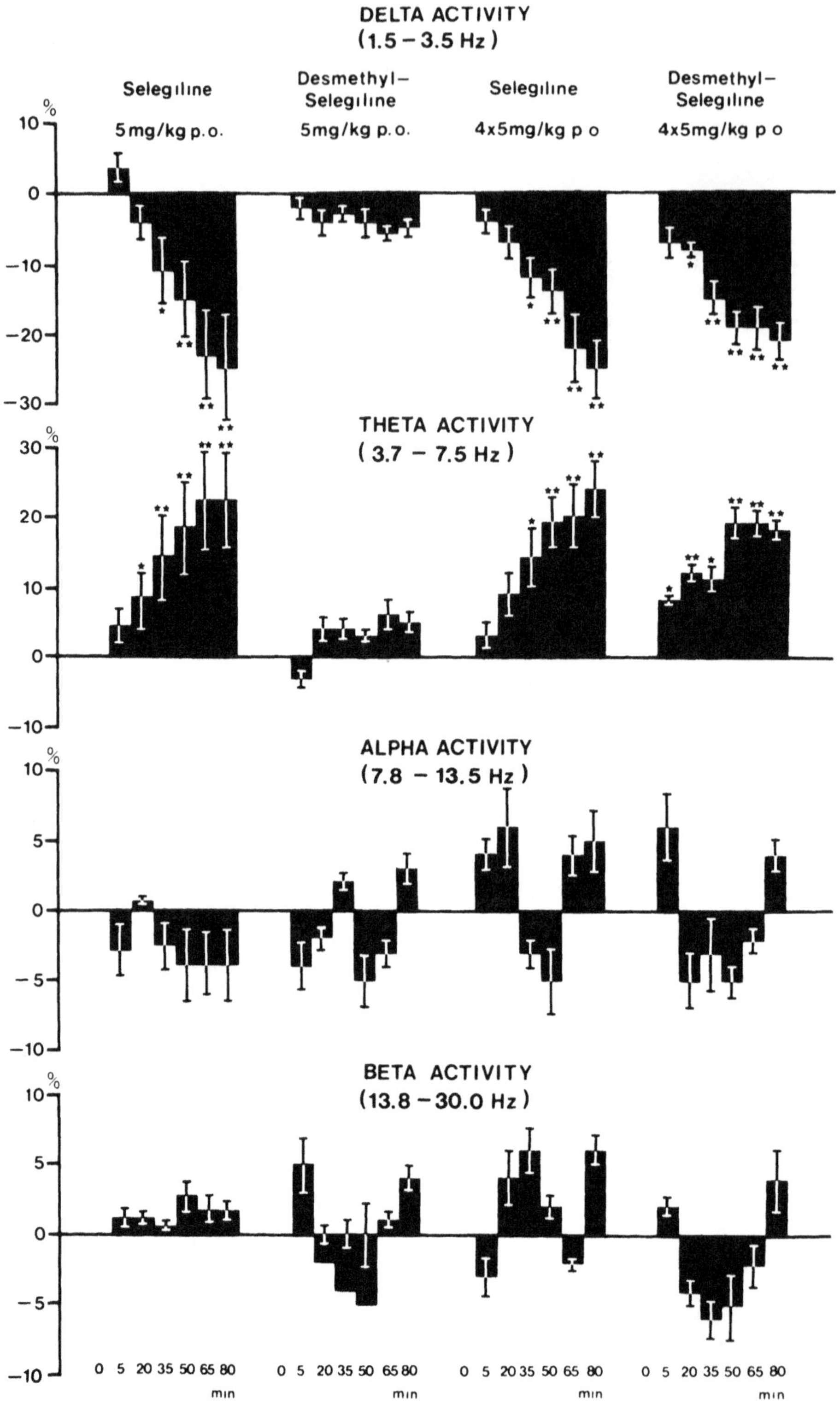

Fig. 1. Drug-induced changes in EEG pattern in rats (n = 6). Mean changes in % to basic activity + / − SEM. Asterisks indicate that EEG frequency changes after drug administration differed significantly from the baseline value obtained before drug application (* $p < 0.05$; ** $p < 0.01$)

After repeated oral treatment on four consecutive days selegiline and desmethyl-selegiline induced comparable changes, consisting in decreases of mean power values in delta and increases in theta frequency bands. By which there were in the opposite to desmethyl-selegiline no differences in the power spectrum changes after a single dose of selegiline or after four days treatment (Table 1 and Fig. 1).

The first signs of EEG changes, caused by both compounds after repeated administration, were already observed during the first 15 min recording period after oral administration of the drugs. The changes lasted for at least 165 min, with a peak between 30 and 45 min after application.

Both compounds caused only small EEG frequency changes in alpha and beta frequency bands (Table 1 and Fig. 1).

Discussion

Recording the cortical electrical activity by means of chronically implanted electrodes from freely moving conscious rats enable the processing of nearly artefact-free signals by the computer. Moreover, the so-called quantitative EEG analysis has been shown to be a valuable tool to establish differences or similarities between different drugs in animals (Nickel and Szelenyi, 1989). There are relatively few data with regard to comparisons of EEG changes induced by a definite drug in humans and animals (e.g. rats). Recently, we have compared the EEG changes induced by flupirtine, a novel non-opioid centrally acting analgesic, in healthy volunteers with those evoked by this drug in rats. There was a clearcut similarity between the findings observed in humans and rats (Kobal et al., 1990) suggesting the possibility that EEG changes observed in animals (rats) can be projected to human to a certain extent.

To obtain results with predictive character for human use, both pharmacodynamic and particularly pharmacokinetic properties and possible differences should often be taken into consideration. This is, of course, also valid for comparison of EEG changes.

In the present study it could be demonstrated that the desmethyl metabolite of selegiline evoked after repeated administration nearly the same power frequency changes as it has been described for selegiline. These results indicate that the mode of action of desmethyl-selegiline is similar or identical with the mode of action of selegiline. Furthermore, the comparison of the results obtained after single dose of desmethyl-selegiline and after repeated treatment suggest that desmethyl-selegiline is likely an irreversible inhibitor of MAO-B as its parent compound. In restorage experiments of MAO-B activity Borbe et al. (1990) also found the same potency for selegiline and desmethyl-selegiline after repeated administration over three days. Furthermore, they also came to the conclusion that desmethyl-selegiline must be an irreversible inhibitor of MAO-B; nearly equipotent to selegiline. Therefore, it is assumed that desmethyl contributes to the therapeutic efficacy of selegiline in the treatment of P.-D.

References

Birkmayer W, Riederer P, Youdim MBH, Linauer W (1975) The potentiation of the anti-akinetic effect after L-dopa treatment by an inhibitor of MAO-B, deprenyl. J Neural Transm 36:303–326

Borbe HO, Niebch G, Nickel B (1990) Kinetic evaluation of MAO-B activity following oral administration of selegiline and desmethyl-selegiline in the rat. (in preparation)

Dietsch G (1932) Fourier-Analyse von Elektroencephalogrammen des Menschen. Pflügers Arch Ges Physiol 230:106–112

Johnston JP (1968) Some observations upon a new inhibitor of monoamine oxidase in brain tissue. Biochem Pharmacol 17:1285–1297

Knoll J, Magyar K (1972) Some puzzling pharmacological effects of monoamine oxidase inhibitors. In: Costa E, Sandler M (eds) Monoamine oxidase – new vistas. Raven Press, New York, pp 393–408

Kobal G, Nickel B, Hummel Th, Szelenyi I (1990) Comparison of EEG changes induced by flupirtine in man and rats. Neuropharmacology (in press)

Nickel B, Zerrhan H (1987) Pharmaco-electroencephalography in the rat as a method for characterization of different types of analgesics. Postgrad Med J 63 [Suppl 3]:45–47

Nickel B, Szelenyi I (1989) Comparison of changes in the EEG of freely moving rats induced by enciprazine, buspirone and diazepam. Neuropharmacology 28(8): 799–803

Tallarida RJ, Murray RB (1984) Manual of pharmacological calculations with computer programs. Springer, Berlin Heidelberg New York

Zilles K (1985) The cortex of the rat. Springer, Berlin Heidelberg New York

Authors' address: Dr. B. Nickel, Department of Pharmacology, ASTA Pharma AG, Weismüllerstrasse 45, D-6000 Frankfurt/Main, Federal Republic of Germany.

J Neural Transm (1990) [Suppl] 32: 145–148

An enzymatic method for detecting MAO-A and MAO-B inhibitors in plasma and its application in studies with reversible MAO-A selective inhibitors

H. L. White

Division of Pharmacology, Wellcome Research Laboratories, Research Triangle Park, N.C., U.S.A.

Summary. A general method was developed for the simultaneous enzymatic assay of MAO-A and MAO-B inhibitors in plasma of pretreated animals. After extraction of the inhibitor from plasma into an appropriate volatile organic solvent, dried aliquots of the organic layer are incubated with a buffered rat brain MAO preparation and substrates, [^{3}H]serotonin and [^{14}C]phenethylamine. Under conditions employed, the presence in plasma of MAO-A and/or B inhibitors can be detected.

Introduction

The enzymatic procedure described was developed for preliminary testing of reversible MAO inhibitors in order to relate inhibitor levels in plasma to MAO inhibition in brains and livers of rats and other animals. When used in conjunction with more specific HPLC methods, the presence in plasma of active metabolites and the MAO-A/B specificities of such metabolites may be revealed.

Materials and methods

Preparation of standard rat brain MAO (all steps at 0–4 °C)

Whole rat brains were homogenized at a 10-fold tissue weight/buffer volume ratio in 15 mM Tris-Cl, 0.1 mM EDTA, pH 7.4. The homogenate was centrifuged at 800 g for 10 min, and the decanted supernatant was then centrifuged at 8200 g for 10 min to give a mitochondrial pellet. This pellet was resuspended in 1/10 the initial buffer volume and again centrifuged at 8200 g for 10 min. The final pellet, resuspended in 1/20 the initial buffer volume, was stable during storage in small aliquots at − 70 °C for more than a year. Before use in MAO assays, this enzyme suspension was diluted 8 to 20-fold in 0.10 M potassium phosphate, pH 7.4 and preincubated at 37 °C for 10 min before a further 8-fold dilution to give appropriate activities in MAO assays.

Enzymatic assay for detecting MAO-A and B inhibitors in plasma

At designated times after pretreatment of rats with a reversible MAO-A inhibitor, animals were sacrificed by decapitation, and brains, livers, and blood were collected. Whole blood was immediately centrifuged at 3–4 °C and 10,000 g for 10 min to obtain platelet-free plasma. Aliquots of plasma (5–100 µl, depending on range of concentrations expected) were extracted with 0.50 ml of chloroform or methyl 3°-butyl ether by vortexing for 20 sec. After a 5 min centrifugation at 160 g, 0.25 ml of each organic layer was dried under nitrogen in heavy glass centrifuge tubes with tight-fitting screw caps (40 ml size tubes in order to accommodate all further steps in the MAO assay). To each tube the following were added, with vortexing, to give a total assay volume of 300 µl: 200 µl of a standard rat brain extract (approx. 10^{-4} I.U. in 50 mM potassium phosphate, pH 7.4); 50 µl distilled water or 50 µl of 12 mM pargyline (for blanks); and 50 µl of a substrate mixture (stock concentrations: 1.2 mM [^{3}H]serotonin, 5 Ci/mol and 60 µM [^{14}C]β-phenethylamine, 3 Ci/mol). Assay mixtures were incubated for 20 min at 37 °C. Products were extracted by adding 0.2 ml of 2 N HCl + 6 ml of 1:1 ethyl acetate/toluene and shaking vigorously for 10 min, as described by White and Stine (1982). After centrifugation for 10 min at 160 g, aliquots of the organic layers were counted using double-label techniques. Standard plots were prepared by initially adding known amounts of the MAO inhibitor to plasma from vehicle-treated animals.

Ex vivo inhibition of MAO by reversible inhibitors

In order to estimate MAO inhibition in brains and livers of rats pretreated with reversible inhibitors, it was necessary to use an assay procedure that minimized dilution of the inhibitor. Thus, high concentrations of brain tissue homogenates were incubated for very short incubation times. For brain assays, initial tissue was 3-fold diluted into each assay. Substrate concentrations were not saturating, but were chosen relative to K_m values for serotonin and phenethylamine in order to give an estimate of MAO-A and B, respectively. Brains from pretreated rats were homogenized in a buffer consisting of 0.1 M potassium phosphate and 5% sucrose (pH 7.4) at a 1:1 tissue wt/buffer volume ratio, using a motorized Teflon/glass homogenizer. MAO-A and B were determined by incubating 100 µl of tissue homogenate with 50 µl of a double-label substrate mixture to give final concentrations of [^{3}H]serotonin, 0.4 mM (5 Ci/mol); and [^{14}C]β-phenethylamine, 20 µM (3 Ci/mol). For blank assays 100 µl portions of homogenate were pre-incubated for 15 min at 37 °C with pargyline (4 mM) before substrate addition. For brain homogenates, incubations were at 37 °C for 30 sec. Assay mixtures were then acidified with 0.2 ml of 2N HCl and products extracted with 6 ml of ethyl acetate/toluene (1:1 vols) as described above.

Results

A typical standard curve is shown in Fig. 1A for the reversible MAO-A selective inhibitor, BW 616U76. In this experiment the inhibitor concentra-

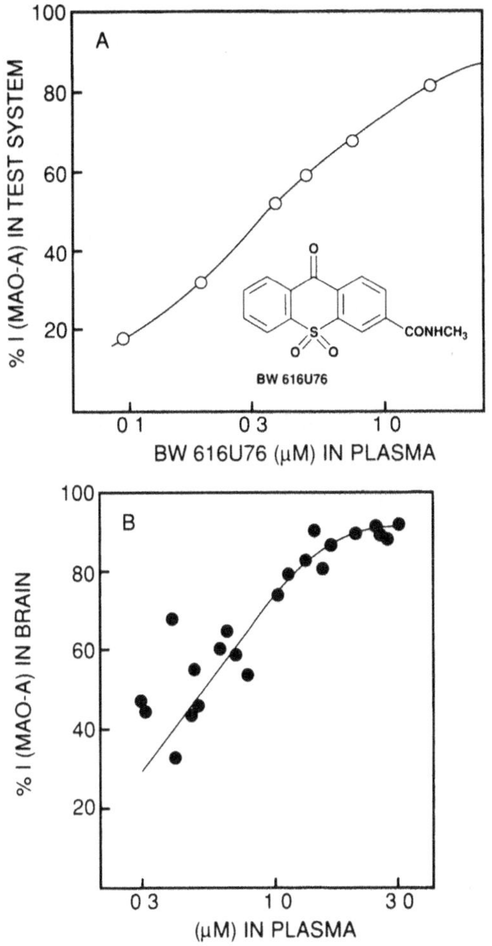

Fig. 1. **A** Standard curve for estimation of BW 616U76 in plasma of pretreated rats. Inhibition conc. in test system was 1/6 of that in original plasma. **B** Plasma conc. related to % I of MAO-A in brain of each rat

tion in each MAO assay was 1/6 of that in the original plasma. Thus 50% inhibition in the test system occurred in the presence of 0.058 μM of BW 616U76. This corresponds with an I_{50} value of 0.055 μM obtained when the compound was added directly to the same rat brain MAO preparation. Plasma concentrations were related to ex vivo inhibition in brains of the same rats in Fig. 1B. Only MAO-A inhibition was observed in the enzyme assays with this compound. Results obtained using the enzymatic method averaged 33 ± 13 % higher than values from HPLC assays of the same plasma samples.

Discussion

This enzymatic procedure for determination of MAO inhibiting activities in plasma requires a suitable extraction procedure, which can generally be developed using knowledge of the structure of the inhibitor and its solubility properties. The standard rat brain mitochondrial enzyme used in the assay contains both MAO forms (A/B ratio = 1.1). At the substrate concentrations employed (approximately twice the K_m concentration for each substrate), MAO-A and MAO-B activities are independent. Product formation from serotonin (MAO-A) was selectively inhibited by 0.1 μM clorgyline, and phenethylamine oxidation (MAO-B) was selectively inhibited by 1 μM deprenyl.

This method has been applied to studies with reversible MAO inhibitors, including BW 616U76 and BW 70C73. It is especially useful during preliminary testing of reversible inhibitors in vivo in order to relate plasma inhibitor levels to MAO inhibitions in brains and livers of rats and other animals without the necessity for development of specific HPLC assays. It is also applicable to human plasma. When used as an adjunct to HPLC, the presence in plasma of active metabolites and the A/B specificities of these metabolites may be revealed.

Fritze et al. (1989) have determined moclobemide in human plasma using an enzymatic method in which human placental MAO-A was used as a test system with tyramine as substrate. Their results indicated that one or more MAO-A inhibiting metabolites, in addition to moclobemide, might be present in plasma. Our studies with plasma of moclobemide-treated rats support a similar conclusion (data not shown).

References

Fritze J, Laux G, Sofic E, Koronakis P, Schoerlin MP, Riederer P, Beckmann H (1989) Plasma moclobemide and metabolites: lack of correlation with clinical response and biogenic amines. Psychopharmacology 99:252–256

White HL, Stine DK (1982) Monoamine oxidases A and B as components of a membrane complex. J Neurochem 38:1429–1436

Authors' address: Dr. Helen L. White, Division of Pharmacology, Wellcome Research Laboratories, Research Triangle Park, NC 27709, U.S.A.

J Neural Transm (1990) [Suppl] 32: 149–154

Serotonin and 5-hydroxyindoleacetic acid in plasma. Potential use as peripheral measures of MAO-A activity

P. Celada, M. J. Sarrias, and **F. Artigas**

Department of Neurochemistry, C.S.I.C., Barcelona, Spain

Summary. We have explored the possibility of using measures of 5-HT (extracellular 5-HT and 5-HIAA in blood) as peripheral indexes of MAO activity. Depressed patients treated with phenelzine for 6 weeks display a dramatic increase of plasma 5-HT (270% of basal values) together with a decrease of plasma 5-HIAA. After treatment with irreversible MAO inhibitors, rat plasma 5-HIAA decreased in a similar fashion than MAO-A activity in liver, lungs and brain, measured using 5-HT as substrate. These results support the use of plasma 5-HT and 5-HIAA in humans to evaluate the action of MAOI on the serotonergic system.

Introduction

MAO inhibitors (MAOI) were among the first drugs clinically used in the treatment of depressive states. Their use declined due to serious side effects ("cheese effect" and hepatotoxicity) (For a review see Cooper, 1989; Zimmerman and Ishak, 1987). However, in recent years, the introduction of new molecules with strong specificity for the A-form of the enzyme and a more accurate control of the risk associated to non-specific MAO inhibitors has led to changing views on this matter (Paykel and White, 1989). Serotonin (5-HT) has been implicated for long in the aetiology of depressive disorders and in the mode of action of antidepressant drugs (Heninger et al., 1987). Most studies have shown a functional decrease of its activity in depressive states (Åsberg et al., 1976; Heninger et al., 1984; Sarrias et al., 1987). Therefore, 5-HT is one of the key targets for the action of MAOI. However, a major problem in the evaluation of the MAO activity that deaminates 5-HT is the lack of reliable peripheral measures to be used in humans after MAOI. Lumbar punctures are ethically restricted in follow-up studies and thus, platelet MAO has been used as a peripheral index of CNS MAO activity. However, no relationship has been found between platelet MAO-B and brain MAO-A or B in humans (Young et al., 1986). Deamination by MAO-A in the liver and the lungs is one of the major elimination pathways of free 5-HT in blood (Gillis, 1985). Therefore, we have examined wether MAOI affect its concentration and that of plasma 5-hydroxyindoleacetic acid (5-HIAA), its main metabolite.

Materials and methods

Patients

We examined eight major depressive patients (four males and four females) admitted to the Department of Psychiatry of the Hospital de Sant Pau, Barcelona. They had been without antidepressant medication for at least 2 months before. Blood sampling (8–10 am) was carried out to fasted subjects in basal conditions and after 6 weeks of treatment with phenelzine (45 mg/daily).

Blood treatment

Blood was collected in several EDTA Venoject tubes, which were immediately mixed by inversion. Platelet-free-plasma (PFP) was obtained by centrifugation according to previously described procedures (Ortiz et al., 1988). Samples of PFP were stored at $-80°C$ until analysis.

Animal treatment

Male Wistar rats (Interfauna, Sant Feliu de Codines, Barcelona) of 220–250 g were used. They were injected i.p. with saline or different doses of tranylcypromine (TCP). 4 h later, the animals were anesthesized with pentobarbitone, had their carotid artery cannulated and PFP was obtained by centrifugation of blood. The animals were killed and their brain, liver and lungs taken out and frozen until use.

Biochemical analyses

The analyses of 5-HT and 5-HIAA in PFP were carried out by HPLC using electrochemical detection as described (Ortiz et al., 1988) with minor modifications for the analysis of these compounds in rat PFP. MAO activity was determined in homogenates of either tissue using 100–200 μM 5-HT with [^{14}C]-5-HT as a tracer. Incubations were carried out for 10 min and the tubes extracted with 4 ml of toluene/ethyl acetate (1:1). Radioactivity was counted in the organic phase. Using these conditions, basal values of MAO-A activity in the tissues analyzed were: 2620 $\pm$ 740 (liver), 520 $\pm$ 95 (lung) and 2460 $\pm$ 400 pmols/mg protein. min ($\bar{X} \pm$ SD, N = 8–10 animals).

Results

Figure 1 shows the effect of 45 mg/day of phenelzine on plasma 5-HT and 5-HIAA. Plasma 5-HT increased significantly after 6 weeks on phenelzine

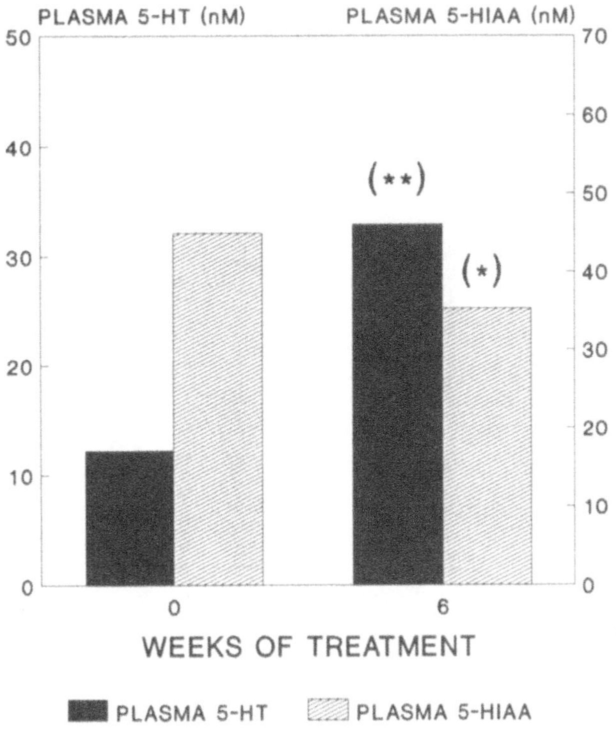

Fig. 1. Mean values of plasma 5-HT and plasma 5-HIAA after 6 weeks of treatment with phenelzine. * $P < 0.07$; ** $P < 0.02$, Wilcoxon t-test for paired data

(270% of basal values, $P < 0.02$, Wilcoxon t-test) while plasma 5-HIAA decreased (– 22%, $P < 0.07$, Wilcoxon t-test).

To study wether these changes induced by MAOI treatment in humans had any relationship to the classical "in vitro" measures of MAO activity, we examined the simultaneous effect of TCP on plasma 5-HT and 5-HIAA and on MAO activity of the liver, lungs and brain of rats. Figure 2 shows the effect of 0.5 and 15 mg/kg of TCP on plasma 5-HIAA and MAO-A activity measured in either tissue. Significant reductions of plasma 5-HIAA to 52 and 15% of control (saline) animals were obtained after these doses ($N = 8$–10 animals per group).

MAO activity in the liver, lungs and brain of the same animals, showed reductions to 20, 18 and 19% of controls (0.5 mg/kg dose) and to 6, 5 and 4% (15 mg/kg dose), respectively. Significant correlations ($r = 0.80$–0.88, $P < 0.01$) were obtained between the plasma 5-HIAA reduction and the MAOI inhibition in either tissue. Analyses of PFP 5-HT were also carried out in some of the animals treated with TCP. Control values were in the range 1–3 ng/ml (1.99 ng/ml, mean of 4 animals) whereas the animals treated with 15 mg/kg of TCP had a mean value of 35.9 ng/ml ($N = 8$). Thus, the decrease of plasma 5-HIAA was accompanied by a large increase of PFP 5-HT.

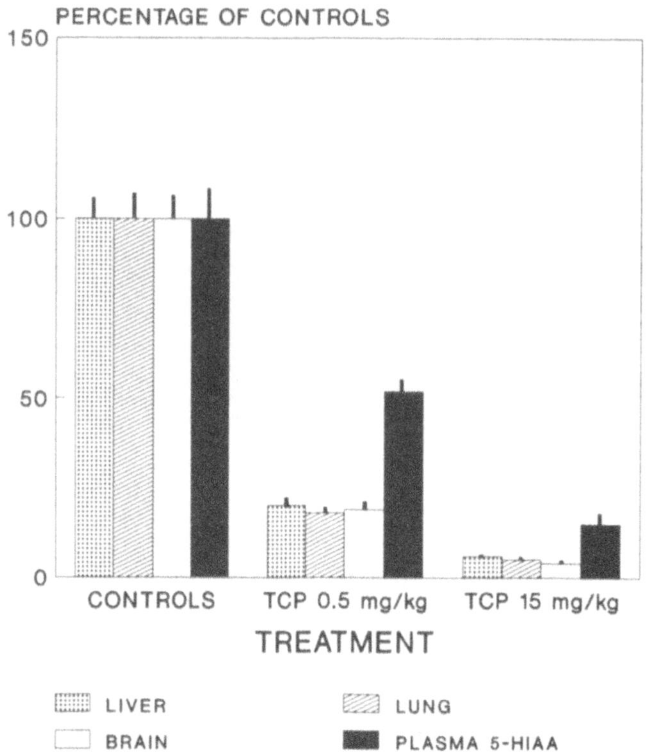

Fig. 2. Effects of 0.5 and 15 mg/kg of TCP on MAO-A activity in liver, lungs and brain and on plasma 5-HIAA. ANOVA showed a significant ($P < 0.001$) effect of the treatment

Discussion

Previous results from this laboratory support the existence of an extracellular pool of 5-HT in blood (Artigas et al., 1985; Ortiz et al., 1988). With some precautions (avoiding high centrifugation forces to damage the platelets, use of EDTA as anticoagulant, careful pippeting of plasma), the 5-HT contained in PFP can be taken as a good estimate of the extracellular pool of 5-HT in blood. The concentration of 5-HT in PFP is usually in the low nM range (mean value in humans is in the range 5–15 nM), while that stored in platelets is much higher (2–4 μM) (Ortiz et al., 1988). Both pools are differentially affected by drugs inhibiting the high-affinity 5-HT uptake (Sarrias et al., 1987; Artigas et al., 1989) because uptake (by the lungs and platelets) is a well-known mechanism of removal of free 5-HT from plasma. However, deamination by MAO in liver and in the vascular endothelium is perhaps the most important mechanism of inactivation of extracellular 5-HT in blood (Gillis, 1985). Accordingly, it could be predicted that MAOI drugs should increase plasma 5-HT and decrease plasma 5-HIAA (see Fig. 3). The present results seem to confirm such a hypothesis. Marked changes of these variables have been observed after two different non-

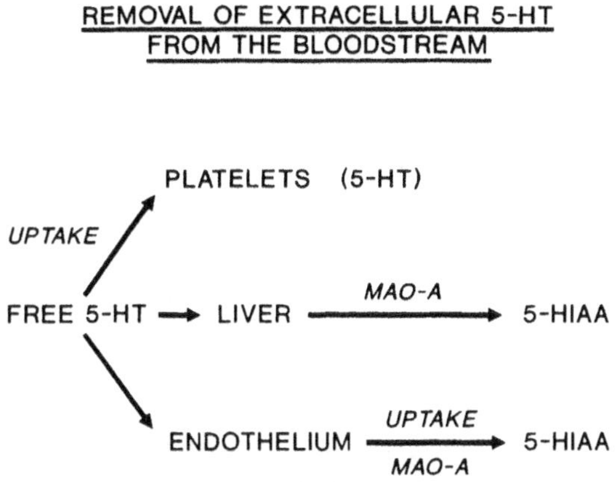

Fig. 3. Extracellular 5-HT is removed from blood through two different mechanisms: uptake (endothelial cells and platelets) and MAO deamination in liver and lung endothelium. Antidepressant drugs like uptake inhibitors and MAOI alter the balance between extracellular 5-HT, platelet 5-HT and 5-HIAA in blood

specific MAO inhibitors. Subchronic treatment with phenelzine yielded large increases of PFP 5-HT that can result from the actual inhibition of MAO activity by this drug. Also, concurrent decreases of plasma 5-HIAA have been observed. Platelet MAO is of B-type (Donnelly and Murphy, 1977). Therefore, the changes of platelet MAO after MAOI drugs may have little relationship with the actual changes of the MAO-A activity deaminating 5-HT, especially when specific MAOI-A drugs are used.

The results obtained in rats treated with TCP support the above interpretation of human data. The inhibition of MAO-A activity in liver and lungs – the two organs removing most 5-HT from the bloodstream – and in brain paralleled a dramatic decrease of the concentration of circulating 5-HIAA. This clearly indicated that the reduction of plasma 5-HIAA can be taken as an "in vivo" index of MAO inhibition elicited by TCP in the rat. Given the similarities of the effects of TCP – in rats – and phenelzine – in humans – , it is reasonable to presume that the changes observed in psychiatric patients are due to the same effect.

Taken together, these preliminary results support the use of plasma 5-HT and 5-HIAA to study the effects of MAOI on the serotonergic system. Given the lack of reliable estimates of MAO-A activity in humans, these measures can prove valuable in clinical practice. Work is in progress to study the effects of MAO specific inhibitors on these variables.

Acknowledgements

Thanks are given to Drs. E. Alvarez and C. Udina for the supply of blood samples of treated patients. The skillful technical assistance of M. Figueras is also acknowledged. This work has been supported by a grant from the Fondo de Investigaciones Sanitarias (89/0387). P. Celada and M. J. Sarrias are fellowship recipients from the FIS and the Ministry of Education and Science, respectively.

References

Artigas F, Sarrias MJ, Martínez E, Gelpí E (1985) Serotonin in body fluids. Characterization of human plasmatic and cerebrospinal fluid pools by means of a new HPLC method. Life Sci 37:441–447

Artigas F, Sarrias MJ, Martínez E, Gelpí E, Alvarez E, Udina C (1989) Increased plasma serotonin but unchanged platelet serotonin in bipolar patients treated chronically with lithium. Psychopharmacology 99:328–332

Åsberg M, Thoren P, Traskman L, Bertilsson L, Reingberger V (1976) "Serotonin depression" – A biochemical subgroup within the affective disorders? Science 191:478–480.

Cooper AJ (1989) Tyramine and irreversible monoamine oxidase inhibitors in clinical practice. Br J Psychiatry 155:38–45

Donnelly CH, Murphy DL (1977) Substrate and inhibitor-related characteristics of human platelet monoamine oxidase. Biochem Pharmacol 26:852–858

Gillis CN (1985) Peripheral metabolism of serotonin. In: Vanhoutte PM (ed) Serotonin and the cardiovascular system. Raven Press, New York, pp 27–36

Heninger GR, Charney DS, Sternberg DE (1984) Serotonergic function in depression: prolactin response to intravenous tryptophan in depressed patients and healthy subjects. Arch Gen Psychiatry 41:398–402

Heninger GR, Charney DS (1987) Mechanism of action of antidepressant treatments: implications for the etiology and treatment of depressive disorders. In: Meltzer HY (ed) Psychopharmacology: the third generation of progress. Raven Press, New York, pp 535–544

Ortiz J, Artigas F, Gelpí E (1988) Serotonergic status in human blood. Life Sci 43:983–990

Paykel ES, White JL (1989) A European study of views on the use of monoamine oxidase inhibitors. Br J Psychiatry 155:9–17

Sarrias MJ, Artigas F, Martínez E, Gelpí E, Alvarez E, Udina C, Casas M (1987) Decreased plasma serotonin in melancholic patients: a study with clomipramine. Biol Psychiatry 22:1429–1438

Young WF, Laws ER, Sharbrough FW, Weinshilboum RM (1986) Human monoamine oxidase. Lack of brain and platelet correlation. Arch Gen Psychiatry 43:604–609

Zimmerman HJ, Ishak KG (1987) The hepatic injury of monoamine oxidase inhibitors. J Clin Psychopharmacol 7:211–213

Authors' address: Dr. F. Artigas, Department of Neurochemistry, C.S.I.C., Jordi Girona 18-26, 08034 Barcelona, Spain.

J Neural Transm (1990) [Suppl] 32: 155–159

Role of monoamine oxidase A and B in the deamination of newly-formed dopamine in the rat kidney

M. Helena Fernandes and **P. Soares-da-Silva**

Laboratório de Farmacologia, Faculdade de Medicina, Porto, Portugal

Summary. The present study has examined the effects of two selective inhibitors of monoamine oxidase (MAO) type A and B, respectively Ro 41-1049 and Ro 19-6327, on the deamination of newly-synthesized dopamine in kidney slices incubated with exogenous L-DOPA (50 and 100 µmol/l). Ro 41-1049 (50, 100 and 250 nmol/l) was found to produce a concentration-dependent increase of newly-formed dopamine (36–56% increase) and reduced DOPAC formation (45–86% reduction). Ro 19-6327 (50, 100 and 250 nmol/l), was found not to affect the accumulation of newly-formed dopamine at 50 µmol/l L-DOPA in the medium, but significantly reduced the formation of DOPAC. At the concentration of 100 µmol/l of L-DOPA, Ro 19-6327 (100 and 250 nmol/l) significantly increased (by 32% and 132%, respectively), the dopamine tissue levels in kidney slices and decreased DOPAC formation. It is concluded that both MAO-A and MAO-B are important in the metabolism of newly-formed dopamine in kidney slices incubated with exogenous L-DOPA.

Introduction

In kidney, dopamine may be formed in tubular epithelial cells by decarboxylation of filtered 3,4-dihydroxyphenylalanine (DOPA) (Baines and Chan, 1980; Lee, 1982). This is a quantitatively important process since tubular epithelial cells are endowed with a high aromatic L-amino acid decarboxylase (AAAD) activity (Adam et al., 1986) and plasma levels of DOPA can attain concentrations up to 2.5 ng/ml (Cuche, 1988). The large capacity of tubular epithelial cells to synthesize dopamine has been confirmed in in vitro studies and it has also been shown that a substantial amount of newly-formed amine is deaminated to 3,4-dihydroxyphenylacetic acid (DOPAC) (Soares-da-Silva and Fernandes, 1990). This would agree with the evidence suggesting that renal tissues are endowed with one of the largest monoamine oxidase (MAO) activities in the body and that deamination represents a major pathway for the inactivation of catecholamines (Kopin, 1985).

The aim of the present work is to study the role of MAO-A and MAO-B in the deamination of dopamine and formation of DOPAC in renal tissues under in vitro experimental conditions. For that purposes rat kidney slices were incubated with exogenous L-DOPA and the influence of two new selective MAO-A and MAO-B inhibitors, respectively Ro 41-1049 and Ro 19-6327 (Da Prada et al., 1990), on the formation of both dopamine and DOPAC studied.

Materials and methods

Male Wistar rats (Biotèrio do Instituto Gulbenkian de Ciência, Oeiras, Portugal) aged 45–60 days and weighing 200–280 g were used in the experiments. Renal slices approximately 1.5 mm thick were prepared and preincubated during 60 min in warm (37 °C) and gassed (95% O_2 and 5% CO_2) 10 ml Krebs solution. After preincubation, renal slices were incubated for 15 min in gassed and warm Krebs solution with added L-DOPA (50 and 100 μmol/l). The Krebs solution had the following composition (mmol/l): NaCl 118, KCl 4.7, $CaCl_2$ 2.4, $MgSO_4$ 1.2, $NaHCO_3$ 25, KH_2PO_4 1.2, EDTA 0.4, ascorbic acid 0.57 and glucose 11; 1-α-methyl-p-tyrosine (50 μmol/l) and copper sulphate (10 μmol/l) were also added to the Krebs solution. The preincubation and incubation were carried out in glass vials, continuously shaken throughout the experiment. In studies of MAO inhibition, the inhibitor, Ro 41-1049 (50, 100 and 250 nmol/l) or Ro 19-6327 (50, 100 and 250 nmol/l), was present throughout the experiment, during the preincubation and incubation periods. After the incubation, renal slices were collected, washed for 30 min in ice cold Krebs solution, blotted with filter paper, weighed and placed in 2 ml of 0.2 mol/l perchloric acid. The assay of L-DOPA, dopamine, noradrenaline and DOPAC in renal tissues was performed by means of HPLC-ECD (Soares-da-Silva and Fernandes, 1990).

Ro 41-1049 ((N-(2-aminoethyl)-5-(m-fluoro-phenyl)-4-thiazole carboxamide hydrochloride) and Ro 19-6327 ((N-(2-aminoethyl)-5-chloro-2-pyridine carboxamide hydrochloride) were kindly donated by Prof. Mosè Da Prada from Hoffmann-La Roche (Basel, Switzerland).

Results

Incubation of renal slices with L-DOPA (50 and 100 μmol/l) for 15 min resulted in a concentration-dependent accumulation of dopamine (Fig. 1). The accumulation of DOPAC, the deaminated metabolite of dopamine, was also found to be dependent on the concentration of L-DOPA used and reached values as high as 4500 ng/g when 100 μmol/l L-DOPA is used. Increasing concentrations (50, 100 and 250 nmol/l) of the selective MAO-A inhibitor Ro 41-1049 were found to produce a concentration-dependent increase of newly-formed dopamine (36–56% increase) in kidney slices

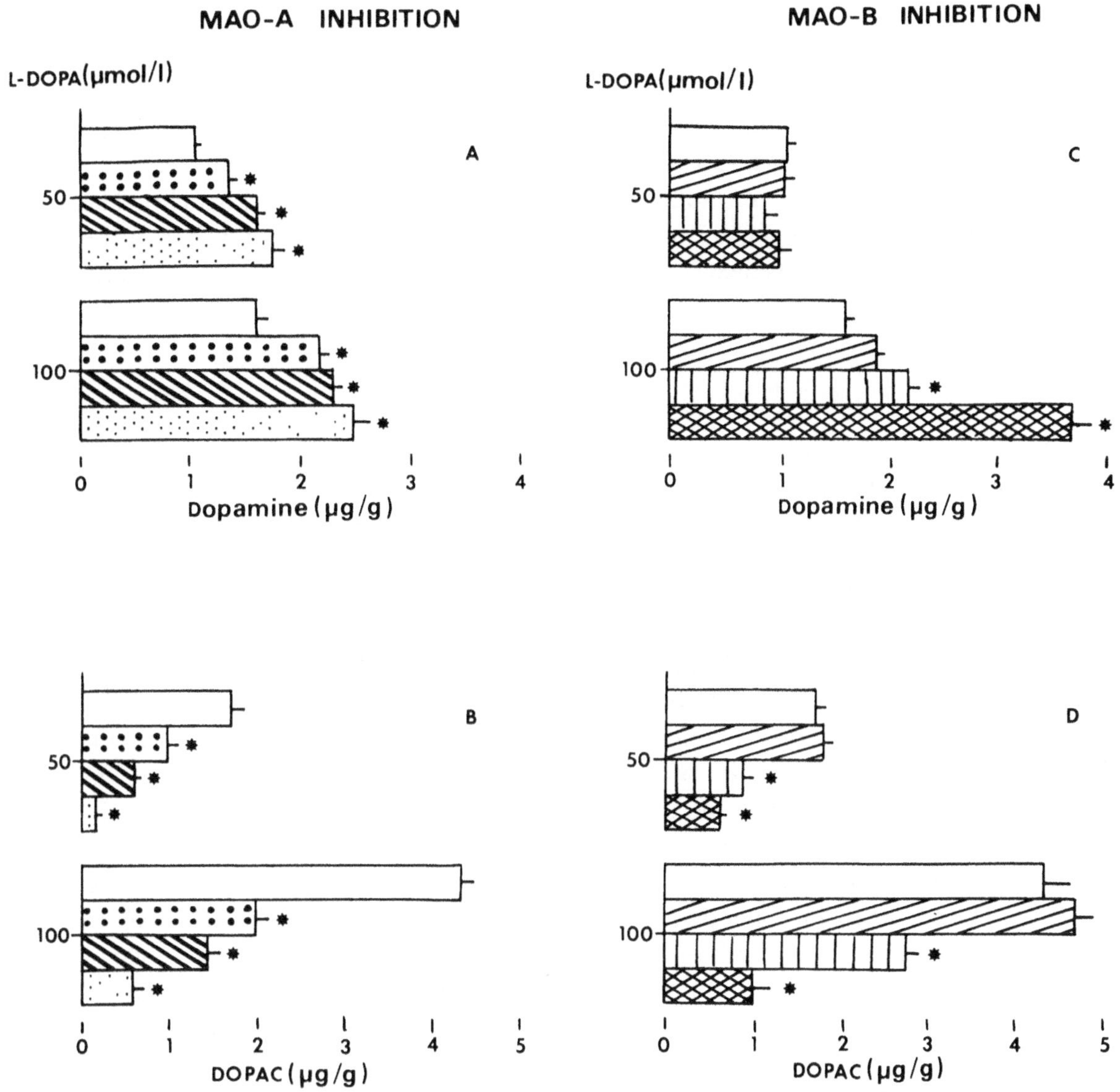

Fig. 1. Effects of MAO-A and MAO-B inhibition on tissue levels (in μg/g) of dopamine and DOPAC in kidney slices incubated with 50 and 100 μmol/l L-DOPA. Each column represents the mean of five to six experiments per group; lines show SEM. Control, ▭; Ro 41-1049 (50 nmol/l), ▭; Ro 41-1049 (100 nmol/l), ▭; Ro 41-1049 (250 nmol/l), ▭; Ro 19-6327 (50 nmol/l), ▭; Ro 19-6327 (100 nmol/l), ▭; Ro 19-6327 (250 nmol/l), ▭

incubated with L-DOPA (Fig. 1A). This effect was accompanied by a decrease (45–86% reduction) of DOPAC tissue content (Fig. 1B). Incubation of kidney slices in the presence of the MAO-B selective inhibitor, Ro 19-6327, was found not to affect the accumulation of newly-formed dopamine for the 50 μmol/l L-DOPA in the medium; however, at the concentration of 100 μmol/l L-DOPA, Ro 19-6327 (100 and 250 nmol/l) significantly increased by 37% and 132%, respectively, the dopamine tissue levels in kidney slices (Fig. 1C). At 50 nmol/l, Ro 19-6327 did not affect DOPAC formation, but at higher concentrations (100 and 250 nmol/l) it decreased DOPAC accumulation (30–70% reduction); this effect was observed even when kidney slices were incubated with 50 μmol/l L-DOPA (Fig. 1D). The amount of L-DOPA accumulated in renal slices was the same either in the presence or in the absence of compounds Ro 41-1049 and Ro 19-6327 (data not shown).

Discussion

The results presented here show that a substantial amount of newly-formed dopamine is deaminated to DOPAC and also suggest that the ability of tubular cells to synthesize dopamine does not exceed the deamination capacity of MAO. It appears, however, that the compartment where dopamine synthesis takes place has a limited size and the results further suggest that there are at least two different compartments where dopamine can be deaminated. Evidence supporting this suggestion is that provided by the results in experiments in which the influence of the two new selective MAO-A and MAO-B inhibitors on the formation of dopamine and DOPAC in kidney slices has been studied. Ro 41-1049 was found to produce a concentration-dependent increase in the accumulation of newly-formed dopamine; this could be observed at 50 and 100 μmol/l L-DOPA and was accompanied by a concentration-dependent reduction in the accumulation of DOPAC in the tissues. On the other hand, incubation of kidney slices in the presence of the MAO-B selective inhibitor Ro 19-6327 did not affect the accumulation of dopamine at 50 μmol/l L-DOPA in the medium, but decreased the formation of DOPAC. However, when kidney slices were incubated with 100 μmol/l of L-DOPA the MAO-B inhibitor was found to increase the accumulation of newly-formed dopamine in a concentration-dependent manner. These results might suggest that most of the MAO located inside tubular cells (where dopamine synthesis takes place) is of the A type, whereas MAO-B would be mainly located outside this compartment and would be responsible for the deamination of dopamine which leaks from renal tubular cells. This is probably why MAO-B inhibition does not increase dopamine accumulation but reduces DOPAC formation.

In conclusion, the results presented here show that newly-formed dopamine under in vitro experimental conditions can be deaminated by both MAO-A and MAO-B. The results might also suggest that MAO-A is mainly located inside the compartment where dopamine synthesis occurs and MAO-B outside this compartment.

Acknowledgements

This work was supported by a grant from Instituto Nacional de Investigaçäo Cientifica (FmP1).

References

Adam WR, Culvenor AJ, Hall J, Jarrot B, Wellard MR (1986) Aromatic L-amino acid decarboxylase: histochemical localization in rat kidney and lack of effect of dietary potassium or sodium loading on enzyme distribution. Clin Exp Pharmacol Physiol 13:47–53

Baines AD, Chan W (1980) Production of urine free dopamine from dopa: a micropuncture study. Life Sci 26:253–259

Cuche JL (1988) Sources of circulating dopamine: a personal view. In: Bell C, MacGrath BP (eds) Peripheral actions of dopamine. Macmillan, London, pp 1–23

Da Prada M, Ketler R, Keller HH, Cesura AM, Richards JG, Marti JS, Muggli-Maniglio D, Kyburz E, Imhof R (1990) From moclobemide to Ro 19-6327 and Ro 41-1049: the development of a new class of reversible, selective MAO-A and MAO-B inhibitors. J Neural Transm [Suppl] 29:279–292

Kopin IJ (1985) Catecholamine metabolism: basic aspects and clinical significance. Pharmacol Rev 37:333–364

Lee MR (1982) Dopamine and the kidney. Clin Sci 62:439–448

Soares-da-Silva P, Fernandes MH (1990) Synthesis and metabolism of dopamine in the kidney: effects of sodium chloride, monoamine oxidase inhibitors and alpha-human atrial natriuretic peptide. Am J Hypert ('in press').

Authors' address: Dr. P. Soares-da-Silva, Laboratório de Farmacologia, Faculdade de Medicina, 4200 Porto, Portugal.

J Neural Transm (1990) [Suppl] 32: 161–164

Monoamine oxidase activity in blood platelets of migraine patients

M. M. Caramona[1], **M. D. Cotrim**[1], **C. Fontes Ribeiro**[2], and **T. Macedo**[2]

[1] Laboratório de Farmacologia, Faculdade de Farmácia, [2] Instituto de Farmacologia e Terapêutica Experimental, Faculdade de Medicina, Coimbra, Portugal

Summary. Biochemical changes in platelets of migraine patients during the attacks have been reported before, however there are some conflicting results. In an attempt to define the biochemical lesion in the platelets, we have carried out a survey of platelets monoamine oxidase in migraine patients *with* and *without aura.* Platelet MAO activity in platelets from migraine patients was significantly reduced when compared with normal platelets.

Introduction

Monoamine Oxidase (MAO) plays an important role in the oxidative deamination of a wide variety of biogenic amines like catecholamines and serotonin. Based on substrate specificities and susceptibilities towards inhibitors, MAO can be divided in two forms, namely MAO-A and MAO-B. In human platelets the MAO is present in the MAO-B type (Donnelly and Murphy, 1977).

Interest in MAO activity in migraine patients was stimulated by the observation that the "cheese reaction", (the hypertensive reaction to certain foods in patients receiving MAO inhibitors), was similar to the dietary migraine episode; as this reaction is thought to be due to tyramine, it was suggested that patients with dietary migraine might be sensitive to tyramine-containing foods by virtue of a deficiency of MAO (Hanington and Harper, 1968).

Platelet MAO activity has been found to be significantly decreased during migraine attacks when compared with activity in the crises-free period (Glover et al., 1977, 1982). It was also found that mean platelet MAO activity between attacks in males with classical migraine, with tension headaches and with cluster headaches was significantly lower than that in a control group of normal men (Glover et al., 1981). However, conflicting results have been reported possibly due to differences in technical procedures used or to the heterogenity of the migraine patients (Thomas, 1982).

Having in mind the headache classification of the National Institute of Neurological Diseases and Blindness (USA-1988) and in an attempt to define more clearly the biochemical lesion of migraine, we have carried out a survey of platelet MAO activity in migraine patients *with* and *without aura* using only platelet samples from subjects in interval periods, i.e. between attacks.

Materials and methods

The migraine patients were selected from a group of females and males presenting themselves without an attack at Coimbra University Hospital (Headache Group of Neurology Service). The subgroups *with* and *without aura* were classified on the basis of the criteria of the National Institute of Neurological Diseases and Blindness (USA-1988).

The normal subjects belong to the Blood Donnors Service from the same hospital and all of them were healthy volunteers.

All the subjects were in diet (phenylethylamine and tyramine absent), without medication for at least 14 days and not depressed (Back Scale). The migraine patients had not experienced an attack during at least the last five days.

The platelets were prepared using a modificated method described by Fowler et al. (1979). Twenty ml blood samples were obtained by venipuncture and collected in standard heparine/lithium tubes; the subjects were aged between 20 and 60 years and the blood obtained, in the morning (9–10 a.m.). After being gently mixed, the blood was centrifuged at 200 g for 10 min. to remove the red cells. The cells were washed twice with 0.15 M NaCl and the combined supernatants further centrifuged at 3,000 g for 15 min. to produce a platelet preparation. The pellet was resuspended in 2 ml of 0.01 M potassium phosphate (pH 7.4) and stored at − 80 °C. Before use the suspension was sonicated in a Tissumizer Ultrasonic Desintegrator at low power for 2 min., to produce a more homogeneous preparation.

Monoamine Oxidase Assay: MAO activity was determined using ^{14}C-Phenylethylamine HCl (PEA, 50 mCi/mmole, New England Nuclear) as a preferential substrate for MAO type B. The reaction mixture containing 50 µl platelets suspension and 50 µl of 0.01 phosphate buffer (pH 7.4) with the substrate was incubated during 20 min.; the deaminated product was extracted and measured by liquid scintillation counting (Caramona, 1982).

To obtaine information about the K_m and V_{max} in the human platelets the PEA concentrations used were 2, 4, 8, 16, 32, 64 µM. The individual measurements were performed with PEA 100 µM. MAO activity is expressed in nanomoles of substrate metabolized per milligram of protein per hour of incubation.

The protein content of the platelet suspensions (milligram of protein per ml of suspension) was determined by the method of Lowry et al. (1951) with bovine serum albumin as a standard.

Results

The rates of deamination in platelets from controls and in those of migraine

experiments. The mean values of K_m obtained were 1.21 µM for the control platelets and 1.19 µM for the migraine patients platelets. Hence the concentration of ^{14}C-PEA selected was 100 µM, i.e. more than 3 times the respective K_m (Caramona, 1982).

The results are summarized in Table 1. Platelet MAO activity in 6 control males was not significantly different from that found in 8 females (age-matched, 20–40 years old), according to the Student's t test; however females showed a slightly higher mean activity than males. On the basis of age, considering the 20–40 years old and the 50–60 years old groups, the MAO activity is significantly different. The mean value of the younger group are 30% higher than the corresponding one for the older controls.

In platelets of migraine patients, MAO activity was significantly reduced when compared with normal platelets. The reduction was of about 70% in migraine patient *with aura* platelets and of 50% in the migraine patient *without aura*.

Discussion

The most important point in the discussion of these results is related to the definition of the two types of migraine, according to the accepted classification of the Institute of Neurological Diseases and Blindness, in migraine *with* and *without aura*; the results published until now have never reported this separation and we have shown that MAO activity decreases in the migraine patients *with* and *without aura* platelets but more so in the former.

Another point is related to the MAO activity not only during the attacks (Glover et al., 1977; Thomas, 1982; Summers et al., 1982) but also during the intervals of the attacks; in fact we used platelets from migraine patients which had not experienced an attack for at least five days.

It is also important to note that the two groups are compared with control groups which are age-matched, because Hanington (1978) described changes in platelet MAO activity related to age. The number of experiments performed by us until now is not enough to conclude definitely about differences between males and females, although the data appears to point to a slightly higher activity in females than in males.

Low platelet MAO activity reported in schizophrenia (Wyatt, 1979; Murphy, 1984), alcoholism (Major and Murphy, 1978), depression (Agarwal et al., 1979), demential disorders (Murphy, 1984; Regland, 1988) also is shown to occur in attack-free periods in migraine *with* and *without aura*. The biochemical changes that occur during the migrainous episode process may help us to understand the pathophysiology of migraine.

References

Agarwal DP, Goedde HW, Schrappe O (1979) Blood platelet monoamine oxidase activity in schizophrenia, affective disorders and alcoholism. In: Singer TP, Von

Korff R, Murphy DL (eds) Monoamine oxidase: structure, function and altered functions. Academic Press, New York, pp 397–402
Caramona MM (1982) Monoamine oxidase of types A and B in the saphenous vein and mesenteric artery of the dog. Naunyn-Schmiedbergs Arch Pharmacol 319:121–124
Donnelly CH, Murphy DL (1977) Substrate and inhibitor-related characteristics of human platelet monoamine oxidase. Biochem Pharmacol 26:853–858
Fowler CJ, Ekstedt B, Egashira T, Kinemuchi H, Oreland L (1979) The interaction between human platelet monoamine oxidase, its monoamine substrates and oxygen. Biochem Pharmacol 28:3063–3068
Glover V, Sandler M, Grant E, Rose FC (1977) Transitory decrease in platelet monoamine oxidase activity during migraine attacks. Lancet i:391–393
Glover V, Peatfield R, Zammit-Pace R (1981) Platelet monoamine oxidase activity and headache. J Neurol Neurosurg Psychiatry 44:786–790
Glover V, Littlewood J, Sandler M, Peatfield R, Summers K, Craig I, Rose FC (1982) Why is platelet monoamine oxidase activity low in some headache patients? In: Rose FC (ed) Advances in migraine research and therapy. Raven Press, New York, pp 127–132
Hanington E (1978) Migraine a blood disorder. Lancet ii:501–502
Hanington E, Harper M (1968) The role of tyramine in the aetiology of migraine and related studies on the cerebral and extracerebral circulations. Headache 8:84–97
Lowry OH, Rosebrough NJ, Farr AL, Randall RJ (1951) Protein measurement with the Folin phenol reagent. J Biol Chem 193:265–275
Major LF, Murphy DL (1978) Platelet and plasma amineoxidase activity in alcoholic individuals. Br J Psychiatry 132:548–554
Murphy DL (1984) Are there diseases attributable to monoamine oxidase abnormalities? In: Tipton KF, Dostert P, Strolin Benedetti M (eds) Monoamine oxidase and disease. Prospects for therapy with reversible inhibitors. Academic Press, New York, pp 321–332
Regland B, Gottfries CG, Oreland L, Svennerholm L (1988) Low B12 levels related to high activity of platelet MAO in patients with dementia disorders. A retrospective study. Acta Psychiatr Scand 78:451–457
Summers KM, Brown GK, Craig IW (1982) Platelet monoamine oxidase: specific activity and turnover number in migraine. Clin Chim Acta 121:139–146
Thomas DV (1982) Platelet monoamine oxidase in migraine. Adv Neurol 33:279
Wyatt RJ, Potkin SG, Murphy DL (1979) Platelet monoamine oxidase activity in schizophrenia: a review of data. Am J Psychiatry 136:337–385

Authors' address: Dr. Maria Margarida Caramona, Laboratório de Farmacologia, Faculdade de Farmácia da Universidade de Coimbra, Largo de D. Dinis, 3000 Coimbra, Portugal.

J Neural Transm (1990) [Suppl] 32: 165–170

Monoamine oxidase inhibition by moclobemide and 2-Amino-ethyl carboxamide derivatives: mode of action and kinetic characteristics

A. M. Cesura, D. Muggli-Maniglio, G. Lang, R. Imhof, and **M. Da Prada**

Pharmaceutical Research Department, F. Hoffmann-La Roche Ltd, Basel, Switzerland

Summary. The selective, reversible inhibitors of monoamine oxidase (MAO) moclobemide and Ro 41-1049 (selective for MAO-A), as well as of Ro 16-6491 and Ro 19-6327 (selective for MAO-B) inhibited the enzyme with an initial competitive phase, followed by a time-dependent inhibition of MAO. Ro 41-1049, Ro 16-6491 and Ro 19-6327, being activated by MAO into reversible adducts, fit into the classification as mechanism-based inhibitors. Conversely, since no product formation was observed after incubation of tissue homogenates with moclobemide, this drug probably belongs to the class of the "slow-binding" MAO inhibitors.

Introduction

Since the discovery that monoamine oxidase (MAO) exists as two isoenzymes (Johnston, 1968), i.e. MAO-A and MAO-B, and the recent demonstration that the two isoforms have different amino acid sequences (Bach et al., 1988), several selective inhibitors with different chemical structures have been developed (Dostert et al., 1989). The search for new inhibitors selective towards MAO-A or MAO-B has been largely stimulated by the clinical evidence that MAO-A inhibitors have antidepressant activity (Fowler and Ross, 1984), whereas MAO-B inhibitors are beneficial in the therapy of Parkinson's disease (Tertrud and Langston, 1989).

The study on the mechanism of action and on the metabolism of the MAO-A inhibitor antidepressant moclobemide (Aurorix®) led to the discovery of a new class of potent and reversible MAO inhibitors possessing a very high selectivity towards MAO-A or MAO-B (Da Prada et al., 1990). Among these novel compounds Ro 16-6491 and Ro 19-6327 behave as highly selective MAO-B inhibitors. By chemical modification of the aromatic moiety of these molecules, Ro 41-1049 was identified, which is a

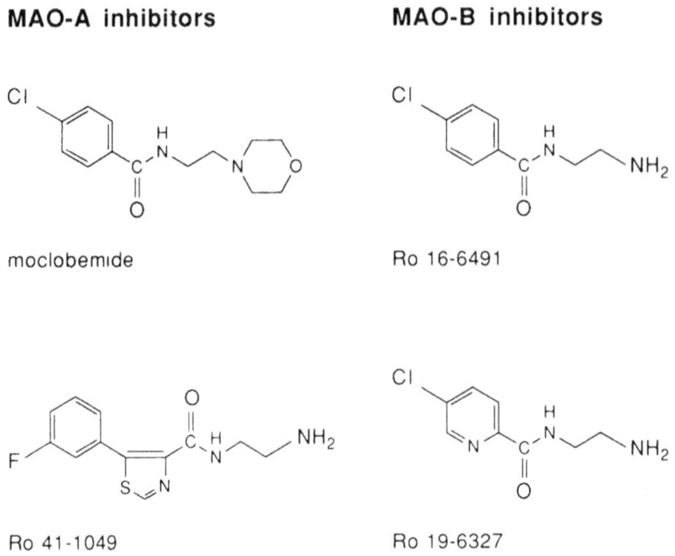

Fig. 1. Chemical structures of reversible MAO inhibitors

reversible, highly selective inhibitor of MAO-A (for chemical structures, see Fig. 1).

In the present paper we describe some characteristics of the mode of inhibition of these MAO-A and MAO-B blockers. In these experiments the enzyme sources were rat brain homogenates and MAO enzymes partially purified from human platelets and placenta.

Materials and methods

MAO activity was measured radioenzymatically (Keller et al., 1987) using [^{14}C]5-hydroxytryptamine (5-HT, 200 μM) as substrate for MAO-A and [^{14}C]-phenylethylamine (PEA, 20 μM) for MAO-B. MAO-B was purified from human platelets by affinity chromatography (Cesura et al., 1990a). Human placenta MAO-A was purified by affinity chromatography using as affinity ligand a Ro 41-1049 derivative coupled to AH-Sepharose 4B (unpublished method), followed by ion-exchange HPLC on a DEAE-TSK-3SW column. The specific activity of the final enzyme preparations were about 500-fold (MAO-B) and 50-fold (MAO-A) higher than in the crude homogenate.

The metabolism of Ro 16-6491 and moclobemide was studied by incubating crude membrane preparations from human platelets (MAO-B) and placenta (MAO-A) with the radiolabeled inhibitors. These incubations were carried out at 37 °C, for different times and using different concentrations of [^{3}H]Ro 16-6491 and tritiated or [^{14}C]moclobemide. Proteins were precipitated in the presence of 0.3 M $HClO_4$ (final concentration). Unlabeled Ro 16-6491, together with its aldehyde derivative Ro 19-7731 (0.5 mM each) or alternatively unlabeled moclobemide were added to the samples. After centrifugation, an aliquot of the supernatant was injected into a C18 reverse-phase HPLC column. The column was eluted isocratically (1 ml/min) with 0.125 M ammonium acetate buffer, containing different concentrations of CH_3CN. The radioactivity present in the eluate was either counted directly by means of a Flo-One/Beta radioactivity flow detector or counted by adding scintillation fluid to fractions collected during 1 min.

Results

Enzymatic studies

When the inhibition of MAO-A and MAO-B activity of rat brain homogenates was measured without exposure of the enzyme preparation to the inhibitors prior to adding the substrates, moclobemide, Ro 41-1049, Ro 16-6491 and Ro 19-6327 behaved as competitive inhibitors. Replotting the apparent K_m values (obtained by Lineweaver-Burk plot) versus the inhibitor concentrations, low K_i values for Ro 41-1049 (0.17 μM), Ro 16-6491 (0.21 μM) and Ro 19-6327 (0.18 μM) were obtained. In contrast moclobemide had a remarkably high K_i (400 μM) for rat brain MAO-A. In contrast, already after 5 min incubation of the homogenates with the inhibitors, the inhibition pattern for all drugs changed from competitive to mixed competititve/non-competitive inhibition. All the compounds tested behaved as time-dependent inhibitors. The reversible inhibition of purified MAO-B by Ro 16-6491, by its α,α′-dideuterated derivatives, and by Ro 19-6327 (all at 0.1 μM) followed a pseudo-first order kinetic (Fig. 2, left panel), reaching equilibrium (maximum inhibition) after various preincubation times. The

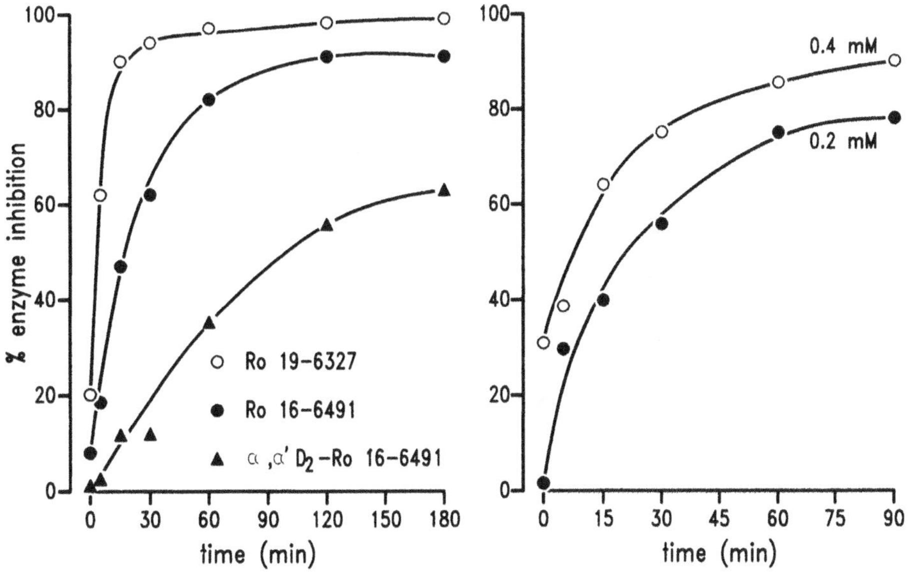

Fig. 2. Time-dependent inhibition of human platelet purified MAO-B by various inhibitors (0.1 μM, left panel) and of MAO-A purified from human placenta by moclobemide (right panel)

inactivation of MAO-B by the deuterated derivative of Ro 16-6491 was slower than that of Ro 16-6491, indicating a marked isotope effect of the deuterated derivative. The inhibition of purified MAO-A by moclobemide (Fig. 2, right panel) showed first a rapidly equilibrating phase (probably corresponding to the competitive interaction) which was followed by an apparent pseudo first-order kinetic enzyme inhibition. Notably, the IC_{50} values of moclobemide in inhibiting the deamination of 5-HT by MAO-A were about 10-fold higher after 5 min than after 60 min exposure of the enzyme to moclobemide (IC_{50} 300 and 33 μM, respectively).

Inhibitor metabolism

After incubation of [^{3}H]Ro 16-6491 at various concentrations with human platelet membranes, the compound was metabolized by MAO-B into its aldehyde derivative. The deamination of Ro 16-6491 followed Michaelis-Menten kinetics (Fig. 3), with a high affinity ($K_m = 0.12$ μM) but with a velocity which was 1000-fold lower ($V_{max} = 1.94$ pmol mg protein^{-1} min^{-1}) than that observed for PEA ($V_{max} = 1.43$ nmol mg protein^{-1} min^{-1}). As expected, the presence of selegiline in the incubation mixture completely blocked the deamination of Ro 16-6491. Conversely, even after prolonged incubation (48 h at 37 °C) of human placenta membranes with radiolabeled moclobemide at various concentrations (5 nM up to 50 μM), no formation of moclobemide derivatives was observed by HPLC analysis.

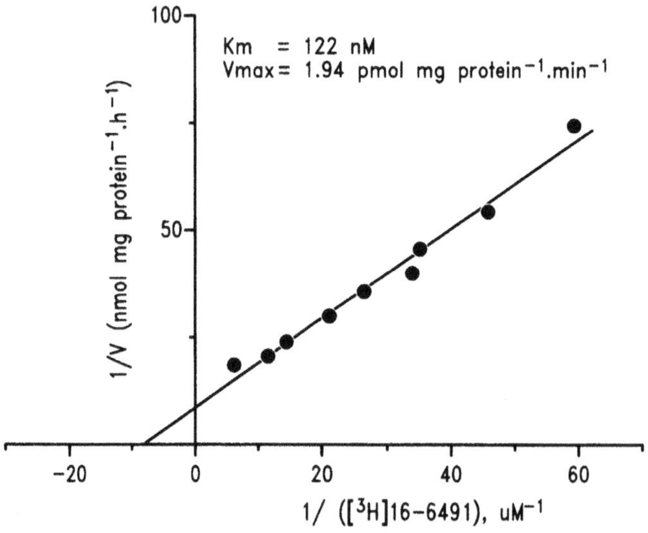

Fig. 3. Lineweaver-Burk plot of [^{3}H]Ro 16-6491 deamination by human platelet MAO-B. Incubation was performed for 1 h at 37 °C. The tritiated aldehyde product was separated by HPLC

Discussion

The present results, together with previously published data (Cesura et al., 1989, 1990b), support the view that 2-amino-ethyl-carboxamide derivatives can be considered as reversible, mechanism-based enzyme inhibitors. This is clearly indicated by the saturation kinetics, by the time-dependent inhibition and by the formation of adducts or products (Palfreyman et al., 1987). Thus, Ro 41-1049, Ro 16-6491 and Ro 19-6327 are substrates for MAO-A and MAO-B, and after their oxidation they are activated into intermediates which remain tightly bound to the enzyme active site.

In previous reports we suggested that moclobemide could inhibit MAO-A by a mechanism-based process. However, until now we have been unable to detect moclobemide adducts or metabolites generated by direct interaction of this drug with MAO-A. Therefore, we have to conclude that moclobemide should not be included in the class of mechanism-based inhibitors. Rather, moclobemide might belong to the class of the "slow-binding" inhibitors (Williams and Morrison, 1979). Hence, after an initial competitive interaction, moclobemide inhibits MAO-A by slow binding kinetics. If so, the moclobemide-MAO-A interaction might trigger conformational changes in one or both of the species leading to a relatively stable inhibitor-enzyme complex.

References

Bach AWJ, Lan NC, Johnson DL, Abell CW, Bembenek ME, Kwan S-W, Seeburg PH, Shih JC (1988) cDNA cloning of human liver monoamine oxidase A and B: molecular basis of difference in enzymatic properties. Proc Natl Acad Sci USA 85:4934–4938

Cesura AM, Galva MD, Imhof R, Kyburz E, Picotti GB, Da Prada M (1989) [^{3}H]Ro 19-6327: a reversible ligand and affinity labeling probe for monoamine oxidase-B. Eur J Pharmacol 162:457–465

Cesura AM, Imhof R, Muggli-Maniglio D, Picotti GB, Da Prada M (1990a) Characterization of [^{3}H]Ro 16-6491 binding to digitonin solubilized monoamine oxidase-B and purification of the enzyme from human platelets by affinity chromatography. Biochem Pharmacol 39:216–220

Cesura AM, Bös M, Galva MD, Imhof R, Da Prada M (1990b) Characterization of the binding of [^{3}H]Ro 41-1049 to the active site of human monoamine oxidase-A. Mol Pharmacol 37:358–366

Da Prada M, Kettler R, Keller HH, Cesura AM, Richards JG, Saura Marti J, Muggli-Maniglio D, Wyss P-C, Kyburz E, Imhof R (1990) From moclobemide to Ro 19-6327 and Ro 41-1049: the development of a new class of reversible, selective MAO-A and MAO-B inhibitors. J Neural Transm [Suppl] 29:279–292

Dostert PL, Strolin-Benedetti M, Tipton KF (1989) Interactions of monoamine oxidase with substrates and inhibitors. Med Res Rev 9:45–89

Fowler CJ, Ross SB (1984) Selective inhibitors of monoamine oxidase A and B: biochemical, pharmacological, and clinical properties. Med Res Rev 4:323–358

Johnston JP (1968) Some observations upon a new inhibitor of monoamine oxidase in brain tissue. Biochem Pharmacol 17:1285–1297
Keller HH, Kettler R, Keller G, Da Prada M (1987) Short-acting novel MAO inhibitors: in vitro evidence for the reversibility of MAO inhibition by moclobemide and Ro 16-6491. Naunyn-Schmiedebergs Arch Pharmacol 335:12–20
Palfreyman MG, Bey P, Sjoerdsma A (1987) Enzyme-activated/mechanism-based inhibitors. Essays in Biochemistry 23:28–81
Tertrud JW, Langston JW (1989) The effect of deprenyl (selegiline) on the natural history of Parkinson's disease. Science 245:519–522
Williams JW, Morrison JF (1979) The kinetic of reversible tight-binding inhibition. Meth Enzymol 63:337–467

Authors' address: Prof. Dr. M. Da Prada, F. Hoffmann-La Roche Ltd, Pharmaceutical Research Department, CH-4002 Basel, Switzerland.

J Neural Transm (1990) [Suppl] 32: 171–175

Does moclobemide stimulate melatonin synthesis as the other selective MAO-A inhibitors do?

G. F. Oxenkrug[1], **P. J. Requintina**[1], and **A. Yuwiler**[2]

[1] Brown University, VAMC, Providence, Rhode Island, and
[2] University of California at Los Angeles/Brentwood, VAMC, Los Angeles, California, U.S.A.

Summary. It has been reported that selective MAO-A inhibitors, clorgyline and brofaromine, but not the MAO-B inhibitors, deprenyl and pargyline, stimulated rat pineal melatonin synthesis in humans and animals. Recent studies, however, found no effect of moclobemide, a selective MAO-A inhibitor, on human plasma melatonin levels.

Present study found that in vitro moclobemide produced very weak stimulation of rat pineal NAT activity. However, in vivo moclobemide induced a significant increase of rat pineal NAS and melatonin content, and a dramatic decrease of 5-HIAA content (HPLC-fluorimetric procedure). Moclobemide's effect on melatonin and related indoles could be detected as early as 30 min after the injection and lasted, at least, for 2 h.

The possible reasons for discrepancies between human and animal data are discussed.

Introduction

We have reported at the first Amine Oxidase Club meeting in Cambridge, UK (1984) that single dose of clorgyline, a selective monoamine oxidase-A type (MAO-A) inhibitor but not deprenyl, a selective MAO-B inhibitor, increased rat pineal N-acetylserotonin (NAS) and melatonin content (Oxenkrug et al., 1984). Further studies confirmed that selective MAO-A inhibitors, clorgyline and brofaromine, but not MAO-B inhibitors, deprenyl and pargyline, stimulate rat pineal melatonin synthesis in vivo (Oxenkrug et al., 1985) and in vitro (Oxenkrug et al., 1988), and increase plasma melatonin levels in normal human volunteers (Bieck et al., 1988). Chronic administration of MAO inhibitors have also demonstrated the ability of MAO-A (but not MAO-B) inhibitors to stimulate melatonin synthesis (Murphy et al., 1986; Golden et al., 1988).

Two recent studies (Bieck et al., 1989; Scheinin et al., 1990), however, found no effect of moclobemide, a new selective MAO-A inhibitor (Da Prada et al., 1989), on human plasma melatonin levels.

We have studied moclobemide effect on rat pineal melatonin synthesis in vitro and in vivo.

Methods

F 344 male rats (3 month old, 200–250 g b.w.) have been housed under 12 h light 12 h dark schedule with the food and water ad libitum. N-acetyltransferase (NAT) activity in freshly exized pineals (5 h of incubation) has been studied as described elsewhere (Oxenkrug et al., 1988). All drugs were injected i.p. and rats were decapitated. Pineal melatonin and related indoles: serotonin. N-acetylserotonin (NAS) and 5-hydroxyindoleacetic acid (5-HIAA), were determined by HPLC-fluorimetric procedure (Oxenkrug et al., 1986b).

Results

In vitro moclobemide produced much weaker stimulation of rat pineal NAT activity than clorgyline and brofaromine (Table 1).

In vivo moclobemide (50 and 100 mg/kg, i.p.) induced significant increase of NAS and melatonin content, and dramatic decrease of 5-HIAA pineal content (Table 2). Moclobemide effect on melatonin and related indoles could be noticed as early as 30 min. after the injection and lasted, at least, for 2 h.

Discussion

The weak effect of moclobemide on the pineal NAT activity in vitro is in line with the weak effect of this drug on MAO activity in vitro (Da Prada et al., 1989). However, in vivo moclobemide produced dramatic (comparable with

Table 1. The effect of MAO-A inhibitors on NAT activity

The NAT activity for 5 hours of acute culture for the MAO inhibitors are as follows:

	0	1 μM	10 μM	100 μM
None	(3) 0.29 ± 0.09			
Moclobemide		(4) 0.17 ± 0.02	(4) 0.37 ± 0.06	(4) 3.22 ± 0.21
Clorgyline		(4) 7.08 ± 0.18	(3) 7.70 ± 0.85	(3) 9.87 ± 0.89
Brofaromine		(4) 3.40 ± 0.74	(4) 7.88 ± 1.13	(4) 14.05 ± 0.60

Table 2. Moclobemide effect on rat pineal melatonin and related indoles

	NAS	Melatonin	Serotonin	5-HIAA
Saline (13)	< 0.15	0.15 ± 0.04	98.85 ± 7.46	8.97 ± 1.19
MBM (25 mg/kg)				
120 min (4)	1.69 ± 0.52[a]	1.20 ± 0.12[b]	98.5 ± 8.24	4.17 ± 0.73[c]
MBM (50 mg/kg)				
60 min (4)	0.44 ± 0.06	0.83 ± 0.06	109.31 ± 12.57	1.75 ± 0.29
120 min (5)	3.32 ± 0.40*	1.33 ± 0.14*[b]	79.93 ± 14.27**	1.74 ± 0.38[b]
MBM (100 mg/kg)				
30 min (4)	0.15 ± 0.05*	0.51 ± 0.10*	97.48 ± 13.8	2.57 ± 0.29*
60 min (8)	0.47 ± 0.18	0.84 ± 0.14	109.61 ± 9.33	1.42 ± 0.16
120 min (4)	2.66 ± 0.82*	2.15 ± 0.26*	95.19 ± 14.08	1.76 ± 0.30***

Male Fisher 344N rats, 10–12 week old, about 200 g b.w. were injected with moclobemide (MBM) between 8 and 9 a.m., i.p.: mean ± S.D. (ng/pineal). Number of animals in group in brackets. Time – between injection and decapitation;
All saline groups (except serotonin) are statistically significant (Student t-test) vs all other groups ($p < 0.001$):
* $p < 0.0001$, ** $p < 0.001$ and *** $p < 0.02$ vs the same dose at 60 min;
[a] $p = 0.06$, [b] $p < 0.008$ vs the same time, 100 mg/kg;
[c] $p < 0.0005$ vs the same time, 50 mg/kg

clorgyline) increase of NAS and melatonin content and decrease of 5-HIAA content.

Our results indicate that in vivo moclobemide is a powerful stimulator of the pineal melatonin synthesis. It is quite possible that moclobemide might stimulate human melatonin synthesis as well. Negative results of human studies might be caused by some methodological problems such as, for example, moclobemide interference with the RIA procedures of melatonin plasma determinations.

Present results together with the literature data (for rev., see Oxenkrug et al., 1990) suggest that all, so far known selective MAO-A inhibitors might stimulate melatonin biosynthesis. We have previously proposed (Oxenkrug et al., 1986a) that this ACUTE effect (if repeated) might change the circadian rhythm (CR) of melatonin, and this, in its turn, would normalize the other CRs important for the maintaining of the normal affect. According to our hypothesis, the time lag between the beginning of MAO inhibitors administration and the occurrence of the clinical antidepressant effect depends upon the relative resistance of the endogenous CR to the changes of the 24 h environmental light/dark cycle induced by the MAO-A inhibitors.

In addition, stimulation of melatonin synthesis might be responsible for the antihypertensive effect of MAO inhibitors (Oxenkrug et al., 1986b). Moclobemide does decrease blood pressure in spontaneously hypertensive rats (Da Prada et al., 1989). Therefore, melatonin synthesis stimulation by moclobemide might contribute to both antidepressive and antihypertensive effects of moclobemide.

Acknowledgements

Prof. Da Prada and Prof. Haefely's encouragement and supply of moclobemide are very much appreciated.

References

Bieck PR, Antonin KH, Balon R, Oxenkrug GF (1988) Brofaromine and pargyline effect on human plasma melatonin. Prog Neuropsychopharmacol Biol Psychiatry 12:93–101

Bieck PR, Muhlbauer B, Antonin KH (1989) Differential effects of brofaromine and moclobemide on melatonin secretion. In: Stefanis CN, Soldatos CR, Rabavilas AD (eds) Psychiatry today. Abstracts of the 7th World Congress of Psychiatry, Athens, 1989. Excerpta Medica, International Congress Series 899, p 503, abstract 1946

Da Prada M, Kettler R, Keller HH, Burkard W, Muggli-Maniglio D, Haefely WE (1989) Neurochemical profile of moclobemide, a short-acting and reversible inhibitor of monoamine oxidase type A. J Pharmacol Exp Ther 248:400–414

Golden RN, Markey SP, Risby ED, Rudorfer MV, Cowdry RW, Potter WZ (1988) Antidepressants reduce whole-body norepinephrine turnover while enhancing 6-hydroxymelatonin output. Arch Gen Psychiatry 45:150–154

Murphy DL, Tamarkin L, Sunderland T, Garrick NA, Cohen R (1986) Human plasma melatonin is elevated during treatment with the monoamine oxidase inhibitors clorgyline and tranylcypromine but not deprenyl. Psychiatry Res 17:119–127

Oxenkrug G, McCauley R, McIntyre I, Filipowicz C (1984) Effect of clorgyline and deprenyl on rat pineal melatonin. J Pharm Pharmacol 36:55W

Oxenkrug G, McIntyre I, McCauley R, Filipowicz C (1985) Selective inhibition of MAO-A but not MAO-B activity increases rat pineal melatonin. J Neural Transm 61:265–270

Oxenkrug GF, Balon R, Jain AK, McIntyre IM, Appel D (1986a) Single dose of tranylcypromine increases human plasma melatonin. Biol Psychiatry 21: 1085–1089

Oxenkrug GF, McCauley RB, Fontana DJ, McIntyre IM, Commissaris RL (1986b) Possible melatonin involvement in the hypotensive effect of MAO inhibitors. J Neural Transm 66:271–280

Oxenkrug G, McIntyre I, McCauley R, Yuwiler A (1988) The effect of selective MAO inhibitors on rat pineal melatonin synthesis in vitro. J Pineal Res 5:99–109

Oxenkrug GF (1990) The acute effect of monoamine oxidase inhibitors on serotonin conversion to melatonin. In: Sandler M, Coppen A, Harnett S (eds) 5-Hydroxytryptamine in psychiatry: spectrum of ideas. Oxford University Press, pp 98–109

Scheinin M, Koulu M, Vakkuri O, Vuorinen J, Zimmer H (1990) Moclobemide, an inhibitor of MAO-A, does not increase daytime melatonin levels in normal humans. Prog Neuropsychopharmacol Biol Psychiatry 14:73–82

Authors' address: Prof. Dr. G.F. Oxenkrug, Department of Psychiatry, VAMC, Davis Park, Providence, RI 02908, U.S.A.

J Neural Transm (1990) [Suppl] 32: 177–184

Efficacy and safety of moclobemide compared with imipramine in the treatment of major depressive disorder. Double-blind Multicenter Study, Austria*

G. F. Hebenstreit[1], M. Loidl[2], U. Baumhackl[3], B. Gallhofer[4], Ch. Geretsegger[5], E. Radmayr[6], M. Saletu[7], W. Schöny[8], and M. Stabl[9]

[1] NÖ Landes-Nervenklinik, Amstetten/Mauer, [2] Hoffmann-La Roche Wien, Ges. m.b.H., [3] A.Ö. Krankenhaus, St. Pölten,
[4] Psychiatrisch Neurologische Universitätsklinik, Graz, [5] Landes-Nervenklinik, Salzburg,
[6] Facharzt für Psychiatrie und Neurologie, Dornbirn, [7] Anton-Proksch-Institut, Wien-Kalksburg, and
[8] Wagner-Jauregg-Krankenhaus des Landes Oberösterreich, Linz, Austria
[9] F. Hoffmann-La Roche AG, Basel, Switzerland

Summary. In a double-blind, 4-week, prospective, randomized multicenter (17 centers) study we checked on the efficacy, tolerability and safety of moclobemide (300–600 mg/d) compared to imipramine (100–200 mg/d) in parallel groups of patients with a Major Depressive Episode (DSM III).

The mean % reduction of the HAMD at the end of treatment was 51.7 in the moclobemide group and 52.1 in the imipramine group. The percentage of patients in whom efficacy was globally judged as "good" or "very good" was 62% in the moclobemide group and 60% in the imipramine group.

---

* "Further Investigators"

H. Donat, Kaiser-Franz-Josef-Spital, Wien
R. Fartacek, Landes-Nervenklinik, Salzburg
K. Fellerer, NÖ Landes-Nervenklinik, Amstetten/Mauer
R. Fischbach, Landes-Nervenklinik, Salzburg
P. König, Landes-Nervenklinik, Valduna/Rankweil
H. Pfolz, Psychiatrisches Krankenhaus der Stadt Wien-Baumgartner, Höhe
H. Pietschmann, Pflegeheim der Stadt Wien, Lainz
T. Platz, Landes-Krankenhaus, Klagenfurt
L. Rieder, Pflegeheim der Stadt, Wien-Lainz
G. Schnaberth, Neurologisches Krankenhaus, Wien-Rosenhügel
H. Schubert, Landes Nervenklinik, Hall in Tirol
G. Wieselmann, Psychiatrisch-Neurologische Univ.-Klinik, Graz
W. Wolf, Psychiatrische Univ.-Klinik Wien, Austria

There was no statistically significant difference in the efficacy in both groups but in some factors there was a trend for a better amelioration favoring moclobemide.

The final overall physician's judgement of tolerability was "good" or "very good" in 83% of moclobemide patients and in 74% of imipramine patients. Adverse events were reported or observed in 56% of moclobemide patients and in 69% of imipramine patients. The number of mild, moderate and severe adverse events was higher in the imipramine group with a total of 286 versus 189.

There was a statistically high significant difference considering the tolerability favoring moclobemide again.

In this project the basic goal to find a substance with at least the same efficacy but a much better tolerability for sure got fulfilled.

Introduction

Moclobemide is a reversible MAOI, which both in vitro and in vivo, preferentially inhibits type-A MAO. In rodents moclobemide is active in most models of depression and appears to be devoid of anticholinergic activity. The hypothesis according to which the risk of a hypertensive crisis should be less with reversible MAOIS, appears to hold true for moclobemide. Extensive studies performed in human volunteers examining the interactions of moclobemide with tyramine, have shown only minimal potentiation of the pressor effects of tyramine.

Objectives and hypotheses

The objective of this study was to compare in a 4-week trial the efficacy and the tolerability of flexible doses of moclobemide (300–600 mg/day) to those of flexible doses of imipramine (100–200 mg/day) for the treatment of a Major Depressive Episode (DSM III). It was hypothesized that the efficacy of moclobemide would be comparable to that of imipramine, but that moclobemide would have advantages over imipramine with regards to tolerability (in particular, a lower incidence of anticholinergic side-effects). It was also hypothesized that moclobemide would have a faster onset of antidepressant action than imipramine.

Inclusion criteria

Patients considered for participation were men and non-pregnant, non-lactating women over 18 years of age (the protocol specified an age range of

18–80 years; for two geriatric centers no upper limit was fixed). Patients could be either in- or out-patients, and they were enrolled on a diagnosis of a Major Depressive Disorder as defined by the DSM-III criteria. Additionally participants were required to have a minimum baseline score of 17 on the 21-item Hamilton Depression Rating Scale (HAMD). Patients with psychotic symptomatology were eligible only if the psychotic symptoms were mood-congruent.

Exclusion criteria

Patients with poorly controllable, marked suicidal intent, with mood-incongruent delusions and/or hallucinations, with acute confusional states, with symptoms of present alcohol- or drug-abuse or dependency; as well as patients who had received MAO inhibitors within 7 days prior to study entry were excluded from the trial.

Duration of the treatment and dosage of trial medication

The protocol specified that the duration of treatment with trial drug was to be 28 days and that the target daily was to be 400 mg for moclobemide and 133 mg for imipramine. To guarantee double-blind conditions, trial medication was blister-packaged. The initial daily dose of moclobemide was 300 mg and the initial daily dose of imipramine was 33.3 mg. This dose was further increased step by step until day 6. The maximal allowed daily dose was 600 mg for moclobemide and 200 mg for imipramine if necessary.

Concomitant medication

Throughout the trial concomitant psychotropic medication was prohibited, with the exception of oral benzodiazepines (the investigators agreed on either nitrazepam or diazepam) for the treatment of severe insomnia or agitation, if deemed necessary by the investigator.

Results

This study provides evidence that the efficacy of moclobemide in the treatment of Major Depression (DSM-III) is comparable to that of imipramine according to the measures provided by the Hamilton Depression Rating Scale, the Investigator's Final Global Assessment of Efficacy, and the

Zung Self-Rating Depression Scale. The tolerability of moclobemide however, appears to be superior to that of imipramine.

Upon study entry both treatment groups were comparable for demographic and illness characteristics with approximately 60% of patients in each group diagnosed as suffering from endogenous depression. Efficacy was judged primarily on the Hamilton Depression Rating Scale and on the Investigator's Final Global Assessment of Efficacy. No differences between the two treatment groups were observed, and as the number of patients enrolled in each group is high (189 in the moclobemide group and 192 in the imipramine group) one can assume that the difference, if any, in the antidepressant efficacy of moclobemide and imipramine is small (Fig. 1). The percentage of patients whose total Hamilton score decreased by 50% or more as compared to baseline, approximated 60% in both treatment groups, a result which is in keeping with what is usually reported for imipramine. The Investigator's Final Global Assessment of Efficacy gave comparable results, and efficacy was judged to be "good" or "very good" in approximately 60% of patients in each treatment group (Fig. 2). In this study, no relevant difference in onset of activity was observed between moclobemide and imipramine.

First generation MAOIs were considered not to be effective in patients with endogenous-type depression (West and Dally, 1959; Robinson et al., 1978; Davidson et al., 1978; Paykel et al., 1971). For instance Paykel (1971) found that phenelzine was effective in only 18% of patients with endogen-

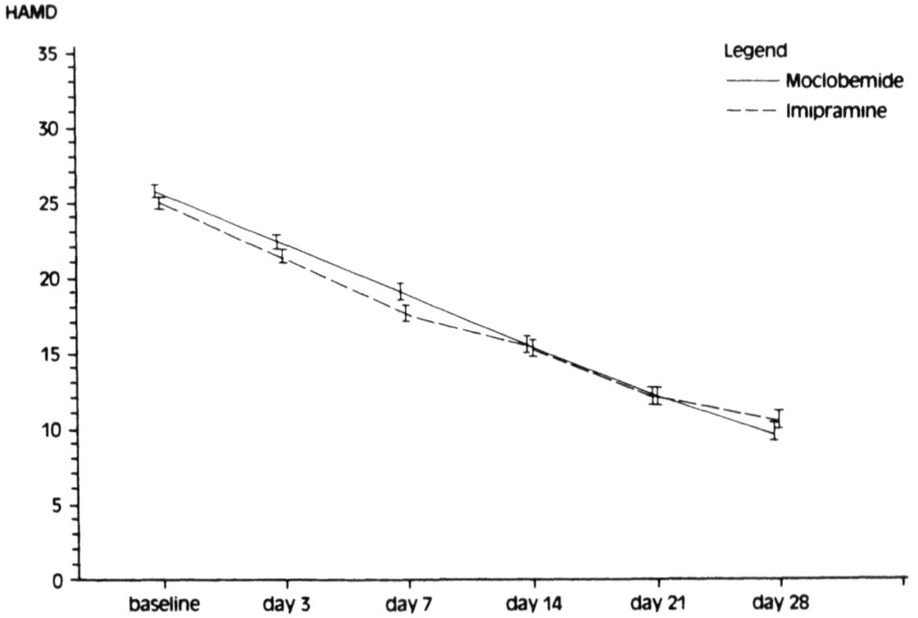

Fig. 1. Mean HAMD-scores (17 items) ± SEM. Drop-outs included as long as in study

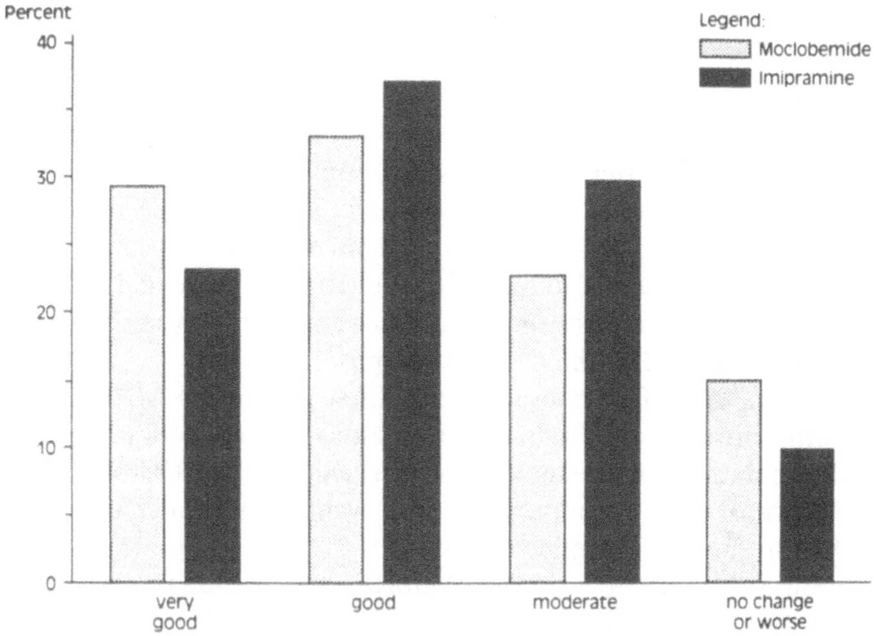

Fig. 2. Overall assessment of efficacy

ous-type depression, but was effective in 70% of patients with depressive neurosis. In the present study, approximately 60% of the patients in each treatment group were judged to suffer from endogenous-type depression, yet the response rate in the moclobemide group was comparable to the response rate in the imipramine group. This was also true, when subgroup analysis was performed, and when only patients with endogenous-type depression were considered. These results suggest that, unlike first generation MAOIs, which irreversibly inhibited both type-A and type-B MAO, moclobemide, (a reversible and selective type-A MAO), possesses an antidepressant efficacy which is comparable to that of imipramine. The dose of imipramine used in this study was a little low, and one could argue that if a higher dose had been used, moclobemide would have been found to be less effective than imipramine. However, it should be noted that a recent W.H.O. collaborative study (1986) failed to show any difference in the antidepressant activity of imipramine 75 mg/day and imipramine 150 mg/day. Moreover, the percentage of patients whose total Hamilton score decreased by 50% or more in the imipramine group is of 60%, a figure comparable to that which is usually reported for TCAs (e.g. Angst, 1986). Thus the findings observed in this study may be considered as valid, and the classical notion according to which MAOIs are less effective than TCAs for the treatment of depression does not seem to hold true for moclobemide.

Treatment tolerability and safety was judged on reported and observed adverse events, vital signs, physical examination, ECG, and a laboratory panel. The number of patients presenting with adverse events, as well as the total number of adverse events were greater in the imipramine than in the moclobemide treatment group. This difference is mainly due to the fact that imipramine induced clearly more anticholinergic adverse events than moclobemide (Fig. 3). Therapy with first generation MAOIs has been reported to induce a particular pattern of CNS side effects characterized by insomnia, irritability, agitation, motor restlessness and hypomania (for review see e.g. Kline and Cooper, 1980). In this study moclobemide appeared to induce mainly insomnia, but only in a small proportion of patients (7%), and the incidence of the above-mentioned CNS adverse events was comparable in the moclobemide and in the imipramine group.

Autonomic side effects observed with first generation MAOIs included dry mouth, constipation, dizziness, orthostatic hypotension, impotence and delayed ejaculation (for review see e.g. Kline and Cooper, 1980). In this study the incidence of these autonomic adverse events was greater in the imipramine than in the moclobemide group.

The cardiovascular tolerability of moclobemide tended to be superior to that of imipramine, although one hypertensive episode was reported in a hospitalized patient treated with moclobemide. It should be stressed however, that the investigator judged this adverse event to be of moderate severity and that the observed blood pressure increase was not life-threatening. No symptomatic treatment was given, the patient continued to receive moclobemide, and blood pressure values normalized in 24 hours. Postural hypotension, a frequent and often severe side effect of first generation

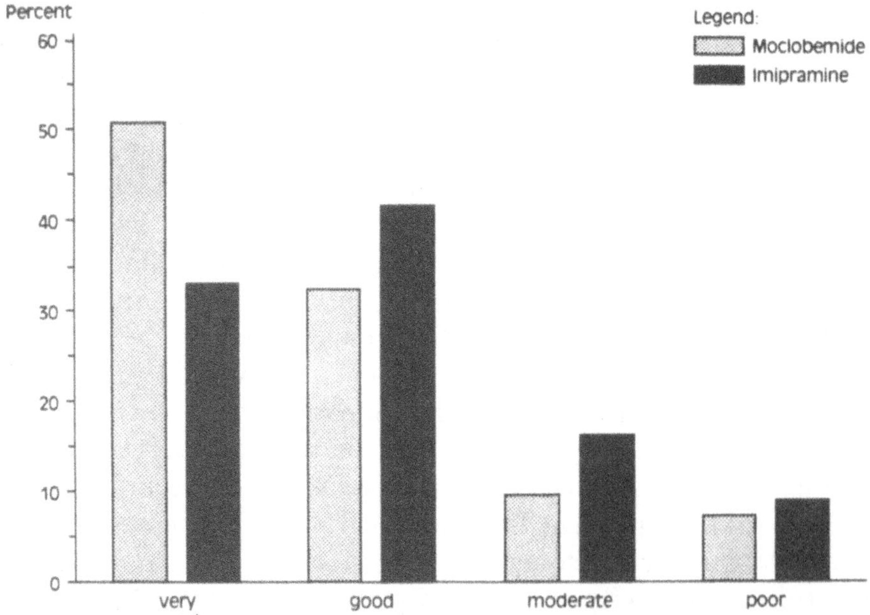

Fig. 3. Overall assessment of tolerance

MAOIs, was reported in 1.5% of patients treated with moclobemide and in 3% of patients treated with imipramine.

Thus, unlike first generation MAOIs, moclobemide appears to be as effective as imipramine for the treatment of depression. Moreover, moclobemide does not seem to induce the characteristic side effects profile of first generation MAOIs, and seems to be better tolerated than imipramine.

Conclusions

This report describes the results obtained in a double-blind, 4-week, prospective, randomized multicenter (17 centers) study. The efficacy, tolerability and safety of moclobemide (300–600 mg/d) were compared to those of imipramine (100–200 mg/d) in parallel groups of patients with a Major Depressive Episode (DSM-III). In addition, participants were required to have a minimum baseline total score of 17 on the 21-item Hamilton Depression Rating Scale (HAMD). Upon inclusion (no run-in), 381 male or female in- or out-patients (aged over 18 years) were randomly allocated to moclobemide t.i.d. (n = 189) or to imipramine t.i.d. (n = 192). The demographic and illness characteristics were comparable in both treatment groups upon study entry. During the study the patients were not required to avoid tyramine-rich foods but they were advised to take trial medication after meals. Patients were assessed for efficacy and tolerability on days 3, 7, 14, 21 and 28. Efficacy was judged primarily on the HAMD, the investigator's final overall assessment, and on the number of premature terminations associated with insufficient efficacy. No significant difference was observed between both groups. The mean % reduction of the HAMD at the end of treatment was 51.7 in the moclobemide group and 52.1 in the imipramine group. The percentage of patients in whom efficacy was globally judged as "good" or "very good" was 62% in the moclobemide group and 60% in the imipramine group. 8% of moclobemide patients and 4% of imipramine patients did not complete in association with insufficient efficacy. The final overall physician's judgement of tolerability was "good" or "very good" in 83% of moclobemide patients and in 74% of imipramine patients. Adverse events were reported or observed in 56% of moclobemide patients and in 69% of imipramine patients. The number of mild, moderate and severe adverse events was higher in the imipramine group with a total of 286 versus 189. This difference was mainly due to the higher incidence of anticholinergic adverse events with imipramine. Poor tolerability was given as the reason for premature termination for 5% of patients on moclobemide and for 4% of patients on imipramine. Cardiovascular tolerability (blood pressure, heart rate, ECG, cardiovascular adverse events) was satisfactory in both groups and tended to be better with moclobemide. Hematology, clinical chemistry and urinalysis were not affected in a clinically significant fashion in either group. It is concluded that moclobemide is as effective as imipramine for the treatment of a Major Depressive Episode, but is better tolerated.

References

Angst J (1986) Allgemeine Aspekte zur Behandlung mit Antidepressiva. In: Schmid-Burgk W Koella W (Hrsg) Neue Aspekte der Therapie mit Antidepressiva. Fischer, Stuttgart New York

Baumhackl U, Biziere K Fischbach R, Geretsegger Ch, Hebenstreit G, Radmayr E, Stabl M (1990) Efficacy and tolerability of moclobemide compared with imipramine in depressive disorder (DSM III): an Austrian double-blind, multicentre study. Br J Psychiatry 155 [Suppl] 6:78–83

Davidson J, McCleod MN, Blum MR (1978) Acetylation phenotype, platelet monoamine oxidase inhibitor and the effectiveness of phenelzine in depression. Am J Psychiatry 135:467–469

Hebenstreit GF (1988) Efficacy and safety of moclobemide compared to imipramine in patients with major depression: results of an Austrian multicenter double-blind trial. Neurologija 37 [Suppl] 1:25

Hebenstreit GF, et al. (1990) Efficacy, safety and tolerability of moclobemide compared with imipramine in the treatment of major depressive disorder. WPA-Regional Symposium, book of abstracts, p 114

Kline NS, Cooper TB (1980) Monoamine oxidase inhibitors as antidepressants. In: Hoffmeister F, Stille G (eds) Psychotropic agents, part I. Antipsychotics and antidepressants. Springer, Berlin Heidelberg New York, pp 369–397

Paykel ES (1971) Classification of depressed patients: a cluster analysis derived grouping. Br J Psychiatry 188:275–288

Robinson DS, Nies A, Ravaris CL, Ives JO, Bertlett D (1978) Clinical psychopharmacology of phenelzine: MAO activity and clinical response. In: Lipton MA, DiMascio A, Killam KF (eds) Psychopharmacology: a generation of progress. Raven Press, New York

West ED, Dally LJ (1959) Effects of iproniazid in depressive syndromes. Br Med J 1:1491–1494

WHO Collaborative Study (1986) J Affect Dis 2:1–67

Authors' address: Prim. Prof. Dr. G. F. Hebenstreit, NÖ Landesnervenklinik, A-3362 Mauer/Amstetten, Austria.

J Neural Transm (1990) [Suppl] 32: 185–188

Psychometric alterations in treatment with the MAO-A-inhibitor moclobemide

W. Classen and **G. Laux**

Department of Psychiatry, University of Würzburg, Federal Republic of Germany

Summary. Two groups of depressed patients were treated either with a selective MAO-A-inhibitor, moclobemide (n = 13) or a tetracyclic antidepressant, maprotiline (n = 18), in a 28 days lasting double blind investigation. Before and after treatment psychopathologic symptoms were rated, motor performance was proven, and acoustic and visual sensomotoric performance were investigated. Deterioration of psychomotor performance were seen in patients without amelioration of their psychopathologic symptoms, especially when treated with moclobemide. These findings were regarded as a hint that possibly the therapeutic agent interacts with the wrong transmitter system and perhaps this is the reason for the deterioration of psychomotor functions.

Introduction

Affective psychosis are often accompanied by cognitive impairments (Robertson and Taylor, 1985) and sensorimotor disturbances (Cohen et al., 1982). It remains unclear however, which functional system (Brand and Joles, 1987) and whether dysfunction of the nondominant right hemisphere (Coffey, 1987) are affected.

The aim of the study is to find whether a substituted benzamide, the MAO-A-inhibitor moclobemide as compared to the tetracyclic agent, maprotiline, differentially alter sensorimotor functions in major depressive patients.

Patients and methods

After giving informed consent patients received substances [mean doses/d: 300 mg moclobemide (13 patients), 150 mg maprotiline (18 patients)] in a double blind design. No differences in duration of illness and illness episodes, age, gender or therapy response between both groups could be shown (Table 1). Sensorimotor acoustic and visual reaction times and motor acts (right and left hand) were investigated by Bettendorf Reaktiometer T 96 and the "Motorische Leistungs-Serie (MLS)" (Hamster, 1980) two times before drug intake (1) and after 28 days of therapy (2).

Results

Regarding treatment groups no significant differences could be demonstrated with respect to motor and sensorimotor performance. Looking at therapy-response (50% reduction of HAM-D) vs therapy-non-response groups once more no significant differences in sensorimotor reaction times could be demonstrated. However, precise motor acts (line tracing, aiming and pegboard-task) were impaired in nonresponders (NR). Especially in NR to moclobemide therapy motor performance was worsened after treatment ($p < .01$–$p < .05$).

Table 1. Demographic, nosological and therapeutic data of patients

		Moclobemide n = 13	Maprotiline 18	
Male		6	6	$x^2 = 1.03$
Female		7	14	$p > 0.05$
Duration of episodes	<	6	9	$x^2 = 0.00$
(mdn = 17.5 weeks)	>	6	10	$p > 0.05$
Duration of illness	<	6	9	$x^2 = 0.00$
(mdn = 4.5 years)	>	9	10	$p > 0.05$
Type of illness (according to DSM III)				
unipolar		7	14	
bipolar		3	0	$x^2 = 6.02$
neurotic		1	3	$p > 0.05$
atypical		2	1	
Age	mean	48.62	48.67	$t = 0.01$
	s	11.78	12.48	$p > 0.05$
Mean dose/d		300 mg	150 mg	
Hamilton-D				
before treatment	mean	28.3	28.1	time-course
	s	6.9	4.3	effect:
after treatment	mean	14.4	15.9	$p < 0.01$
		7.7	11.4	
Therapy response	yes	8	10	$x^2 = 0.16$
	no	5	8	$p > 0.05$

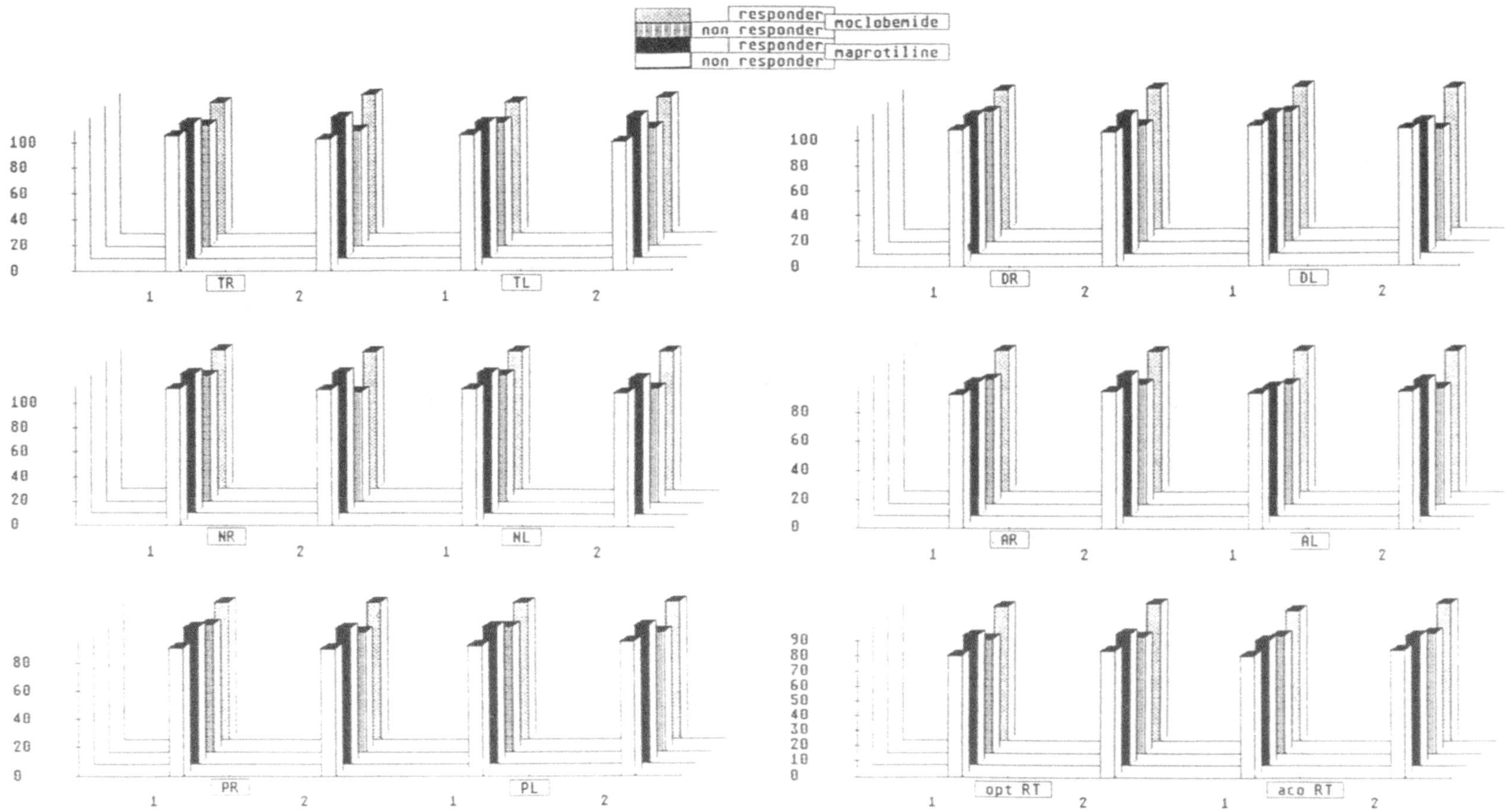

Fig. 1. Mean standard scores of performance in therapy-response- and therapy-non-response subgroups before (*1*) and after (*2*) treatment with moclobemide or maprotiline. *TR* duration time right hand, *TL* duration time left hand, *DR* duration of faults right hand, *DL* duration of faults left hand, *NR* number of faults right hand, *NL* number of faults left hand, *AR* aiming right hand, *AL* aiming left hand, *PR* pegboard right hand, *PL* pegboard left hand opt *RT* visual reaction time, *aco RT* acoustic reaction time

Discussion

In line with findings of Posner and McLeod (1982) motor performance was impaired in depressed patients and improved only with therapy response.

Pharmacological properties of both substances may cause these effects. Whereas maprotiline acts as a noradrenergic and serotonergic agent, moclobemide exclusively inhibits noradrenaline metabolism. Psychomotor activation of moclobemide without antidepressive effect in NR may produce arousal in patients, which is followed by impairments in attention and motor functions. Deterioration of psychomotor function may be a hint that the therapeutic agent in NR-patients interacts with the wrong transmitter system.

References

Brand N, Jolles J (1987) Information processing in depression and anxiety. Psychol Med 17:145–153

Cohen RM, Weingartner SA, Smallberg A, Pickar D, Murphy DL (1982) Effort and cognition in depression. Arch Gen Psychiatry 39:593–597

Coffey CE (1987) Cerebral laterality and emotion: the neurology of depression. Compr Psychiatry 28:197–219

Posner MI, McLeod P (1982) Information processing models in search of elementary operations. Annu Rev Psychol 33:477

Robertson G, Taylor PJ (1985) Some cognitive correlates of affective disorders. Psychol Med 15:297–309

Authors' address: Dr. W. Classen, Department of Psychiatry, University of Würzburg, Füchsleinstrasse 15, D-8700 Würzburg, Federal Republic of Germany.

J Neural Transm (1990) [Suppl] 32: 189–195

Clinical, biochemical and psychometric findings with the new MAO-A-inhibitors moclobemide and brofaromine in patients with major depressive disorder

G. Laux, W. Classen, E. Sofic, T. Becker, P. Riederer, K. P. Lesch, M. Struck, and **H. Beckmann**

Department of Psychiatry, University of Würzburg, Würzburg, Federal Republic of Germany

Summary. N = 53 inpatients with major depressive disorder have been treated with the reversible, selective MAO-A-inhibitors moclobemide (double-blind versus maprotiline) and brofaromine (open study), respectively. Clinically, significant improvement of depression and an activating profile of action could be observed, typical side effects were sleep disturbances, agitation and weight loss. The neurobiochemical data showed an increase of noradrenaline plasma concentrations under treatment with moclobemide. Visual reaction times improved with antidepressant treatment. MAO-A inhibitors proved to be effective antidepressants in the treatment of hospitalized patients with predominantly endogenous depressions.

Introduction

Renewed therapeutic interest in monoamine oxidase inhibitors (MAOIs) has followed the advent of a new generation of reversible, selective MAO-A-inhibitors. Inhibition of MAO-A is likely to be the decisive factor in antidepressant action of MAOIs since noradrenaline is almost exclusively catabolized by MAO-A. Lack of hepatotoxicity and reduced risk of hypertensive crisis ("cheese effect") facilitate clinical use (no diet restrictions), they are likely to be due to reversibility of MAO inhibiton. In recent years, controlled studies have provided evidence that MAOIs are very effective in the treatment of so-called therapy-resistant depressions (i.e., tricyclic-nonresponders), atypical depressions (subgroup with hypochondriac, histrionic features), phobias, panic disorders and anxiety states (Nies and Robinson, 1982; Pare, 1985; Sheehan et al., 1980). In addition, low cardiotoxicity, absence of anticholinergic side effects and mood-activating properties may be factors in favor of MAOIs as compared to tricyclic antidepressants.

Moclobemide, a substituted benzamide, is a novel MAOI with predominant inhibition of MAO-A, characterized by reversible binding, short half-life (1–2 hrs) and lack of hepatotoxicity (Da Prada et al., 1983). For review of clinical data see Laux and Riederer (1990).

Brofaromine is a reversible and selective MAO-A inhibitor with a plasma half-life of 12 hours (Waldmeier et al., 1983).

Patients and methods

A double blind study with moclobemide and maprotiline, the most selective noradrenaline reuptake inhibitor available at present has been conducted in n = 40 severely depressed inpatients suffering from predominantly endogenous depressions ($\bar{x}$ age = 48.1 years, 26 females, $\bar{x}$ HAMD day 0 = 28.2).

During treatment blood samples were drawn three hours after the morning dose for pharmacokinetic and neurobiochemical analyses. The patients were not subject to food restrictions (tyramine-free diet); drugs were, however, ingested at the end of meals.

Brofaromine was given to 13 inpatients with a major depressive disorder ($\bar{x}$ age = 47.5 years, 4 females) at a dosage of 100 to 150 mg daily over 4 weeks.

Informed written consent was obtained for participation in the respective trials. Blood samples were drawn at the same intervals, the patient remaining supine with an i.v. drip during the procedure. Drug level measurements were done by HPLC with a detection limit of 10 to 20 ng/ml.

Biogenic amines and their metabolites were determined using HPLC-coulometric analysis. Statistical analysis comprised one- and two-tailed Student's t-test.

Results

In the moclobemide group the trial was interrupted in 5 cases. Three drop-outs were due to agitation, delirious state and ineffectiveness, and were considered drug-related. The mean doses of moclobemide and maprotiline at the end of the study were 327 mg and 169 mg, respectively.

The results of the Clinical Global Impressions (CGI) are presented in Fig. 1: change in CGI revealed a tendency favoring moclobemide on day 21. The global profile of action of moclobemide can be summarized as primarily activating and anxiolytic to some extent. Figure 2 shows the mean total scores of the Hamilton Rating Scale for Depression (HAMD, 21 items). Overall tolerance was rated as good or very good in the majority of patients in both treatment groups. Type and frequency of side effects are presented in Table 1. Typical side effects were sedation and anticholinergic symptoms with maprotiline, and disturbed sleep, agitation, and pruritus with moclobemide (Laux et al., 1989).

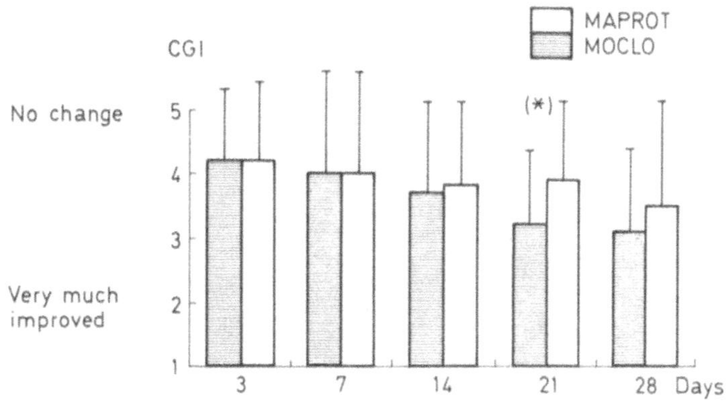

		3	7	14	21	28
Moclobemide	x̄	4.2	4.0	3.7	3.2	3.1
	SD	(1.1)	(1.4)	(1.4)	(1.1)	(1.2)
Maprotiline	x̄	4.2	4.0	3.8	3.9	3.5
	SD	(1.2)	(1.4)	(1.3)	(1.2)	(1.3)

Fig. 1. Clinical Global Impressions (CGI) and profile of action during treatment with moclobemide vs. maprotiline

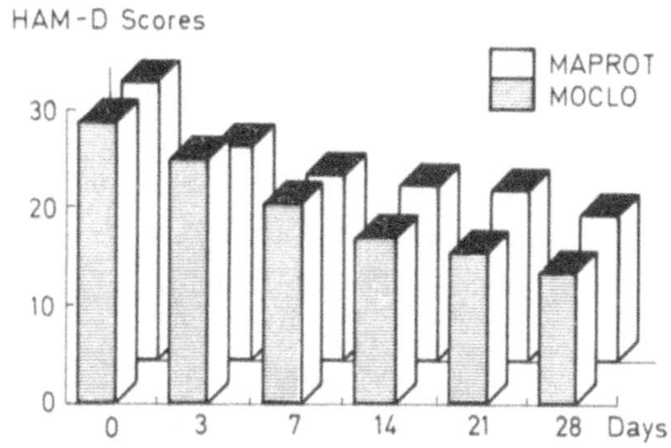

		0	3	7	14	21	28
Moclobemide	x̄	28.9	25.0	20.0	16.9	15.4	13.3
	SD	(6.1)	(9.3)	(9.0)	(9.2)	(9.5)	(7.7)
		n = 20	n = 20	n = 20	n = 18	n = 17	n = 15
Maprotiline	x̄	28.3	21.8	18.8	17.8	17.3	14.8
	SD	(4.9)	(6.0)	(7.9)	(8.2)	(9.5)	(11.4)
		n = 20	n = 20	n = 20	n = 20	n = 20	n = 20

Fig. 2. Mean values (x̄) of Hamilton Rating Scale for Depression (HAMD) scores during treatment with moclobemide vs. maprotiline. *SD* standard deviation

Table 1. Side effects under treatment with moclobemide and maprotiline

	Moclobemide (n = 20)	Maprotiline (n = 20)
Sedation	1	5
Dizziness	2	2
Nausea	3	2
Restlessness	4	–
Dry mouth	4	11
Disturb. micturition	–	5
Pruritus/Paraesthesia	4	1
Dysarthria, Tremor	2	5
Hypotension (> 20 mmHg)	3	3
Hypertension (> 20 mmHg)	3	3

Nine subjects completed the study with brofaromine, 2 males dropped due to increasing agitation, one female and one male due to lacking efficacy. Nevertheless the group as a whole significantly improved; more than 50% final reduction of HAMD-scores was reached by 7 patients in two of whom, however, loss of efficacy could be observed during trial. When base-line values were compared to final HAMD-ratings (either day 28 or day of drop out) decrease in the brofaromine group was from 24 ± 3 to 16 ± 9. Typical side effects seen with brofaromine were sleep disturbances, weight loss and restlessness.

Results of the neurobiochemical data showed an increase of noradrenaline (NA) plasma concentrations on day 14, a significant decline in homovanillic acid (HVA) plasma concentration on day 28 of treatment with moclobemide (Table 2).

In the patient receiving a single dose of brofaromine most plasma concentrations of catecholamines and major amine metabolites were and remained in the normal range. There was some transient increase of plasma adrenaline, 5-HIAA fell slightly in plasma and was below the normal range by the time of final sampling. There appears to have been some decrease of HVA in plasma.

Regarding the psychometric results, visual reaction times improved with antidepressant treatment. So, activating properties of moclobemide could be confirmed. Reduction of reaction time was more extensive in responders (defined as reduction of HAMD scores more than 50%) (Fig. 3) (see p. 194).

In conclusion, the reversible MAO-A inhibitors moclobemide and brofaromine proved to be effective antidepressants in the treatment of hospitalized patients suffering predominantly from endogenous depressions.

Table 2. Plasma catecholamines and major amine metabolites of inpatients with major depressive disorder before and under treatment with moclobemide

	NA (pg/ml) $\bar{x} \pm SD$	A (pg/ml) $\bar{x} \pm SD$	DA (pg/ml) $\bar{x} \pm SD$
Day 0	327.9 ± 205.7 (n = 19)	47.8 ± 36.4 (n = 12)	41.4 ± 49.9 (n = 9)
Day 14	401.6 ± 201.9 (n = 20) *	40.3 ± 37.5 (n = 15)	37.3 ± 33.7 (n = 15)
Day 28	372.6 ± 232.1 (n = 16)	57.5 ± 36.4 (n = 12)	33.8 ± 22.2 (n = 5)
Normal values	100–650 pg/ml	0–100 pg/ml	0–100 pg/ml

NA Noradrenaline; *A* Adrenaline; *DA* Dopamine
* $p < .05$ (Student's t-Test)

	5-HIAA (ng/ml) $\bar{x} \pm SD$	DOPAC (ng/ml) $\bar{x} \pm SD$	HVA (ng/ml) $\bar{x} \pm SD$
Day 0	6.1 ± 2.9 (n = 20)	0.9 ± 0.4 (n = 4)	9.4 ± 5.5 (n = 19)
Day 14	6.4 ± 4.2 (n = 19)	0.8 ± 0.4 (n = 4)	8.3 ± 4.9 (n = 20) *
Day 28	5.1 ± 2.7 (n = 17)	1.3 ± 1.3 (n = 3)	7.0 ± 5.7 (n = 17)
Normal values	4–12 ng/ml	1–4 ng/ml	2–10 ng/ml

5-HIAA 5-Hydroxyindoleacetic acid; *DOPAC* Dihydroxyphenylacetic acid; *HVA* Homovanillic acid
* $p < .10$ (Student's t-Test)

Selectivity of action has been confirmed by neurobiochemical data, activating properties involve facilities and limitations regarding clinical profile and therapeutic use.

References

Da Prada M, Kettler R, Keller HH, Haefely WE (1983) Neurochemical effects in vitro and in vivo of the antidepressant Ro 11-1163, a specific and short-acting MAO-A inhibitor. In: Beckmann H, Riederer P (eds) Monoamine oxidase and

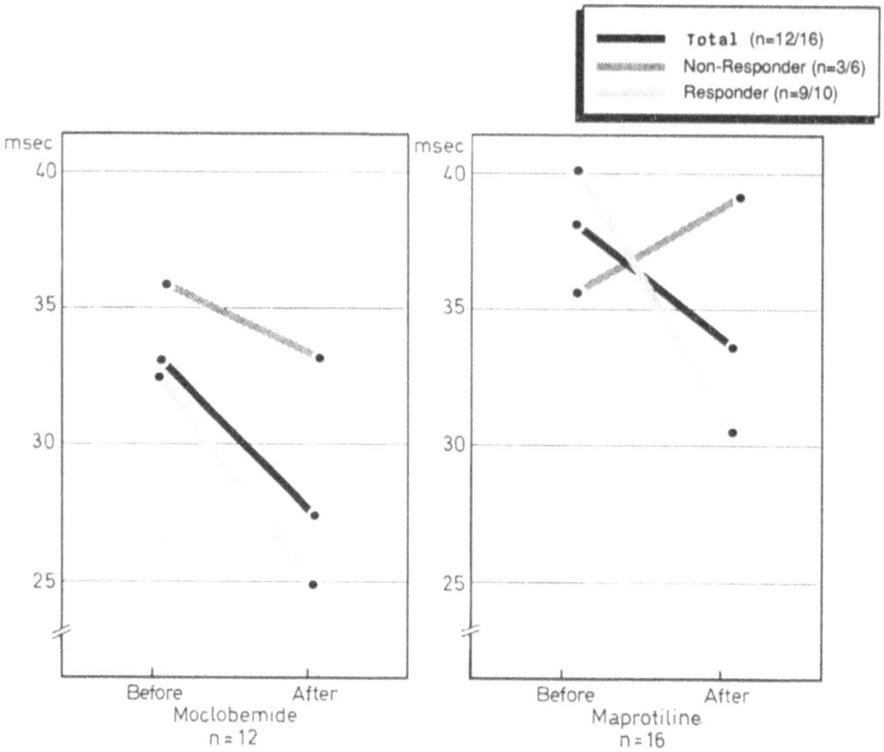

Fig. 3. Visual reaction times in endogenous depressed inpatients before and after treatment with moclobemide vs. maprotiline

its selective inhibitors. Modern problems of pharmacopsychiatry. Karger, Basel, pp 231–245

Laux G, Beckmann H, Classen W, Becker T (1989) Moclobemide and maprotiline in the treatment of inpatients with major depressive disorder. J Neural Transm [Suppl] 28:45–52

Laux G, Riederer P (eds) (1990) Clinical approaches to moclobemide, a new reversible inhibitor of MAO-A (RIMA). Psychiat Prax 17:1–30

Nies A, Robinsons DS (1982) Monoamine oxidase inhibitors. In: Paykel ES (ed) Handbook of affective disorders. Guilford, New York, pp 246–261

Pare CMB (1985) The present status of monoamine oxidase inhibitors. Br J Psychiatry 146:576–584

Sheehan DV, Ballenger J, Jacobsen G (1980) Treatment of endogenous anxiety with phobic, hysterial and hypochondrial symptoms. Arch Gen Psychiatry 37:51–59

Waldmeier PC, Baumann PA, Delini-Stula A, Bernascont R, Sigg K, Buech O, Felner AE (1983) Characterization of a new, short-acting and specific inhibitor of type A monoamine oxidase. In: Beckmann H, Riederer P (eds) Monoamine oxidase and its selective inhibitors. Modern problems of pharmacopsychiatry. Karger, Basel, pp 31–52

Authors' address: Dr. G. Laux, Department of Psychiatry, University of Würzburg, Füchsleinstrasse 15, D-8700 Würzburg, Federal Republic of Germany.

J Neural Transm (1990) [Suppl] 32: 197–201

Brofaromine (CGP 11 305 A): estimation of plasma concentrations by a biologic technique as compared to liquid chromatography

J. Fritze[1], **Th. Becker**[1], **V. Ziegler**[1], **G. Laux**[1], **P. Bieck**[2], **E. Sofic**[1], and **P. Riederer**[1]

[1] Department of Psychiatry and Clinical Neurochemistry, School of Medicine, University of Würzburg, and
[2] Institute of Human Pharmacology, Ciba-Geigy, Tübingen, Federal Republic of Germany

Summary. In an open clinical trial 13 depressives significantly improved under the reversible and selective type-A monoamine oxidase (MAO) inhibitor brofaromine. The inhibitory potency of deproteinated plasma on a crude MAO preparation from human placenta was measured as a parameter for plasma brofaromine. There were no significant differences in plasma MAO inhibitory potency between responders (improvement $> 50\%$; $n = 5$) and non-responders. MAO inhibitory potency significantly ($p < 0.05$) increased parallel to the increase of the dosage from 50 mg b.i.d. to t.i.d. confirming the validity of this technique. The biologic assay, however, overestimated brofaromine by a factor of two in acute kinetic experiments with healthy volunteers as compared to a chromatographic technique, although both methods significantly correlated ($r = 0.928$).

Introduction

The relationship of the plasma concentrations of antidepressants with clinical efficacy is far from clarified (Brotman et al., 1987). To control pharmacokinetic sources of variance, all biologically active forms of a drug should be quantified (Rudorfer and Potter, 1987). For the putative antidepressant brofaromine (CGP 11 305 A) (Schiwy et al., 1990), a reversible and selective inhibitor on monoamine oxidase (MAO-A) (Waldmeier et al., 1983), it seemed worthwhile to establish a procedure for estimating the MAO inhibitory potency of patients' plasma rather than the definite compound.

Patients and methods

Nine males and 4 females entered the study after giving their informed consent. Their age was 47.2 ± 11 years (range 28–67). They all fulfilled DSM-III criteria for major depression, 4 subjects being bipolars, 3 subjects recurrent monopolars. In 6 patients the depressive episode was superimposed to a dysthymic disorder. 6 patients were of the melancholic subtype. 6 patients had been refractory to at least two standard antidepressants for a minimum of 4 weeks each. 3 patients were on lithium which was continued. Flunitrazepam was allowed as needed. In the interest of tolerability the initial dosage of brofaromine was 50 mg b.i.d. orally maintaining 50 mg t.i.d. from day 14 to day 28. Psychopathology was rated before (day 0) and on days 1, 3, 7, 14, 21 and 28 during treatment using the Hamilton Depression Scale (HAMD) (Hamilton, 1960). Blood was drawn in EDTA-tubes on days 0, 7, 14, 21 and 28 at 10.00 a.m., i.e. 3 h after the morning drug administration.

A crude mitochondrial fraction of human placenta was prepared according to Nakano et al. (1986) and suspended by sonification (10 sec) in 100 mM sodium phosphate buffer pH 7.4 yielding a final activity of 2 nmoles H_2O_2/ml*min. MAO activity was determined by the spectrophotometric assay of Szutowicz et al. (1984). Tyramine (1 mM) was used as substrate. The mechanism of MAO inhibition by brofaromine was studied using 10 μm to 1 mM tyramine with or without preincubating MAO for various periods (15 to 60 min) at 37 °C in the presence or absence of 1 nM to 20 nM brofaromine. Moreover, the dependency of the IC_{50} of brofaromine on the time of preincubation was studied. Data were analysed by linear regression in logit-log plots. Initial experiments revealed that native plasma interfered with the spectrophotometric assay. Deproteination with 9 volumes of 0.16% uranylacetate completely removed this interference. The inhibition of MAO by deproteinated plasma from patients was transformed to nM and ng/ml concentrations of brofaromine in original plasma by means of a calibration curve from plasma of known brofaromine concentrations linearized as above. All laboratory experiments were performed in triplicate and blind to the clinical data. To test its validity, the biological assay was applied to plasma from kinetic experiments in two healthy volunteers where brofaromine was estimated additionally by gas-liquid chromatography (Hengen et al., 1984). The clinical data were analysed by the Freedman-Test and Wilcox-Wilcoxon-Tests. Drop-outs were considered by end-point analysis. Differences in plasma MAO inhibitory potency between responders (HAMD-score on day 28/end-point $< 50\%$ of day 0) and non-responders were analysed by two-factor ANOVA with repeated measures. Linear correlations were calculated to test the validity of both assays for brofaromine.

Results

Nine subjects completed the study. Two males dropped out on day 7/11 due to increasing agitation, one female and one male on day 14/23 due to lacking efficacy. Nevertheless, the group as a whole significantly ($p < 0.01$) improved, 5 patients by $> 50\%$. 4 of these responders had an underlying dysthymic disorder.

The mechanism of MAO inhibition by brofaromine changed from a competetive to a non-competetive mode with preincubation. Thus, its IC_{50}

changed from 37 nM to 9 nM after 30 min of preincubation at 37 °C in the presence of brofaromine. There was no further change with longer periods of preincubation. Although the recovery rate in deproteinated plasma was only 60% the sensitivity was high yielding a lower limit of detectability of ≈ 5 nM (≈ 2 ng/ml). The inter- and intraassay coefficients of variation were < 5%. In the healthy volunteers, both measures of brofaromine significantly correlated ($r = 0.928$; $p \ll 0.001$), although the biological assay overestimated brofaromine by a factor or two.

'Brofaromine' as estimated biologically significantly increased over time ($F = 3.31$; df = 3.30; $p < 0.05$) parallel to the increase of the daily dosage (Fig. 1) with no differences between responders and non-responders. Plasma from day 0 showed no MAO inhibitory potency. Plasma 'brofaromine' did not correlate to age.

Discussion

The kinetic properties of brofaromine measured in placenta mitochondria correspond to previous findings (Keller et al., 1987; Waldmeier et al., 1983). Concentrations of brofaromine as low as ≈ 2ng/ml could be detected with an acceptable reliability. Thus, the biologic procedure reaches the sensitivity of gas-liquid-chromatographic (GLC) procedures (Hengen et al., 1984). The reason for the biological assay overestimating brofaromine remains to be clarified. Although brofaromine is extensively metabolized, only part of the

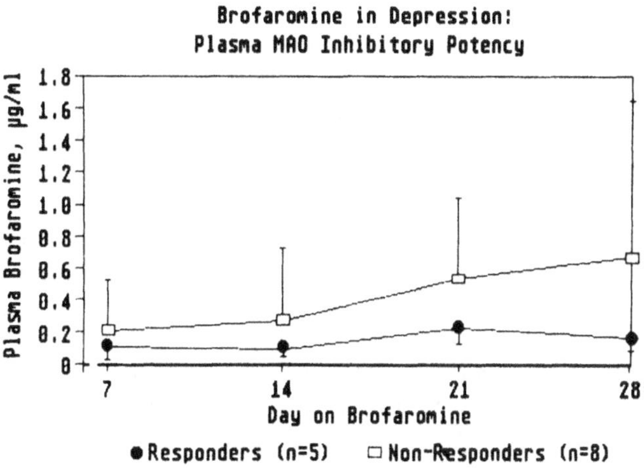

Fig. 1. Plasma brofaromine (means ± SD) as estimated by the inhibition of a standard MAO-A preparation from human placenta by deproteinated plasma from depressives treated by 50 mg brofaromine b.i.d. till day 14 and 50 mg t.i.d. from day 15 to day 28. ANOVA: no significant differences between responders and non-responders ($F = 0.93$; df = 3.30); significant increase over time parallel to the increase in dosage ($F = 3.31$; df = 3.30; $p < 0.05$) in the group as a whole

metabolites have been identified. Active metabolites might have contributed to the overestimation. The increase of the plasma MAO inhibitory potency parallel to the increase of the dosage (Fig. 1) is a further indicator of the validity of the biological assay. The lack of MAO inhibitory potency on day 0 signifies that there was no interference with endogenous MAO inhibitors (Giambalvo, 1984; Glover et al., 1980). The plasma levels calculated from the calibration curves correspond to those described by Steiger et al. (1987). Despite the considerable amount of non-responders and drop-outs the present open trial adds to the evidence of antidepressant properties of brofaromine (Schiwy et al., 1990). The outcomes described might be biased by the selection of patients refractory to previous treatment with standard antidepressants. Interestingly, and possibly corresponding to the findings of Steiger et al. (1987), subjects with dysthymic disorder preferably responded. This interpretation would be in line with previous (Quitkin et al., 1979) although controversial results with irreversible MAO inhibitors. The plasma levels as estimated by plasma MAO inhibitory potency were unrelated to the clinical response (Fig. 1) corresponding to the findings of Steiger et al. (1987).

References

Brotman AW, Falk WE, Gelenberg AJ (1987) Pharmacologic treatment of acute depressive subtypes. In: Meltzer HY (ed) Psychopharmacology: the third generation of progress. Raven Press, New York, pp 1031–1040

Giambalvo CT (1984) Purification of endogenous modulators of monoamine oxidase from plasma. Biochem Pharmacol 33:3929–3932

Glover V, Reveley MA, Sandler M (1980) A monoamine oxidase inhibitor in human urine. Biochem Pharmacol 29:467–470

Hamilton M (1960) A rating scale for depression. J Neurol Neurosurg Psychiatry 23:56–62

Hengen N, Jedrychowski ME, Hoffmann E (1984) Pharmacokinetics of CGP 11 305A in man after acute and prolonged oral treatment. In: Tipton KF, Dostert P, Strolin-Benedetti M (eds) Monoamine oxidase and disease. Academic Press, London, pp 185–191

Keller HH, Kettler R, Keller G, Da Prada M (1987) Short-acting novel MAO inhibitors: in vitro evidence for the reversibility of MAO inhibition by moclobemide and Ro 16-6491. Naunyn-Schmiedebergs Arch Pharmacol 335:12–20

Nakano T, Saito S, Higashida H, Kojima K, Nagatsu T (1986) Assignment of A and B types of monoamine oxidase in NCB20 hybrid cells to those of the parental cells by peptide mapping. J Neurochem 46:686–694

Quitkin F, Rifkin A, Klein DF (1979) Monoamine oxidase inhibitors. Arch Gen Psychiatry 36:749–760

Rudorfer MV, Potter WZ (1987) Pharmacokinetics of antidepressants. In: Meltzer HY (ed) Psychopharmacology: the third generation of progress. Raven Press, New York, pp 1353–1363

Schiwy W, Heath WR, Delini-Stula A (1990) Therapeutic and side-effect profile of a selective and reversible MAO-A inhibitor, brofaromine. Results of dose finding trials in depressed patients. J Neural Transm [Suppl] 28:33–44

Steiger A, Holsboer F, Gerken A, Demisch L, Benkert O (1987) Results of an open clinical trial of brofaromine (CGP 11 305 A), a competetive, selective, and short-acting inhibitor of MAO-A in major endogenous depression. Pharmacopsychiatry 20:262–269

Szutowicz A, Kobes RD, Orsulak PJ (1984) Colorimetric assay for monoamine oxidase in tissues using peroxidase and 2,2′-azinodi-(3-ethyl-benzthiazoline-6-sulfonic acid) as chromogen. Anal Biochem 138:86–94

Waldmeier PC, Baumann PA, Delini-Stula A, Bernasconi R, Sigg K, Buech O, Felner AE (1983) Characterization of a new, short-acting and specific inhibitor of type A monoamine oxidase. Mod Probl Pharmacopsychiatry 19:31–52

Authors' address: Dr. J. Fritze, Department of Psychiatry, University of Würzburg, Füchsleinstrasse 15, D-8700 Würzburg, Federal Republic of Germany.

J Neural Transm (1990) [Suppl] 32: 203–209

A double-blind, placebo-controlled study of the tolerability and effects on platelet MAO-B activity of single oral doses of MDL 72.974A in normal volunteers

C. Hinze, D. Harland, M. Zreika, B. Dulery, and **J. Hardenberg**

Merrell Dow Research Institute, Strasbourg, France

Summary. MDL 72.974A [(E) 4-fluoro-beta-fluorethylene benzene butanamine] has been shown in animal studies, both in vitro and in vivo, to be a potent, selective, enzyme-activated irreversible inhibitor of MAO-B (Zreika et al., 1989). This compound is under clinical development for the treatment of Parkinson's disease.

In this double blind, randomized, placebo-controlled normal volunteer study the tolerability, effects on platelet MAO-B activity and associated pharmacokinetics of increasing single oral doses of MDL 72.974A (0.1–12 mg) were assessed.

MDL 72.974A was extremely well tolerated and no treatment-related changes in vital signs or the adjectival check-list (EWL-N) occurred. The compound caused significant dose-dependent inhibition of platelet MAO-B activity at all dose levels with a return to baseline values by day 14. The mean (± S.D.) elimination half-life of parent compound was 51 ± 26 min and mean (± S.D.) urinary excretion was 0.54 ± 0.26% of the administered dose. These results, long action on platelet MAO-B and short elimination half-life, demonstrate MDL 72.974A to be a potent, irreversible inhibitor of MAO-B in man.

Introduction

Monoamine oxidase (MAO) enzymes metabolize a wide variety of amines including monoamine neurotransmitters. There are two forms of mammalian brain MAO which can be classified by a number of criteria, including substrate specificity. (Fowler and Ross, 1984; Denney et al., 1985). According to current nomenclature, noradrenaline and 5-HT are preferentially metabolized by MAO type A (MAO-A) whereas benzylamine and phenylethylamine are preferred substrates for type B (MAO-B). Dopamine is a good substrate for both forms of MAO (Fowler et al., 1984). However, brain homogenate studies suggest that MAO-B predominates in human brain and it appears that dopamine is largely metabolized by this form of the enzyme in man (Glover et al., 1980; Riederer et al., 1982).

MDL 72.974A [(E)-4-fluoro-beta-fluorethylene benzene butanamine] is a potent, selective, enzyme-activated irreversible inhibitor of monoamine oxidase type B (MAO-B). Comparison of the in vitro IC_{50} values for inhibition of rat brain mitochondrial MAO-A and MAO-B by MDL 72.974A indicates a selectivity ratio of 190 in favour of MAO-B (Zreika et al., 1989). Unlike the prototype MAO-B inhibitor, L-deprenyl, MDL 72.974A has only minor indirect sympathomimetic effects in rats.

L-deprenyl has been reported to be a valuable adjunct to conventional L-dopa plus dopa decarboxylase inhibitor treatment of Parkinson's disease, particularly for patients who exhibit fluctuations in response to L-dopa (for review, see Heinonen and Rinne, 1989).

The results of animal studies indicate that MDL 72.974A is a more potent and selective inhibitor of MAO-B than L-deprenyl and, as such may prove to be an effective treatment for Parkinson's disease. The present study involved the first administration to man of MDL 72.974A and was designed to evaluate the tolerability, biochemical effects and pharmacokinetics of single rising doses of this compound.

Materials and methods

The study was carried out as a double-blind, placebo-controlled randomized trial. It was approved by a local ethics committee. Forty-eight healthy male volunteers who had given written informed consent received single oral doses of 0.1, 0.25, 0.5, 1, 2, 4, 8, or 12 mg of MDL 72.974A or placebo. Each dose group consisted of 6 subjects (5 active, 1 placebo). Medication was supplied as 0.1, 0.25, 0.5, 1, 2 and 4 mg capsules of MDL 72.974A with matching placebo capsules. Concomitant medication was not allowed from 2 weeks before to 2 weeks following administration of study medication. No tyramine-containing foodstuffs or alcoholic beverages were permitted from admission to the Clinical Pharmacology Unit until 2 weeks following administration of MDL 72.974A. A full physical examination was performed at the screening visit and on day 14 (last day). Haematology, biochemistry and urinalysis tests were performed, resting heart rate and blood pressure was measured and a 12-lead ECG was recorded at pre-determined intervals. The subjects completed an adjectival check-list (EWL-N, Janke and Debus, 1978) at baseline and 2, 4, 12 and 24 h following administration. Blood samples were taken for the measurement of MDL 72.974A, and subsequent determination of pharmacokinetic parameters at baseline and then at 15, 30, 45 min, 1, 1.5, 2, 3, 4, 6, 8, 12, 24 and 48 h post administration. The separated plasma was stored at − 18 °C until it was assayed. All urine voided from 0 to 24 h, 24 to 48 h, from day 6 to 7 and from day 13 to 14 was collected into plastic containers to which 10 ml of 1 N HCl had been added. The volume was measured and 50 ml aliquots were taken and frozen at − 18 °C for the subsequent determination of concentrations of MDL 72.974A and its metabolites. Plasma concentrations and urinary excretion were determined using an analytical method based on automatic sample preparation with disposable solid phase extraction columns followed by gas chromatography-mass spectrometry. MDL 71.946A [1-fluoromethyl-2-(4-chlorobenzene)ethanamine, hydrochloride] was used as the internal standard. Mass spectrometric analysis (Finnigan TSQ 46) was carried out in the positive chemical ionization mode with ammonium and in the selected ion monitoring mode.

Blood samples (10 ml) were taken for the preparation of platelet-rich plasma by clean venipuncture into a SARSTEDT-Monovette^R containing potassium EDTA before (baseline, 0 h) and at intervals after (1, 3, 6, 12, 24, 48 h, days 4, 7, 11, 14) administration of MDL 72.974A.

The samples were centrifuged at 300 × gravity for 10 minutes and the platelet-rich plasma fraction was removed using an Eppendorf pipette. Platelet counts (standard Coulter method, $\times 10^3/mm^3$) were performed on an aliquot of each fresh platelet rich plasma sample prior to freezing at $-18\,^{\circ}C$. Platelet MAO-B activity was determined by incubating 200 µl of platelet rich plasma for 30 minutes at 37 °C in a shaking water bath with 800 µl of 0.1 M phosphate buffer, pH 7.2 and 20 µl of ^{14}C-labelled substrate (1.9 ml 0.1 M phosphate buffer pH 7.2, 50 µl 10^{-2} M phenylethylamine, 50 µl ^{14}C-phenylethylamine, 80.2 mCi/mmol). The reaction was stopped by adding 300 µl of 2N HCl. The products of the reaction were extracted into 7 ml of toluene:ethyl acetate (1 : 1) by shaking for 10 minutes and then centrifuged for 10 minutes at 1000 rpm to separate the two phases. 5 ml aliquots of the organic layer were taken and placed into scintillation vials for counting after the addition of 10 ml Econofluor. The MAO-B activity was then expressed as $\mu mol/hour/10^9$ platelets.

Results

A total of 48 healthy male subjects entered and completed the study. They had a mean age of 27.6 years, a mean weight of 73.0 kg and a mean height of 176.8 cm. There were no dropouts or withdrawals and data from all subjects were analyzed for safety and efficacy.

There were no drug-related changes in ECG, heart rate or systolic and diastolic blood pressure when these parameters were recorded at intervals up to 24 hours post dosing. There were also no significant differences between active drug and placebo for any of these parameters. EWL-N tests (adjectival check list, self assessment) did not reveal any drug- or dose-related differences for any of the 17 subtests recorded. Results of routine haematology, biochemistry and urinalysis revealed no drug-attributable changes in any parameters. All values outside the normal laboratory range were considered to be of no clinical significance.

Two cases of headache, 1 mild and 1 moderate, following administration of, respectively, 2 mg and 8 mg of MDL 72.974A, were the only unwanted effects reported.

MDL 72.974A caused significant inhibition of platelet MAO-B activity compared with placebo at each of the doses administered.

Enzyme activity was inhibited to a greater extent following administration of 0.25 mg and 0.5 mg of MDL 72.974A than following administration of 0.1 mg (Fig. 1). In the 0.1 mg dose group, enzyme activity was inhibited by a maximum of 74% 12 hours post-dosing. Inhibition was greater than 90% within 3 hours of dosing in the groups receiving 0.25 mg and 0.5 mg of MDL

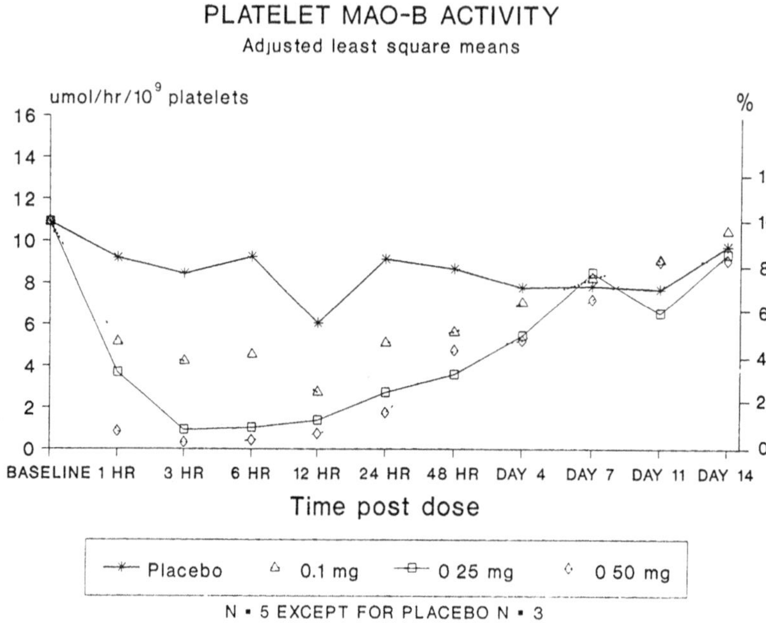

Fig. 1

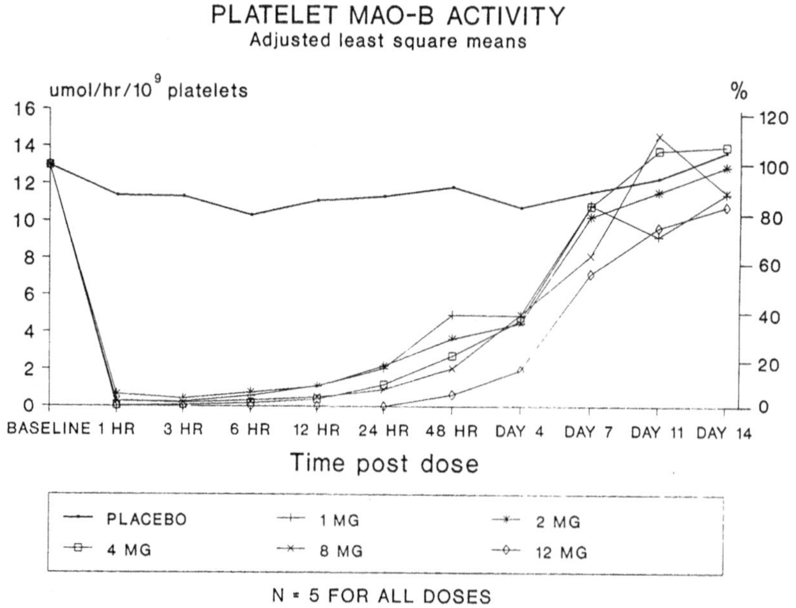

Fig. 2

72.974A and was still significant ($p < 0.05$) 48 h following administration of 0.1 to 0.5 mg. Following administration of doses of 1 mg and above, enzyme inhibition was greater than 95% within 1 hour of dosing (Fig. 2) and was still significant 4 days following administration. Enzyme activity had returned to baseline levels by day 14 in all groups. Plasma concentrations of parent drug were low. Thus, with a lower limit of detection of 50 pg/ml, pharmacokinetic parameters could only be determined for the 8 mg (Table 1) and 12 mg (Table 2) dose groups. As the absolute oral bioavailability of MDL 72.974A is unknown, values of CL_{total} and aV_d are uncorrected for bioavailability.

Mean ($\pm$ S.D.) peak plasma concentrations were 4.4. $\pm$ 2.5 ng/ml after 8 mg and 6.7 $\pm$ 3.4 ng/ml after the 12 mg dose. Time to peak (t_{max}) ranged from 0.5 to 1.5 h. The mean terminal half-life for both groups combined was 51 min. Very little unchanged drug was excreted with the urine, i.e. for both dose groups, mean 0–24 h urinary excretion corresponded to 0.54 $\pm$ 0.26% of the administered dose (range 0.25–0.98%). No parent drug was detected in the 24–48 h urines.

Table 1. Pharmacokinetic parameters of MDL 72.974A after an oral dose of 8 mg MDL 72.974A

Subject	19	20	21	22	23	MEAN	S.D.
C_{max}(ng/ml)	8.77	2.54	3.70	4.04	2.78	4.37	2.54
t_{max}(h)	0.75	1.0	0.5	1.5	1.0	1.0	0.4
AUC (ng.h/ml)	15.87	5.37	8.43	8.70	7.50	9.17	3.97
$t_{1/2}$ (min)	102	37	81	54	54	66	26
Cl_{tot} (1/min)	8.4	24.8	15.8	15.3	17.7	16.4	5.9
V_d (1)	1250	2390	1858	1205	4450	2231	1333
U_E (% dose)	0.35	0.47	0.67	0.25	0.50	0.45	0.16

Table 2. Pharmacokinetic parameters of MDL 72.974A after an oral dose of 12 mg MDL 72.974A

Subject	26	27	28	29	30	MEAN	S.D.
C_{max}(ng/ml)	2.04	8.40	4.31	8.62	10.30	6.73	3.43
t_{max}(h)	1.5	0.5	0.75	1.5	1.5	1.2	0.5
AUC (ng.h/ml)	3.65	6.88	4.94	13.24	17.12	9.17	5.77
$t_{1/2}$ (min)	44	22	17	33	62	36	18
CL_{tot} (1/min)	54.8	29.0	40.5	15.1	11.7	30.2	17.9
V_d (1)	3480	908	987	720	1037	1426	1154
U_E (%dose)	0.32	0.85	0.26	0.98	0.78	0.64	0.33

Discussion

The results of this study suggest that MDL 72.974A is a potent inhibitor of platelet MAO-B activity in man. MDL 72.974A was extremely well tolerated. There were no significant effects on cardiovascular parameters or laboratory safety parameters at doses which proved to be 12-fold those required to inhibit platelet MAO-B activity by more than 95% (i.e. 12 mg compared with 1 mg). Only 2 cases of headache, (1 mild, 1 moderate) with a possible relation to drug treatment were reported. No subjective effects were revealed by the EWL-N tests.

Inhibition of platelet MAO-B activity was dose-dependent over the dose range of 0.1 mg to 0.5 mg. The profile of enzyme inhibition following administration of doses of 1 mg and above, with a maximal inhibition of enzyme activity occuring within 1 hour of dosing and gradual recovery to baseline levels by day 14, is characteristic of the irreversible nature of enzyme inhibition produced by this compound. Comparison with the inhibition of platelet MAO-B activity produced by single doses of the prototype MAO-B inhibitor L-deprenyl in man, (Elsworth et al., 1978; Koulu et al., 1989) suggests that MDL 72.974A is 10- to 20-fold more potent than L-deprenyl.

It is interesting to contrast the long duration of action of MDL 72.974A on platelet MAO-B activity with the low detectable amount of unchanged drug in plasma and its short elimination half-life. These results suggest that low plasma concentrations of MDL 72.974A are sufficient to irreversibly inhibit platelet MAO-B activity, which is restored only following the regeneration of platelets and thus the synthesis of new enzyme.

References

Denney RM, Denney CB (1985) An update on the identity crisis of monoamine oxidase: new and old evidence for the independence of MAO-A and -B. Pharmacol Ther 30:227–259

Elsworth JD, Glover V, Reynolds GP, Sandler M, Lees AJ, Phuapradit P, Shaw KM, Stern GM, Kumar P (1978) Deprenyl administration in man: a selective monoamine oxidase B inhibitor without the "cheese effect". Psychopharmacology 57:33–38

Fowler CJ, Ross SB (1984) Selective inhibitors of monoamine oxidase A and B: biochemical, pharmacological and clinical properties. Med Res Rev 4:323–358

Fowler CJ, O'Carroll AM, Tipton KF (1984) Deamination of dopamine by monoamine oxidase -A and -B in the rat and man. In: Tipton KF, Benedetti MS (eds) Monoamine oxidase and disease. Academic Press, London, pp 393–402

Glover V, Elsworth JD, Sandler M (1980) Dopamine oxidation and its inhibition by (−)-deprenyl in man. J Neural Transm [Suppl] 16:163–172

Heinonen EH, Rinne U (1989) Selegiline in the treatment of Parkinson's diesease. Acta Neurol Scand 126:103–111

Janke W, Debus G (1978) Die Eigenschaftswörterliste. Hogrefe, Göttingen

Koulu M, Schernin M, Kaartinen A, Kallio J, Pykko K, Vuorinen J, Zimmer RM (1989) Inhibition of monoamine oxidase by moclobemide: effects on monoamine metabolism and secretion of anterior pituitary hormones and cortisol in healthy volunteers. Br J Clin Pharmacol 27:243–255

Riederer P, Reynolds JP, Youdim MBH, Jellinger K (1982) In vitro tests of MAO inhibitors in human brain tissue: chemical structure and pharmacologic action. In: Kamijo K, Usdin E, Nagatsu T (eds) Monoamine oxidase, basic and clinical frontiers. Excerpta Medica, Amsterdam, pp 345–350

Zreika M, Fozard JR, Dudley MW, Bey P, McDonald IA, Palfreyman MG (1989) MDL 72.974: a potent and selective enzyme-activated irreversible inhibitor of monoamine oxidase type B with potential for use in Parkinson's disease. J Neural Transm [PD-Sect] 1:243–254

Authors' address: Dr. C. Hinze, Merrell Dow Research Institute, Department of Clinical Research, 2, rue de Stockholm, F-67000 Strasbourg, France

J Neural Transm (1990) [Suppl] 32: 211–214

MAO-B inhibition in rabbit tissues and in human platelets by Ro 19-6327 shows similar time-course

R. Kettler[1], **A. M. Cesura**[1], **J. Dingemanse**[2], and **M. Da Prada**[1]

[1] Pharmaceutical and [2] Clinical Research Departments, F. Hoffmann-La Roche Ltd, Basel, Switzerland

Summary. In rabbits, 3 mg/kg Ro 19-6327 p.o., induced complete monoamine oxidase type B (MAO-B) inhibition in striatum, cerebral cortex, liver and platelets for about 12 h. In healthy human volunteers complete inhibition of platelet MAO was obtained with only 5 mg Ro 19-6327. After 40 and 200 mg Ro 19-6327 the duration of complete MAO-B inhibition was prolonged to 12 and 24 hours, respectively. Changes in brain MAO-B in humans after Ro 19-6327 are likely to be faithfully reflected by platelet MAO activity.

Introduction

In clinical trials with the irreversible monoamine oxidase type B (MAO-B) inhibitor, selegiline (L-deprenyl), inhibition of human platelet MAO-B has often been used as a reliable biochemical index of the degree of blockade of brain MAO-B (Riederer et al., 1978; Oreland et al., 1983). Platelets from most laboratory animals have little or no MAO-B activity, in contrast to those from humans (Kettler and Da Prada, 1989). For this reason comparative time-course studies on MAO-B inhibition in brain and platelets of laboratory animals have not been reported.

The effect of the reversible and highly selective MAO-B inhibitor Ro 19-6327 (Da Prada et al., 1988; for chemical structure see Cesura et al., these Proceedings) on MAO-B activity was investigated in two rabbit brain regions (frontal cortex and striatum) as well as in extracerebral tissues (platelets and liver). This animal species has been chosen because rabbit platelets' MAO-B activity can be measured radiochemically (Kettler and Da Prada, 1989). For comparison, the time-course of the inhibition of MAO-B was also measured in human platelets obtained from healthy subjects administered Ro 19-6327 at various single doses.

Materials and methods

Blood was collected and platelets were isolated from plasma and washed once with Tyrode as previously described (Kettler and Da Prada, 1989). MAO-B activity was measured radiochemically (Da Prada et al., (1989) and proteins were measured spectrophotometrically (BCA* Protein Assay Reagent; Pierce, Rockford, Ill., USA).

ED_{50} values with 95% confidence limit for MAO-B inhibition in platelets were calculated using values for five different doses of Ro 19-6327 and a computer program for curve fitting.

Results

The time-course of MAO-B inhibition in platelets, liver, striatum and frontal cortex of rabbits after 3 mg/kg Ro 19-6327 p.o. is shown in Fig. 1. At this dose Ro 19-6327 markedly inhibited (about 80%) MAO-B activity in all peripheral and central tissues investigated already 30 min after dosing. Sixteen hours after administration MAO-B activity in platelets and liver as well as in striatum and frontal cortex had recovered to about half-maximum activity. Even 24 h after administration a certain degree of MAO-B inhibition (20–30% inhibition) was observed. After lower doses (0.3 and 1 mg/kg p.o., not shown) Ro 19-6327 induced pronounced MAO-B inhibition (70–80%) in the periphery and in the CNS 1 h after dosing. After these low

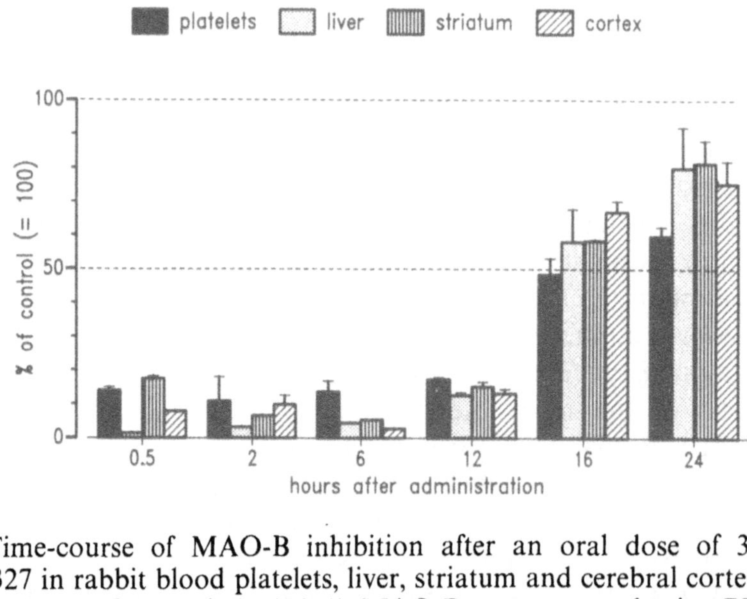

Fig. 1. Time-course of MAO-B inhibition after an oral dose of 3 mg/kg of Ro 19-6327 in rabbit blood platelets, liver, striatum and cerebral cortex. Data are given in percent of controls (= 100%). MAO-B was measured using PEA (20 μM) as substrate. Absolute MAO-B activities in untreated controls (N = 17) in blood platelets: 1.8 ± 0.1 nmoles/h/mg protein; in liver, striatum and cerebral cortex: 44.5 ± 3.3, 4.7 ± 0.2 and 4.6 ± 0.3 nmoles/h/mg wet tissue, respectively

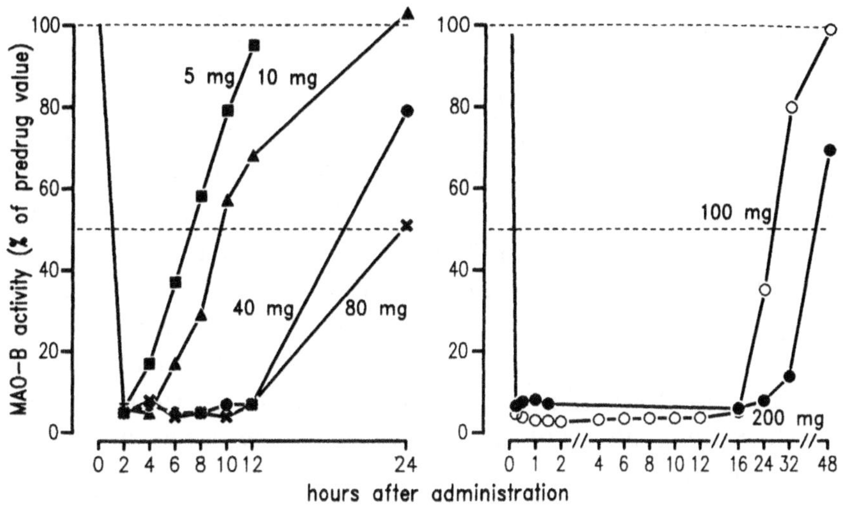

Fig. 2. Time-course of MAO-B inhibition in blood platelets of young volunteers (N = 6 per group) administered single oral doses of 5, 10, 40, 80, 100 and 200 mg Ro 19-6327 as tablets. Predrug MAO-B activities in these groups were: 22.6 ± 4.0, 17.9 ± 0.9, 20.5 ± 1.0, 18.5 ± 2.7, 16.2 ± 1.6 and 16.5 ± 2.2 nmol/h/mg protein, respectively

doses complete recovery of MAO-B activity occurred 12 and 16 h after drug administration, respectively (not shown).

The dose-response curves for human platelet MAO-B inhibition (Fig. 2) show that a single dose of 5 mg Ro 19-6327 p.o. produced complete inhibition (> 90%) 2 h post-dosing. The duration of virtually complete platelet MAO-B inhibition increased with incremental doses and, at the highest dose shown here (200 mg), marked inhibition was maintained for at least 24 h. A comparison of the ED_{50} values (mg/kg p.o., 12 h) measured in rabbit [$ED_{50} = 0.695$ (0.692–0.698)] and human [$ED_{50} = 0.202$ (0.196–0.207)] platelets indicate that Ro 19-6327 is about 3.5 times more potent in man than in rabbit.

Discussion

The main findings of this study are a) that the time-course of MAO-B inhibition in the rabbit is virtually the same in platelets and liver and b) that the extracerebral MAO-B activity closely paralleled the time-course of MAO-B inhibition in two brain areas (striatum and frontal cortex).

The fact that at low doses of Ro 19-6327 (0.3 mg/kg p.o.) the MAO-B inhibition was more pronounced in platelets and liver than in the CNS indicates that at this low dose, Ro 19-6327 (being highly hydrophilic) penetrates with a certain delay into the brain.

As shown in Fig. 2, the duration of complete platelet MAO-B inhibition in man increases dose-dependently and, at 40 mg, marked MAO-B inhibition lasted for at least 12 h. Positron emission tomography investigations using labeled selegiline showed that in one healthy subject Ro 19-6327 at about 0.7 mg/kg p.o. (50 mg) markedly inhibited (> 90%) brain MAO-B activity 12 h after dosing. Interestingly, at the same time, complete platelet MAO-B inhibition was found in the platelets (Price et al., 1990). Therefore, it can be concluded that the time-course of MAO-B inhibition in human brain probably parallels platelet MAO-B inhibition when 50 mg or higher doses of Ro 19-6327 are administered. Clinical trials have demonstrated an excellent tolerability for Ro 19-6327 up to 200 mg. Therefore Ro 19-6327 seems to be an attractive and safe tool as adjuvant in the long-term therapy of Parkinson's disease.

References

Da Prada M, Kettler R, Keller HH, Burkard WP (1988) Ro 19-6327, a reversible, highly selective inhibitor of type B monoamine oxidase, completely devoid of tyramine-potentiating effects: comparison with selegiline. In: Sandler M, Dahlström A, Belmaker RH (eds) Progress in catecholamine research, part B. Central aspects. Alan R Liss, New York, pp 359–363

Da Prada M, Kettler R, Keller HH, Burkard WP, Muggli-Maniglio D, Haefely WE (1989) Neurochemical profile of moclobemide, a short-acting and reversible inhibitor of monoamine oxidase type A. J Pharmacol Exp Ther 248:400–414

Kettler R, Da Prada M (1989) Platelet MAO-B activity in humans and stumptail monkeys: in vivo effects of the reversible MAO-B inhibitor Ro 19-6327. In: Przuntek H, Riederer P (eds) Early diagnosis and preventive therapy in Parkinson's disease. Springer, Wien New York, pp 213–219

Oreland L, Johanson F, Ekstedt J (1983) Dose regimen of deprenyl (selegiline) and platelet MAO activities. Acta Neurol Scand [Suppl] 95: 87–89

Price GW, Bench CJ, Cremer JC, Luthra SK, Turton DR, Lammertsma AA, Kettler R, Da Prada M, Wood N, Jamieson V, McClelland G, Frackowiak RSJ (1990) Inhibition of human brain monoamine oxidase-B by Ro 19-6327 – in vivo measurement using positron emission tomography. Eur J Pharmacol 183: 166

Riederer P, Youdim MBH, Rausch WD, Birkmayer W, Jellinger K, Seemann D (1978) On the mode of action of L-deprenyl in the human central nervous system. J Neural Transm 43: 217–226

Authors' address: Prof. M. Da Prada, F. Hoffmann-La Roche Ltd, Pharmaceutical Research Department, CH-4002 Basel, Switzerland

Oxidative stress and autooxidation

J Neural Transm (1990) [Suppl] 32: 217–227

Monoamine oxidase and the bioactivation of MPTP and related neurotoxins: relevance to DATATOP

R. E. Heikkila, I. Terleckyj, and **B. A. Sieber**

Department of Neurology, UMDNJ-Robert Wood Johnson Medical School, Piscataway, New Jersey, USA

Summary. The DATATOP study is a clinical trial in which deprenyl, a selective inhibitor of monoamine oxidase-B (MAO-B), is being given to newly diagnosed Parkinsonian patients in an attempt to halt the progression of their disorder. In part, this is being done because of the working hypothesis than an MPTP-like molecule may be the cause of Parkinsonism, and deprenyl is known to protect against MPTP-induced dopaminergic neurotoxicity in experimental animals. In the present study we point out that several analogs of MPTP are good substrates not only for MAO-B but also for MAO-A. In addition, we point out that with long-term administration to rodents, deprenyl loses its selectivity as an inhibitor of MAO-B and also inhibits MAO-A. We believe that these observations have relevance for the DATATOP study.

Introduction

The self-administration of 1-methyl-4-phenyl-1,2,3,6-tetrahydropyridine (MPTP) by several young drug abusers led to the rapid development of a Parkinsonian state (Davis et al., 1979; Langston et al., 1983). These individuals suffered from tremor, rigidity and akinesia, which are hallmark symptoms of Parkinsonism. These behavioral deficits were reversed by drugs such as 1-dopa which are used to treat Parkinson patients. At about the same time, it was shown that the administration of MPTP to non-human primates led to behavioral deficits similar to those observed in Parkinson patients and those who had self-administered MPTP (Burns et al., 1983). Moreover, these MPTP-treated primates exhibited a severe dopamine deficiency and a loss of cells in the zona compacta of the substantia nigra, which are the major biochemical and pathological deficits observed in Parkinson patients (Burns et al., 1983).

Surprisingly, it was originally shown that rats were unaffected by MPTP administration (Chiueh et al., 1984), although subsequently it was shown that treatment with large and repeated doses of MPTP caused some dopaminergic neurotoxicity in rats (Fuller and Steranka, 1985; Jarvis and Wagner, 1985). Mice treated with MPTP were found to exhibit severe dopaminergic deficits (Heikkila et al., 1984a; Sundstrom et al., 1987), although the doses of MPTP required in mice were considerably higher than doses which were effective in primates. As an example, C57 black mice treated with 4 intraperitoneal injections of MPTP at 2 hr intervals at 20 mg/kg/injection (80 mg/kg total dose) exhibited approximately a 90% decrement in their neostriatal content of dopamine, a parallel loss in neostriatal tyrosine hydroxylase activity and in the capacity of neostriatal synaptosomes to take up ^{3}H-dopamine, slightly lesser decrements in their neostriatal content of dopamine metabolites, and a marked loss of cells in the zona compacta of the substantia nigra (Heikkila et al., 1984a; Sonsalla and Heikkila, 1986; Mayer et al., 1986). Monkeys treated once daily with 3 to 4 injections of MPTP at 0.25 to 0.35 mg/kg (total dose of 0.75 to 1.40 mg/kg) exhibited similar dopaminergic deficits (Burns et al., 1983; Langston et al., 1984).

It was shown by Markey et al. (1984) that the 1-methyl-4-phenylpyridinium species (MPP^+) was the major metabolite of MPTP found in the brain of experimental animals. It was subsequently shown (Castagnoli et al., 1985; Chiba et al., 1984) that the conversion of MPTP to its 4-electron oxidation product MPP^+ occurred in vitro in rat brain preparations via the formation of the 2-electron intermediate 1-methyl-4-phenyl-2,3-dihydropyridinium ($MPDP^+$). These investigators also demonstrated that monoamine oxidase-B (MAO-B) inhibitors including deprenyl and pargyline were potent inhibitors of MPTP oxidation while in contrast, the MAO-A inhibitor clorgyline was ineffective (Chiba et al., 1984). It was also shown that MPTP was a good substrate for purified MAO-B in vitro (Salach et al., 1984). After learning of the above-mentioned work by Castagnoli, Trevor and colleagues, we determined that pretreatment of mice with inhibitors of MAO-B (i.e. deprenyl, pargyline, nialamide, tranylcypromine) prior to MPTP administration, provided protection against all of the dopaminergic deficits normally caused by MPTP (Heikkila et al., 1984b). Similar protective effects against the neurotoxicity of MPTP were subsequently obtained in primates with inhibitors of MAO-B (Cohen et al., 1984; Langston et al., 1984). It was also shown that MAO-B inhibitors were able to markedly attenuate MPP^+ formation from MPTP in vivo in mice and in monkeys (Langston et al., 1984; Markey et al., 1984). These observations suggested that MPP^+ formation was intimately involved with the neurotoxicity observed after MPTP administration, and that MPP^+ itself was the actual neurotoxic entity.

MPP^+ is an effective cytotoxin, having been developed by Gulf Oil several years ago as a herbicide. It is now well-known that MPP^+ is a potent inhibitor of mitochondrial respiration (Hoppel et al., 1987; Mizuno et al., 1987a,b; Nicklas et al., 1985; Ofori et al., 1989; Ramsay et al., 1986; Ramsay and Singer, 1986; Vyas et al., 1986). It has been suggested that the inhibition of respiration, occurring after the accumulation of MPP^+ by the dopamine transport system and its subsequent accumulation by mitochondria within the dopaminergic neuron, is the ultimate mechanism responsible for the neurotoxicity observed after MPTP administration (see e.g. Youngster et al., 1989 a,b). Interestingly, the pesticide rotenone which, inhibits respiration at the same site in the electron transport chain as MPP^+, namely Complex I, also was found to cause dopaminergic neurotoxicity after its stereotoxic administration to rats (Heikkila et al., 1985).

Steve Youngster, a colleague of ours who is trained both as a medicinal chemist and as a neuropharmacologist, has synthesized over 75 analogs of MPTP and close to 50 analogs of MPP^+. In the present study, we will report on some of our observations with several of the MPTP analogs. We will further discuss the relevance of these and other observations to the DATATOP study (Deprenyl and Tocopherol Antioxidative Therapy of Parkinsonism). The DATATOP study is a large prospective study in which deprenyl and/or tocopherol are being given to newly diagnosed Parkinson patients in an attempt to halt the progression of their disease (Parkinson Study Group, 1989 a,b; Tetrud and Langston, 1989).

Materials and methods

The capacity of several analogs of MPTP to be oxidized by MAO was determined in a crude mitochondrial preparation as previously described (Youngster et al., 1987). Briefly, the whole brain was removed from male, Swiss-Webster mice (Taconic Farms) and rinsed in ice cold saline. The brain was then homogenized in 9 volumes of 0.32 M sucrose, followed by centrifugation at 1500 × g for 10 min in a Sorvall RC-5B refrigerated centrifuge. The resulting supernatant was layered over 15 ml of 0.88 M sucrose and the samples centrifuged at 18,000 × g for 20 min. The resulting pellet was dispersed in 0.05 M phosphate buffer at pH 7.4, containing 3 mM sodium azide. This latter addition was made to inactivate brain catalase, since the principle of the assay is the measurement of the MAO-dependent formation of H_2O_2, which is decomposed by catalase. Then the MPTP analogs were added (10 μM to 1000 μM concentration) to 1 ml of the buffer containing 30 mg of original weight of tissue and the samples incubated with shaking at 37 °C for 30 min. At this time 0.5 ml of a peroxidase chromogen color reagent (Boehringer-Mannheim) and 0.5 ml of 6% Triton X-100 were added and the absorbancy determined at 520 nM. MAO-A activity was determined in the presence of 0.25 μM deprenyl and MAO-B activity was determined in the presence of 0.25 μM clorgyline. Blanks were run with no substrate and with a combination of the above concentrations of deprenyl and clorgyline. These concentrations of the inhibitors are selective and complete for inhibition of MAO-B and MAO-A respectively. Known substrates for MAO-A (i.e. tryptamine) and MAO-B (i.e. benzylamine) were run in each experiment for comparison. MAO activity was expressed as nmoles of H_2O_2 formed/g tissue/hr.

In other experiments, various MAO inhibitors including deprenyl were given to mice (male Swiss-Webster, Taconic) and to rats (male, Sprague-Dawley) for periods ranging from 1 day to 28 days. At 16–18 hours after the last injection, the animals were sacrificed, their brains were removed and MAO activity was determined as described previously (Kindt et al., 1988). Briefly, the tissue was homogenized at 5 mg/ml of 0.05 M phosphate buffer at pH 7.4. Then radioactive ^{3}H-5-hydroxytryptamine (MAO-A substrate) and ^{14}C benzylamine (MAO-B substrate) were added simultaneously to 0.5 ml of the tissue preparation. Non-radioactive substrates were included so that the final substrate concentration ranged from 10 to 1000 μM (usually 250 μM). The samples were incubated with shaking for 30 min at 37 °C and the reaction stopped by addition of 0.3 ml of 2N HCl. The oxidation products were extracted into 5 ml of ethyl acetate: toluene (1:1) by vortexing. After centrifugation at low speed, the samples were kept at − 70 °C for 15 min to freeze the aqueous layer. The amount of radioactivity in the upper layer (organic) containing the oxidation products was then determined. The data were calculated as nmoles of product formed/g tissue/hr. Blanks containing deprenyl and clorgyline or 2N HCl were run in each experiment.

Results and discussion

For purposes of discussion, we will focus on MPTP and three of its analogs, all of which are substituted in the 2′ position of the phenyl ring. These MPTP analogs include 2′methyl-MPTP, 2′ethyl-MPTP and 2′n-propyl-MPTP. As stated above, MPTP was found to be a good substrate for MAO-B. As may be seen in Table 1, there was a concentration-dependent oxidation of MPTP by MAO-B. These values for MPTP oxidation were similar to those obtained with benzylamine, a commonly used substrate for MAO-B (data not shown). In contrast, MPTP was a poor substrate for MAO-A (Table 1). Both 2′methyl-MPTP and 2′ethyl-MPTP were good substrates for MAO-B; in fact 2′methyl-MPTP was considerably better than even benzylamine. In contrast to MPTP, both of these MPTP analogs were also good substrates for MAO-A. It may be seen that at 25 to 100 μM, 2′ethyl-MPTP was oxidized to a greater extent by MAO-A than by MAO-B, while 2′ methyl-MPTP was oxidized to a greater extent by MAO-B than by MAO-A at all concentrations studied.

2′Methyl-MPTP and 2′ethyl-MPTP are, like MPTP, effective dopaminergic neurotoxins in mice (Youngster et al., 1989a,b). In fact, these two analogs of MPTP are even more potent than MPTP itself. Based on the fact that 2′methyl-MPTP was such a good substrate for MAO-B (see Table 1), we anticipated that deprenyl and other MAO-B inhibitors would provide protection against 2′methyl-MPTP induced neurotoxicity as they did against MPTP. We were quite surprised to find that deprenyl failed to protect. Clorgyline alone, as expected, also failed to protect. However,

Table 1. The oxidation of several MPTP analogs by MAO

Substrate	Conc. (μM)	MAO-A activity	MAO-B activity	Proportion A activity
		(nmoles H_2O_2/g tissue/h)		
MPTP	25	0	246	0
	50	9	347	0.03
	100	0	526	0
	250	0	675	0
2′Me-MPTP	25	269	756	0.26
	50	348	1293	0.21
	100	408	1791	0.19
	250	428	2089	0.17
2′Et-MPTP	25	328	149	0.69
	50	507	269	0.65
	100	627	358	0.64
	250	726	746	0.49
2′nPr-MPTP	25	140	18	0.89
	50	254	26	0.91
	100	298	35	0.89
	250	368	61	0.86

Experiments were done in mouse brain preparations (Swiss-Webster Mice, Taconic Farms). MAO-A activity is defined in the presence of 0.25 μM deprenyl and MAO-B activity in the presence of 0.25 μM clorgyline. Data are from a representative experiment. The experiment was repeated several times with similar results

a combination of deprenyl and clorgyline was found to be protective against the dopaminergic neurotoxicity of 2′methyl-MPTP (Heikkila et al., 1988). In parallel with the observed protection, only the combination of deprenyl and clorgyline effectively decreased the neostriatal content of 2′methyl-MPP^+ after the administration of 2′methyl-MPTP to mice (Heikkila et al., 1988).

In the same series of experiments, deprenyl alone failed to provide protection against the dopaminergic neurotoxicity of 2′ethyl-MPTP. Surprisingly, clorgyline alone provided substantial protection, while the combination of deprenyl and clorgyline provided complete protection (Heikkila et al., 1988). Again, there was a parallel between the protection against the dopaminergic neurotoxicity and the attenuation of pyridinium formation by the inhibitors of MAO. Clorgyline alone substantially decreased the neostriatal content of 2′ethyl-MPP^+ after 2′ethyl-MPTP administration to mice, while the combination of deprenyl and clorgyline was even more effective. All of these observations suggest that the corresponding pyridinium species formed by MAO-A and/or MAO-B acting on the tetrahydropyridine analogs of MPTP are the ultimate dopaminergic neurotoxins.

Of the neurotoxic analogs of MPTP that we have studied, all have been good substrates for MAO-A and/or MAO-B. Additionally, the pyridinium species which would be the theoretical 4-electron oxidation products have been, like MPP^+, good substrates for the dopamine transport system and also effective inhibitors of mitochondrial respiration. This has led to the working hypothesis that all three of these properties are important determinants of neurotoxicity for MPTP and structurally related compounds (Youngster et al., 1989a,b).

After realizing that both 2′methyl-MPTP and 2′ethyl-MPTP were such good substrates for MAO-A, we synthesized 2′n-propyl-MPTP. We anticipated that it would be a good substrate for MAO-A and that it might be a poor substrate for MAO-B. Our hope was to find a dopaminergic neurotoxin which would be bioactivated exclusively by MAO-A. As it turned out, 2′n-propyl-MPTP proved to be a good substrate for MAO-A but not for MAO-B (see Table 1). Unfortunately it was not a dopaminergic neurotoxin under experimental conditions in which MPTP is effective in mice. This lack of dopaminergic neurotoxicity with 2′n-propyl-MPTP and several other MPTP analogs which were good substrates for MAO has led to the concept that oxidation by MAO is a necessary, but not sufficient, criterion for a compound to be an MPTP-like dopaminergic neurotoxin. After further study, we discovered that the pyridinium species of 2′n-propyl-MPTP, namely 2′n-propyl-MPP^+, was a very weak substrate for the dopamine transport system. This explains its lack of neurotoxic activity.

As stated above, deprenyl is being given to newly diagnosed Parkinson patients in the DATATOP study in an attempt to halt the progression of the disease. There are two theoretical considerations which form the basis for the use of deprenyl in this study:

1) The bioactivation of an MPTP-like molecule by MAO-B plays a role in the pathogenesis of Parkinsonism.
2) The H_2O_2 formed as a result of the deamination of dopamine by MAO-B plays a role in the pathogenesis of Parkinsonism.

If either of these two mechanisms were responsible for causing Parkinsonism, it is obvious that deprenyl should halt the progression of the disorder. Deprenyl would prevent the MAO-B catalyzed formation of an MPP^+-like molecule from MPTP and the resulting deleterious effects. It would further prevent any oxidative stress resulting from the H_2O_2 formed during the MAO-B catalyzed deamination of dopamine, which is primarily a substrate for MAO-B in humans (Glover et al., 1977; Oreland et al., 1983). Furthermore, tocopherol, is an effective free radical scavenger, theoretically should provide protection against H_2O_2 dependent lipid peroxidation. It should also be emphasized that deprenyl has no known toxicity and, in fact, it was previously reported that Parkinson patients receiving deprenyl plus l-dopa had an increased life expectancy compared to Parkinson patients receiving l-dopa alone (Birkmayer et al., 1985). Moreover, it was also shown that patients in whom l-dopa had lost its effectiveness appeared to regain the beneficial response to l-dopa after receiving deprenyl (Birkmayer et al., 1983, 1985).

When we first heard of the DATATOP study, we had some initial concerns, based in part on our own experience. For example, it is possible that both an MPTP analog and MAO could play an important role in the pathogenesis of the disorder, and still deprenyl therapy might be ineffective in halting the progression of Parkinsonism. This could happen if the actual neurotoxin were, like 2′ethyl-MPTP, bioactivated exclusively or primarily by MAO-A. A similar concern can be raised about the effectiveness of deprenyl in combatting oxidative stress. If the H_2O_2 which caused Parkinsonism were formed via an MAO-A dependent oxidation, deprenyl theoretically would be ineffective in halting the progression of Parkinsonism. Another concern is that tocopherol, while being an effective inhibitor of lipid peroxidation, is ineffective against other types of oxidative stress. Thus deprenyl and tocopherol might be ineffective in halting the progression of Parkinsonism, and yet both MAO and oxidative stress could still be involved in the pathogenesis of the disorder. Our fear was that most investigators might disregard a role for MAO and for oxidative stress in the pathogenesis of Parkinsonism when, in actuality, either or both might be extremely important.

Deprenyl is an extremely selective inhibitor of MAO-B, both in vitro and in vivo after a single injection of 0.2 or 2.0 mg/kg to mice or rats (data not shown). However, deprenyl effectively inhibits not only MAO-B activity but also MAO-A activity after one injection per day to mice or rats for 28 consecutive days. This loss of selectivity of deprenyl, which was previously pointed out by Ekstedt et al. (1979), is particularly evident in rats (Table 2). Ekstedt et al. (1979) found that after only a single injection of 5.25 mg/kg,

Table 2. The effects of deprenyl administration on MAO activity in rats and mice

Species	Deprenyl dose (mg/kg)	% Inhibition of MAO-A	MAO-B
S.W. mouse	0.2	6	85
	2.0	23	95
S.D. rat	0.2	22	89
	2.0	79	95

Deprenyl was given s.c. once daily for 28 consecutive days at 0.2 or 2 mg/kg to male Sprague-Dawley rats and i.p. to male Swiss-Webster mice (both from Taconic Farms). The animals were sacrificed one day after the last injection and MAO activity determined. Data are the mean % inhibition for 3 animals in each group

deprenyl had lost its selectivity. Moreover, after receiving deprenyl for seven consecutive days at 1.0 mg/kg/injection/day, deprenyl had lost its selectivity for inhibiting MAO-B that was observed after a single injection of 1.0 mg/kg. In general, it is evident that deprenyl becomes markedly less selective as an inhibitor of MAO-B with increasing dose and with an increasing number of injections.

If this crossover phenomenon were to occur in humans, it is quite possible that deprenyl would start to inhibit MAO-A activity at some point during its repeated administration to Parkinson patients. That this is possible and even likely has recently been suggested by the observation that the administration of deprenyl to human volunteers at 20 mg daily for 13 to 16 days led to an enhanced pressor response to tyramine (Schulz et al., 1989). The authors concluded, not unreasonably, that this increased pressor response was due to the inhibition of MAO-A by the dosing regimen with deprenyl. If 20 mg/day for 13–16 days can result in an inhibition of MAO-A activity, what is the effect of 10 mg/day (the dose used in the DATATOP study) for weeks and even months? Furthermore, Riederer and Youdim (1986) have shown that deprenyl at 10 mg/day for as little as six days led to extensive inhibition of MAO-A activity in various regions of human Parkinson brain obtained after autopsy. In this study, the inhibition of the oxidation of 5-HT (a preferred substrate for MAO-A) after deprenyl administration ranged from 38% in the putamen to 84% in the hypothalamus. The inhibition of the oxidation of dopamine (a substrate for both MAO-A and MAO-B) in these same brain areas was 88% and 90% respectively.

Preliminary results from the DATATOP study have been encouraging (Parkinson Study Group, 1989a; see also Tetrud and Langston, 1989). Patients on deprenyl have taken considerably longer than patients not on deprenyl to reach "end-point", the time at which it is determined that they require l-dopa therapy. The nature of this effect is at present unclear. Is there, as many believe, a slowing down of the progression of Parkinsonism by deprenyl? Or is there some other explanation for this fascinating observation? Only time will tell. And what is the mechanism for this effect? Is it due to the inhibition of MAO-B by deprenyl? Or rather, is it due to its inhibition of MAO-A? Or is the effect due to the relatively large amounts of methamphetamine or of amphetamine which are formed during the metabolism of deprenyl? Is the effect in some way due to the fact that the nigrostriatal content of dopamine is higher in those patients on deprenyl therapy because of the inhibition of dopamine catabolism by deprenyl (by its inhibition of MAO-A or of MAO-B)? It should be pointed out that many consider that dopamine can be harmful because of its capacity to autoxidize. However, it cannot be ignored that the patients on deprenyl therapy in the DATATOP study must have had a higher brain dopamine content than those not receiving deprenyl. One may thus want to reevaluate the concept of withholding l-dopa from newly diagnosed Parkinson patients because of fear of the deleterious effects of the excess dopamine that would be formed from l-dopa.

It is our hope that careful and systematic studies similar to the DATATOP study will be carried out with other MAO inhibitors and with other experimental treatments. Until such studies are carried out, it will be difficult to further understand the nature of deprenyl's apparent protective effect and the fundamental mechanism for the pathogenesis of Parkinsonism. Not to proceed aggressively at this time with careful, thoughtful and new experimentation would be unfortunate.

References

Birkmayer W, Knoll J, Riederer P, Youdim MBH (1983) (−)-Deprenyl leads to prolongation of l-dopa efficacy in Parkinson's disease. Mod Probl Pharmacopsychiatry 19:170–176

Birkmayer W, Knoll J, Riederer P, Youdim MB, Hars V, Marton J (1985) Increased life expectancy resulting from addition of l-deprenyl to madopar treatment in Parkinson's disease: a longterm study. J Neural Transm 64:113–127

Burns RS, Chiueh CC, Markey SP, Ebert MH, Jacobowitz DM, Kopin IJ (1983) A primate model of parkinsonism: selective destruction of dopaminergic neurons in the pars compacta of the substantia nigra by N-methyl-4-phenyl-1,2,3,6-tetrahydropyridine. Proc Natl Acad Sci USA 80:4546–4550

Castagnoli N Jr, Chiba K, Trevor AJ (1985) Potential bioactivation pathways for the neurotoxin 1-methyl-4-phenyl-1,2,3,6-tetrahydropyridine (MPTP). Life Sci 36:225–230

Chiba K, Trevor A, Castagnoli N Jr (1984) Metabolism of the neurotoxic tertiary amine, MPTP, by brain monoamine oxidase. Biochem Biophys Res Commun 120:574–578

Chiueh CC, Markey SP, Burns RS, Johannessen JN, Jacobowitz DM, Kopin IJ (1984) Neurochemical and behavioral effects of 1-methyl-4-phenyl-1,2,3,6-tetrahydropyridine (MPTP) in rat, guinea pig, and monkey. Psychopharmacol Bull 20:548–553

Cohen G, Pasik P, Cohen B, Leist A, Mytilineou C, Yahr MD (1984) Pargyline and deprenyl prevent the neurotoxicity of 1-methyl-4-phenyl-1,2,3,6-tetrahydropyridine (MPTP) in monkeys. Eur J Pharmacol 106:209–210

Davis GC, Williams AC, Markey SP, Ebert MH, Caine CD, Reichert CM, Kopin IJ (1979) Chronic Parkinsonism secondary to intravenous injection of meperidine analogues. Psychiatry Res 1:249–254

Ekstedt B, Magyar K, Knoll J (1979) Does the B form selective monoamine oxidase inhibitor lose selctivity by long term treatment? Biochem Pharmacol 28:919–923

Fuller RW, Steranka LR (1985) Central and peripheral catecholamine depletion by 1-methyl-4-phenyl-tetrahydropyridine (MPTP) in rodents. Life Sci 36:243–247

Glover V, Sandler M, Owen F, Riley GJ (1977) Dopamine is a monoamine oxidase B substrate in man. Nature 265:80–81

Heikkila RE, Hess A, Duvoisin RC (1984a) Dopaminergic neurotoxicity of 1-methyl-4-phenyl-1,2,5,6-tetrahydropyridine in mice. Science 224:1451–1453

Heikkila RE, Manzino L, Cabbat FS, Duvoisin RC (1984b) Protection against the dopaminergic neurotoxicity of 1-methyl-4-phenyl-1,2,5,6-tetrahydropyridine by monoamine oxidase inhibitors. Nature 311:467–469

Heikkila RE, Nicklas WJ, Vyas I, Duvoisin RC (1985) Dopaminergic toxicity of rotenone and the 1-methyl-4-phenylpyridinium ion after their stereotaxic administration to rats: implication for the mechanism of 1-methyl-4-phenyl-1,2,3,6-tetrahydropyridine toxicity. Neurosci Lett 62:389–394

Heikkila RE, Kindt MV, Sonsalla PK, Giovanni A, Youngster SK, McKeown KA, Singer TP (1988) Importance of monoamine oxidase A in the bioactivation of neurotoxic analogs of 1-methyl-4-phenyl-1,2,3,6-tetrahydropyridine. Proc Natl Acad Sci USA 85:6172–6176

Hoppel CL, Grinblatt D, Kwok HC, Arora PK, Singh MP, Sayre LM, Greenblatt D (1987) Inhibition of mitochondrial respiration by analogs of 4-phenylpyridine and 1-methyl-4-phenylpyridinium cation (MPP^+), the neurotoxic metabolite of MPTP. Biochem Biophys Res Commun 148:684–693

Jarvis MF, Wagner GC (1985) Neurochemical and functional consequences following 1-methyl-4-phenyl-1,2,5,6-tetrahydropyridine (MPTP) and methamphetamine. Life Sci 36:249–254

Kindt MV, Youngster SK, Sonsalla PK, Duvoisin RC, Heikkila RE (1988) Role for monoamine oxidase-A (MAO-A) in the bioactivation and nigrostriatal dopaminergic neurotoxicity of the MPTP analog, 2′Me-MPTP. Eur J Pharmacol 146:313–318

Langston JW, Ballard P, Tetrud JW, Irwin I (1983) Chronic Parkinsonism in humans due to a product of meperidine- analog synthesis. Science 219:979–980

Langston JW, Irwin I, Langston EB, Forno LS (1984) Pargyline prevents MPTP-induced Parkinsonism in primates. Science 225:1480–1482

Markey SP, Johannessen JN, Chiueh CC, Burns RS, Herkenham MA (1984) Intraneuronal generation of a pyridinium metabolite may cause drug-induced Parkinsonism. Nature 311:464–467

Mayer RA, Walters AS, Heikkila RE (1986) 1-Methyl-4-phenyl-1,2,3,6-tetrahydropyridine (MPTP) administration to C57-black mice leads to parallel decrements in neostriatal dopamine content and tyrosine hydroxylase activity. Eur J Pharmacol 120:375–377

Mizuno Y, Sone N, Saitoh T (1987) Effects of 1-methyl-4-phenyl-1,2,3,6-tetrahydropyridine and 1-methyl-4-phenylpyridinium ion on activities of the enzymes in the electron transport system in mouse brain. J Neurochem 48:1787–1793

Mizuno Y, Sone N, Suzuki K, Saitoh T (1988) Studies on the toxicity of 1-methyl-4-phenylpyridinium ion (MPP^+) against mitochondria of mouse brain. J Neurol Sci 86:97–110

Nicklas WJ, Vyas I, Heikkila RE (1985) Inhibition of NADH-linked oxidation in brain mitochondria by 1-methyl-4-phenyl-pyridine, a metabolite of the neurotoxin, 1-methyl-4-phenyl-1,2,5,6-tetrahydropyridine. Life Sci 36:2503–2508

Ofori S, Heikkila RE, Nicklas WJ (1989) Attenuation by dopamine uptake blockers of the inhibitory effects of 1-methyl-4-phenyl-1,2,3,6-tetrahydropyridine and some of its analogs on NADH-linked metabolism in mouse neostriatal slices. J Pharmacol Exp Ther 251:258–266

Oreland L, Arai Y, Stenstrom A (1983) The effect of deprenyl (selegiline) on intra- and extraneuronal dopamine oxidation. Acta Neurol Scand [Suppl] 95:81–85

Parkinson Study Group (1989a) Effect of deprenyl on the progession of disability in early Parkinson's disease. N Engl J Med 321:1364–1371

Parkinson Study Group (1989b) DATATOP: a multicenter controlled clinical trial in early Parkinson's disease. Arch Neurol 46:1052–1060

Ramsay RR, Singer TP (1986) Energy-dependent uptake of N-methyl-4-phenyl-pyridinium, the neurotoxic metabolite of 1-methyl-4-phenyl-1,2,3,6-tetrahydropyridine, by mitochondria. J Biol Chem 261:7585–7587

Ramsay RR, Salach JI, Singer TP (1986) Uptake of the neurotoxin 1-methyl-4-phenylpyridine (MPP^+) by mitochondria and its relation to the inhibition of the mitochondrial oxidation of NAD^+-linked substrates by MPP^+. Biochem Biophys Res Commun 134:743–748

Riederer P, Youdim MBH (1986) Monoamine oxidase activity and monoamine metabolism in brains of parkinsonian patients treated with l-deprenyl. J Neurochem 46:1356–1365

Salach JI, Singer TP, Castagnoli N Jr, Trevor A (1984) Oxidation of the neurotoxic amine 1-methyl-4-phenyl-1,2,3,6-tetrahydropyridine (MPTP) by monoamine oxidases A and B and suicide inactivation of the enzymes by MPTP. Biochem Biophys Res Commun 125:831–835

Schulz R, Antonin K-H, Hiffmann E, Jedrychowski M, Nilsson E, Schick C, Bieck PR (1989) Tyramine kinetics and pressor sensitivity during monoamine oxidase inhibition by selegiline. Clin Pharmacol Ther 46:528–536

Sonsalla PK, Heikkila RE (1986) The influence of dose and dosing interval on MPTP-induced dopaminergic neurotoxicity in mice. Eur J Pharmacol 129:339–345

Sundstrom E, Stromberg I, Tsutsumi T, Olson L, Jonsson G (1987) Studies on the effect of 1-methyl-4-phenyl-1,2,3,6-tetrahydropyridine (MPTP) on central catecholamine neurons in C57BL/6 mice. Comparison with three other strains of mice. Brain Res 405:26–38

Tetrud JW, Langston JW (1989) The effect of deprenyl (selegiline) on the natural history of Parkinson's disease. Science 245:519–522

Vyas I, Heikkila RE, Nicklas WJ (1986) Studies on the neurotoxicity of 1-methyl-4-phenyl-1,2,3,6-tetrahydropyridine: inhibition of NAD-linked substrate oxidation by its metabolite, 1-methyl-4-phenylpyridinium. J Neurochem 46: 1501–1507

Youngster SK, Sonsalla PK, Heikkila RE (1987) Evaluation of the biological activity of several analogs of the dopaminergic neurotoxin 1-methyl-4-phenyl-1,2,3,6-tetrahydropyridine. J Neurochem 48:929–934

Youngster SK, Sonsalla PK, Sieber BA, Heikkila RE (1989a) Structure activity of the mechanism of 1-methyl-4-phenyl-1,2,3,6-tetrahydropyridine (MPTP) induced neurotoxicity I. Evaluation of the biological activity of MPTP analogs. J Pharmacol Exp Ther 249:820–828

Youngster SK, Nicklas WJ, Heikkila RE (1989b) Structure-activity study of the mechanism of 1-methyl-4-phenyl-1,2,3,6-tetrahydropyridine (MPTP)-induced neurotoxicity II. Evaluation of the biological activity of the pyridinium metabolites formed from the monoamine oxidase-catalyzed oxidation of MPTP analogs. J Pharmacol Exp Ther 249:829–835

Authors' address: Dr. R. E. Heikkila, Department of Neurology, UMDNJ-Robert Wood Johnson Medical School, Piscatway, NJ 08854, U.S.A.

J Neural Transm (1990) [Suppl] 32: 229–238

Monoamine oxidase and oxidative stress at dopaminergic synapses

G. Cohen

Department of Neurology and Neurobiology Center, Mount Sinai School of Medicine, New York, U.S.A.

Summary. Increased oxidation of dopamine by monoamine oxidase (MAO) in the striatum is associated with an oxidant stress, expressed as a rise in the level of oxidized glutathione. Oxidation of glutathione is suppressed by MAO inhibitors, such as deprenyl and clorgyline. These observations relate to Parkinson's disease and to the clinical trial of deprenyl as an agent that may retard progression of the disease.

Introduction

The oxidative deamination of catecholamines by monoamine oxidase (MAO) produces hydrogen peroxide, which is a known cellular toxin. Therefore, the turnover of dopamine (DA) by MAO can, in theory, be associated with an oxidative stress (Cohen, 1983). The purpose of these studies was to assess one aspect of oxidant stress, namely, the conversion of reduced glutathione (GSH) to glutathione disulfide (GSSG) during increased synthesis and turnover of DA (Equations 1–3).

$$\text{L-DOPA} \xrightarrow{\text{DOPA-decarboxylase}} \text{DA} + CO_2 \quad (1)$$

$$\text{DA} + O_2 + H_2O \xrightarrow{\text{MAO}} \text{3,4-dihydroxyphenylacetaldehyde} + H_2O_2 + NH_3 \quad (2)$$

$$H_2O_2 + 2\text{GSH} \xrightarrow{\text{GSH peroxidase}} \text{GSSG} + H_2O \quad (3)$$

$$\text{GSSG} + \text{NADPH} + H^+ \xrightarrow{\text{GSSG reductase}} 2\text{GSH} + \text{NADP}^+ \quad (4)$$

GSH is broadly distributed in relatively high concentration (0.5–10 mM) in body tissues and serves, in part, to set the redox status of tissues. Levels in rat brain range from 1.4 to 3.3 mM (Slivka et al., 1987). GSSG, on the other hand, is maintained at very low levels and constitutes only 0.5–1.2% of the total glutathione in rat, monkey, and human brain (Slivka et al., 1987). Low GSSG levels are also the rule in peripheral organs, such as liver (Adams et al., 1983; Akerboom and Sies, 1981). The efficiency of GSSG reductase, which utilizes NADPH as cofactor (Equation 4), is responsible for maintaining low GSSG levels in cells, even during exposure to peroxides. However, an elevation in the steady-state level of GSSG can reflect a change in redox status. When elevations in GSSG do occur in various organs and tissues, generally during exposure to redox cycling cell toxins or other sources of H_2O_2, biologic sequelae follow: In liver, for example, alterations in calcium homeostasis, proteolytic activity, enzymatic activity (SH-dependent enzymes) and cell structure (blebs) have been observed. The nervous system, however, has not been well studied.

It is known that surviving nigrostriatal neurons in Parkinson patients exhibit increased turnover of DA. This is evident from the increase in the HVA/DA ratio at autopsy (e.g., Hornykiewicz and Kish, 1986). Similar observations have been made in rats with unilateral lesions of the nigrostriatal tract, particularly when the lesions encompass 80% or more of the DA neurons (Altar et al., 1987; Hefti et al., 1980). Since increased DA turnover is accompanied by increased generation of H_2O_2, Parkinson's disease may be associated with a peroxidative stress. When DA turnover is elevated, there is a greater risk of peroxidative damage at both pre- and post-synaptic sites. In addition, accumulation of GSSG can lead to changes in activities of various SH-dependent enzymes and changes in structural proteins via mixed disulfide formation. The latter may be responsible for subtle changes in neuronal function.

The concept of a peroxidative stress in Parkinson's disease, emanating from increased turnover of DA by MAO, was a prime consideration in the recent DATATOP trial of deprenyl as a treatment to delay the progression of Parkinson's disease (Parkinson Study Group, 1989a).

Methods

For in vitro experiments, tissue slices and synaptosomal preparation (P2 pellet) were prepared from the striata of male Sprague-Dawley rats (Spina and Cohen, 1988). The slices and synaptosomes were incubated in a Krebs-Ringer-phosphate buffer (pH 7.4) which contained glucose (5.5 mM), ascorbic acid (0.1 mM), and diethylenetriaminepentaacetic acid (0.1 mM, metal chelator). Drugs were added at the concentrations indicated in the text.

For in vivo experiments, male Swiss-Webster mice were used. Mice received i.p. injections of either reserpine (10 mg/kg) or haloperidol (1 mg/kg). Some animals had been pretreated (18 hrs) with MAO inhibitors at the doses indicated in the text.

GSSG was measured by a modification (Slivka et al., 1987) of the enzymatic recycling procedure of Tietze (1969). The method utilizes GSSG reductase and it is a particularly sensitive assay for GSSG. GSH can also be measured by this method; however, GSH was first completely removed by reaction with N-ethylmaleimide, as described by Adams et al. (1983). Dopamine, DOPAC and HVA were measured by HPLC with electrochemical detection.

Results

The studies described here summarize previously reported experiments (Spina and Cohen, 1988, 1989; Cohen and Spina, 1989), with some additional information.

1. In vitro studies: L-DOPA and reserpine

Experiments were carried out with both striatal slices and synaptosomal preparations from rat striatum (Spina and Cohen, 1988) in order to assess the effects of L-DOPA, with and without reserpine, on tissue levels of GSSG. Since endogenous GSSG levels are maintained at very low levels due to the effectiveness of GSSG reductase, we first examined the effect of added peroxide to see if CNS tissue was capable of responding with an elevation in GSSG. t-Butyl hydroperoxide, a known substrate for glutathione peroxidase, was used. The use of an organic hydroperoxide, rather than reagent H_2O_2, avoided interference from catalase within erythrocytes in the tissue slices.

The results presented in Table 1 show that GSSG levels rose when slices of rat striatum were incubated with increasing concentrations of t-butyl hydroperoxide. Glucose was present to provide for metabolic production of NADPH to sustain GSSG reductase activity. Table 1 shows that the steady

Table 1. A model experiment: incubation of slices of rat striatum with t-butyl hydroperoxide

t-BuOOH added (μM)	GSSG in tissue (μM, pmoles/mg tissue)	Rise in GSSG
0 (control)	6.0	–
10	10.9	4.9
50	23.6	17.6
100	38.5	32.5

Rat striatal slices (1.0 × 0.4 × 0.4 mm; 25 mg) were incubated at 37 °C for 30 min in 2.0 ml Krebs-Ringer phosphate buffer, pH 7.4, in the presence of glucose (1 mg/ml) and the indicated concentrations of t-butyl hydroperoxide (t-BuOOH). Assays for GSSG were conducted after 30 min

state level of GSSG in brain tissue can become elevated despite a very active GSSG reductase system. The elevation in GSSG was proportional to the concentration of added t-butyl hydroperoxide. In additional experiments, H_2O_2 was used and tissue catalase was inhibited with sodium azide (50 μM). Addition of 100 μM H_2O_2 resulted in an elevation of GSSG from 1.0 ± 0.2 μM (control) to 21.8 ± 0.3 μM (experimental) at 30 min., confirming that protection by GSSG reductase could be overwhelmed by exposure to peroxides.

In subsequent experiments, synaptosomes were used because they represent a relative purification of nerve terminals and, moreover, they can be easily pipetted into replicate aliquots. Reserpine (10 μM) was added in most experiments to prevent vesicular storage of DA. Incubation of striatal synaptosomes with L-DOPA and reserpine resulted in a rise in GSSG (within the tissue pellet), which was concentration-dependent in the range of 0.04–1.0 mM L-DOPA. With 1.0 mM L-DOPA, GSSG was elevated on the average by 38.0 ± 4.5% compared to control. In general, reserpine amplified the increase in GSSG levels when L-DOPA was present at lower concentrations (0.04 mM), but the difference was small (less than 10%) and not statistically significant with 1.0 mM L-DOPA.

The effects of MAO inhibitors and DOPA decarboxylase inhibitors were tested. In the presence of 10 μM clorgyline, the rise in GSSG was suppressed by 88–92% (Table 2). Clorgyline also suppressed the formation of DOPAC and elevated the level of DA. Similar results were obtained when 10 μM pargyline was used in place of 10 μM clorgyline. The concentrations of MAO inhibitors employed in these in vitro experiments do not produce selective inhibition of MAO-A or MAO-B. The powerful inhibitory action of

Table 2. GSSG levels after treatment of synaptosomes with reserpine (10 μM) plus L-DOPA (1 mM), with and without clorgyline (10 μM)

Expt	Control (A)	Res/L-DOPA (B)	Clorgyline + Res/L-DOPA (C)	Increase in GSSG (B-A)	(C-A)
	(pmoles GSSG/mg striatal tissue ± SEM)			(pmoles/mg tissue)	
1	15.6 ± 0.2	22.7 ± 0.2	16.4 ± 0.4	7.1	0.8[a]
2	20.2 ± 1.2	27.0 ± 0.5	20.3 ± 0.2	6.8	0.1[a]
3	21.9 ± 0.8	31.8 ± 1.2	22.9 ± 0.3	9.9	1.0[a]
			Mean ± SEM	7.9 ± 1.0	0.6 ± 0.3[a]

A resuspended P2 pellet, prepared from rat striatum, was incubated at 37 °C in Krebs-phosphate buffer (pH 7.4) containing 5.5 mM glucose (n = 3 samples/group). The rise in GSSG after incubation with L-DOPA and reserpine (column B compared to column A) was significant ($p < 0.01$) in each experiment.

[a] $p < 0.01$ compared (B-A)

MAO inhibitors excludes autoxidation of L-DOPA as a prominent source of H_2O_2. NSD-1055 and carbidopa, two inhibitors of DOPA-decarboxylase, were also examined. The decarboxylase inhibitors suppressed the rise in GSSG by 50–60% ($p < 0.01$).

The resuspended P2 pellet contains MAO that is associated with both intrasynaptosomal mitochondria and free mitochondria in the medium. In order to distinguish between the two sites for H_2O_2 production, catalase (2 μg/ml) was added to the medium in order to intercept the H_2O_2 produced in the medium by free mitochondria. Catalase suppressed the rise in GSSG from 9.1 ± 0.7 pmoles/mg to 4.6 ± 0.4 pmoles/mg in 3 experiments. Therefore, approximately 50% of the GSSG was derived from the penetration of externally-generated H_2O_2 into the synaptosomes.

These experiments show that the synthesis of DA from L-DOPA in vitro, with subsequent catabolism of DA by MAO, can evoke a significant rise in the level of GSSG, which reflects the oxidant stress associated with MAO activity.

2. *In vivo studies: effects of reserpine or haloperidol*

Administration of reserpine and haloperidol represent two ways to experimentally increase the turnover of striatal DA by MAO in experimental animals. Reserpine produces an "artificial" increase in DA turnover by blocking the uptake and storage of DA by synaptic vesicles. Reserpine exposes primarily presynaptic DA to metabolic degradation by MAO. Haloperidol provides a more physiologic approach since it depends upon an increased firing rate of nigrostriatal neurons, with increased flux and metabolism of DA. Increased secretion of DA from nigrostriatal nerve terminals in mice treated with haloperidol would be expected to place relatively greater emphasis on post-synaptic metabolism of DA by MAO.

Reserpine (10 mg/kg) was administered i.p. and, two hours later, the striatum was assayed for GSSG. L-DOPA was not injected. In 8 independent experiments with 3–4 mice per group, GSSG in the striatum rose from 6.8 ± 0.9 μM (pooled mean $\pm$ SEM) to 12.9 ± 1.1 μM, constituting a doubling of the GSSG within the striatum ($p < 0.001$). Because reserpine interferes with the storage of DA by synaptic vesicles in vivo, levels of DA fall and acid metabolites rise, reflecting increased activity of MAO. In these experiments, the mean level of DA in the striatum fell from 50.9 μM to 1.4 μM ($n = 6$), while the steady-state levels of the acidic metabolites DOPAC and HVA doubled from 4.8 μM to 8.8 μM (DOPAC) and from 5.9 to 12.3 μM (HVA). These results confirm the expected increase in MAO activity. The mean rise in GSSG (6.1 μM) is readily accommodated by the H_2O_2 that would be generated during the observed loss of DA (49.5 μM). Moreover, the observed loss in DA fails to account for metabolism of newly synthesized DA during exposure to reserpine.

If the oxidation of GSH to GSSG is the result of H_2O_2 generated in vivo by MAO, then MAO inhibitors should suppress the rise in GSSG. The results in Table 3 show experiments with clorgyline (2.5 mg/kg). This dose of clorgyline produces a selective inhibition of MAO-A in the CNS of Swiss-Webster mice (Kindt et al., 1988). In each of 3 experiments, reserpine elevated the mean level of GSSG in the striatum ($p < 0.001$). Clorgyline, given 18 hrs earlier, suppressed the rise by 93.1% ± 6.3%. Hence, the oxidant stress derives in large measure from the activity of MAO A, which is the form present within DA nerve terminals.

The striatum is richly innervated by DA nerve terminals, whereas the cortex receives a much sparser innervation by catecholamine terminals: The level of DA in the striatum is in the range of 10 μg/g, whereas the level of catecholamines (DA plus norepinephrine) in the cortex is less that 1 μg/g, reflecting the relative difference in innervation by catecholamine terminals. If the rise in GSSG reflects the innervation by monoamine nerve terminals, a much larger effect is expected in the striatum, which is more densely innervated than the cortex. In 3 independent experiments comparing striatum to cortex, the mean rise in GSSG was 5.0 ± 0.3 μM in the striatum, but − 0.3 ± 0.6 μM in the cortex (Spina and Cohen, 1989). Therefore, the rise in GSSG reflects the degree of innervation by catecholamine nerve terminals.

Whereas reserpine induces a wasteful metabolism of DA, haloperidol, on the other hand, increases the firing rate of nigrostriatal neurons and simultaneously promotes the activity of tyrosine hydroxylase. Experiments with haloperidol provide a model for the increased turnover of DA seen in Parkinson's disease.

When mice were injected with haloperidol (1 mg/kg), striatal levels of GSSG rose (Cohen and Spina, 1989). In 8 independent experiments

Table 3. Levels of oxidized glutathione (GSSG) in the striatum after treatment of mice with reserpine alone or reserpine plus clorgyline

Expt	Control	Reserpine	Reserpine + clorgyline	% Suppression by clorgyline
	(pmoles GSSG/mg tissue; μM)			
1	12.8 ± 0.2	17.1 ± 0.0^a	13.5 ± 0.3^b	83.7%
2	6.1 ± 0.6	12.4 ± 0.1^a	6.7 ± 0.0^b	90.5%
3	5.1 ± 0.4	10.4 ± 0.2^a	4.9 ± 0.3^b	105.1%

Swiss-Webster mice received i.p. injections of reserpine (10 mg/kg). Some animals had been treated 18 hrs earlier with clorgyline (2.5 mg/kg); this dose produces a selective inhibition of MAO-A in brain (Kindt et al., 1988). GSSG was studied at two hours after reserpine. Values are the mean ± SEM for n = 3–4 mice per group. [a] $p < 0.001$ compared to control; [b] $p < 0.001$ compared to reserpine alone

(n = 3–4 mice/group), mean levels of GSSG rose from 6.6 ± 1.0 μM for controls to 19.1 ± 1.9 μM for haloperidol-injected mice at one hour ($p < 0.001$). The mean increase in GSSG after haloperidol (12.5 μM) was double that seen with reserpine (6.1 μM). Levels of deaminated DA metabolites were also more greatly elevated after haloperidol compared to reserpine, while levels of DA did not fall due to replacement of DA via biosynthesis from available tyrosine.

If the oxidation of GSH to GSSG after injection of haloperidol derives from the H_2O_2 generated by MAO, the formation of GSSG should be suppressed by inhibition of MAO. The results in Table 4 (Cohen and Spina, 1989) are from 3 experiments with deprenyl (2.5 mg/kg). This dose of deprenyl produces a selective inhibition of MAO-B in the CNS of Swiss-Webster mice (Kindt et al., 1988). In each experiment, haloperidol elevated the mean level of GSSG in the striatum ($p < 0.001$). Deprenyl, given 18 hrs earlier, suppressed the rise by 71.0% ± 4.9%. The results indicate that the rise in GSSG is mediated by formation of H_2O_2 by MAO, with a major contribution by MAO-B.

In separate experiments with rats, unilateral lesions, of the substantia nigra were made by stereotaxic injection of 6-hydroxydopamine. The control (left) striatum was compared to the denervated (right) striatum. Results are shown in Table 5. The extent of the lesion (−90.5%) was verified by measurement of DA. The rise in GSSG induced by haloperidol (1 mg/kg) was significantly suppressed on the lesioned side (−88.7%). Therefore, the participation of DA neurons in the response to haloperidol is confirmed.

In recent experiments (unpublished), deprenyl and clorgyline were compared in the reserpine and haloperidol models of increased DA turnover. Although both MAO inhibitors suppressed the rise in GSSG in each

Table 4. Levels of oxidized glutathione (GSSG) in the striatum after treatment of mice with haloperidol alone or haloperidol plus deprenyl

Expt	Control	Haloperidol	Deprenyl + Haloperidol	% Suppression by deprenyl
	(pmoles GSSG/mg tissue; μM)			
1	4.5 ± 0.4	17.3 ± 0.8[a]	7.3 ± 0.5[b]	78.2%
2	5.8 ± 0.1	17.5 ± 0.5[a]	8.7 ± 0.2[b]	75.2%
3	8.4 ± 1.8	23.0 ± 0.9[a]	13.9 ± 1.8[b]	62.4%

Swiss-Webster mice received i.p. injections of haloperidol (1 mg/kg). Some animals had been treated 18 hrs earlier with deprenyl (2.5 mg/kg); this dose produces a selective inhibition of MAO-B in brain (Kindt et al., 1988). GSSG was studied at one hour after haloperidol. Values are the mean ± SEM for n = 3 mice per group. [a] $p < 0.001$ compared to control; [b] $p < 0.001$ compared to haloperidol alone

Table 5. Levels of oxidized glutathione (GSSG) in the striatum of rats after treatment with haloperidol: effect of unilateral right nigrotomy

Tissue	Dopamine (pmol/mg)	GSSG (pmole/mg)	Increase in GSSG (pmole/mg)
Left striatum	42.0 ± 2.0	9.7 ± 0.3	7.1
Right striatum	4.0 ± 1.3[a] (−90.5%)	3.4 ± 0.9[a]	0.8 (−88.7%)

Sprague-Dawley rats were lesioned by stereotaxic injection of 6-OHDA into the right substantia nigra, and allowed to recover for one week. Rats received injections of haloperidol (1 mg/kg, i.p.). GSSG and DA were measured in the control (left) and denervated (right) striatum at one hour. Values are the mean ± SEM for n = 9 rats. Unlesioned rats, used as controls, showed a mean GSSG level of 2.6 pmoles/mg (n = 3). The increase in GSSG is the observed mean value minus the control value in unlesioned rats. [a] $p < 0.01$ compared to the unlesioned side

experimental model, their relative efficacy was reversed in the two models. Whereas clorgyline was more effective in the reserpine experiments, deprenyl was more effective in the haloperidol model. Most likely this reflects the predominant role of presynaptic deamination (MAO-A) in the reserpine model, with a shift to greater dependence on post-synaptic activity (MAO-B plus MAO-A) in the haloperidol model. In addition, both forms of MAO may contribute to GSSG formation at some anatomical sites (e.g., glia).

Discussion

The main conclusion is that a significant peroxidative stress is associated with increased turnover of DA. The rise in GSSG signifies a shift in redox state. The question arises: How great is the redox shift within DA nerve terminals and in the immediate post-synaptic environment? And how much of an amplification does it represent over that associated with the *normal* turnover of DA? A broadly distributed increase in the average GSSG level in the striatum might be readily tolerated; however, a large and selective change at DA synapses would be expected to lead to biologically relevant sequelae. Control studies showed that the normal turnover of DA is *not* associated with detectable increases in MAO-generated GSSG: neither deprenyl nor clorgyline had any effect on endogenous GSSG levels in control mice. Therefore, formation of H_2O_2 during normal DA turnover is tolerated, but places the cells close to a "trigger zone" where increased turnover can overcome repair by GSSG reductase. When DA turnover is increased several-fold, the steady state levels of GSSG rise. The major protective effects of MAO inhibitors (Tables 3 and 4) indicate that the rise in GSSG must be relatively large within DA terminals and in the immediate post-synaptic environment.

The key to making the observations described here lay in the very low levels of GSSG normally found in brain. If it were not for these low endogenous levels of GSSG, the additional GSSG made by enhanced metabolic activity at DA synapses would not have been readily visible. It is estimated that DA neurons contribute only 1% or less to the mass of the striatum. Therefore, the rise in the steady state level of GSSG in the striatum reflects a contribution from a markedly limited mass of tissue.

When DA turnover is elevated, as in Parkinson's disease and, perhaps, during the natural aging process as well, there is a greater risk of peroxidative damage at both pre- and post-synaptic sites in the vicinity of DA neurons. Direct damage by H_2O_2 is possible. In addition, mixed disulfide formation with various enzymes, structural proteins, and receptors may alter neuronal function. In Parkinson's disease, it is even possible that some of these events can be exacerbated by treatment with L-DOPA. Therefore, treatment with L-DOPA needs to be looked on with caution, even though no direct evidence exists for damage mediated by L-DOPA. Some of the motor anomalies seen during treatment with L-DOPA may contain components related to the interaction of GSSG with enzymes and structural proteins.

The apparent success of the DATATOP trial of deprenyl alone (no L-DOPA) in delaying the progression of early Parkinson's disease (Parkinson Study Group, 1989b), as well as prior observations by others on the beneficial effects of combining deprenyl with L-DOPA (Birkmayer et al., 1985), are supported by the results reported here. The experimental data show that an oxidative stress does exist when DA turnover is increased and that administration of deprenyl to rodents can prevent a significant portion of the build-up in GSSG within the striatum. Therefore, treatment of Parkinson patients with deprenyl may diminish the oxidative stress associated with increased turnover of endogenous DA by surviving nigrostriatal neurons, as well as some of the complications of L-DOPA therapy.

References

Adams JD, Lauterberg EH, Mitchell JR (1983) Plasma glutathione and glutathione disulfide in the rat: regulation and response to oxidative stress. J Pharmacol Exp Ther 227:749–754.

Akerboom TPM, Sies H (1981) Assay of glutathione, glutathione disulfide, and glutathione mixed disulfides in biological samples. Meth Enzymol 77:373–382

Altar CA, Marien MR, Marshall JF (1987) Time course of adaptations in dopamine biosynthesis, metabolism, and release following nigrostriatal lesions: implications for behavioral recovery from brain injury. J Neurochem 48:390–399.

Birkmayer W, Knoll J, Riederer P, Youdim M, Hars V, Marton J (1985) Increased life expectancy resulting from the addition of L-deprenyl to Madopar therapy in Parkinson's disease. J Neural Transm 64:113–127

Cohen G (1983) The pathobiology of Parkinson's disease: biochemical aspects of dopamine neuron senescence. J Neural Transm [Suppl] 19:89–103
Cohen G, Spina MB (1989) Deprenyl suppresses the oxidant stress associated with increased dopamine turnover. Ann Neurol 26:689–690
Hefti F, Melamed E, Wurtman RJ (1980) Partial lesions of the dopaminergic nigrostriatal system: biochemical characterization. Brain Res 195:123–137
Hornykiewicz O, Kish SJ (1986) Biochemical pathophysiology of Parkinson's disease. In: Yahr MD, Bergmann KJ (eds) Parkinson's disease. Raven Press, New York (Adv Neurol 45:19–34)
Kindt MV, Youngster SK, Sonsalla PK, Duvoisin R, Heikkila RE (1988) Role for monoamine oxidase-A (MAO-A) in the bioactivation and nigrostriatal dopaminergic neurotoxicity of the MPTP analog, 2'Me-MPTP. Eur J Pharmacol 146:313–318
Parkinson Study Group (1989a) DATATOP: a multicenter controlled clinical trial in early Parkinson's disease. Arch Neurol 46:1052–1060
Parkinson Study Group (1989b) Effect of deprenyl on the progression of disability in early Parkinson's disease. N Engl J Med 321:1364–1371
Slivka A, Spina MB, Cohen G (1987) Reduced and oxidized glutathione in human and monkey brain. Neurosci Lett 74:112–118
Spina MB, Cohen G (1988) Exposure of striatal synaptosomes to L-dopa elevates levels of oxidized glutathione. J Pharmacol Exp Ther 247:502–507
Spina MB, Cohen G (1989) Dopamine turnover and glutathione oxidation: implications for Parkinson's disease. Proc Natl Acad Sci (USA) 86:1398–1400
Tietze F (1969) Enzymic method for quantitative determination of nanogram amounts of total and oxidized glutathione: applications to mammalian blood and other tissues. Anal Biochem 27:502–522

Author's address: Dr. G. Cohen, Box 1137, Mount Sinai School of Medicine, New York, NY 10029, U.S.A.

J Neural Transm (1990) [Suppl] 32: 239–248

The role of monoamine oxidase, iron-melanin interaction, and intracellular calcium in Parkinson's disease

M. B. H. Youdim[1], **D. Ben-Shachar**[1], and **P. Riederer**[2]

[1] Technion – Israel Institute of Technology, Faculty of Medicine, Department of Pharmacology, Haifa, Israel
[2] Department of Psychiatry, University of Würzburg, Würzburg, Federal Republic of Germany

Summary. Recent evidence suggests that iron accumulates in substantia nigra pars compacta of patients with Parkinson's disease (PD). This finding is compatible with changes in the respiratory chain activity, increase of malondialdehyde concentration (a measure of lipid peroxidation), decrease of enzyme activity of enzymes involved in detoxication of hydrogen peroxide and oxygen radical species, increased MAO-B-activity in this brain area etc. All these data suggest that oxidative stress may play a certain role in the pathobiochemistry of PD. In addition to the description of the neuroprotective mechanism of the MAO-B-inhibitor L-deprenyl a new aspect focuses the role of the endogenous MAO-B substrates "polyamines" which occur both in neurons and glia. A further aspect of this review deals with the role of calcium as cellular toxin. Although of major importance it is not decided yet whether these biochemical changes are of primary or secondary importance to the pathogenesis of PD.

Iron neurotoxicity and nigro-striatal dopamine neurons

The mechanism of neurotoxicity leading to the degeneration of nigrostriatal dopamine neurons of zona compacta leading to a deficiency of dopamine in Parkinson's disease remains obscure. On numerous occasions the involvement of endogenously or exogenously produced neurotoxins has been implicated in the progression of this disease. Evidence, however, is lacking, even though synthetic neurotoxins such as MPTP (N-methyl-4-phenyl-1,2,3,6-tetrahydropyridine) and 6-hydroxydopamine producing a Parkinsonian syndrome in man and animals have been discovered (Markey et al., 1986).

It is apparent that during the normal process of aging of human brain there is a loss of melanized nigrostriatal neurons. When about 80% of the neurons are lost, symptoms (tremor and rigidity) of Parkinson's disease (PD) appear. The PD is characterized by an accelerated degeneration of pigmented (melanized) dopamine neurons in the pars compacta of the substantia nigra. The neurons project to the striatum where they regulate dopamine dependent motor activity, where they can synthesize, store, release and catabolize dopamine as their neurotransmitter. The characteristic pigmentation of substantia nigra is related to the formation of neuromelanin as a result of polymerization of autooxidative products of dopamine. Dopamine can also be oxidatively metabolized by the enzyme monoamine oxidase (MAO) A and B, present in highest concentration in the basal ganglia. Both of the above dopamine oxidative processes result in the formation of hydrogen peroxide (H_2O_2). In normal circumstances, H_2O_2 is rather inert and does not accumulate in the tissue, due to a highly active presence of catalase, peroxidase and glutathione peroxidase. However, in the brain the former two enzymes are absent and the H_2O_2 detoxification is handled almost solely by glutathione peroxidase, a glutathione requiring enzyme.

The presence of high lipid content and large deposits of iron in neurotransmitter-rich brain regions (such as substantia nigra, globus pallidus and caudate nucleus) would make these brain regions ideal for oxidative stress, resulting from metal induced lipid peroxidation, in the presence of H_2O_2 (Halliwell and Gutteridge, 1986; Cohen, 1988). Although the function of regional high brain iron content is not known, the homeostasis of brain iron is thought to be necessary for normal brain function (Youdim et al., 1989; Youdim, 1990). Free tissue metal (e.g. Fe^{2+} or Cu^{2+}) can initiate the formation of cytotoxic oxygen free radicals (e.g. hydroxyl radical, $^{\cdot}OH$) resulting from its interaction with H_2O_2 (Fenton reaction) and leading to promotion of membrane lipid peroxides. The mobilization of intracellular and mitochondrial calcium by iron is now thought to be directly involved in cellular toxicity in oxidative stress since calcium antagonists (dihydropyridines) can inhibit iron induced lipid peroxidation. Although lipid peroxidative process was originally identified in vitro, its in vivo occurrence can no longer be questioned. There are numerous examples where iron induced lipid peroxidation and calcium cellular toxicity have been implicated in the pathogenesis of systemic disorders (Braughler, 1988; Halliwell, 1989; Hayashi, 1989). Its application to stroke and trauma induced neurodegeneration and nigro-striatal dopamine neuron degeneration in Parkinson's disease has not escaped scrutiny (Youdim et al., 1989; Hayashi, 1989). These findings highlight the recent reports from a number of laboratories in which a highly selective increase of iron and basal lipid peroxidation and decrease of dismutase activity have been noted in the substantia nigra of Parkinsonian brains (Dexter et al., 1989; Riederer et al., 1989; Saggu et al.,

1989). The increase of iron appears to be associated solely within zona compacta of nigra (Jellinger et al., 1990; Sofic et al., 1991). Iron is accumulated mainly in astrocytes, glia, macrophage and in the capillaries structures. The latter findings are compounded by the decreased ability of Parkinsonian brains to scavenge H_2O_2 due to diminished availability of reduced glutathione (GSH) and glutathione peroxidase activity (Cohen, 1988; Youdim et al., 1989; Riederer et al., 1989). The latter is the main pathway of H_2O_2 breakdown in the brain.

Both oxidative deamination by monoamine oxidase (MAO) and autooxidation of dopamine results in excessive generation of H_2O_2. The inability to dispose of H_2O_2 could result in H_2O_2 accumulated cytotoxicity (Cohen et al. 1974) and its interaction with Fe^{2+} may promote the Fenton reaction.

$$Fe^{2+} + H_2O_2 \rightarrow Fe^{3+} + OH^- + {}^{\cdot}OH$$

Iron induced oxidative stress and lipid peroxidation can proceed optimally with either Fe^{2+} or Fe^{3+} provided mechanisms exist to facilitate the interconversion of iron between its redox states. Fe^{3+} formed can be converted to Fe^{2+} in the presence of endogenous chelators such as ADP, melanin and ferritin. Such an interaction makes Fe^{3+} even more reactive as an initiator of ${}^{\cdot}OH$. ADP is a weak iron chelator and of limited capacity to chelate iron. If, however, we are to implicate iron in the neurodegeneration of substantia nigra it is in this brain region we must look for an endogenous chelator within the substantia nigra which would have a high affinity for iron.

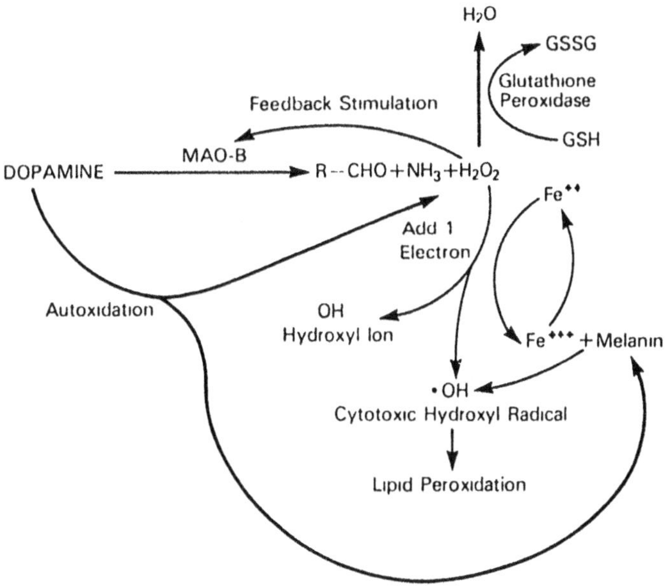

Fig. 1. The pathway of dopamine deamination and autooxidation leading to the formation of melanin with significant generation of H_2O_2. Accumulation of H_2O_2 as a result of decreased GSH-glutathione can lead to increased formation of oxygen free radicals via interaction of H_2O_2 with Fe^{2+} (Fenton Reaction)

The substantia nigra is unique among the brain regions for having excessive amounts of neuromelanin formed from auto-oxidation of dopamine. It is the pigmented melanin containing dopamine neurons of pars compacta which are thought to degenerate in Parkinson's disease (Hirsch et al., 1988). Although melanin has the capacity to bind high amounts of metals, in normal circumstances it is thought to participate in scavenging metal induced free radicals, wherever it is formed. The best example is the melanin in the skin which acts as protection against ultraviolet light induced damage to the skin. Recently we have demonstrated that melanin has high affinity binding sites for Fe^{3+} with K_D values of 1.3 and 200 nM respectively. In this condition melanin-iron interaction results in potentiation of iron induced lipid peroxidation in the presence of rat brain cortical synaptosomal preparation (Ben-Shachar and Youdim, 1990; Ben-Shachar et al., 1990) (Fig. 2). The binding of iron to melanin and lipid peroxidation has been attributed to the ability of melanin to convert Fe^{3+} to Fe^{2+} in the presence of H_2O_2 and drive a Fenton like reaction with liberation of cytotoxic hydroxyl radical ($^{\cdot}OH$). This process is thought to take place with Fe^{3+} and not Fe^{2+} (Pilas et al., 1988). The accumulation and alteration in the ratio of Fe^{2+} to Fe^{3+} from 3:1 in control zona compacta of substantia nigra to 1:1 in Parkinsonian substantia nigra is compatible with optimal conditions necessary for induction of lipid peroxidation process (Braughler et al., 1986).

$$Fe^{3+} + H_2O_2 + \text{melanin} \rightarrow Fe^{2+} + {}^{\cdot}OH + \text{melanin}$$

These findings have led us to postulate iron-melanin induced oxidative stress as a possible phenomenon in the etiology of Parkinson's disease

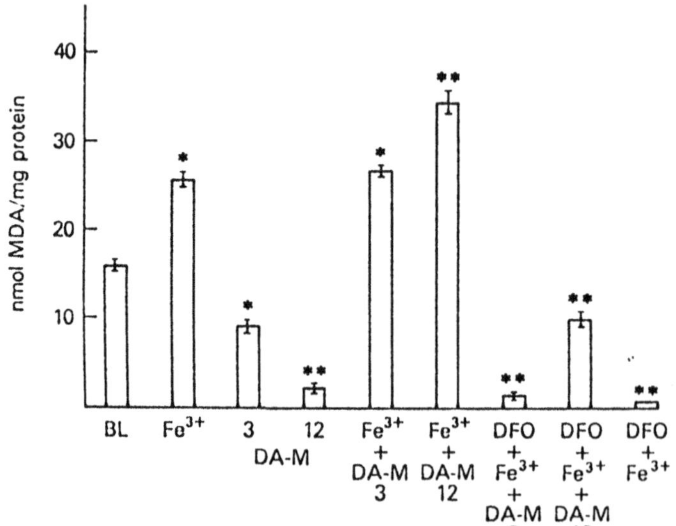

Fig. 2. Lipid peroxidation as induced by iron in the presence and absence of dopamine melanin (DA-M) and its inhibition by desferrioxamine (DFO), Fe^{3+} (100 μM); DA-M (3 or 12 μg/ml); DFO (100 μM)

(Table 1), without involving an endogenously or exogenously derived neurotoxin (Youdim et al., 1989). Accumulation of free tissue iron could act as the neurotoxic agent in Parkinson's disease, similar to what has been reported for the heart in cases of iron overload. However, a number of important questions remain unanswered, namely: (a) how and why does iron accumulate in substantia nigra and why is it further increased in Parkinsonian substantia nigra? (b) what is the mechanism of iron transport into these brain regions low in transferrin receptor? (c) why is it that the regions (globus pallidus, substantia nigra, caudate nucleus and dentate gyrus) with high iron content have extremely low transferrin and transferrin receptor sites? and (d) is iron sequestered and transported from one brain site to another (Table 2). Preliminary studies using the MPTP induced model of Parkinsonism in mice and monkeys show that the degeneration of substantia nigra does not lead to increase of free tissue iron in this region. The possibility of an infection-induced break down of blood brain barrier in this region, which would allow a greater uptake of iron, cannot be ignored, nor the possible genetic basis of the disease. A reduced transport of iron out of the brain may also be contributory since the turnover of iron in the brain is extremely slow (Ben-Shachar et al., 1986).

Table 1. Evidence for oxidative stress in Parkinsonian substantia nigra

1. Increase of iron in pars compacta of SN
2. Reduction of glutathione (GSH)
3. Increase of MAO-B activity with aging
4. Increased basal lipid peroxidation
5. Decreased mitochondrial complex I and III activities
6. Altered Ca^{2+} homeostasis, resulting from its diminished binding
7. Interaction of iron with melanin

Table 2. Possible mechanisms for increase of iron* in Parkinsonian substantia nigra pars compacta

1. Alteration in blood brain barrier at the level of substantia nigra
2. Increased uptake and sequestering of iron
3. Decreased efflux of iron from the brain
4. Genetics
5. Infection
6. Neurodegeneration
7. Neurotoxic agent

* In normal circumstances iron does not cross the blood brain barrier even though transferrin receptors are present on capillary endothelial cells

Intracellular calcium homeostasis

The evidence for the condition of oxidative stress in substantia nigra of Parkinsonian brains has been accumulating. As noted in Table 1, the increases of SN iron and basal lipid peroxidation on the one hand and the demise of mitochondrial electron transport system on the other, resulting in decreased activities of complex I and III (Lestienne et al., 1990), suggest that the process of mitochondrial toxicity in SN may involve other events or neurotoxicity. It is now accepted that alteration in homeostasis of intracellular calcium, especially that in the mitochondria and endoplasmic reticulum, is closely linked with altered cell physiology and death (Fig. 3) (Boobis et al., 1989; Orrenius et al., 1989). Furthermore it has been demonstrated that numerous drugs as well as iron (Braughler, 1988; Youdim et al., 1991) which are known to induce cell toxicity and death are closely associated with altered intracellular calcium due to opening of calcium channels. In this regard the relationship of cellular calcium binding requires examination and indeed several recent studies do indicate that the calcium binding protein calbindin is selectively decreased in substantia nigra of Parkinsonian brains (Iacopino and Chirstako, 1990). This finding could explain both the demise of mitochondrial function as well as neuronal degeneration. The crucial point in this discussion is whether these changes and cell death are the consequence of the altered SN iron metabolism noted. Indeed, recently we (Youdim et al., 1991a) have demonstrated that Fe^{2+} or Fe^{3+} causes a time and concentration dependent opening of dihydropyridine sensitive calcium channels in rat cortical synaptosomes, resulting in parallel increased uptake of $^{45}Ca^{2+}$ and lipid peroxidation as measured by malondialdehyde formation in intact rat cortical synaptosomal preparation. Both the calcium

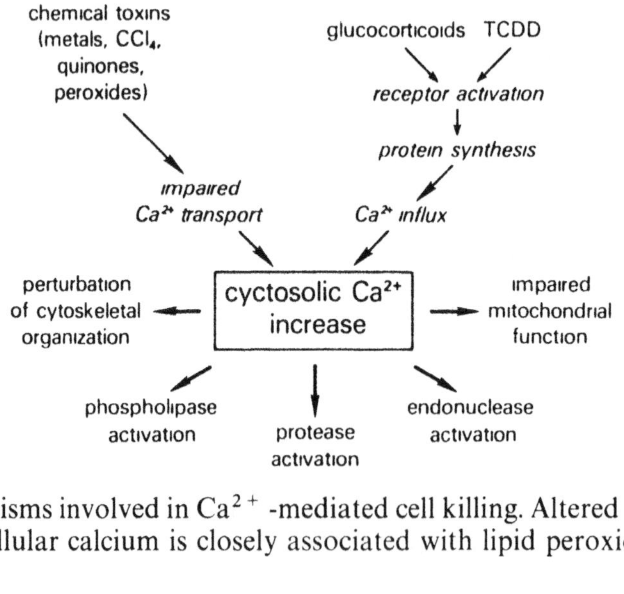

Fig. 3. Mechanisms involved in Ca^{2+} -mediated cell killing. Altered homeostatis of intracellular calcium is closely associated with lipid peroxidation

uptake and lipid peroxidation can be inhibited by dihydropyridines (nifedipine) and iron chelating agents (desferrioxamine and U47500A). The ability of iron salts to induce opening of calcium channels resulting in alteration of intracellular calcium would support the recent hypothesis that iron could be the agent to induce the neurotoxic event in the pars compacta, where it is selectively deposited. This finding also lends support to the link between the siderosis of substantia nigra (Youdim et al., 1989) and oxidative stress.

Mechanism of L-deprenyl neuroprotective activity: the possible involvement of polyamines

Long term l-deprenyl (selective MAO-B inhibitor) treatment of Parkinson's disease is thought to retard the neurodegeneration of dopamine neurons in Parkinson's disease (Birkmayer et al., 1983, 1985; Tetrud and Langston, 1989; Datatop, 1989). However, these results can also be explained by symptomatic action of this drug on increasing striatal dopamine in Parkinsonian brains (Riederer and Youdim, 1986).

On numerous occasions it has been postulated that the mechanism of action of l-deprenyl is associated with the inhibition of striatal MAO-B (Riederer and Youdim, 1986) and prevention of H_2O_2 generation and formation of highly reactive oxygen free radical (e.g. hydroxyl radical) (Birkmayer et al., 1983, 1985; Cohen, 1974, 1988). Indeed recent experiments from our laboratory have shown that if l-deprenyl does have any neuroprotective activity it is not via inhibition of oxygen free radical formation linked to the action of increased SN iron. Neither clorgyline (MAO-A inhibitor) nor l-deprenyl, even at high concentration ($> 10\ \mu M$) were able to inhibit basal and iron H_2O_2 induced lipid peroxidation in vitro. Nor was l-deprenyl able to alter iron induced synaptosomal uptake of $^{45}Ca^{2+}$, a process intimately linked with increased lipid peroxidation of neuronal membrane (Youdim et al., 1991). If l-deprenyl is neuroprotective other actions of this drug must be considered.

Polyamines, such as putrescine, spermine and spermidine are normal constituents of most cells, including neurons and glia (Bolkenius and Seiler, 1986; Seiler, 1989; Canellakis et al., 1989), which are important in cell proliferation and differentiation (Russell, 1980; Heby, 1981; Anon, 1981). Principally, all results support the theory that polyamines are essential for the progression through the cell cycle and lead to cell differentiation. Thus neuronal cells are not excluded since these amines are thought to be also involved in neuronal differentiation, plasticity and viability. The enzyme ornithine decarboxylase is the enzyme responsible for conversion of orthinin to putrescine, which is the *key* regulatory step in the synthesis of all aliphatic polyamines such as spermidine and spermine from putrescine (Russel, 1980; Canellakis et al., 1989).

The main degradating enzyme involved in the oxidative deamination of polyamines is diamine oxidase, an enzyme very low in activity or absent in the mammalian brain (Seiler and Al-Therib, 1974 a, b; Seiler, 1985). Thus the major metabolic pathway in the brain is thought to be N-acetylation by acetyl CoA to N-acetylated derivatives (Seiler, 1985). N-acetylated polyamines which are present in the glia (where MAO-B resides) have a direct strong positive feed back stimulation on ornithine decarboxylase activity for synthesis of polyamine. By this means the high cellular concentrations of polyamines are maintained. In normal circumstances brain cellular concentration of N-acetylated derivatives are relatively low as compared to the polyamines (Bolkenius and Seiler, 1986). The report by Seiler and Al-Therib (1974a, b) demonstrated that N-acetylputrescine is a substrate for MAO. This finding led us to examine the specificity of the latter amine, as well as N-acetylspermidine and N-acetylspermine in relation to their metabolism by MAO-A and MAO-B, using rat brain mitochondrial preparation. Our preliminary results (Youdim, pers. comm.) have clearly indicated that these amines are substrates of MAO-B with relative K_m values close to that for phenylethylamine and benzylamine. This could be one reason why their concentrations are significantly lower than the polyamines in the brain. Furthermore l-deprenyl rather than clorgyline selectively inhibits their oxidation. These data point to N-acetylated polyamines as being an important class of *endogenous* substrates of MAO-B. The question remains as to whether during l-deprenyl monotherapy or in combination with L-dopa, the inhibition of MAO-B in the striatum (Riederer and Youdim, 1986) would result in accumulation of N-acetylated polyamines with consequential positive feed back stimulation of ornithine decarboxylase. With what is known about the physiological role of polyamines, an assessment of their possible involvement as neuroprotective agents cannot be ignored in Parkinson's disease.

References

Anon B (1981) Polyamines as cellular regulators. Med Biol 59 (5–6)

Ben-Shachar D, Riederer P, Youdim MBH (1990) The implications of high affinity binding sites for ^{39}Fe on dopamine-melanin and potentiation of iron induced lipid peroxidation for Parkinson's disease. Eur J Pharmacol (submitted)

Ben-Shachar D, Youdim MBH (1990) Selectivity of melaninized nigrostriatal dopamine neurons to degeneration in Parkinson's disease may depend on iron-melanin interaction. J Neural Transm [Suppl] 29:251–258

Birkmayer W, Knoll J, Riederer P, Hars V, Youdim MBH, Marton J (1985) Improvement of life expectancy due to L-deprenyl addition to Madopar treatment in Parkinson's disease. A long term study. J Neural Transm 64:113–127

Birkmayer W, Knoll J, Riederer P, Youdim MBH (1983) (-)Deprenyl leads to prolongation of L-dopa efficacy in Parkinson's disease. Mol Probl Pharmacopsychiatry 19:170–176

Bolkenius FM, Seiler N (1986) Role of polyamines in the development of gabaergic neurons. Int J Dev Neurosci 4:217–224

Boobis AR, Fawthrop DJ, Davies DS (1989) Mechanism of cell death. TIPS 10:275–280

Braughler JM (1988) Calcium and lipid peroxidation. In: Halliwell B (ed) Oxygen radicals and tissue injury. FASEB, Bethesda, pp 99–106

Braughler JM, Duncan LA, Chase RL (1986) The involvement of iron in lipid peroxidation: importance of ferric to ferrous rat in initiation. J Biol Chem 261:10282–10289

Canellakis ZN, Marsh LIL, Bondy PK (1989) Yale J Biol Med 62:481–491

Cohen G (1988) Oxygen radicals and Parkinson's disease. In: Halliwell B (ed) Oxygen radicals and tissue injury. FASEB, Bethesda, pp 130–138

Cohen G, Heikkila RE, McNamee D (1974) The generation of hydrogen peroxide, superoxide radical and hydroxyl radical by 6-hydroxy-dopamine dialuric acid and related cytotoxic agents. J Biol Chem 249:2447–2459

Dexter DT, Carter CJ, Well FR, Javoy-Agid F, Agid Y, Lee SN, Jenner P, Marsden CD (1989) Basal lipid peroxidation in substantia nigra is increased in Parkinson's disease. J Neurochem 52:381–389

Halliwell B (1989) Oxidants and the central nervous system: some fundamental questions. Acta Neurol Scand 126:35–47

Halliwell B, Gutteridge JM (1986) Oxygen free radicals and iron in relation to biology and medicine: some problems and concepts. Arch Biochem Biophys 246:501–514

Hayashi O (ed) (1989) Oxygen free radicals and medicine. Elsevier, Amsterdam

Heby O (1981) Role of polyamines in the control of cell proliferation and differentiation. Differentiation 19:1–11

Hirsch E, Graybiel AM, Agid YA (1988) Melanized dopaminergic neurons are differentially susceptible to degeneration in Parkinson's disease. Nature 334:345–348

Iacopino AM, Christakos S (1990) Specific reduction of calcium-binding protein (28-kilodalton calbindin-D) gene expression in aging and neurodegenerative diseases. Proc Natl Acad Sci USA 87:4078–4082

Jellinger K, Paulus W, Grundke-Iqbal I, Riederer P, Youdim MBH (1990) Brain iron and ferritin in Parkinson's and Alzheimer's diseases. J Neural Transm 2(4): 327–340

Lestienne P, Nelson J, Riederer P, Jellinger K, Reichmann H (1990) Normal mitochondrial genome in brain from patients with Parkinson's disease and complex I defect. J Neurochem 55:1810–1812

Markey S, Kopin I, Trevor A, Castagnoli N Jr (eds) (1986) MPTP: a neurotoxin producing a Parkinsonian syndrome. Academic Press, London

Orrenius S, McConkey DJ, Bellomo G, Pierluigi N (1989) Role of Ca^{2+} in toxic cell killing. TIPS 10:281–285

Pilas B, Sana T, Kalyanarman B, Swartz HM (1988) The effect of melanin on iron associated decomposition of hydrogen peroxide. Free Rad Biol Med 4: 285–293

Riederer P, Sofic E, Rausch W, Schmidt B, Reynolds GP, Jellinger K, Youdim MBH (1989) Transition metals, ferritin, glutathione and ascorbic acid in Parkinsonian brains. J Neurochem 52:515–521

Riederer P, Youdim MBH (1986) Monoamine oxidase activity and monoamine metabolism in the brains of Parkinsonian patients treated with l-deprenyl. J Neurochem 46:1349–1356

Russel DH (1980) Ornithine decarboxylase as a biological and pharmacological tool. Pharmacology 20:117–126

Saggu H, Cooksey J, Dexter D, Wells FR, Lees A, Jenner P, Marsden CD (1989) A selective increase in particulate superoxide dimutase activity in Parkinsonian substantia nigra. J Neurochem 53:692–698

Seiler N (1985) Acetylpolyamines as substrates of amine oxidases. In: Mondovi B (ed) Structure and function of amine oxidases. CRC Press, Boca Raton, pp 21–37

Seiler N (1989) Metabolism and physiological roles of polyamines. In: Bacharach U, Keimer Y (eds) Physiology of polyamines. CRC Press, Boca Raton, pp 159–179

Seiler N, Al-Therib MJ (1974a) Putrescine catabolism in mammalian brain. Biochem J 144:29–35

Seiler N, Al-Therib MJ (1974b) AcetylCoA: 1,4-diaminobutane N-acetylpolyamines in vertebrate organs and subcellular distribution. Biochim Biophys Acta 354:206–212

Sofic E, Paulus W, Jellinger K, Riederer P, Youdim MBH (1991) Selective increase of iron in substantia nigra zona compacta of Parkinsonian brains. J Neurochem (in press)

Tetrud JW, Langston W (1989) The effect of deprenyl (selegiline) on the natural history of Parkinson's disease. Science 245:519–522

The Parkinson Study Group Datatop (1989) Selegiline and Parkinson's disease. N Engl J Med 321:1364–1371

Youdim MBH (1985) Brain iron metabolism: biochemical and behavioural aspects in relation to dopaminergic neurotransmission In: Lajtha A (ed) Handbook of neurochemistry, vol 10. Plenum, New York, pp 731–756

Youdim MBH (1990) Developmental neuropharmacological and biochemical aspects of iron-deficiency In: Dobbing J (ed) Brain behaviour and iron deficiency. Springer, Berlin Heidelberg New York

Youdim MBH, Ben-Shachar D, Pollard HB (1991) The effects of iron chelators and calcium antagonists on iron induced neuronal peroxidation. Br J Pharmacol (in press)

Youdim MBH, Ben-Shachar D, Riederer P (1989) Is Parkinson's disease a progressive siderosis of substantia nigra resulting in iron and melanin induced neurodegeneration? Acta Neurol Scand 126:47–54

Authors' address: Prof. M. B. H. Youdim, Technion – Israel Institute of Technology, Faculty of Medicine, Department of Pharmacology, Haifa, Israel.

J Neural Transm (1990) [Suppl] 32: 249–257

Oxidation of the indole nucleus of 5-hydroxytryptamine and formation of dimers in the presence of peroxidase and H_2O_2

G. Huether[1], **A. Reimer**[1], **F. Schmidt**[2], **P. Schuff-Werner**[2], and **M. M. Brudny**[3]

[1] Neurochemische Forschungsgruppe der Psychiatrischen Klinik, and
[2] Abteilung für Klinische Chemie, University of Göttingen,
[3] Abteilung für Molekulare Neuroendokrinologie, Max-Planck-Institut für experimentelle Medizin, Göttingen, Federal Republic of Germany

Summary. 5-Hydroxytryptamine (5-HT) is rapidly oxidized in the presence of peroxidase and H_2O_2. The major reaction product was isolated by gel chromatography and analyzed by mass spectroscopy. It is a 5-HT dimer formed under abstraction of two protons, most likely by the reaction at the C(4) position of two phenoxyradicals of 5-HT. In the presence of 5-HT an increased H_2O_2-consumption and a dose dependent reduction of the formation of reactive oxygen metabolites during H_2O_2 degradation was observed. The pattern of 5-HT reaction products separated by TLC was dependent on the H_2O_2 concentrations used. In the presence of albumen, plasma or tissue homogenates, massive binding of the 5-HT oxidation products to proteins was observed.

Introduction

In the last few years there has been an upsurge of interest in the chemistry of biologically active amines, not only because of their possible physiological importance but also because of their potential pathological implications. The main metabolic fate of oxidative deamination of monoamines by MAO has been well established (Erspamer, 1954; Udenfriend et al., 1956). The oxidation of the phenolic hydroxyl-groups of indoles and catechols is, however, poorly elucidated, although their enzymatic oxidation by peroxidaseoxidases (Porter et al., 1957; Martin et al., 1960; Blaschko and Milton, 1960; Blaschko and Levine, 1960; Nelson and Huggins, 1975; Weber and Horita, 1963; Svensson, 1989; Napolitano et al., 1988; Methodiewa et al., 1989) and their nonenzymatic oxidation by oxyhaemoglobin (Blum and Ling, 1959), by ferricytochrome (Alivisatos and Williams-Ashman, 1964) and by inorganic oxidants such as copper, and $AgNO_3$ (Eriksen et al., 1960), ammonium persulfate, potassium ferricyanide and sodium periodate (Napolitano et al., 1988) have been reported. 5-Hydroxytryptamine (5-HT) for example is oxidized by human serum and erythrocyte haemolysates (Martin et al., 1960; Porter et al., 1957; Curzon and Vallet, 1960; Barass et al., 1973; Blum and Ling, 1959) by liver and brain microsomes (Uemura et al., 1979, 1980) and by spleen homogenates (Nelson and Huggins, 1975) to give colored, but poorly identified products. The oxidation of the indole nucleus is claimed to generate radical intermediates, although the fate of these radicals is not known (Uemura et al., 1980; Perez-Reyes and Mason, 1981; Wrona and Dryhurst, 1987). The 5-HT oxidation products are of considerable interest because it has long been speculated that the formation of unusual 5-HT metabolites might be connected with the etiology of mental dysfunctions (Woolley and Shaw, 1954; Gaddum, 1953; Himwich et al., 1967; Woolley, 1962; Kaplan and Mann, 1982). Wrona and Dryhurst (1987, 1988, 1989) investigated the electrochemical oxidation of 5-HT and proposed an oxidative pathway involving the formation of phenoxyl and nitrogen centered radicals which could either couple to give a series of oligomeric products, or lose another electron to generate an unstable quinone imine. The intermediate formation of the same quinone imine was also suggested to occur under the more biomimetic conditions of hydroxyindole oxidation in the presence of peroxidase and H_2O_2, resulting in the ultimate formation of a complex mixture of oligomeric products (Napolitano et al., 1988).

The present communication focuses on the further clarification of the interaction of 5-HT with the peroxidase-catalyzed H_2O_2 degradation. Evidence is presented that 5-HT, in this system, acts like a true radical scavanger under the formation of a 5-HT dimer. Peroxidase is thus protected against inactivation by reactive oxygen metabolites released during H_2O_2 degradation.

Material and methods

Chemicals

5-Hydroxy(side chain-2^{14}C)tryptamine creatine sulphate, specific activity 2.0 GBg/nmol, was obtained from Amersham (Amersham-Buchler) and checked for purity by TLC. Unlabelled 5-hydroxytryptamine (hydrochloride) and horseradish peroxidase (HRP, EC 1.11.1.7, Type I, salt free powder) were from Sigma. Hydrogen peroxide (30%) and all other chemicals used for TLC, chemiluminescence and gel filtration chromatography were obtained at highest available purity from Merck.

Peroxidative conversion of 5-HT

A typical reaction mixture consisted of 2 µl 1 mM 5-HT, 8 µl 0.1 M potassium phosphate buffer, pH 7.0 and 5 µl HRP-stock solution (25 mU/µl). Reaction was started by the addition of 5 µl H_2O_2-stock solution (0.2 mM), and terminated with acetone (– 20 °C, 40 µl). In some experiments, various antioxidants, catalse and supernatants of tissue homogenates in 0.1 M potassium phosphate buffer were added instead of the pure buffer.

The formation of reactive oxygen metabolites during the enzymatic degradation of H_2O_2 was monitored by the luminol-intensified chemiluminescence assay as described elsewhere (Ernst et al., 1983).

The consumption of H_2O_2 during the reaction was measured with the color reagent of the commercially available kit for the enzymatic determination of cholesterol (CHOD-iodide, Merck). The oxidative loss of 5-HT in the reaction mixture was determined by HPLC and electrochemical detection.

Characterization of peroxidative products of 5-HT

Thin layer chromatography on silica gel plates (5 × 10 precoated TLC plates silica gel 60 F-254, Merck) was used to separate the reaction products of 5-HT and to determine the effects of the variation of the incubation mixture described above. The solvent system used contained ethylacetate-isopropanol-25% ammonia-water (45:35:10:10 by volume). Plates were developed at 20 °C in darkness, dried in air and viewed under ultraviolet light. Radioactive areas were visualized by autoradiography using Osray M3 X-ray films (Agfa-Gevaert). The radioactive components were quantitized by scraping off the areas shown up in the autoradiograms for liquid scintillation counting. For liquid chromatographic separation of the reaction products, sephadex LH-20 (Pharmacia) was used as the stationary phase. The mobile phase was water-methanol (9:1, v/v) adjusted to pH 2.5 with HCl, the column size was 20 × 1.0 cm and a flow rate of 20 ml/h was used. The eluant was monitored by UV-detection at 280 nm. Radioactivity was measured in 1 ml fractions.

Mass Spectrometry of the 5-HT peroxidative product was performed using ^{252}Cf-plasma a plasma desorption time-of-flight mass spectrometry (PDMS) as described in the literature (Roepstorff, 1989). The instrument was manufactured by Bio Ian Nordic AB, Sweden. 10 µl of the sample solution were applied to the nitrocellulose surface for adsorption and spin dried after 15 min. Samples were measured at 15 kV accelerating voltage and 5×10^6 spectra were accumulated.

Results

In the presence of HRP and H_2O_2, 5-HT was almost completely oxidized during the first minute of incubation. After 5 min of incubation, authentic 5-HT could no longer be measured by HPLC and ECD. With the addition of increasing amounts of 5-HT to the HRP/H_2O_2 system a dose-dependent suppression of the chemiluminescence signal and an increased consumption of H_2O_2 was observed (Table 1). As seen in autoradiograms of TLC-separations, the loss of 5-HT coincides with the appearance of several oxidation products of 5-HT with smaller Rf-values. The pattern of TLC separations of the reaction products was dependent on the H_2O_2 concentration used. Separation of the reaction products on a Sephadex LH 20 column revealed a major peak representing about 50% of total radioactivity at about half the elution volume of 5-HT. This major oxidation product was identified by plasma desorption mass spectrometry as a 5-HT-dimer which was formed under abstraction of 2 protons (Fig. 1). The most likely structure

Table 1. Effect of increasing concentrations of 5-HT on the luminol-intensified chemiluminescence and on the consumption of H_2O_2 in the HRP/H_2O_2-system

Initial concentration of 5-HT in incubation mixture (μM)	Total chemiluminescence signal measured during 15 min of incubation	H_2O_2 concentration after 15′ of incubation in % of initial H_2O_2 concentration
0	100%	86%
0.1	98%	99%
1	77%	97%
10	41%	91%
100	22%	48%
1000	13%	1%

50 μl HRP (25 mU/μl), 80 μl 0.1 M phosphate buffer (pH 7.0) and 20 μl buffer ± 5-HT were mixed and the reaction was started by the addition of 50 μl H_2O_2-solution (0.2 mM). Even at the highest concentration, 5-HT was completely oxidized during the 15 min of incubation at 37 °C

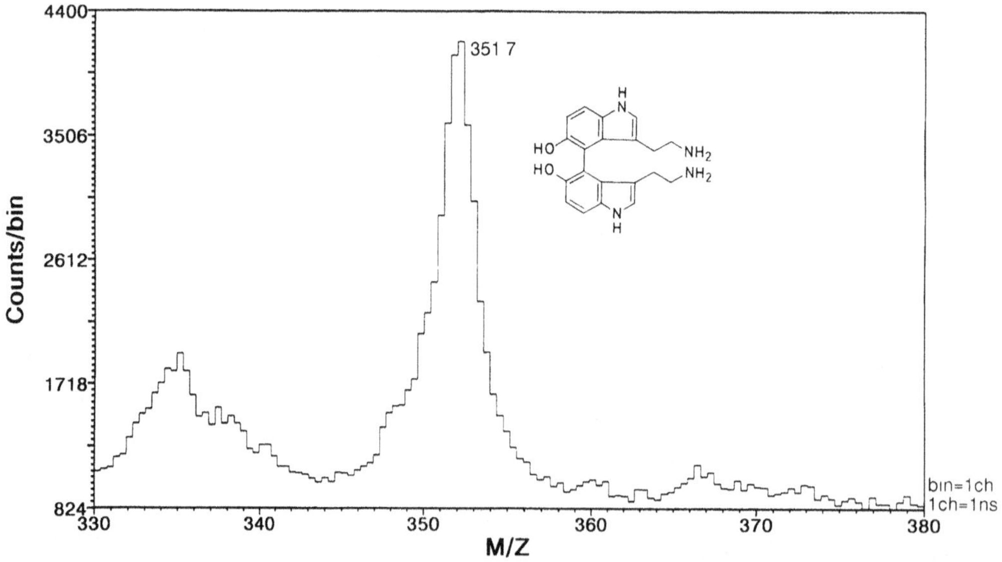

Fig. 1. Plasma desorption mass spectrum of the major oxidation product of 5-HT formed in the presence of HRP and H_2O_2 and isolated by gel chromatography on a LH 20 column as described in Material and methods. Most likely, the reaction product is a 4,4′-bi-5-hydroxytryptamine

Table 2. Effect of the addition of albumin (10 mg/ml), plasma (1:10 v/v) and liver or brain homogenates (1:10 w/v) on the fate of 5-HT and its oxidation products formed in the HRP H_2O_2-system

Addition of		Authentic 5-HT in supernatant	Oxidized 5-HT in supernatant	Protein-bound 5-HT oxidation products
Buffer	a)	0	100	0
	b)	0	100	0
Albumen	a)	4	16	80
	b)	0	44	56
Plasma	a)	3	49	48
	b)	0	67	33
Liver	a)	53	24	23
	b)	0	26	74
Brain	a)	17	13	70
	b)	0	22	78

Incubation mixture: 50 µl HRP (25mU/µl), 50 µl 0.1 M phosphate buffer (pH 7.0), 20 µl ^{14}C-5-HT (1 mM) and 50 µl H_2O_2-solution (0.2 mM). 30 µl of the test-solutions were added either before the reaction was started (a) or after completion of the oxidative conversion of 5-HT, i.e. after 15 min of incubation at 37 °C (b). Proteins were precipitated by cold acetone. Supernatants and pellets were counted for radioactivity and the amount of 5-HT and 5-HT-oxidation products in the supernatants was measured by LSC after TLC-separation. Values are given in % of total radioactivity

of this dimer, as suggested in Fig. 1, is that of a 4,4'-bi-5-hydroxytryptamine.

Almost no oxidation of 5-HT was observed during incubation in the presence of either HRP or H_2O_2 alone. The oxidation of 5-HT in the presence of HRP and H_2O_2 was prevented by the addition of ascorbic acid or $NADH_2$. Glutathione was less effective and gave rise to a conjugate characterized by a very small Rf-value in TLC-separations. In the presence of plasma or homogenates of liver and brain tissue, the rate of 5-HT oxidation was reduced, and the formation of soluble oxidation products was diminished on the expense of considerable binding of oxidized 5-HT to tissue proteins (Table 2). Massive binding of 5-HT-oxidation products was also observed if plasma or tissue homogenates were added at a time when 5-HT was already completely oxidized in the HRP/H_2O_2 system (see Table 2).

Discussion

Peroxidases can, by definition, use H_2O_2 as a cosubstrate for the enzymatic oxidation of other compounds. The capacity of the peroxidase system to oxidize phenoles and indoles has been known for many years, and several attempts have been made to isolate and to identify the reaction products formed. The formation of colored 5-HT oxidation products is well documented and a number of enzyme systems have been implicated including monoamine oxidase (Blaschko and Hellmann, 1953), caeruloplasmin (Eriksen et al., 1960), hydroxyindole oxidase (Blaschko and Levine, 1960), cytochrome oxydase (Weber and Morita, 1963) and pseudoperoxidase (Blum and Ling, 1959). Using various indolic compounds, Nelson and Huggins (1975) clearly established the structural requirement of the C5 substituted hydroxyl group for interaction of indolic compounds with peroxidase systems. From gel chromatography experiments these authors suggested that the primary stable oxidation product was a dimer. This is in agreement with the fact that indoles are known to dimerize (Smith, 1963; Saunders et al., 1964) and that the majority of known hydrogen donors in the peroxidase catalyzed H_2O_2 degradation tend to form dimeric or polymeric products (Saunders et al., 1964; Guilbault et al., 1968; Mason, 1957; Bayes et al., 1972; Napolitano et al., 1988). However, a definite analysis of the 5-HT oxidation chemistry in the HRP–H_2O_2 system has never been presented.

Napolitano et al. (1988) studied the formation of oligomeric products formed from 5-hydroxyindole in the presence of HRP using ^1H and ^{13}C-NMR spectroscopy of the reaction products isolated by TLC after acetylation. From the consideration of the structure of the two dimers and the three trimers analysed, an ionic process was assumed in which the electrophilic 4 position of a transiently generated quinone imine was attacked by the nucleophilic site (C 4) of a 5-hydroxyindole counterpart. More likely, however, is the formation of 5-HT dimers in the HRP–H_2O_2 system by a free radical mechanism involving the one-electron oxidation of 5-HT and formation of semiquinone imine or phenoxy radicals in which the unpaired electron is located at C(4). Coupling of two of such phenoxyl-centered radicals would give the proposed C(4)–C(4′) dimer. The formation of a phenoxylic radical as a precursor of dimerization was also concluded by Wrona and Dryhurst (1987). Based on their studies of the electrochemical oxidation of 5-HT, an oxidative pathway was proposed, involving the initial formation of phenoxyl and nitrogen centered radicals, which could either couple to give C(4)–C(4′) and N(1)–C(4′) dimers. An unstable quinone imine was formed by the abstraction of a second electron, but in contrast to the suggestion of Napolitano et al. (1988), it did not form a dimer but reacted rapidly with water to the tryptamine-4,5-dione.

The formation of dimeric and polymeric oxidation products of 5-HT has long been thought to imply the initial formation of 5-HT radicals (Martin et al., 1960; Eriksen et al., 1960; Wrona and Dryhurst, 1987; Perez-Reyes and Mason, 1989). There is some evidence that 5-HT can be oxidized to a radical e.g. by caeruloplasmin, (Walaas and Walaas, 1961), in the nonenzymatic mediation of the oxidation of pyridine nucleotides by cytochrome C (Alivisatos and Williams-Ashman, 1964), during permanganate oxidation (Borg, 1965) and during autoxidation in alkaline solutions (Perez-Reyes and Mason, 1981). Free radical reactions of the serotonin free radicals with themself, with serotonin and its oligomeric oxidation products, and with nucleophilic sites of tissue proteins may be important in the formation of the melanin like materials observed upon incubation of microsomes in the presence of 5-HT and NADPH (Uemura et al., 1979, 1980). Such reactions would also explain the massive binding of the 5-HT-oxidation products to tissue proteins seen in this study. The suppression of chemiluminescence seen during coincubation of 5-HT with HRP and H_2O_2 indicates that 5-HT, like other hydroxylated indoles and catechols, prevent the formation of reactive oxygen metabolites during H_2O_2 cleavage. The generation of reactive oxygen metabolites during peroxidatic degradation of H_2O_2 seem to be responsible for the substrate induced inactivation of peroxidase (Arnao et al., 1990). That 5-HT, as a radical scavanger and reductant cosubstrate, can prevent this inactivation process is indicated by the increased consumption of H_2O_2 by HRP in the presence of 5-HT.

It is not unlikely that peroxidases use H_2O_2 also in vivo as a cosubstrate for the enzymatic oxidation of 5-HT. In vivo such peroxidatic oxidation can be assumed to occur preferentially at sites where H_2O_2 is generated, e.g. during the respiratory burst of activated leucocytes or at sites of high MAO-activity. Interestingly enough, the local concentration of free 5-HT is higher than anywhere else at sites of inflammation as well as at sites of stimulated synaptic 5-HT-release and presynaptic 5-HT reuptake. This coincidence invites further studies on the possible function of 5-HT as a radical scavanger and the possible consequences of the formation of 5-HT-oxidation products in both, the immune and the nervous system.

Acknowledgements

This research was supported by grants of the Deutsche Forschungsgemeinschaft (Heisenbergprogramm, G. Huether) and of the Land Niedersachsen (Volkswagenstiftung).

References

Alivisatos SGA, Williams-Ashman HG (1964) Serotonin-mediated oxidation of dihydronicotinamide derivatives by cytochrome C. Biochim Biophys Acta 86:392–395

Arnao MB, Acosta M, Del Rio JA, Garcia-Canovas F (1990) Inactivation of peroxidase by hydrogen peroxide and its protection by a reductant agent. Biochim Biophys Acta 1038:85–89

Barass BC, Coult DB, Pinder RM, Skeels M (1973) Substrate specificity of caeruloplasmin for indoles and indole isosteres. Biochem Pharmacol 22:2891–2895

Bayes GS, Michaels AW, Morrison M (1972) The peroxidase-catalyzed oxidation of tyrosine. Biochem Biophys Acta 284:34–42

Blaschko H, Hellmann K (1953) Pigment formation from tryptamine and 5-hydroxytryptamine in tissue: a contribution to the histochemistry of amine oxidases. J Physiol 122:419–427

Blaschko H, Levine WG (1960) A comparative study of hydroxyindole oxidases. Br J Pharmacol 15:625–633

Blaschko H, Milton AS (1960) Oxidation of 5-hydroxytryptamine and related compounds by mytilus gill plates. Br J Pharmacol 15:42–468

Blum JJ, Ling NS (1959) Oxidation of serotonin and 5-hydroxyindoles during the denaturation of oxyhaemoglobin. Biochem J 73:530–535

Borg DC (1965) Transient free radical forms of hormones. EPR spectra from iodothyrosins indoles, estrogens, and insulin. Proc Natl Acad Sci USA 53:829–836

Curzon G, Vallet L (1960) The purification of human caeruloplasmin. Biochem J 74:279–287

Eriksen N, Martin GM, Benditt EP (1960) Oxidation of the indole nucleus of 5-HT and the formation of pigments: isolation and partial characterization of a dimer of 5-HT. J Biol Chem 235:1662–1668

Ernst M, Heberer M, Fischer H (1983) Chemiluminescence measurements of immune cells—a tool in immunobiology and clinical research. J Clin Chem Clin Biochem 21:555–560

Erspamer V (1954) Pharmacology of indolealkylamines. Pharmacol Rev 6:425–487

Gaddum JH (1953) Tryptamine receptors. J Physiol (Lond) 119:363–368

Guilbault GC, Brignac PJ, Zimmer M (1968) Homovanillic acid as a fluorometric substrate for oxidative enzymes. Anal Chem 40:190

Himwich HE, Kety SS, Smythies JR (1967) Amines and schizophrenia. Pergamon, New York

Kaplan RD, Mann JJ (1982) Altered platelet serotonin uptake kinetics in schizophrenia and depression. Life Sci 31:583–588

Martin GM, Benditt EP, Eriksen N (1960) Enzyme oxidation of the indole nucleus of 5-HT: properties of an enzyme in human serum and the products of oxidation. Arch Biochem Biophys 90:208–217

Mason HS (1957) Mechanisms of oxygen metabolism. Adv Enzymol 19:79–233

Metodiewa D, Reszka K, Dunford MB (1989) Evidence for a peroxidatic oxidation of norepinephrine, a catecholamine, by lactoperoxidase. Biochem Biophys Res Commun 160:1183–1188

Napolitano A, D'Ischia M, Prota G (1988) A profile of the oxidation chemistry of 5-hydroxyindole under biomimetic conditions. Tetrahydron 44:7265–7270

Nelson DR, Huggins AK (1974) Interference of 5-HT in the assay of glucose by glucose oxidase: peroxidase—chromagen based methods. Anal Biochem 59:46–53

Nelson DR, Huggins AG (1975) The interaction of 5-HT and related hydroxyindoles with horseradish and mammalian peroxidase systems. Biochem Pharmacol 24:181–192

Perez-Reyes E, Mason RP (1981) Characterization of the structure and reactions of free radicals from serotonin and related indoles. J Biol Chem 256:2427–2432

Porter CC, Titus DC, Sanders BEE, Smith EVC (1957) Oxidation of serotonin in the presence of ceruloplasmin. Science 126:1014–1015

Roepstorff P (1989) Plasma desorption mass spectrometry of peptides and proteins. Acc Chem Res 22:421–427

Saunders BC, Holmes-Siedle G, Stark BP (1964) The properties and uses of a versatile enzyme and of some related catalysts. In: Saunders BC (ed) Peroxidase. Butterworths, London, pp 25–186

Smith GF (1963) The acid catalyzed polymerization of pyroles and indoles. Adv Heterociclic Chem 2:287

Svensson BE (1989) Involvement of cysteine, serotonin and their analogues in peroxidase-oxidase reactions. Chem Biol Interact 70:305–321

Udenfriend S, Weissbach ETH, Peterson RE (1956) Biogenesis and metabolism of 5-OH indole compounds. J Biol Chem 219:333–344

Uemura T, Matsushita H, Ozawa M, Fiori A, Chiesara E (1979) Irreversible binding of 5-HT and 5-HTP metabolites to rat liver microsomal protein. FEBS Lett 101:59–62

Uemura T, Shimazu T, Miura R, Yamano T (1980) NADPH-dependent melanin pigment formation from 5-OH-indolealkylamines by hepatic and cerebral microsomes. Biochem Biophys Res Commun 93:1074–1081

Weber LJ, Horita A (1963) Oxidation of 4- and 5-OH indole derivatives by mammalian cytochrome oxidase. Life Sci 3:44–49

Walaas E, Walaas O (1961) Oxidation of reduced phosphopyridine nucleotides by p-phenylenediamines, catecholamines and serotonin in the presence of ceruloplasmin. Arch Biochem Biophys 95:151–162

Woolley DW (1962) The biochemical bases of psychoses. Wiley, New York

Wrona MZ, Dryhurst G (1987) Oxidation chemistry of 5-HT. 1. Mechanism and products formed at micromolar concentrations. J Organic Chem 52:2817–2825

Wrona MZ, Dryhurst G (1988) Further insights into the oxidation chemistry of 5-HT. J Pharmacol Sci 77:911–917

Wrona HZ, Dryhurst G (1989) Electrochemical oxidation of 5-HT in acidic aqueous-solution. J Organic Chem 54:2718–2721

Authors' address: Priv. Doz. Dr. G. Huether, Max-Planck-Institut für experimentelle Medizin, Hermann-Rein-Strasse 3, D-3400 Göttingen, Federal Republic of Germany

J Neural Transm (1990) [Suppl] 32: 259–267

Reflection of changes in membrane constituents in various regions of Alzheimer brains to differential scanning thermograms

E. Sofic[1], **M. Götz**[1], **L. Frölich**[1], **R. Burger**[1], **S. Heckers**[1], **P. Riederer**[1], **K. Jellinger**[2], and **H. Beckmann**[1]

[1] Clinical Neurochemistry, Department of Psychiatry, University of Würzburg, Federal Republic of Germany
[2] Department of Clinical Neurobiology, Ludwig Boltzmann Institute, Vienna, Austria

Summary. To test whether changes in basic biochemical membrane constituents were reflected in membrane fluidity measurements, protein, total lipids, triglycerides, lecithine, cholesterol and alkaline phosphatase were determined in frontal cortex, hippocampus, putamen and nucleus basalis Meynert (NbM) of DAT brains and controls and compared to differential scanning thermograms. Biochemical changes were most pronounced in the hippocampus, while thermostability was altered in the NbM and the frontal cortex. The results indicate that changes in protein content, but not in lipid composition were reflected in alterations of thermostability of the brain regions examined.

Introduction

In dementia of Alzheimer type (DAT), characteristic neuropathological findings may be observed in cortical and subcortical brain regions (Jellinger, 1987). In the nucleus basalis of Meynert, a selective loss of cholinergic neurons and their projections into the cortex occurs (Coyle et al., 1983). With respect to biochemical alterations, multiple deficits have been observed in post-mortem tissue of various brain regions. Several neurotransmitter systems are altered in the disease (Hardy et al., 1985) with a pronounced deficit in the cholinergic system (Coyle et al., 1983). Deficits in the glycolytic (Bowen et al., 1979) and oxidative (Sorbi et al., 1983; Gibson et al., 1988) glucose metabolism have been observed. Concerning biochemical constituents of nervous tissue, Bowen et al. (1977) in an extensive study of biochemical substrates found that among the 35 parameters analysed, a number of nerve cell indices in the temporal cortex were reduced, including protein content (Bowen et al., 1977). Lipid composition is known to be altered in several brain areas (Gottfries et al., 1983; Crino et al., 1989; Wolfe et al., 1982). All these biochemical changes might reflect membrane alterations and possibly cellular degeneration. Membrane alterations could also be demonstrated in studies of membrane fluidity in platelets of patients with DAT (Zubenko et al., 1987).

Most of the earlier studies have been performed on one or only a few cortical brain areas. Neuropathological analyses have revealed a distinct pattern of distribution of the neuropathological hallmarks in DAT. The hippocampus, the tertiary association areas and some subcortical nuclei appear most affected (Jellinger, 1987; Brun and Englund, 1981; Hyman et al., 1984). However, despite the obvious pathophysiological significance of the biochemical changes, they could not be correlated with the classical topographic distribution of neuropathological changes in DAT.

The morphological and biochemical changes might be reflected in disturbances of the thermostability of different brain regions. Therefore we have spectrophotometrically measured membrane constituents, i.e. total lipids, protein, triglyceride, lecithine, cholesterol, and alkaline phosphatase activity in several brain regions of post-mortem tissue from neuropathologically verified DAT and brains from controls and compared this to differential scanning calorimetry techniques (Perkin Elmer DSC 7).

Materials and methods

Patients

Clinical diagnosis of dementia was made according to DSM-IIIR and included a mental status examination. A full medical examination of each patient was performed to rule out metabolic causes of dementia. The patients fulfilled the diagnostic criteria of the NINCDS/ADRDA for probable dementia of Alzheimer type (McKhann et al., 1984). Control patients were hospitalized and died from somatic disorders. They showed no history of neurological or psychiatric disease according to their records. Drug treatment consisted of cardiovascular active drugs and antibiotics only.

Brain tissue

Brains were removed at autopsy and divided by mid-sagittal section. Post-mortem interval until autopsy in all cases was between 3 and 15 hours. One half brain was deep-frozen in dry ice for biochemical and neurochemical analysis and stored at − 80 °C. The other half brain was fixed in buffered formaldehyde solution for histological examination. Multiple blocks were embedded in paraffin and the preparation of sections followed routine methods (Khachaturian, 1985). The final diagnosis of DAT was established histologically by the presence of numerous plaques and tangles in the cerebral cortex and most pronounced changes in the hippocampus. Cases with moderate or severe vascular changes were excluded. In controls brains, the macroscopic examination of the brain did not show any pathological changes, and the histological examination of the cerebral cortex and the hippocampus did not show more senile plaques and tangles than could be explained by age.

The frozen brains were partially thawed to $-20\,^{\circ}C$ to help in dissection. Brain tissue for biochemical assays were obtained from 11 patients with DAT and 10 controls using an anatomical atlas (Nieuwenhuys et al., 1981). Brain tissue from the frontal cortex, the hippocampus and the putamen was used. Not in every case all brains regions were available, thus the samples consisted of 7–11 cases with DAT, and 8 normal control cases. The brain tissue was put into airtight plastic tubes after dissection and stored again at $-80\,^{\circ}C$. All probes were assayed within 4 weeks. For the thermoscans, brain tissue from 14 patients with DAT and 11 controls was used. The frontal cortex, the hippocampus, the putamen and the nucleus basalis of Meynert were analysed. Again, not in every case all brain regions were available, thus sample number varied between 7–14 for DAT cases and 7–11 for controls.

A known weight (20–500 mg tissue depending on the brain area) of the previously dissected brain tissue was homogenized in ice cold physiological saline solution and used for biochemical analysis.

Biochemical analysis

Protein was determined by the method of Bradford (1976). Determinations of total lipid content in brain tissue was performed using a modified method of Zöllner and Kirsch (1965). Triglycerides, lecithine, cholesterol, alkaline phosphatase were measured using commercially available kits which had been modified (Boehringer Mannheim Test Combination, Behring-Test Combination, and Mercko-test). The accuracy of the cholesterol assay was confirmed by the simultaneous determination of cholesterol with the method of Huang et al. (1961), which gave almost identical results. Calorimetry scans were made in a Perkin-Elmer DSC7 differential scanning calorimeter. Scans were started at $20\,^{\circ}C$, and indium was used to calibrate the apparatus. Samples consisted of 5 mg $\pm$ 0.2 mg native tissue. Heat was increased by $5\,^{\circ}C$ (in case of NbM) or $1.5\,^{\circ}C$ per minute in all other cases.

Statistical analysis

Group differences were analysed using a parametric test (Student's t-Test).

Results

Due to difference in sampling of brain regions, the mean age of controls varied between 70.3 $\pm$ 1.6 and 73.5 $\pm$ 3.3 years among the 6 brain regions investigated. The mean age of the DAT groups varied between 77.7 $\pm$ 0.9 and 78.8 $\pm$ 1.0 years. There were no significant sex differences. Post mortem interval until autopsy in all cases was between 3 and 17 hours. Brain tissue for thermostability experiments consisted of samples with a wider range of post-mortem delay and age. The mean age varied between 56.8 $\pm$ 20.0 to

77.3 ± 7.0 for the control group and 75.6 ± 3.2 to 78.3 ± 0.8 for the DAT group. Post mortem delay until autopsy was between 3 and 17 hours. There was no indication of age-dependent changes of thermostability in these brains. Data for these general parameters are presented in Table 1.

Data of the biochemical substrates are presented in Table 2. Protein was only reduced in the frontal cortex by 29 percent. Alkaline phosphatase was reduced in the frontal cortex by 50 percent only when the activity was related to the weight of the fresh tissue. When the activity was related to the amount of protein in the tissue, a reduction of 27 percent was no longer significant. Reduced levels of triglycerides, lecithine and cholesterol were found in the hippocampus of Alzheimer brains.

Table 1. General characteristics of DAT and control group

Brain region	sex [m/f]	age [years]	post-mortem time [h]
a) Brain tissue for biochemical analyses			
Frontal Cortex			
DAT	4/4	77.8 ± 3.7	3–17 (range)
controls	4/4	70.3 ± 1.6	3–17 (range)
Hippocampus			
DAT	4/7	78.6 ± 1.0	3–17 (range)
controls	5/3	72.8 ± 3.2	3–17 (range)
Putamen			
DAT	3/4	77.7 ± 0.9	3–17 (range)
controls	6/2	73.5 ± 3.3	3–17 (range)
b) Brain tissue for differential scanning calorimetry			
Frontal Cortex			
DAT	5/9	75.6 ± 3.2	3–18 (range)
controls	4/7	77.3 ± 2.7	3–18 (range)
Hippocampus			
DAT	4/6	77.6 ± 1.0	4–14 (range)
controls	5/1	56.8 ± 8.2	16–90 (range)
Putamen			
DAT	4/7	78.3 ± 0.8	4–14 (range)
controls	6/1	61.4 ± 8.3	7–70 (range)
Nucleus basalis Meynert			
DAT	4/3	78.3 ± 3.3	3–17 (range)
controls	4/3	73.3 ± 4.5	4–40 (range)

Results are expressed as mean ± SEM, post-mortem delay as range

Table 2. Biochemical substrates in different brain regions in dementia of Alzheimer type and controls

Brain region		Protein [mg/g f.t.]	Lipids [mg/g f.t.]	Lipids [mg/g prot.]
Frontal C.				
DAT		49.8 ± 4.9*	30.3 ± 3.9	621.4 ± 68.6
controls		70.0 ± 4.3	33.5 ± 2.4	482.6 ± 33.2
Hippocampus				
DAT		72.3 ± 4.9	34.1 ± 2.3*	473.4 ± 30.6*
controls		62.3 ± 5.7	45.1 ± 3.5	754.8 ± 86.6
Putamen				
DAT		84.7 ± 2.4	41.9 ± 2.2	486.1 ± 22.6
controls		81.5 ± 2.8	42.6 ± 1.7	554.0 ± 41.5

	Triglycerides [mg/g f.t.]	Triglycerides [mg/g prot.]	Lecithine [mg/g f.t.]	Lecithine [mg/g prot.]
Frontal C.				
DAT	5.5 ± 0.9	110.5 ± 12.9	15.2 ± 2.3	301.1 ± 29.1
controls	9.8 ± 1.3	139.5 ± 14.2	16.9 ± 1.2	243.9 ± 15.7
Hippocampus				
DAT	9.3 ± 0.4*	128.5 ± 4.8*	14.9 ± 0.9*	206.3 ± 10.3*
controls	11.6 ± 0.9	195.1 ± 17.9	17.5 ± 1.4	287.8 ± 16.4
Putamen				
DAT	13.4 ± 0.9	158.7 ± 10.1	17.9 ± 0.9	214.3 ± 14.9
controls	13.6 ± 1.5	169.2 ± 21.5	19.6 ± 0.8	244.9 ± 9.0

	Cholesterol [mg/g f.t.]	Cholesterol [mg/g prot.]	alkaline phosphatase [U/g f.t.]	alkaline phosphatase [U/g prot.]
Frontal C.				
DAT	9.9 ± 1.0	205.2 ± 20.1	1.6 ± 0.3*	34.6 ± 7.1
controls	11.0 ± 0.9	157.2 ± 8.8	3.2 ± 0.4	47.4 ± 8.2
Hippocampus				
DAT	13.3 ± 0.6*	181.5 ± 6.6*	1.7 ± 0.4	24.1 ± 6.2
controls	18.5 ± 1.6	314.4 ± 39.4	1.5 ± 0.1	25.2 ± 3.7
Putamen				
DAT	17.0 ± 0.7	203.3 ± 12.7	2.1 ± 0.2	25.5 ± 2.8
controls	18.6 ± 1.8	232.5 ± 30.0	1.9 ± 0.3	23.6 ± 3.8

Results are expressed as means ± SEM
Significant differences are indicated by an asterisk
($P < 0.05$ by Student's t-test with Bonferroni correction)
C. Cortex, *f.t.* fresh tissue, *U* μmol substrate metabolized/min

Table 3. Thermostability of native tissue slices in different brain regions in dementia of Alzheimer type and controls

Brain region	DAT	controls
Frontal Cortex	52.7 ± 0.8 (14)*	58.7 ± 1.9 (11)
Hippocampus	57.6 ± 1.4 (10)	56.2 ± 0.8 (6)
Putamen	59.9 ± 1.9 (11)	56.1 ± 2.1 (7)
N. basalis Meynert	64.7 ± 0.7 (7)*	78.8 ± 4.1 (7)

Results are presented as means ± SEM of T_{onset} [°C]
Sample size is given in brackets
significant differences are indicated by an asterisk
($P < 0.05$ by Student's t-test with Bonferroni correction)

Differential scanning calorimetry indicated a reduced thermostability of nucleus basalis Meynert (Table 3, Fig. 1) and of frontal cortex (Table 3, Fig. 2). The other brain areas appeared unaltered.

Discussion

Protein content and lipid composition of the human brain changes during aging (Söderberg et al., 1989). DAT group and controls were of comparable, but not of closely matched age for all the brain regions investigated. In the present study, still no dramatic differences between the DAT group and controls were found with respect to biochemical membrane composition.

A few indiced of cellular degeneration may be found in the frontal cortex. Protein content, which may be regarded an index of cell loss or atrophy (Bowen et al., 1977), was reduced in the frontal cortex only. Consistent with this finding is the reduced activity of alkaline phosphatase, 85 percent of which are located in the external cell membrane (Gennis, 1989), and which may be regarded a sign of cellular degeneration.

Indices of structural membrane alterations could be found in the hippocampus. Lipid composition was changed most strongly in this area with reductions of all, total lipids, triglycerides, lecithine and cholesterol. This alteration of membrane composition may give rise to changes of membrane fluidity and thermostability as has been found in platelets of patients with DAT (Zubenko et al., 1987). The changes in lecithine and cholesterol content is in partial accordance with the observation of Gottfries et al. (1986) who found a reduction of phospholipids and cholesterol in DAT hippocampus. It has been suggested that phosphatidylcholine of neuronal cell membranes might serve as an endogenous substitute in the synthesis of acetylcholine in DAT (Bluztajn and Wurtman, 1983). Thus, a reduction of lecithin ought to be expected in all brain regions with cholinergic degeneration. Since such a

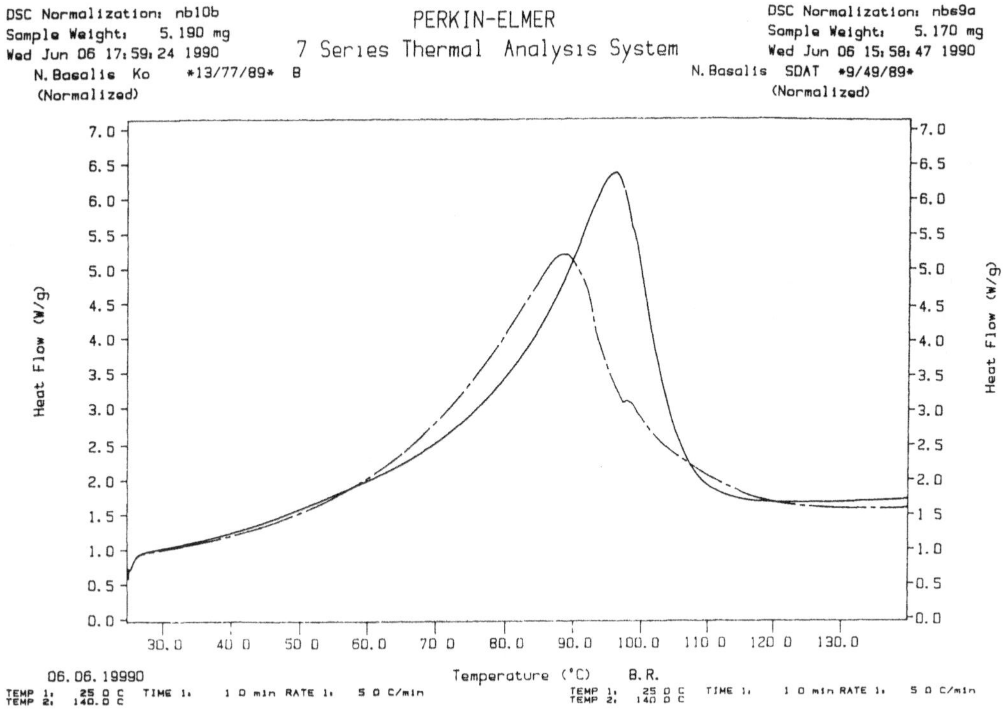

Fig. 1

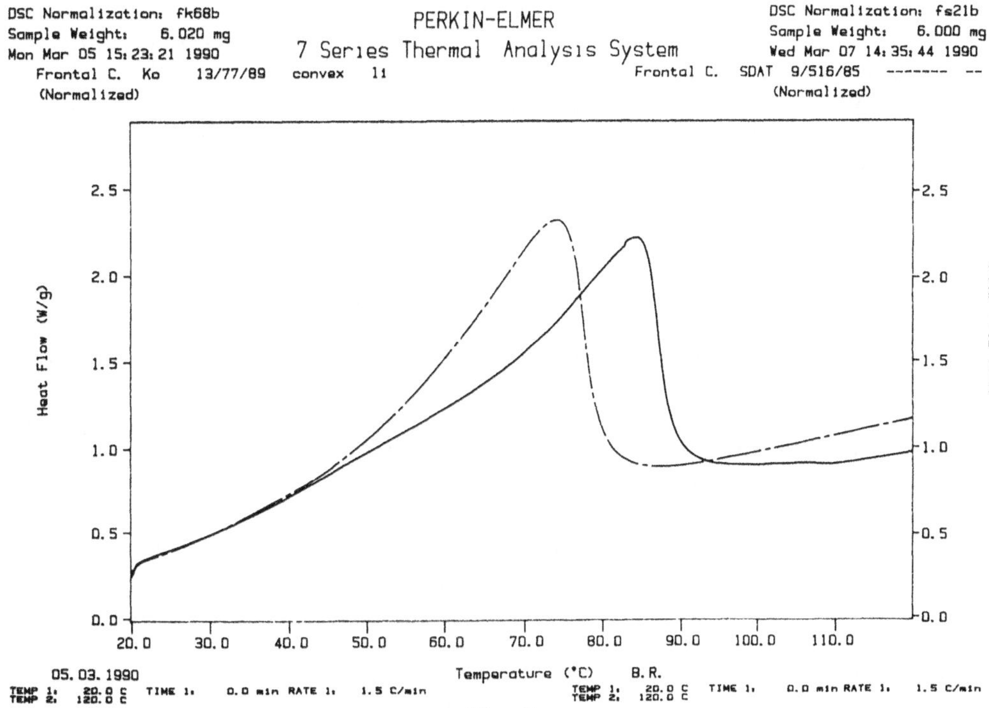

Fig. 2

Fig. 1 and 2. Differential calorimetric scan in dementia of Alzheimer type (— • —) and controls (—). Diagram show thermograms from native tissue slices of nucleus basalis Meynert (Fig. 1) and frontal cortex (Fig. 2) measured against an inert reference. Heat rate 1.5 °C/min for frontal cortex and 5.0 °C/min for nucleus basalis Meynert

reduction could not be demonstrated, this hypothesis is not supported by the present findings.

The observed changes in the thermostability of nucleus basalis Meynert and frontal cortex of SDAT might depend on the decomposition of membrane lipids and the formation of unstable protein-lipid complexes leading to alterations in the fluidity of membranes (Rothman and Lenard, 1977). Neuropathological changes were most pronounced in the hippocampus. The data indicate that only in the hippocampus, the most severely affected brain region, major alterations of membrane constituents occurred. However, these changes did not result in alterations of thermostability in this brain region. Among the neocortical brain regions, only the frontal cortex showed some signs of degradation of cell membranes, which were parallelled by alterations of thermostability. Interestingly, not changes in lipid composition, but changes in protein content were reflected in alterations of thermostability of the brain regions examined.

References

Blusztajn JK, Wurtman RJ (1983) Choline and cholinergic neurons. Science 221:614–620

Bowen DM, Smith CB, White P, Goodhardt MJ, Spillane JA, Flack RHA, Davison AN (1977) Chemical pathology of organic dementias. II Quantitative estimation of cellular changes in post-mortem brains. Brain 100:427–453

Bowen DM, White P, Spillane JA, Goodhardt MJ, Curzon G, Iwangoff P, Meier-Ruge W, Davison AN (1979) Accelerated ageing or selective neuronal death as an important cause of dementia? Lancet i:11–14

Bradford MM (1976) A rapid and sensitive method for the quantitation of microgram quantities of protein utilizing the principle of protein-dye binding. Anal Biochem 72:248–254

Brun A, Englund E (1981) Regional pattern of degeneration in Alzheimer's disease: neuronal loss and histopathological grading. Histopathology 5:549–564

Coyle JT, Price DL, DeLong M (1983) Alzheimer's disease: a disorder of cortical cholinergic innervation. Science 219:1184–1190

Crino PB, Ullman D, Vogt BA, Bird ED, Volicer L (1989) Brain gangliosides in dementia of the Alzheimer type. Arch Neurol 46:398–401

Gennis RM (1989) Biomembranes: molecular structure and function. Springer, Berlin Heidelberg New York Tokyo, p 86

Gibson GE, Sheu KFR, Blass JP, Baker A, Carlson KC, Harding B, Perrino P (1988) Reduced activities of thiamine-dependent enzymes in the brains and peripheral tissues of patients with Alzheimer's disease. Arch Neurol 45:836–840

Gottfries CG, Bartfai T, Carlsson A, Eckernäs A, Svennerholm L (1986) Multiple biochemical dificits in both gray and white matter of Alzheimer brains. Prog Neuropsychopharmacol Biol Psychiatry 10:405–413

Hardy J, Adolfsson R, Alafuzoff I, Bucht G, Marcusson J, Pedahl E, Wester P, Winblad B (1985) Transmitter deficits in Alzheimer's disease. Neurochem Int 7:545–563

Huang TC, Chen PC, Wefler V, Raftery A (1961) A stable reagent for the Liebermann-Burchard reaction. Anal Chem 33:1405–1407

Hyman BT, Hoesen van GW, Damasio Ar, Barnes CL (1984) Alzheimer's disease: cell-specific pathology isolates the hippocampal formation. Science 225: 1168–1170

Jellinger K (1987) Neuropathological substrates of Alzheimer's disease and Parkinson's disease. J Neural Transm [Suppl] 24:109–129

Khachaturian ZS (1985) Diagnosis of Alzheimer's disease. Arch Neurol 42: 1097–1105

McKhann G, Drachman D, Folstein M, Katzman R, Price DL, Stadlan EM (1984) Clinical diagnosis of Alzheimer's disease: report of the NINCDS-ADRDA work group under the auspices of department of health and human services task force on Alzheimer's disease. Neurology 34:939–944

Nieuwenhuys R, Voogd J, van Huijzen C (1981) The human central nervous system. A synopsis and atlas. Springer, Berlin Heidelberg New York

Rothman JE, Lenard J (1977) Membrane asymmetry. Science 195:743–753

Sofic E (1986) Untersuchung von biogenen Aminen, Metaboliten, Ascorbinsäure und Glutathion mittels HPLC-ECD und deren Verhalten in ausgewählten Lebensmitteln und im Organismus von Tier und Mensch. Doctoral Thesis, Technical University Vienna, Austria

Sorbi S, Bird ED, Blass JP (1983) Decreased pyruvate dehydrogenase complex activity in Huntington and Alzheimer brain. Ann Neurol 13:72–78

Söderberg M, Edlund C, Kristensson K, Dallner G (1990) Lipid compositions of different regions of the human brains during aging. J Neurochem 54:415–423

Wolfe LS, NgYingKin NMK, Palo J, Haltia M (1982) Raised levels of cerebral cortex dolichols in Alzheimer's disease. Lancet ii:99

Zubenko GS, Cohen BM, Reynolds CF, Boller F, Malinakova I, Keefe N (1987) Platelet membrane fluidity in Alzheimer's disease and major depression. Am J Psychiatry 144:860–868

Zöllner N, Kirsch K (1965) Untersuchung und Bestimmung der Lipide im Blut. Springer, Berlin Heidelberg New York, p 258

Authors' address: Dr. L. Frölich, Clinical Neurochemistry, Department of Psychiatry, University of Würzburg, Füchsleinstrasse 15, D-8700 Würzburg, Federal Republic of Germany.

J Neural Transm (1990) [Suppl] 32: 269–275

Effects of L-deprenyl and amantadine in an MPTP-model of parkinsonism

W.-D. Rausch[1], **E. Schallauer**[1], **W. W. Chan**[1], **P. Riederer**[2], and **M. Weiser**[1]

[1] Institute for Medical Chemistry, Veterinary Medical University Vienna, Austria
[2] Clinical Neurochemistry, Department of Psychiatry, University of Würzburg, School of Medicine, Würzburg, Federal Republic of Germany

Summary. Mongolian gerbils of both sexes received a single daily dose of 40 mg/kg of 1-methyl-4-phenyl-1,2,3,6-tetrahydropyridine (MPTP) over 4 consecutive days. On the fifth day the animals were treated with 15 mg/kg i.p. of L-deprenyl or amantadine or the combination of both drugs. At different time intervals (1, 2, 5 hours) the animals were sacrificed. In the caudate nuclei dopamine (DA), 3,4-dihydroxyphenylacetic acid (DOPAC), serotonin (5-HT), 5-hydroxyindoleacetic acid (5-HIAA) and homovanillic acid (HVA) were measured by an HPLC technique.

MPTP affected the dopaminergic (HVA − 25%) as well as the serotoninergic system (5-HT − 54%, 5-HIAA − 31%). L-deprenyl and amantadine accumulated DA and 5-HT in the MPTP affected caudates. Synergistic effects of the drug combination could be proven.

Introduction

The incorporation of MPTP, a neurotoxin with selective action on the nigrostriatal pathway has been shown to have potent effects in humans and primates (Chiueh et al., 1985). Though most rodents appear to be less susceptible, strains as C57B1 mice are also affected, though in much higher dosages of the drug (Hallman et al., 1985).

The toxicity of MPTP is based primarily on it's conversion to 1-methyl-4-phenylpyridinium ion (MPP^+) by monoamine oxidase (MAO)-B. Thus the action of MAO-B inhibitors as L-deprenyl can prevent the toxic action of MPTP (Cohen et al., 1985). This may in part explain it's clinical efficacy in the treatment of Parkinson's disease.

Amantadine (1-aminoadamantane), as well acts effectively on Parkinsonian symptoms (Parkes et al., 1970). In mice models amantadine increases the locomotor activity and particularly antagonizes reserpine induced catalepsy (Maj et al., 1972). The pharmacological effects of the drug are explained by it's capacity to increase membrane fluidity being integrated into cellular membranes. However additionally dopamine agonistic effects, reuptake inhibition of serotonin and MAO-A inhibition (Wesemann and Ekenna, 1982) have been demonstrated.

This study has been conducted to study effects of repeated MPTP treatment on the striatal dopaminergic and serotoninergic system of the Mongolian gerbil. Moreover it has been investigated how dose and time dependent treatment wiht L-deprenyl, amantadine or a combination of both could influence these parameters.

Materials and methods

Adult Mongolian gerbils (3–6 months, 70–80 grams) of both sexes were used for the experiments. A 12 hours dark and light cycle was maintained. The animals were fed a standard diet (Altromin R pellets) and water ad libitum.

At 9 a.m. the animals received a single daily dose of 40 mg/kg MPTP.HCl i.p. in 0.5 ml saline (placebo only saline) over 4 consecutive days. At the morning of the fifth day either L-deprenyl, amantadine · HCl (both Asta Pharma, Germany) or a combination of both in physiological saline in a concentration of 15 mg/kg were given i.p. 1, 2 or 5 hours later the animals were sacrificed. The skulls were opened, the brains removed and the caudate nuclei dissected.

The samples were kept frozen at $-80\,^{\circ}C$ until further processing. Samples (20 mg) were sonicated in 0.6 ml of cold 0.4 M $HClO_4$. The homogenates were centrifuged at 7.000 × g for 2 minutes. The clear supernatants were directly injected into an HPLC-system (Constametric III-pump connected to a Tacussel electrochemical detector). DA, DOPAC, 5-HT, 5-HIAA and HVA were separated by a method of Sperk (1982). The potential of the electrochemical detector was + 700 mV. Standards of the individual substances (Sigma Chem.) were used for reference. Statistical significances compared to the MPTP-group were calculated using the Welsh t-test.

The animal studies were approved by the Austrian Commission for Animal Experiments, Ministry for Science and Research No. 68 205/307-12/86.

Results

MPTP impairs striatal dopaminergic and serotoninergic systems in the gerbil

MPTP-treatment for four days affected the dopaminergic and serotoninergic systems. A 20% reduction in the DA-levels was not significantly different from placebo treated animals. DOPAC levels were virtually unchanged. However HVA was decreased significantly ($p < 0.01$) (Table 1).

Table 1. Changes of DA, 5-HT and their metabolites in caudate nucleus due to MPTP

	Placebo-treated (17)	MPTP-treated (16)
DA	11.65 ± 4.66	9.50 ± 2.15
DOPAC	5.86 ± 1.08	5.81 ± 2.08
HVA	3.26 ± 0.14	2.43 ± 0.05**
5-HT	0.26 ± 0.14	0.12 ± 0.05**
5-HIAA	0.55 ± 0.16	0.38 ± 0.18**
DA/DOPAC	2.11 ± 0.25	1.84 ± 0.17
DA/HVA	3.56 ± 0.17	4.08 ± 0.31
5-HT/5-HIAA	0.50 ± 0.08	0.22 ± 0.01**

Values for transmitter concentrations are expressed as μg/g caudate nucleus and represent means ± S.E.M. Statistically significant differences from placebo values are expressed as ** $p < 0.01$

The effects on the serotoninergic system in the caudate nucleus were more pronounced. Reductions of 5-HT (– 54%, $p < 0.01$) and it's acid metabolite 5-HIAA (– 31%, $p < 0.01$) were measured (Table 1).

The turnover rate of DA remained unchanged by MPTP, while the metabolism of 5-HT was increased (Table 1).

L-deprenyl, amantadine and their combination affect striatal dopamine turnover

Different time (1–5 hrs) and dosage regimens were tested for their efficacy in MPTP treated gerbils (data not shown). However higher dosages (15 mg/kg) were needed to reverse the effect of MPTP. MAO-inhibition by L-deprenyl led to a twofold increase in DA concentrations within two hours. Amantadine increased DA concentrations though to a more moderate extent. Only at five hours the combination reached the same efficacy as L-deprenyl alone.

L-deprenyl significantly reduced DOPAC levels after 5 hours. Amantadine decreased DOPAC formation at all time points measured though shorter lasting and to a smaller extent.

Combined treatment enhanced the effects of L-deprenyl leading to an extensive reduction (– 86%, $p < 0.01$) of DOPAC within 2 hours. Extraneuronally L-deprenyl exerted inhibitory action not before 5 hours as could be seen from the HVA levels. Amantadine and the combination depressed HVA formation 1–5 hours after administration. At 2 hours the combination appeared superior to both of the single drugs (Table 2).

All drugs slowed down the intraneuronal formation of DOPAC (Fig. 1). The synergistic effects of the drug combination became more obvious when

Table 2. Effects of L-deprenyl, amantadine and their combination on dopaminergic and serotoninergic metabolism in caudate nucleus

	L-Deprenyl	Amantadine	Both
DA			
1 hr (4)	17.69 ± 2.42*	13.53 ± 0.92*	11.57 ± 0.80
2 hrs (4)	19.77 ± 1.49**	12.64 ± 0.38**	13.86 ± 0.11**
5 hrs (5)	15.73 ± 1.40**	12.53 ± 0.73*	16.72 ± 1.64**
DOPAC			
1 hr (4)	2.68 ± 0.45	2.06 ± 0.20**	3.11 ± 0.68*
2 hrs (4)	2.98 ± 0.41	3.34 ± 0.22**	1.16 ± 0.06**†
5 hrs (5)	1.15 ± 0.16**	4.05 ± 0.44*	0.84 ± 0.03**
HVA			
1 hr (4)	2.37 ± 0.25	1.87 ± 0.12**	1.92 ± 0.18*
2 hrs (4)	2.11 ± 0.01	2.49 ± 0.08	1.30 ± 0.09**,††
5 hrs (5)	1.76 ± 0.12**	1.71 ± 0.09**	1.69 ± 0.11**
5-HT			
1 hr (4)	0.41 ± 0.02**	0.31 ± 0.02**	0.36 ± 0.07*
2 hrs (4)	0.38 ± 0.04**	0.66 ± 0.11**	0.39 ± 0.04**
5 hrs (5)	0.42 ± 0.03*	0.32 ± 0.05*	0.50 ± 0.04**
5-HIAA			
1 hr (4)	0.36 ± 0.09	0.32 ± 0.04	0.41 ± 0.04
2 hrs (4)	0.50 ± 0.11	0.81 ± 0.11*	0.24 ± 0.04
5 hrs (5)	0.33 ± 0.08	0.49 ± 0.02*	0.25 ± 0.02*

Values are expressed as μg/g caudate nucleus and represent means ± S.E.M. Statistically significant differences from 1) MPTP values (see Table 1). ** $p < 0.01$, *$p < 0.05$, 2) from monotherapy groups $^{++}$ $p < 0.01^{+}$, $p < 0.05$

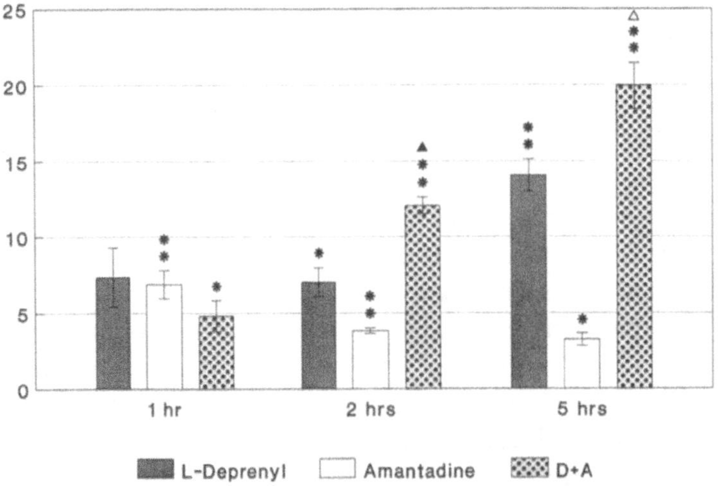

Fig. 1. Ratio DA versus DOPAC concentrations. Values are expressed as means ± S.E.M. Statistical significances 1) ** $p < 0.01$, * $p < 0.05$ from MPTP group ratios, 2) ▲ $p < 0.01$, △ $p < 0.05$ from monotherapy groups

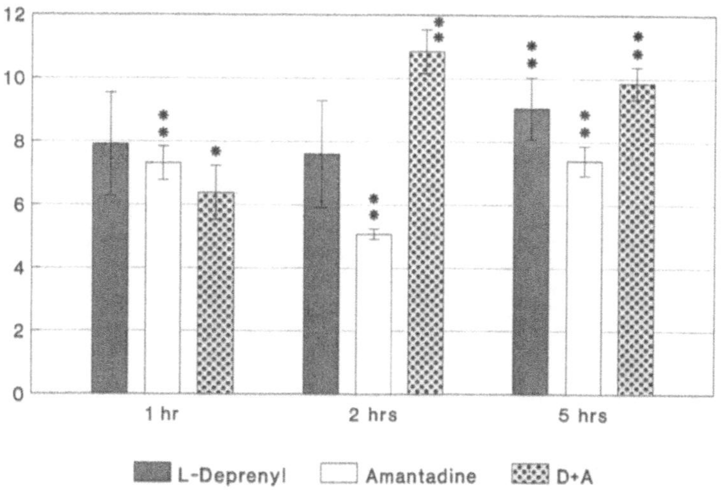

Fig. 2. Ratio DA versus HVA concentrations. Values are expressed as means ± S.E.M. Statistical significances ** $p < 0.01$, * $p < 0.05$ from MPTP group ratios

turnover ratios were compared. L-deprenyl and particularly the combination within two to five hours acted two to seven fold more potent than amantadine alone (Fig. 1). Extraneuronally both drugs decreased the HVA formation. Maximally a 2.6 fold accumulation could be reached by the combined drug regimen after two hours (Fig. 2).

L-Deprenyl, amantadine and their combined effects on striatal 5-HT-turnover

Generally 5-HT levels were increased at all time points and by all drugs (Table 2). Elevations after two hours by amantadine were even higher than placebo treated groups. Amantadine alone produced towfold elevation of 5-HT concentrations within two hours. This effect could not be found in combination with L-deprenyl. 5-HIAA formation in contrast appeared little changed though a maximal action particularly with amantadine at 2 hours may be hypothesized (Table 2).

MPTP alone stimulated the turnover rate to 5-HIAA (Fig. 3). All drugs significantly inhibited the metabolism of 5-HT. A synergistic effect became apparent at 2 hours (Fig. 3).

Discussion

MPTP in Mongolian gerbils next to strong peripheral toxicity also affected the striatal transmitter levels. It is interesting to note that the changes of the

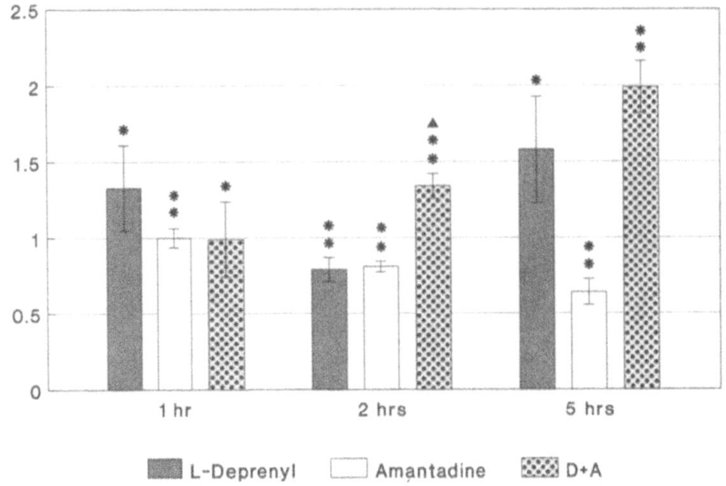

Fig. 3. Ratio 5-HT versus 5-HIAA concentrations. Values are expressed as means ± S.E.M. Statistical significances 1) $^{**}p < 0.01$, * $p < 0.05$ from MPTP group ratios, 2) ▲ $p < 0.01$ from monotherapy groups

dopaminergic system were not as profound as that of the serotoninergic system. This may represent some species difference as the contrary seems to be the case for C57/B1 mice (Adams et al., 1989). MAO inhibitory actions seem to hold responsible for the accumulation of DA and 5-HT in the MPTP challenged neurons. The selective inhibitor of MAO-B, L-deprenyl proved to be most effective for DA. On the contrary amantadine acted more effectively on serotoninergic neurons, which may be attributed to MAO-A inhibition (Wesemann and Ekenna, 1982). The combination appeared to prolong the onset of the drug effect.

Interestingly the co-administration of both drugs at different times appeared more effective than the single drugs. Thus other mechanisms of action should as well be considered. It is still unclear in which brain regions amantadine exerts it's antiparkinsonian effect, however an interesting mechanism of action has been postulated recently. Amantadine may exert NMDA-receptor antagonistic action at therapeutic levels (Kornhuber et al., 1989). In an animal model of Parkinsonism NMDA-antagonists lead, independently of the dopaminergic system, to stimulation of motoric activity (Carlsson and Carlsson, 1989). Possibly an overactive glutamatergic system is involved in the Parkinson syndrome (Albin et al., 1989). Provided the same mechanism exists in MPTP treated animals NMDA antagonists may help to overcome a dopaminergic-glutamatergic imbalance, thus sparing DA at the synapse. In fact combined treatment with the dopamimetic substance L-deprenyl and the glutamatergic antagonist amantadine is superior to monotherapy and leads to improve the impaired motor loop (Albin et al., 1989) at different brain regions.

References

Adams JD, Kalivas PW, Miller CA (1989) The acute histopathology of MPTP in the mouse CNS. Brain Res Bull 23:1–17

Albin RL, Young AB, Penney JB (1989) The functional anatomy of basal ganglia. Trends Neurosci 12:366–375

Carlsson M, Carlsson A (1989) The NMDA antagonist MK-801 causes marked locomotor stimulation in monoamine-depleted mice. J Neural Transm 75:221–226

Chiueh CC, Burns RS, Markey SP, Jacobowitz DM, Kopin IJ (1985) Primate model of Parkinsonism: selective lesion of nigrostriatal neurons by 1-methyl-4-phenyl-1,2,3,6-tetrahydropyridine produces an extrapyramidal syndrome in rhesus monkeys. Life Sci 36:213–218

Cohen G, Pasik, P, Cohen B, Leist A, Mytilineou C, Yahr MD (1985) Pargyline and deprenyl prevent the neurotoxicity of 1-methyl-4-phenyl-1,2,3,6-tetrahydropyridine (MPTP) in monkeys. Eur J Pharmacol 106:209–210

Hallman H, Lange J, Olson L, Stromberg I, Jonsson G (1985) Neurochemical and histochemical characterization of neurotoxic effects of 1-methyl-4-phenyl-1,2,3,6-tetrahydropyridine on brain catecholamine neurons in the mouse. J Neurochem 44:117–127

Kornhuber J, Bormann J, Retz W, Hübers M, Riederer P (1989) Memantine displaces (^{3}H) MK-801 at therapeutic concentrations in postmortem human frontal cortex. Eur J Pharmacol 166:589–590

Maj J, Sowinska H, Baran L (1972) The effect of amantadine on motor activity and catalepsy in rats. Psychopharmacologia 24:296–307

Parkes JD, Zilkha KJ, Calver DM, Knill-Jones RP (1970) Controlled trial of amantadine hydrochloride in Parkinson's disease. Lancet i:259–262

Sperk G (1982) Simultaneous determination of serotonin, 5-HIAA, DOPAC and HVA with ECD. J Neurochem 38:840–843

Wesemann W, Ekenna O (1982) Effect of 1-aminoadamantanes on the MAO activity in brain, liver and kidney of the rat. Arzneimittelforschung/Drug Res 32:1241–1243

Author's address: Univ. Doz. Dr. W.-D. Rausch, Institute for Medical Chemistry, Veterinary Medical University Vienna, Linke Bahngasse 11, A-1030 Vienna, Austria.

Diamine oxidase – semicarbazide sensitive amine oxidase – polyamine oxidase

J Neural Transm (1990) (Suppl] 32: 279–290

Some aspects of the pharmacology of semicarbazide-sensitive amine oxidases

B. A. Callingham, A. Holt, and **J. Elliott**

Department of Pharmacology, University of Cambridge, United Kingdom

Summary. Semicarbazide-sensitive amine oxidase enzymes (SSAO) are found in animals, plants, fungi and bacteria. In vertebrates, their distribution in tissues and blood plasma varies between species. Studies of the SSAO enzymes have concentrated on their biochemical identities separate from those of MAO. Attention is now being paid to their possible physiological and pharmacological significance. These may include, besides the scavenging of circulating amines, functions dependent upon the hydrogen peroxide these enzymes produce. Modulation, by SSAO, of blood vessel tone may be due to the control of amine concentration itself or to actions of released peroxide. In the plasma the activity of SSAO may be susceptible to hormonal control as well as being an indicator of copper status of the animal. However, SSAO may convert xenobiotics to more toxic metabolities. Use of highly selective SSAO inhibitors, such as procarbazine and B24 should enable these preliminary observations to be examined further.

Introduction

Like monoamine oxidase (EC 1.4.3.4), semicarbazide-sensitive amine oxidases (SSAO) are widely distributed, in animals, plants, fungi and bacteria (see Zeller, 1963; Blaschko, 1974). In the vertebrate kingdom, the distribution of SSAO enzymes in tissues and blood plasma varies between species (see Zeller, 1963; Buffoni, 1966; Kapeller-Adler, 1970). Until recently, studies of the SSAO enzymes were concerned mainly with molecular aspects (see Knowles and Yadav, 1984; Finazzi-Agro, 1989) and less attention has been paid to their physiological functions and possible pharmacological significance (see Callingham and Barrand, 1987). All the SSAO enzymes except lysyl oxidase are classified as EC 1.4.3.6 with the systematic name, "Amine: oxygen oxidoreductase (deaminating) (copper containing)" and the trivial name: "Copper containing amine oxidases" (Enzyme nomenclature, 1984).

However, in view of thier common sensitivity to carbonyl reagents typified by semicarbazide and the fact that not all have been shown to contain copper (see Callingham and Barrand, 1987), the abbreviation, "SSAO" has now gained a reasonable acceptance although it is clearly time that a more rational classification of this group of enzymes was undertaken.

There are many differences between SSAO and MAO enzymes, such as subcellular location, co-factor requirement and substrate and inhibitor selectivity (see Callingham and Barrand, 1987). The association with copper of many SSAO enzymes has become more certain since their measured copper content has risen in line with improved methods of purification (Yamada and Yasunobu, 1962; Mondovi et al., 1963; Buffoni and Blaschko, 1964; Harris et al., 1974). In those SSAO enzymes where copper has been unequivocally demonstrated, there is a strong case for the presence of one copper atom per subunit, with most enzymes having two subunits (see Knowles and Yadav, 1984). Practically all copper-containing enzymes are oxidoreductases but, to date, reduction of Cu(II) has not been seen in the SSAO catalytic cycle (see Mondovi and Riccio, 1989). Recent controversy has surrounded the identity of the organic co-factor of these enzymes. Pyridoxal phosphate (PLP), thought to be the most likely candidate, (see Blaschko, 1974) has a rival in pyrroloquinoline quinone (PQQ; see Duine and Frank, 1981). Some of the evidence for PQQ is weakened by the demonstration of Buffoni (1988) that some samples of pronase, used to hydrolyse enzymes to release fragments containing the co-factor, was contaminated with PQQ. However, Raman resonance spectroscopy favours PQQ against PLP (Moog et al., 1986; Knowles et al., 1987). All this may turn out to be somewhat esoteric in view of the recent, detailed and compelling evidence that the cofactor is most likely to be 6-hydroxydopa (TOPA; Janes et al., 1990). The fact that TOPA, a known neurotoxin, could be the cofactor will provide much scope for hypothesis.

The reaction catalysed by SSAO enzymes, while being double-displacement (ping-pong), appears to be of the amino-transferase type, to produce aldehyde, ammonia and hydrogen peroxide, i.e.,

$$E\text{–}CHO + R\text{–}CH_2NH_2 = E\text{–}CH_2NH_2 + RCHO$$

$$E\text{–}CH_2NH_2 + O_2 + H_2O = E\text{–}CHO + NH_3 + H_2O_2$$

Anaerobic release of aldehyde but not of ammonia and the kinetics of product inhibition are evidence in favour of this form of reaction (Oi et al., 1970; see Knowles and Yadav, 1984). PQQ has been implicated in this reaction with SSAO from bovine plasma (Klinman et al., 1989).

Some SSAO enzymes

At present, no satisfactory means of classifying the SSAO enzymes exists, and reliance is placed on such devices as tissue and subcellular distribution and substrate selectivity. Use of an allegedly preferred substrate, irrespective of any physiological significance, to name an enzyme, while convenient to distinguish it from MAO, can lead to great confusion (see Zeller, 1979; Lewinsohn, 1984). There are few selective inhibitors of SSAO enzymes and substrate selectivities of the enzymes vary between species and between tissues of the same species (see Kapeller-Adler, 1970; Callingham and Barrand, 1987). In the present context, two groups of SSAO enzymes will be considered, 1, plasma enzymes and 2, tissue-bound enzymes.

Plasma amine oxidases

A group of SSAO enzymes, which have a different substrate profile from diamine oxidase (DAO, histaminase), has been found circulating freely in the blood plasma of many species (see Blaschko, 1974; Callingham and Barrand, 1987). While most enzymes restrict their activities to the deamination of primary monoamines, the first to be discovered, in the plasma of ruminants by Hirsch (1953) was capable, in addition, of deaminating the polyamines, spermine and spermidine through an action on the primary amine group (see Morgan, 1985). In most, but not all cases, benzylamine (BZ), which is not apparently a physiologically important amine, is a preferred substrate. It seems likely that more than one SSAO enzyme may be found in the plasma. In the horse, with benzylamine as substrate, two activities with K_m values of about 30 and 106 μM were found (Williams and Callingham, 1987) and, in the sheep, with the same substrate, the values were 2.4 and 946 μM (Callingham et al., 1988).

Tissue-bound SSAO enzymes

At present no suitable collective name for these enzymes has been agreed upon. Benzylamine oxidase has been used by many authors (see Lewinsohn, 1984) but MAO also deaminates BZ. Clorgyline- or pargyline-resistant amine oxidase has also been used as a name, but all the SSAO enzymes are relatively resistant to these inhibitors, at least in vitro (Zeller, 1979), while in vivo, SSAO enzymes may even be inhibited irreversibly by metabolites of the acetylenic inhibitors (Elliott et al., 1989a). Often, it is the tissue or organ in which it is being studied or from which it has been extracted, that serves to identify the enzyme. For simplicity, the singular term SSAO will be used from now on without any implication that it will refer to just a single enzyme. Not enough is known to say for certain that a tissue only contains one SSAO and it is the overall activity that is being examined.

Physiological and pharmacological significance of SSAO

Plasma SSAO

Rather little is known of the function of plasma SSAO. Substantial differences exist between the substrate selectivities and amounts of these enzymes between the various species, which makes the task of apportioning functional significance all the harder. The ability of ruminant plasma SSAO to deaminate spermine and spermidine has been shown to parallel the development of the rumen, both phylogenetically and during maturation of the young animal (Blaschko and Hawes, 1959). This has led to the suggestion that this SSAO serves to protect against polyamines formed by cellulose fermentation in the rumen. However, the appearance of plasma SSAO in goats was unaffected by dietary alterations that prevented the development of the rumen (Blaschko and Bonney, 1962). Moreover, the products of polyamine oxidation, the aminoaldehydes, are more toxic than the parent compounds (Tabor et al., 1964; Byrd et al., 1977; Morgan, 1987). While the importance of plasma SSAO is yet to be found, its activity can change with the condition of the animal. For example, in some recent experiments (Elliott et al., 1990), plasma SSAO activity in Welsh Mountain ewes has been measured colorimetrically, with spermine as substrate. Some were made diabetic with alloxan, and a number of these, together with normal control ewes, were then allowed to become pregnant. Induction of diabetes mellitus caused a rise in plasma SSAO activity of 60% which was reached after 70 days. This change was seen in both pregnant and non-pregnant animals. In normal pregnant ewes, SSAO activity remained constant for the first 100 days but declined by 50% during the final month before parturition. These results would suggest the presence of hormonal influences on the plasma SSAO. In addition, there is also the possibility that, during the period before birth, there is competition between fetus and mother for the available supplies of copper. It is possible that the plasma SSAO activity could be a sensitive indicator of the copper status of sheep.

Tissue-bound SSAO

The contractile responses of rat aorta to applied doses of tryptamine (TRP) may be modified by the combined actions of MAO and SSAO. In the rabbit aorta, these contractions are due, in the main, to an action of TRP on 5-HT receptors, although α-adrenoceptors may be involved as well (Stollak and Furchgott, 1983). While SSAO of the rat aorta is responsible for a significant proportion of [^{3}H]-TRP metabolism, selective inhibition of SSAO potentiates the action of TRP only after prior inhibition of MAO, with MAO-A apparently the more important of the enzymes involved (Lyles and Taneja, 1987; Taneja and Lyles, 1988). Inhibition of MAO in the rabbit aorta does not affect the contractile potencies of TRP and 5-HT to the same degree, and this may be due to a combination of TRP's adrenergic potency and its susceptibility to metabolism by SSAO (Stollak and Furchgott, 1983; see Bradley et al., 1985).

Further evidence suggesting that blood vessel SSAO may influence the actions and fate of circulating amines has been obtained from use of the isolated perfused mesenteric arterial bed of the rat. Pretreatment of rats with MDL 72145 (E-2-(3′,4′-dimethoxyphenyl)-3-fluoroallylamine), which is a potent inhibitor of SSAO in these vessels, reduced the amount of metabolites following the addition of BZ (25 μM) or TYR (100 μM) to the perfusing fluid by 83% and 53%, respectively. While cocaine (3 μM) increased the amount of metabolites from TYR, inactivation of MAO-A with clorgyline (10 μM) had little effect (Elliott et al., 1989b). When the effects of these agents were examined on the pressor responses of the perfused arterial bed, neither inhibition of SSAO nor of MAO-A on their own caused any change in the maximum or the area under the curve of the response to administered TYR (Elliott et al., 1989c). However, inhibition of both amine oxidases did lead to substantial potentiation, suggesting that SSAO and MAO-A in blood vessels may operate in concert to deaminate their common substrates. Thus, in those tissues that contain high amounts of SSAO such as blood vessels and brown adipose tissue (BAT; see Callingham and Barrand, 1987), a scavenging rôle is possible, which may be important under conditions when this enzyme is inhibited by drugs.

Quite apart from any useful physiological function which SSAO may or may not possess, it is now well documented that metabolism of certain xenobiotics by SSAO can give rise to products more harmful than the parent compounds. Allylamine (3-aminopropene) is used in a number of industrial processes, for example the manufacture of pharmaceuticals and the vulcanisation of rubber. It is also a cardiovascular toxin, chronic exposure to which can result in lesions which mimic acute myocardial necrosis and atherosclerosis (see Boor and Hysmith, 1987). It is now known that allylamine is metabolised by SSAO to its aldehyde, acrolein (2-propenal). This compound is able to alkylate the thiol group on glutathione transferase, thereby depleting glutathione and rendering the tissue susceptible to peroxidative damage (Haenen et al., 1988a). Initial damage takes the form of intimal smooth muscle cell proliferation, with an astonishing predilection for heart and aortic tissue. In the rat, this can be explained when one considers the high SSAO activity associated with these cells, resulting in a localised production of acrolein. Inhibitors of SSAO, administered before exposure of aortic smooth muscle cells to allylamine, can prevent such damage (Hysmith and Boor, 1988).

The need for selective inhibitors of SSAO

Without highly selective inhibitors of SSAO, it may prove very difficult to resolve the relative importance of these enzymes in the presence of MAO. This is of particular importance if experiments are to be carried out in vivo. As was seen above, it was only possible to employ MDL 72145 as a selective inhibitor of SSAO in the rat mesenteric arterial bed because MAO-B was virtually absent. However, in tissues where this enzyme is present, such as the blood vessels of the horse, MDL 72145 cannot be used. In an attempt to overcome this problem we have recently re-investigated the suitability of procarbazine (N-isopropyl-α-(2-methylhydrazino)-*p*-toluamide HCl; PCZ) as a selective inhibitor of SSAO. PCZ is a carcinostatic agent used, in combination with other drugs, in the treatment of Hodgkin's disease. Its anti-tumour activity is probably due to its azoxy metabolites (Tweedie et al., 1987), with both cytochrome P_{450} and MAO involved in their production (Coomes and Prough, 1983). Further metabolism to free radicals may also contribute to anti-cancer activity (Dost and Reed, 1967; Prough et al., 1985; Tweedie et al., 1987). Administration of PCZ occasionally results in characteristic CNS toxicity (Pfefferbaum et al., 1989), and this, along with its hydrazine structure, led to the demonstration that it was a weak inhibitor of MAO (De Vita et al., 1965).

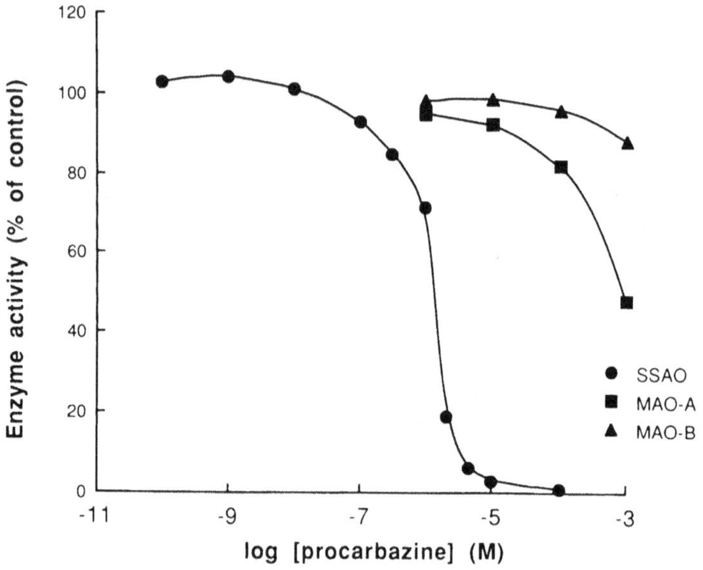

Fig. 1. Inhibition of amine oxidase activities of rat brown adipose tissue and liver following incubation with procarbazine. Volumes of crude homogenates of rat brown adipose tissue (SSAO) and of rat liver (MAO-A and -B) were incubated with various concentrations of procarbazine for 30 min before the addition of 5-hydroxytryptamine (250 μM, final concentration for MAO-A) or benzylamine (250 μM, 10 μM, for MAO-B and SSAO). The resulting enzyme activity was compared with that of an homogenate without inhibitor

This agent was suggested to us as a possible selective inhibitor of SSAO by Dr. Kettler of Hoffmann-La Roche, Basel, as a result of earlier studies undertaken in his department and as a result of the original report by Lewinsohn et al. (1978), who demonstrated that PCZ was an inhibitor of SSAO in various human tissues, displaying a potency greater than compounds such as semicarbazide and benserazide.

With SSAO from the interscapular BAT of the rat we have shown that PCZ had an IC_{50} value of about 2 μM with very little effect on either MAO-A or -B (Fig. 1). The degree of inhibition depended upon the period of pre-incubation of enzyme and inhibitor before the addition of benzylamine as substrate. Dialysis of enzyme, pre-incubated with PCZ until maximum inhibition had occurred, caused partial recovery of activity to an extent that increased with temperature. This might be due to some ageing of the original enzyme-inhibitor bond similar to that seen with organophosphorus anticholinesterases, or partial conversion from tight-binding to covalent interaction.

The metabolic route to the azoxy metabolites of PCZ is via azoprocarbazine (APCZ; Fig. 2) Preliminary studies have been made into the possibility that APCZ is the active inhibitor or that it shares with PCZ an ability to inhibit SSAO in a highly selective manner. APCZ was synthesised by the method of Swaffar et al. (1989). The resulting product turned out to possess an IC_{50} value of approximately 20 nM, and should further studies demonstrate a high degree of selectivity over MAO, APCZ could prove to be a useful tool.

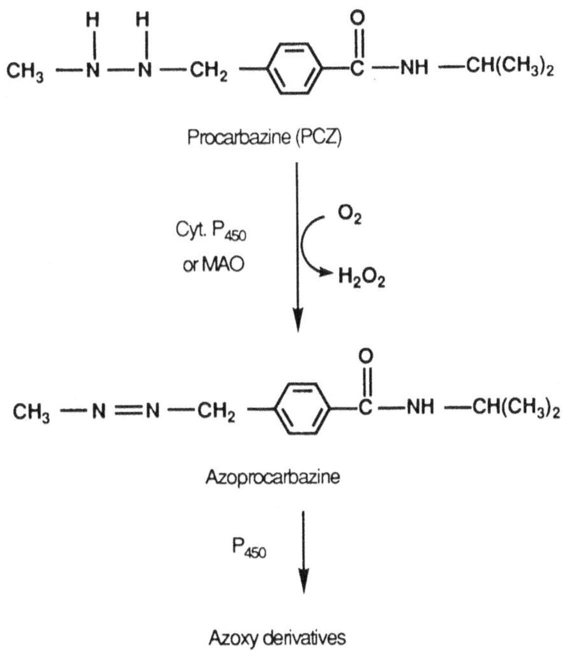

Fig. 2. The metabolic conversion of procarbazine to azoprocarbazine

Another compound with potential is the 4-picolylamine derivative, B24 (3,5-ethoxy-4-aminomethylpyridine. 2HCl; Banchelli et al., 1990). It is a highly selective, reversible inhibitor of SSAO, with a potency similar to that of PCZ. The compound would seem also to be a substrate for plasma-borne, but not tissue-bound SSAO, and displays non-competitive kinetics.

A physiological role for reaction products?

Although the identification of an endogenous substrate for SSAO is of paramount importance, the fate of the reaction products is also worthy of our attention. Hydrogen peroxide production invariably accompanies that of the aldehyde, independent of the substrate used, and while usually thought of as a toxic by-product with its rapid removal essential (see Chance et al., 1979), recent evidence points towards a role for H_2O_2 in a number of physiological systems (see Ramasarma, 1982). For instance, the effects of insulin on glucose transport in fat cells can be mimicked by H_2O_2 (Czech et al., 1974a, b; see Callingham and Barrand, 1987) and while it appears unlikely that H_2O_2 acts as a second messenger in this system, the activity of the insulin receptor kinase can be stimulated by peroxide, suggesting a modulatory role for H_2O_2 (Yu et al., 1987). This would require a localised production of H_2O_2, and a membrane bound SSAO enzyme in close proximity to the receptor would be ideally suited to this task.

The ability of peroxide to oxidise sulphydryl groups also renders adenylate cyclase, guanylate cyclase and a number of ATPase enzymes, which contain such a group, susceptible to modulation, and this again suggests that a proximal SSAO enzyme could affect a cell's response on the binding of an agonist to a receptor (Braughler, 1982; Wright and Drummond, 1983; Bellomo et al., 1983, 1987; Waldman and Murad, 1987; Haenen et al., 1988b). Finally, H_2O_2 has been implicated in catecholamine-stimulated prostaglandin (PG) biosynthesis, its action being to stimulate the synthesis of endoperoxides, which in turn are transformed non-enzymatically into $PGF_{2\alpha}$ using catecholamines as hydrogen donors (Seregi et al., 1982, 1983).

Although existing information mainly supports the view that SSAO in the smooth muscles of blood vessels may be a scavenger of pharmacological concentrations of circulating amines, we have little idea of its physiological importance both in blood vessels and in adipose tissue. In blood plasma, it seems that our knowledge is even more rudimentary. It may be that, with these newer selective inhibitors, there is now a real chance that the mysteries surrounding these enzymes will be solved.

Acknowledgements

A. H. is a Medical Research Council Scholar. We wish to thank Prof. Da Prada and Dr. Kettler of Hoffmann-La Roche & Co. Ltd., Basel, for all their help and for the procarbazine and the Horserace Betting Levy Board for financial support.

References

Banchelli G, Buffoni F, Elliott J, Callingham BA (1990) A study of the biochemical pharmacology of 3,5-ethoxy-4-aminomethylpyridine (B24), a novel amine oxidase inhibitor with selectivity for tissue bound semicarbazide-sensitive amine oxidase enzymes. Neurochem Int 17:215–221

Bellomo G, Mirabelli F, Richelmi P, Orrenius S (1983) Critical role of sulfhydryl group(s) in ATP-dependent Ca^{2+} sequestration by the plasma membrane fraction from rat liver. FEBS Lett 163:136–139

Bellomo G, Thor H, Orrenius S (1987) Alterations in inositol phosphate production during oxidative stress in isolated hepatocytes. J Biol Chem 262:1530–1534

Blaschko H (1974) The natural history of amine oxidases. Rev Physiol Biochem Pharmacol 70:83–148

Blaschko H, Bonney R (1962) Spermine oxidase and benzylamine oxidase. Distribution, development and substrate specificity. Proc Roy Soc (Lond) B156:268–279

Blaschko H, Hawes R (1959) Observations on spermine oxidase of mammalian plasma. J Physiol 145:124–131

Boor PJ, Hysmith RM (1987) Allylamine cardiovascular toxicity. Toxicology 44:129–145

Bradley PB, Humphrey PPA, Williams RH (1985) Tryptamine-induced vasoconstrictor responses are mediated predominantly via 5-hydroxytryptamine receptors. Br J Pharmacol 84:919–925

Braughler JM (1982) Involvement of sulfhydryl groups in the oxidative modulation of particulate lung guanylate cyclase by nitric oxide and N-methyl-N′-nitro-N-nitrosoguanidine. Biochem Pharmacol 31:1239–1244

Buffoni F (1966) Histaminase and related amine oxidases. Pharmacol Rev 18: 1163–1199

Buffoni F (1988) On the nature of the organic cofactor of pig plasma benzylamine oxidase. Pharmacol Res Commun 20 [Suppl] 4:159–160

Buffoni F, Blaschko H (1964) Benzylamine oxidase and histaminase: purification and crystallization of an enzyme from pig plasma. Proc Roy Soc (Lond) B161:153–167

Byrd WJ, Jacob DM, Amos MS (1977) Synthetic polyamines added to cultures containing bovine sera reversibly inhibit in vitro parameters of immunity. Nature 267:621–623

Callingham BA, Barrand MA (1987) Some properties of semicarbazide-sensitive amine oxidases. J Neural Transm [Suppl] 23:37–54

Callingham BA, Elliott J, Williams RB (1988) Amine oxidase interactions in the cardiovascular system. In: Dahlström A, Belmaker RH, Sandler M (eds) Progress in catecholamine research, part A. Basic aspects and peripheral mechanisms. Alan R Liss, New York (Neurology and neurobiology, vol 42A, pp 109–113)

Chance B, Sies H, Boveris A (1979) Hydroperoxide metabolism in mammalian organs. Physiol Rev 59:527–605

Czech MP, Lawrence JC, Lynn WS (1974a) Hexose transport in isolated brown fat cells. A model system for investigating insulin action on membrane transport. J Biol Chem 249:5421–5427

Czech MP, Lawrence JC, Lynn WS (1974b) Evidence for the involvement of sulfhydryl oxidation in the regulation of fat cell hexose transport by insulin. Proc Natl Acad Sci USA 71:4173–4177

Coomes MW, Prough RA (1983) The mitochondrial metabolism of 1,2-disubstituted hydrazines, procarbazine and 1,2-dimethylhydrazine. Drug Metab Dispos 11:550–555

De Vita VT, Hahn MA, Oliverio VT (1965) Monoamine oxidase inhibition by a new carcinostatic agent, N-isopropyl-α-(2-methylhydrazino)-*p*-toluamide (MIH). Proc Soc Exp Biol Med 120:561–565

Dost FN, Reed DJ (1967) Methane formation in vivo from, N-isopropyl-α-(2-methylhydrazino)-*p*-toluamide hydrochloride, a tumor-inhibiting methylhydrazine derivative. Biochem Pharmacol 16:1741–1746

Duine JA, Frank J Jn (1981) Quinoproteins: a novel class of dehydrogenases. Trends Biochem Sci 6:278–280

Elliott J, Callingham BA, Barrand MA (1989a) In vivo effects of (E)-2-(3′,4′-dimethoxyphenyl)-3-fluoroallylamine (MDL 72145) on amine oxidase activities in the rat. Selective inhibition of semicarbazide-sensitive amine oxidase in vascular and brown adipose tissue. J Pharm Pharmacol 41:37–41

Elliott J, Callingham BA, Sharman DF (1989b) Metabolism of amines in the isolated perfused mesenteric arterial bed of the rat. Br J Pharmacol 98:507–514

Elliott J, Callingham BA, Sharman DF (1989c) The influence of amine metabolizing enzymes on the pharmacology of tyramine in the isolated perfused mesenteric arterial bed of the rat. Br J Pharmacol 98: 515–522

Elliott J, Fowden AL, Callingham BA, Sharman DF, Silver M (1990) Physiological and pathological influences on sheep plasma amine oxidase: effect of pregnancy and experimental alloxan-induced diabetes mellitus. Res Vet Sci (submitted)

Finazzi-Agro (1989) Copper-containing amine oxidases. In: Jongejan JA, Duine JA (eds) PQQ and quinoproteins. Kluwer Academic Publishers, London, pp 279–282

Haenen GRMM, Vermeulen NPE, Tai Tin Tsoi JNL, Regetli HMN,Timmerman H, Bast A (1988a) Activation of the microsomal glutathione-s-transferase and reduction of the glutathione dependent protection against lipid peroxidation by acrolein. Biochem Pharmacol 37:1933–1938

Haenen GRMM, Van Dansik P, Vermeulen NPE, Timmerman H, Bast A (1988b) The effect of hydrogen peroxide on beta-adrenoceptor function in the heart. Free Radic Res Commun 4:243–249

Harris ED, Gonnerman WA, Savage, JE, O'Dell BL (1974) Connective tissue amine oxidase. II. Purification and partial characterization of lysyl oxidase from chick aorta. Biochim Biophys Acta 341:332–344

Hirsch JG (1953) Spermine oxidase: an amine oxidase with specificity for spermine and spermidine J Exp Med 97:345–355

Hysmith RM, Boor PJ (1988) Role of benzylamine oxidase in the cytotoxicity of allylamine toward aortic smooth muscle cells. Toxicology 51:133–145

Janes SM, Mu D, Wemmer D, Smith AJ, Kaur S, Maltby D, Burlingame AL, Klinman JP (1990) A new redox cofactor in eukaryotic enzymes: 6-hydroxydopa at the active site of bovine serum amine oxidase. Science 248:981–987

Kapeller-Adler R (1970) Amine oxidases and methods for their study. Wiley Interscience, London

Klinman JP, Hartman C, Janes SM (1989) Mechanistic probes of copper amine oxidases. In: Jongejan JA, Duine JA (eds) PQQ and quinoproteins. Kluwer Academic Publishers, London, pp 297–305

Knowles PF, Yadav KDS (1984) Amine oxidases. In: Lontie R (ed) Copper proteins and copper enzymes, vol II. CRC Press, Boca Raton Fl, pp 103–129

Knowles PF, Pandeya KB, Rius FX, Spencer CM, Moog RS, McGuirl MA, Dooley DM (1987) The organic cofactor in plasma amine oxidase: evidence for pyrroloquinoline quinone and against pyridoxal phosphate. Biochem J 241: 603–608

Lewinsohn R (1984) Mammalian monoamine-oxidizing enzymes, with special reference to benzylamine oxidase in human tissues. Braz J Med Biol Res 17:223–256

Lewinsohn R, Böhm K-H, Glover V, Sandler M (1978) A benzylamine oxidase distinct from monoamine oxidase B-widespread distribution in man and rat. Biochem Pharmacol 27: 1857–1863

Lyles GA, Taneja DT (1987) Effects of amine oxidase inhibitors upon tryptamine metabolism and tryptamine-induced contractions of rat aorta. Br J Pharmacol 90:16P

Mondovi B, Riccio P (1989) On the role of copper in PQQ amine oxidases. In: Jongejan JA, Duine JA (eds) PQQ and quinoproteins. Kluwer Academic Publishers, London pp 289–295

Mondovi B, Rotilio G, Costa MT (1963) The purification of hog kidney diamine oxidase. In: Snell EE, Fasella PM, Braunstein A, Rossi-Fanelli A (eds) Chemical and biological aspects of pyridoxal catalysis. Pergamon Press, Oxford, pp 415–428

Moog RS, McGuirl MA, Cote CE, Dooley DM (1986) Evidence for methoxatin (pyrroloquinoline quinone) as the cofactor in bovine plasma amine oxidase from resonance Raman spectroscopy. Proc Natl Acad Sci USA 83:8435–8439

Morgan DML (1985) Polyamine oxidases. Biochem Soc Trans 13:322–326

Morgan DML (1987) Oxidized polyamines and the growth of human vascular endothelial cells. Prevention of cytotoxic effects by selective acetylation. Biochem J 242:347–352

Oi S, Inasumu M, Yasunobu KT (1970) Mechanistic studies of beef plasma amine oxidase. Biochemistry 9:3378–3383

Pfefferbaum B, Pack R, Van Eys J (1989) Monoamine oxidase inhibitor toxicity. J Am Acad Child Adolesc Psychiatry 28:954–955

Prough RA, Brown MI, Moloney SJ, Wiebkin P, Cummings SW, Spearman ME, Guengerich FP (1985) The activation of hydrazines to reactive intermediates. In: Boobis AR, Caldwell J, De Matteis F, Elcombe CR (eds) Microsomes and drug oxidations. Taylor and Francis, London, p 330

Ramasarma T (1982) Generation of H_2O_2 in biomembranes. Biochim Biophys Acta 694: 69–93

Seregi A, Serfözö P, Mergl Z, Schaefer A (1982) On the mechanism of the involvement of monoamine oxidase in catecholamine – stimulated prostaglandin biosynthesis in particulate fraction of rat brain homogenates: role of hydrogen peroxide. J Neurochem 38:20–27

Seregi A, Serfözö P, Mergl Z (1983) Evidence for the localization of hydrogen peroxide-stimulated cyclooxygenase activity in rat brain mitochondria: a possible coupling with monoamine oxidase. J Neurochem 40:407–413

Stollak JS, Furchgott RF (1983) Use of selective antagonists for determining the types of receptors mediating the actions of 5-hydroxytryptamine and tryptamine in the isolated rabbit aorta. J Pharmacol Exp Ther 224:215–221

Swaffar DS, Horstman MG, Jaw JY, Thrall BD, Meadows GG, Harker WG, Yost GS (1989) Methylazoxyprocarbazine, the active metabolite responsible for the anticancer activity of procarbazine against L1210 leukemia. Cancer Res 49:2442–2447

Tabor CW, Tabor H, Bacharach U (1964) Identification of the aminoaldehydes produced by the oxidation of spermine and spermidine with purified plasma amine oxidase. J Biol Chem 239:2194–2203

Taneja DT, Lyles GA (1988) Further studies on the interactions between amine

oxidase inhibitors and tryptamine-induced contractions of rat aorta. Br J Pharmacol 93:253P
Tweedie DJ, Erikson JM, Prough RA (1987) Metabolism of hydrazine anti-cancer agents. Pharmacol Ther 34:111–127
Waldman SA, Murad F (1987) Cyclic GMP synthesis and function. Pharmacol Rev 39:163–196
Williams RB, Callingham BA (1987) MAO and haemoglobin: interaction or artefact? Pharmacol Toxicol 60 [Suppl] 1:52
Wright M, Drummond GI (1983) Inactivation of the β-adrenergic receptor in skeletal muscle by dithiols. Biochem Pharmacol 32:509–515
Yamada H, Yasunobu KT (1962) Monoamine oxidase II. Copper, one of the prosthetic groups of plasma monoamine oxidase. J Biol Chem 237:3077–3082
Yu KT, Khalaf N, Czech MP (1987) Insulin stimulates the tyrosine phosphorylation of a Mr = 160,000 glycoprotein in rat adipocyte plasma membranes. J Biol Chem 262:7865–7873
Zeller EA (1963) Diamine oxidase. In: Boyer PD, Lardy H, Myrbäck K (eds) The enzymes, vol 8, 2nd edn. Academic Press, London, pp 313–335
Zeller EA (1979) Classification and nomenclature of monoamine oxidases and other amine oxidases. In: Singer TP, Von Korff RW, Murphy DL (eds) Monoamine oxidases: structure, function, and altered functions. Academic Press, London, pp 531–537

Authors' address: Dr. B. A. Callingham, Department of Pharmacology, University of Cambridge, Tennis Court Road, Cambridge, CB2 1QJ, U.K.

J Neural Transm (1990) [Suppl] 32: 291–314

Intestinal diamine oxidases and enteral-induced histaminosis: studies on three prognostic variables in an epidemiological model

J. Sattler[1] and **W. Lorenz**[2]

[1] Department of Theoretical Surgery, II. Surgical Clinic, University of Innsbruck, Austria
[2] Institute of Theoretical Surgery, Centre of Operative Medicine I, University of Marburg, Federal Republic of Germany

Summary. The danger of luminal histamine administered orally or formed in the intestinal fluid by bacteria has long been neglected. However, the demonstration of blocking intestinal diamine oxidase (DAO) by a variety of common drugs has revived the discussion and has created a new disease concept: enteral-induced histaminosis. In an animal model the three central prognostic variables of this disease concept (large amounts of histamine in food to make the individual ill, blocking of DAO by commonly used drugs, and the relationship between increased plasma histamine levels and disease manifestation by exogenous histamine application) were tested with randomized trials in vivo and biochemical tests in vitro using semipurified enzymes from pig and man.

In the first trials authentic histamine in quantities similar to that in normal amounts of food or cheese bought from a supermarket produced lifethreatening reactions if the DAO was inhibited by pretreatment with aminoguanidine. In the second series of experiments in vitro a numerous commonly used drugs was shown to inhibit both the porcine and human enzyme. Some of the inhibitors were really strong, such as dihydralazine, chloroquine, pentamidine, cycloserine, clavulanic acid, dobutamine, pancuronium and others. The type of inhibition was sometimes competitive as in the case of dihydralazine and pancuronium, sometimes non competitive (e.g. pentamidine) which may be important for long-term treatment. In the third group of experiments a relationship between the dose of i.v. injected histamine and the elevation in plasma histamine levels and clinical symptoms in pigs was demonstrated. Hence, elevated plasma histamine in pigs acts as a pathogenetic factor for the disease manifestation. It is concluded that after modelling enteral-induced histaminosis in an animal the trias of variables shown in this study should be consequently investigated in man.

Introduction

Similar to other biogenic amines histamine is widely accepted to act as a vasoactive mediator in severe surgical disease states (Neugebauer et al., 1987). It is involved in cardiovascular disturbances such as shock, myocardial ischaemia and arrhythmias (Lorenz and Doenicke, 1985; Lorenz et al., 1990) especially if the normal regulatory (homoeostatic) mechanisms of the organism are altered by age, disability or considerable stress (Lorenz and Doenicke, 1985; Lorenz et al., 1990; Neugebauer and Lorenz, 1981). Histamine formation and inactivation are initiated and controlled enzymatically by histidine decarboxylase (EC. 4.1.1.22) (Schayer, 1966; Maslinski, 1975a), histamine methyltransferase (EC 2.1.1.8) (Schayer, 1959) and diamine oxidase (EC 1.4.3.6) (Kim et al., 1969; Maslinski, 1975b; Lorenz et al., 1975). Hence it was not only subject of many investigations whether histamine was released (Lorenz and Doenicke, 1986; Lorenz et al., 1990; Neugebauer and Lorenz, 1981), but also whether changes in the activity of these enzymes were associated with the previously mentioned important clinical events (Neugebauer and Lorenz, 1981).

In the intestinal tract diamine oxidase (DAO) has been found to be present in high activities in almost all mammalian species investigated. It was mainly located in the mucosa of all these species (Kusche et al., 1975) and was intracellularly found predominantly in the cytoplasm (Shakir et al., 1977; Bieganski et al., 1983). Similarities in the enzyme characteristics could be established since the purification was successful with almost the same procedures. Only the stability of the enzymes after purification showed great differences in various species. The lowest stability was found in human subjects with only a few days (Bieganski et al., 1983) or several weeks at optimum conditions (Sattler et al., 1985). The enzyme from pig and dog intestine, however, was stable for 4–14 months (Kusche et al., 1975; Sattler et al., 1985).

Although monoamine oxidases were also present in the intestinal mucosa histamine was only deaminated by DAO (Kusche et al., 1975). The enzymes acted preferentially on short chain aliphatic diamines, like putrescine and cadaverine. In the case of the human enzyme it was very surprising that also N^{τ}-methylhistamine as the product of histamine methylation by histamine methyltransferase (HMT) was deaminated with a high reaction velocity (Bieganski et al., 1980, 1983) (Fig. 1). Intestinal diamine oxidases from other species, e.g. the dog and pig, also deaminated this monoamine, but to much lesser extent (Kusche et al., 1975). Hence, especially in humans intestinal diamine oxidase can play a central role in histamine catabolism as a bottleneck which is determining the inactivation of histamine either passing the intestinal wall from the lumen or being released by immunological and non-immunological stimuli in the intestinal mucosa (Fig. 1).

Intestinal diamine oxidases differed in their pH-optima for deamination of various substrates (Kusche et al., 1975). However, Michaelis constants

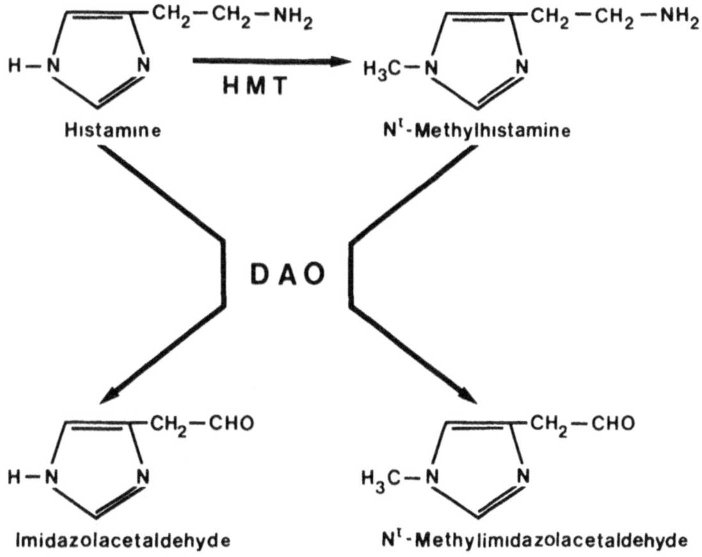

Fig. 1. Central role of intestinal diamine oxidase (*DAO*) as a bottleneck in human histamine catabolism. *HMT* histamine methyltransferase

were quite similar for the most interesting substrates such as putrescine (8.3×10^{-5} M), histamine (1.9×10^{-5} M) and N^{τ}-methylhistamine (9.7×10^{-5} M) (Bieganski et al., 1983). In all cases a strong substrate inhibition was observed at supraoptimum concentrations. This may be of physiological and pathophysiological significance in the case of histamine release from mast cells within the intestinal mucosa (locally very high concentrations around the degranulating mast cells) or in the case of a considerable luminal histamine load.

Intestinal diamine oxidases from all species investigated (Kusche et al., 1975; Bieganski et al., 1983) were strongly inhibited by aminoguanidine (IC_{50} 2–8 $\times 10^{-7}$ M). This uniform behaviour of the enzymes led to the important finding that DAO inhibition aggravated considerably the circulatory insufficiency in mesenteric ischaemia. Plasma histamine was elevated in that disease state (Kusche et al., 1977, 1981, 1986; Wollin et al., 1980). However, not only aminoguanidine inhibited the human and other mammalian enzymes. From nearly 500 drugs used in surgery, anaesthesia and intensive care medicine about 100 inhibited both the canine and human intestinal DAO in vitro in concentrations which could achieve pathophysiological significance (Sattler et al., 1985)

This interesting finding in vitro created the hypothesis whether inhibition of intestinal DAO by drugs could influence the fate of histamine in the human gut. Using the standard disease concept in epidemiology (Fletcher et al., 1982) histamine loading in the lumen could be obtained either in food, by blood after gastrointestinal bleeding or by plasma proteins in ileus and sepsis after decarboxylation of histidine by bacteria. The disease concept of an "enteral-induced histaminosis" (Fig. 2) was created (Sattler and

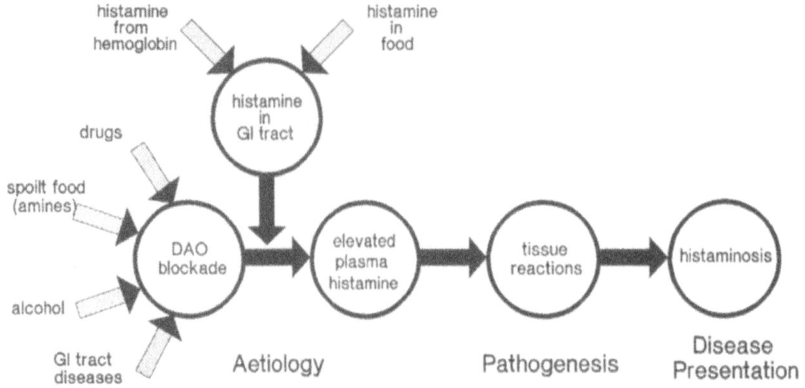

Fig. 2. Scheme for the disease concept of enteral-induced histaminosis: histamine in the gastrointestinal tract (from food or upper GI bleeding) can only enter the circulation in case of DAO blockade (risk factor). Elevated plasma histamine leads to the wellknown disease presentation (Lorenz et al., 1982) which is denominated here as histaminosis. This model is originally based on epidemiological terms (Fletcher et al., 1982) and was modified by Sattler and Lorenz (1987) for enteral-induced histaminosis

Lorenz, 1987; Sattler et al., 1988, 1989) in which three important prognostic variables were especially important for the final disease manifestation:

(1) Presence of high amounts of histamine in the lumen which was derived either from histamine contaminated food, from decarboxylated histidine in haemoglobin splitting products or from plasma secreted into the lumen in the case of ileus, intestinal ischaemia, peritonitis or other septic conditions.
(2) Presence and action of DAO blocking drugs which were given as therapeutics to treat haemorrhagic or septic shock.
(3) Increased plasma histamine concentrations after histamine load and DAO blockade which achieved such high levels as they were observed after histamine injection or histamine release by drugs and which have already been shown to cause severe cardiovascular disturbances in man (Lorenz et al., 1981, 1982, 1990; Lorenz and Doenicke, 1985).

All these three prognostic variables in the epidemiological model (Fig. 2) were investigated in three different randomized studies in pigs and were proven to be finally successful.

Material and methods

Experimental animals

For all 3 separately performed randomized animal studies 40 pigs were used from a local breeder (cross-breed German landrace/Pietrain). Weight (18–30 kg) and age (7–9 weeks) were of similar distribution in all three trials, both sexes were used in approximately equal proportions. Before the experiment the animals were adapted to their environment for a week with a standardized diet (Hemo 111) and starved for the last 24 h before the experiments with tap water ad libitum.

Origin of enzymes

Tissue samples from the terminal ileum from 3 pigs (150 kg, approximately 1 year old) were obtained from the slaughterhouse 10–20 min after death by exsanguination and rapidly used for further purification (see section below).

The healthy intestinal tissue from 5 human subjects, undergoing intestinal and colorectal surgery (age 47–69, male) were collected and used for enzyme preparation.

Drugs and chemicals

486 drugs from all relevant therapeutic groups were purchased or generously obtained from various companies as chemically pure substances. From all water-soluble drugs 210 were tested for DAO inhibition on human and 124 on porcine enzyme respectively. All chemicals used for the histamine assay were the same as described by Lorenz et al. (1987). Aminoguanidine was purchased from Schuchard, histamine from Fluka (Buchs).

General experimental protocol

The following procedure was identical in all three animal studies. The pigs were anaesthetized first by injecting pentobarbital (Nembutal^R, CEVAG), (30 mg/kg b.w.) intraperitoneally. Following oral intubation (Magill tubus with ID 6.0, OD 8.0 mm) with the aid of an appropriate spatula pentobarbital was further administered via a braunula in an ear vein until sufficient anaesthesia was obtained under spontaneous breathing.

After an interval of 10 min for observation 3 femoral catheters (1 arterial, 2 venous) were inserted for blood pressure measurement (Statham), blood withdrawal for plasma samples and an extra line for drug application (diamine oxidase blockers, or saline). For continuous monitoring of the heart rate 3 ECG electrodes were applied to thorax and abdomen to obtain leads with appropriate signals. Ventilation frequency (breaths/min) was monitored via a tension gauge at the intubation equipment. All physical parameters were registered by a Hellige Apparatus (SMV 111).

Following an interval of 20 minutes to obtain homoeostasis all animals in the three trials received histamine either orally (trial 1 and 2) or intravenously (trial 3) as outlined separately in a following section. Thereafter the physical parameters were continuously registered, the animals were observed for clinical symptoms (erythema, bronchospasm, vomiting etc.) according to a defined protocol by two observers and blood for plasma histamine was taken at defined intervals. In trial 1 and 2 blood was taken immediately before and every 20 minutes after oral histamine challenge for two hours. In trial 3 samples were withdrawn again immediately before and 1, 3, 5, 7, 10 and 20 min after the intravenous histamine bolus injection. Plasma preparation in pigs was a critical step in the measurement procedure and was performed according to Sattler et al. (1988). Histamine was assayed according to Lorenz et al. (1972) using the newest elaboration of this fluorometric method (Lorenz and Neugebauer, 1990). The reliability was demonstrated recently in the European external quality control study (Neugebauer et al., 1990).

Administration of histamine in the three trials

Trial 1: In this randomized trial with 2 × 5 pigs half of them received aminoguanidine (100 mg/kg b.w.) for DAO blockade at the end of the preparation period and 20 min before oral instillation of the histamine solution. Then authentic histamine prepared from pure substance in saline was given first in a dose of 60 mg base/100 ml saline and after 2 hours in a second dose of 600 mg/100 ml saline using a stomach tube.
Trial 2: In this randomized study with 2 × 5 animals half of them received again 100 mg/kg aminoguanidine at the end of the preparation period and 20 min before application of a histamine rich cheese into the duodenum. The second half received the same volume of saline instead of DAO blocking drug. Original Swiss cheese was bought from a supermarket and was found to contain 65 mg histamine base/100 g using the fluorometric technique (Lorenz and Neugebauer, 1990). This cheese was suspended in a mixture of saline and 10% ethyl alcohol by homogenization with an Ultraturrax homogenizer (100 g/150 ml). This procedure was choosen because of the fatty consistence of the suspension and the intention to adapt the experimental design as much as possible to the daily custom of alcohol consumption simultaneously with cheese intake. From pilot experiments in our laboratory as well as from data in the literature (Cortot et al., 1981; Meyer et al., 1986; Meyer, 1987) it was recognized that the very high fat content of the cheese and the size of the homogenized particles produced a very long lasting stay of the food in the stomach after feeding through the tube. In addition, the animals were anaesthetized and did not respond normally to food (e.g. salivation, bile production, gut movement etc.). Therefore, the cheese mixture was administered intraduodenally. After laparatomy and incision of the duodenum a ligation was placed orally to a catheter which was inserted into the duodenum through the incision to administer the cheese suspension. Then the duodenal and abdominal wound was closed and the registration of physical parameters and clinical signs and blood collection started as previously described.
Trial 3: In this randomized study with 2 × 10 animals half of them received either 100 mg/kg aminoguanidine or the same volume of saline intravenously at the end of the preparation period. All animals received authentic histamine prepared freshly from the pure substance intravenously through the femoral catheter. The study started with the placebo infusion (saline solution). Blood was taken at the previously defined time points. 30 min later 1 μ/kg histamine, 30 min later 5 μ/kg and again 30 min later 10 μ/kg histamine were applied through the same catheter. Blood withdrawal and measurement of the physical parameters were done as previously defined. Ringer solution was infused with a speed of 1l/h to maintain fluid and electrolyte homoeostasis as far as possible.

In vitro studies with semipurified diamine oxidase from porcine and human intestine

Purification of the enzymes

About 30 g of mucosa was scraped off and the crude enzyme preparation was purified by homogenization, centrifugation and ion exchange chromatography on DEAE Sephacel (Pharmacia) as described previously by Bieganski et al. (1983) and Sattler et al. (1985). The specific activity of the 15-fold purified human enzyme in

0.05 M phosphate buffer was 6.7 mU/mg biuret protein. The porcine DAO was purified 11-fold with a specific activity of 50.5 mU/mg protein. Both enzyme preparations remained sufficiently stable when stored at $-70\,^{\circ}C$ for several weeks (man) or months (pig) and were then used in the in vitro experiments with the inhibitory drugs.

Assay of DAO activity

For the detection of inhibitory capacity of the drugs (^{14}-C) putrescine assay for DAO activity was performed according to Kusche et al. (1973) in the absence and presence of three different concentrations of the drugs under investigation (1×10^{-3}, 1×10^{-4} and 1×10^{-5} M). Inhibition was expressed as loss of DAO activity in percent compared to controls without the inhibitor.

To obtain further information about the type of inhibition, putrescine concentrations were varied in 10 different steps (see Fig. 7) and the inhibitory substance ranged from 1×10^{-6} to 2×10^{-6} M (dihydralazine), 1×10^{-5} to 5×10^{-5} M (pentamidine), 1×10^{-4} to 3×10^{-4} M (pancuronium), 1×10^{-3} to 3×10^{-3} mol/l (pirenzepine, acetylcysteine, isoniazid). Because of the higher dilution of the substrate compared to the standard assay (^{3}-H)-putrescine (specific activity 1430 GBq/m M) (New England Nuclear, Boston MA) was used in this test system. The test mixture was incubated for 20 minutes at $37\,^{\circ}C$ after addition of the substrate. Preliminary studies showed that, under these conditions, the reaction velocity was constant for at least 30 min. The analysis of the type of inhibition was carried out by the classical graphical display according to Lineweaver and Burk (1934).

Statistical analysis

The descriptive analysis of the data was performed both with the median-percentile system and the mean-S.D. system as well as by rates and proportions. Clinical data (flush, vomiting, death etc.) were submitted to inferential statistic using the Chi Square Test. For ordinal data, e.g. the mean arterial pressure (MAP) and the plasma histamine level, non parametric statistics were used (Mann Whitney U-test). The design of trial 3 (repetitive injection of histamine in various concentrations) demanded an analysis of variance (ANOVA) after logarithmic transformation. By this way a comparison between the different groups was possible. Finally, kinetic data for enzyme inhibition were submitted to non linear regression analysis using the sums of least squares.

Results

Luminal histamine load by oral administration of authentic histamine or food

1. Low dose authentic histamine

Receiving 60 mg pure histamine in 100 ml saline by oral route all pigs whose DAO was blocked by aminoguanidine reacted with at least one of the

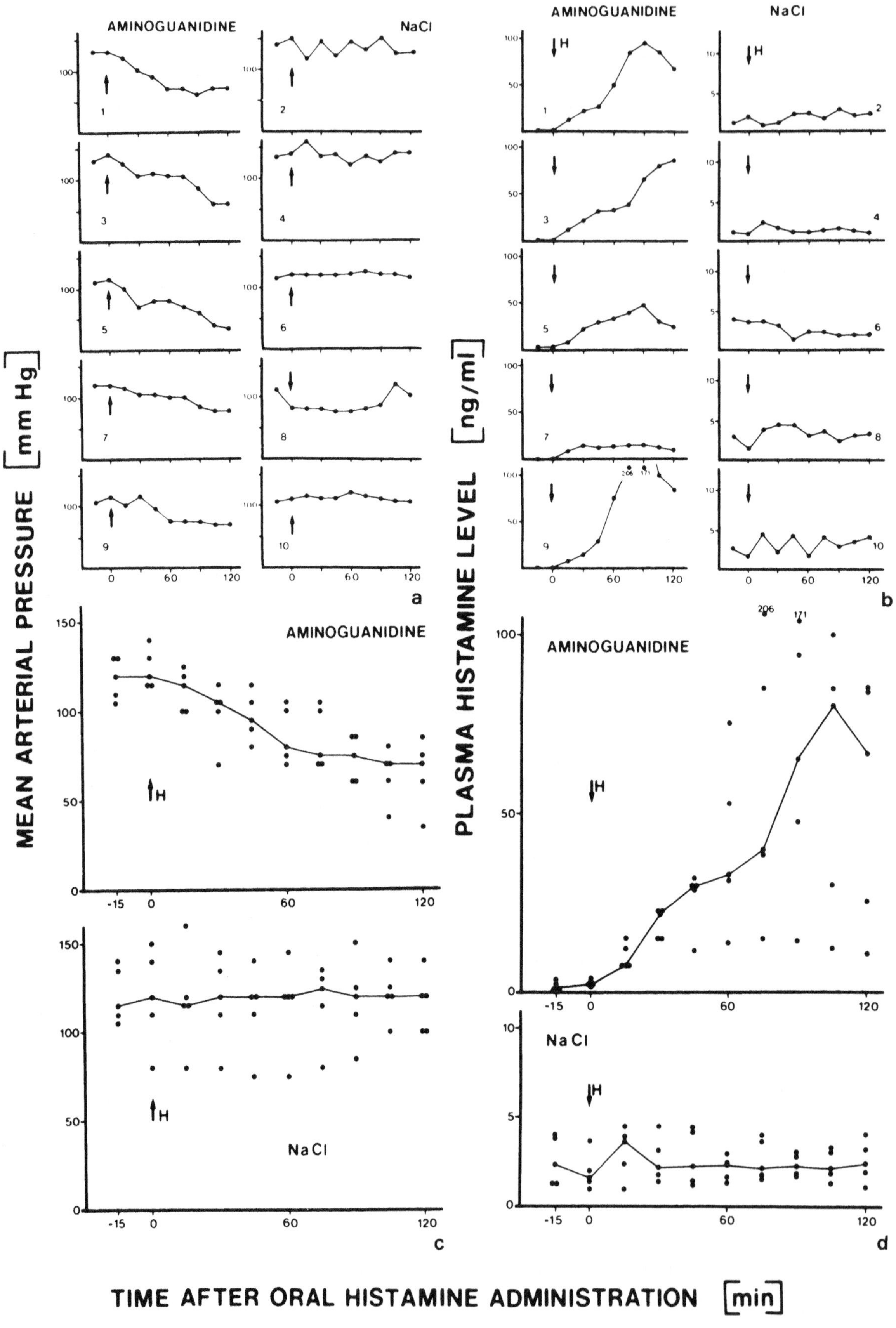

AMINOGUANIDINE
NaCl
MEAN ARTERIAL PRESSURE [mm Hg]
PLASMA HISTAMINE LEVEL [ng/ml]
H
a
b
c
d
206
171
TIME AFTER ORAL HISTAMINE ADMINISTRATION [min]

clinical symptoms erythema, bronchospasm, vomiting, salivation and with physically measured biological reactions. Among them hypotension was the most impressive response as illustrated in Fig. 3 (a and c). There was a good relation in time between maximum hypotension and maximum plasma histamine increase after 90 min (Fig. 3a–d) which reflected the resorption of luminal histamine. The median increase of plasma histamine (Fig. 3d) was 75 ng/ml, (range 14.8 – 206 ng/ml). The one pig which died because of the severity of symptoms had the highest plasma histamine concentrations (No 9 in Fig. 3b and top values in Fig. 3d). No symptoms at all could be observed without DAO blockade (NaCl control). Only 1 animal (No 8 in Fig. 3b) showed a slight plasma histamine elevation up to 5 ng/ml. (Notice the tenfold higher scale in Fig. 3b and d for plasma histamine in control animals (NaCl)!). Single values are shown in the figure 3 because doubts in resorption of histamine from the intestinal lumen raised from variability of the findings (Sattler and Lorenz, 1987).

2. High dose authentic histamine

Tenfold more histamine (600 mg pure base, again in 100 ml saline) given to the same animals as with low dose histamine after 2 hours, led to nearly the same multiplication in histamine resorption (Fig. 4) and to a dangerous aggravation of symptoms and reactions. Such histamine amounts can occur easily in spoilt tuna fish (Sattler and Lorenz, 1987). As a consequence 3 of the remaining 4 animals from the first two hours of the experiment died in shock and the onset of reactions occurred earlier than before – within only 5 and 10 min after application. Fig. 4 (upper curve) illustrates as an example the repetitive start of the plasma histamine elevation, the maximum value immediately before death (1798 ng/ml!!) and the time of death in one pig (No 9 in Fig. 3). Animals without DAO-blockade generally tolerated better the 600 mg doses of histamine. 3 pigs had no detectable histamine resorption (increase in plasma histamine) but animal No 4 (Fig. 4, lower curve) showed both plasma histamine elevation (up to 200 ng/ml!!) and hypotension combined with tachycardia and clinical symptoms. However, all these reactions largely disappeared 180 min after high dose histamine challenge whereas only 1 of the DAO blocked animals survived the observation period and

◀ **Fig. 3.** Effect of oral administration of 60 mg histamine (low dose) to 10 pigs on mean arterial pressure (MAP) (A and C) and plasma histamine concentration (B and D). The upper part shows the individual values of all animals whereas medians of MAP (C) and plasma histamine (D) are given below. Animals without DAO blockade (NaCl) developed no increase in plasma histamine, so the scale in (B) and (D) is tenfold increased in those pigs than in aminoguanidine treated ones. Arrows represent the start of histamine challenge via the stomach tube

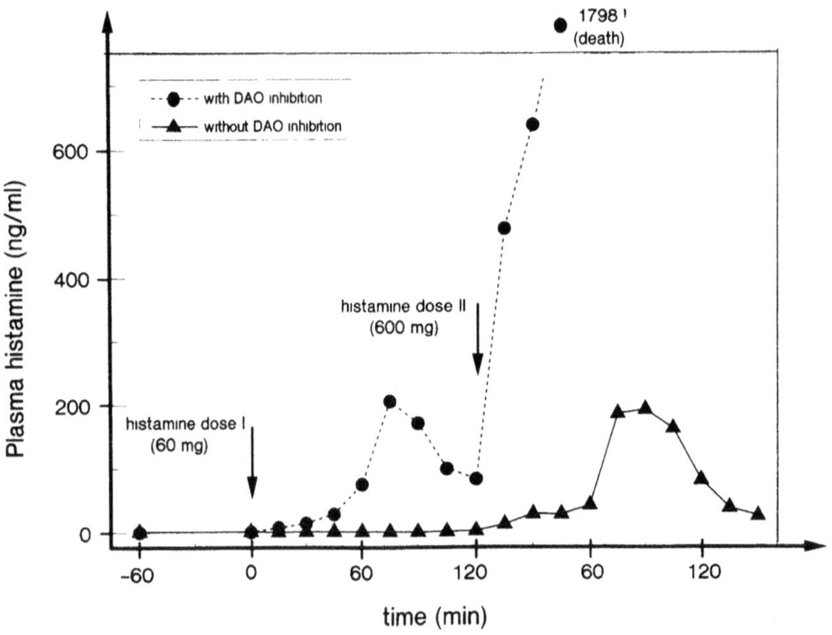

Fig. 4. Effect of oral administration of 60 mg (low dose I) and 600 mg histamine (high dose II) to pigs on plasma histamine concentration. Two animal of each group are shown exemplarily. The upper curve belongs to animal No 9 (see Fig. 3B) with DAO blockade (test group); the lower curve to No 8 without DAO blockade (NaCl)

even in this animal symptoms continued and histamine increase remained at a high level over the whole time of experiment.

3. Histamine-rich food intake

A 100 g portion of histamine-rich cheese administered intraduodenally had fatal consequences in all pigs under DAO-blockade (Fig. 5a) However, without enzyme inhibition nothing happened in the animals with respect to either plasma histamine levels or biological reactions and clinical symptoms (Fig. 5b). Although the amount of histamine in the cheese corresponded to the low histamine dose (60 mg) in the previous experiments the death rate was 4 of 5 animals and hence corresponded to pigs treated with 600 mg histamine via the normal oral route. Figure 5a,b illustrates only one example for each of 5 animals under study. All animals with DAO inhibition reacted in the same massive way with a rapid onset of hypotension, plasma histamine elevation and clinical symptoms (general redness, vomiting, bronchospasm and death). The plasma histamine concentration reached levels of about 100 ng/ml within 60 minutes. For documentation of these dramatic observations the data are compiled in Table 1.

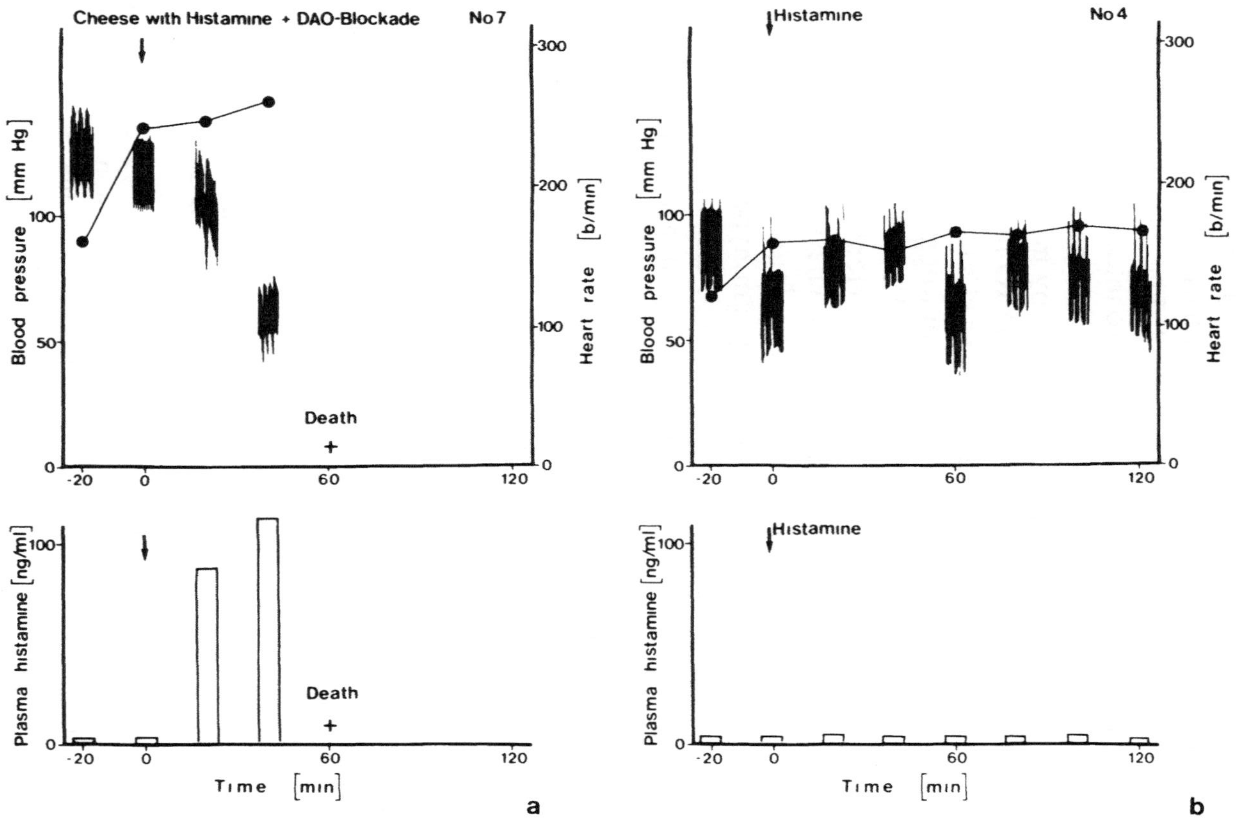

Fig. 5. Effects of histamine rich (65 mg) cheese feeding to pig No 7 under DAO blockade (a) on blood pressure (■ ■), heart rate (●—●—●) (upper part) and plasma histamine concentration (below). The same amount of histamine administered without DAO blockade in animal No 4 (b) yielded neither biological reactions nor any plasma histamine increase

Inhibition of porcine and human intestinal DAO in vitro by commonly used drugs

1. Number and efficacy of in vitro DAO inhibitors

Search for DAO inhibitors included commonly used drugs from all therapeutic groups. Some of the 486 compounds collected which inhibited porcine and/or human DAO are listed in Table 2. The 28 substances in each columns are organized according to their in vitro inhibitory potency against partially purified intestinal DAO. To date 94 substances -all water soluble-inhibited the human enzyme and 49 that of pig by more than 20%.

A comparison of the inhibitory efficacy of DAO inhibitors between pig and man is necessary to justify the transfer of results from the animal model to the human situation. As already known from dog experiments (Sattler et al., 1985) most of the substances listed in Table 2 inhibited diamine oxidases of several species in vitro, but to a different degree. There were, however, a few drugs which inhibited only the enzyme from one species (Table 2; Sattler et al., 1985). Hence, it was very important to find quite a group of drugs in all therapeutic domains, e.g. antihypertensives, antimalarial drugs, antibiotics etc., which inhibited both enzymes quite strongly. Again, like food, some of them are given to patients in g amounts and over a period of weeks or even much longer.

Chloroquine is widely distributed as an antimalarial drug and was shown to be one of the most potent DAO blockers in all species tested. Therefore, a more detailed dose-inhibition curve was established and is presented in Fig. 6. For different drugs different slopes of the curves were

Table 1. Comparison of maximum histamine increase, maximum hypotension and clinical symptoms after histamine-rich cheese application (orally) under the influence of DAO blockade in 10 pigs. Half of them received saline (control group without DAO blockade) instead of aminoguanidine (test group with DAO blockade)

	Test group with DAO blockade				Control group without DAO blockade		
Pig No	Histamine increase (ng/ml)	Hypotension (mmHg)	Clinical symptoms (n)	Pig No	Histamine increase (ng/ml)	Hypotension (mmHg)	Clinical symptoms (n)
1	93	55	E, B	1	5	20	–
2	99	50	E, S, D	2	4	5	–
3	77	45	E, B, S, D	3	6	20	–
4	59	55	E, D	4	2	20	–
5	109	70	E, S, D	5	2	20	–

E erythema, *B* bronchospasm, *S* salivation, *D* death, *V* vomiting

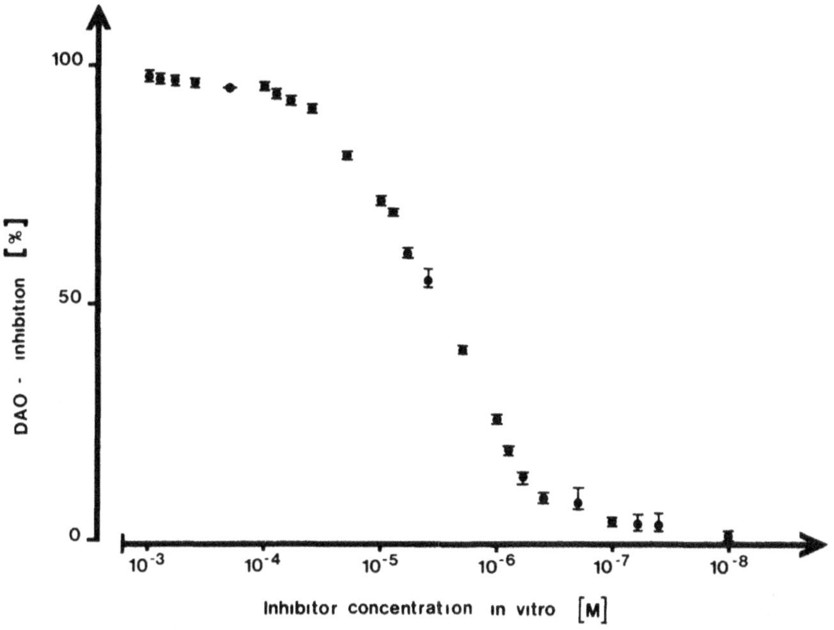

Fig. 6. Dose-inhibition curve of chloroquine tested on semipurified intestinal porcine diamine oxidase in vitro. Inhibitor concentration ranged from 1×10^{-3} to 5×10^{-8} M. Mean values $\pm$ SD of 3 experiments for each inhibitor concentration. Graphical evaluation of IC_{50} at 50% inhibition (see Table 3)

found. From them the IC_{50} was calculated as shown in Table 3. It was most interesting to find pentamidine, a drug used in palliative treatment of AIDS among the strong DAO inhibitors.

2. Kinetic analysis for the type of DAO inhibition

The selection of drugs for a study on the kinetic type of inhibition followed the criteria of high efficacy, clinical relevance in intensive care medicine and high doses of usual drug administration. The results of Lineweaver-Burk type experiments are listed in Table 4. The neuromuscular blocking drug pancuronium and the antihypertensive drug dihydralazine apparently acted as competitive inhibitors when tested with putrescine as substrate, whereas acetylcysteine, pirenzepine and pentamidine behave as non competitive DAO blockers.

As one example, the Lineweaver Burk plot with dihydralazine is shown in Fig. 7. All 3 slopes intersect at the y-axis thus favouring a competitive type of inhibition. The Km value of 3×10^{-4} M for putrescine as substrate is in good agreement with those reported earlier for porcine DAO (Kusche et al., 1975). The finding of non-competitive inhibition for pentamidine is in good agreement with the results of Karvonen (1987) who used also putrescine in the same concentration range, but DAO from small intestine of rats.

Table 2. Drugs with inhibitory effects on porcine and human intestinal DAO listed according to their potency at the highest concentration ($1 + 10^{-3}$ M), Values for DAO inhibition are given as means from 3 experiments per individual drug and represent the percentage of inhibition in the presence of the drug at 3 different concentrations (1×10^{-3}, 1×10^{-4} and 1×10^{-5} M)

Pig				Man			
Drugs (inn)	% Inhibition at drug concentration (m)			Drugs (inn)	% Inhibition at drug concentration (m)		
	1×10^{-3}	1×10^{-4}	1×10^{-5}		1×10^{-3}	1×10^{-4}	1×10^{-5}
Dihydralazine	100	99	96	Dihydralazine	100	100	96
Chloroquine	100	96	75	Cycloserine	100	99	63
Pentamidine	100	92	64	Pentamidine	100	95	79
Cycloserine	100	71	6	Distigmine	96	82	48
Distigmine	94	58	20	Tubocurarine	95	82	51
Clavulanic acid	91	37	3	Imipenem	95	2	0
Dobutamine	85	41	10	Paromomycin	93	74	28
Pancuronium	81	31	9	Chloroquine	90	69	45
Carbocromen	81	29	4	Pancuronium	88	63	29
Reproterol	80	39	7	Dobutamine	84	48	9
Propafenone	77	25	1	Clomethiazole	82	20	17
Tubocurarine	75	23	10	Clavulanic acid	80	23	3
Clomipramine	74	24	8	Propafenone	77	34	15
Colistin	73	19	7	Salazosulfapyridine	75	47	6
Promethazine	70	26	5	Dilazep	73	49	19
Tetroxoprim	70	25	4	Guanethidine	73	34	8

Table 2. Continued

Pig				Man			
Drugs (inn)	% Inhibition at drug concentration (m)			Drugs (inn)	% Inhibition at drug concentration (m)		
	1×10^{-3}	1×10^{-4}	1×10^{-5}		1×10^{-3}	1×10^{-4}	1×10^{-5}
Amiphenazole	70	11	1	Carbocromen	65	27	2
Thiamazole	55	18	1	Verapamil	61	16	9
Pramiverine	53	24	19	Isoniazid	56	5	0
Antazoline	51	11	5	Pirenzepine	55	20	1
Guanethidine	47	10	9	Acetylcysteine	55	11	4
Triflupromazine	44	11	2	Tetroxoprim	54	10	0
Epinephrine	44	3	1	Amiphenazole	52	8	0
Isoniazid	42	11	10	Aminophylline	40	8	2
Aminophylline	36	8	5	Orciprenaline	39	12	7
Verapamil	35	5	1	Metoclopramide	36	9	0
Cefuroxime	34	9	4	Cefuroxime	33	11	7
Dipyrone	33	10	4	Dipyrone	30	3	1

INN international non-proprietary name of the drugs

Table 3. Efficacy of DAO-inhibitors on porcine intestinal DAO in vitro. The values were calculated as IC_{50} from dose-inhibition curves (see Fig. 6)

INN	IC_{50}(mM)
Chloroquine	0.002
Pentamidine	0.004
Dobutamine	0.2
Pancuronium	0.3
Acetylcysteine	1.0
Verapamil	2.0

INN international non-proprietary name of the drugs

Table 4. Apparent type of DAO-inhibition of several commonly used drugs investigated on porcine intestinal DAO in vitro

INN	competitive	non competitive
Pancuronium	+	
Acetylcysteine		+
Pirenzepine		+
Dihydralazine	+	
Pentamidine		+

INN international non proprietary name of the drugs

Plasma histamine levels in pigs after intravenous histamine administration: comparison with those after food intake

The crucial question in the epidemiological model of an enteral-induced histaminosis is Koch-Dale criterion 3: can the disease be elicited by exogenous administration of the chemical factor? (Sattler et al., 1990). This was tested in the third group of experiments. Exogenous histamine given (i.v.) as a bolus produced in all 20 pigs studied a dose dependent increase in plasma histamine levels. This is demonstrated by medians in Fig. 8 and by parameters of variation including the results of an analysis of variance in Table 5. The animals pretreated with a DAO-blocking drug always showed significantly higher increases than the controls with placebo solution. The rapid onset in the first minute and the immediate clearance from the circulation within the following 3 or 5 min are in agreement with data in man (Lorenz et al., 1982). Clinical symptoms and biological reactions followed the course of the plasma histamine curve. Already at a dose of 1 μg/kg 7 of 10 pigs showed

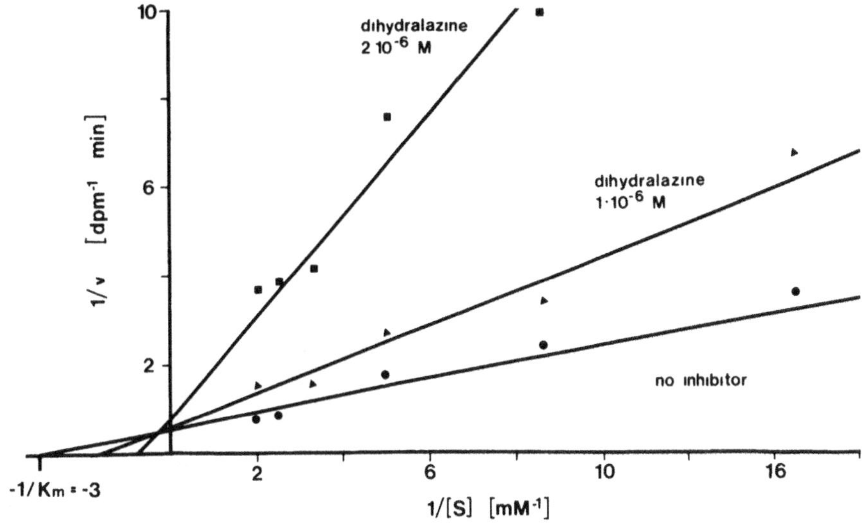

Fig. 7. Kinetics of inhibition of porcine intestinal diamine oxidase by the antihypertensive drug dihydralazine. The substrate concentration ranged from 0.07 to 0.5×10^{-3} M

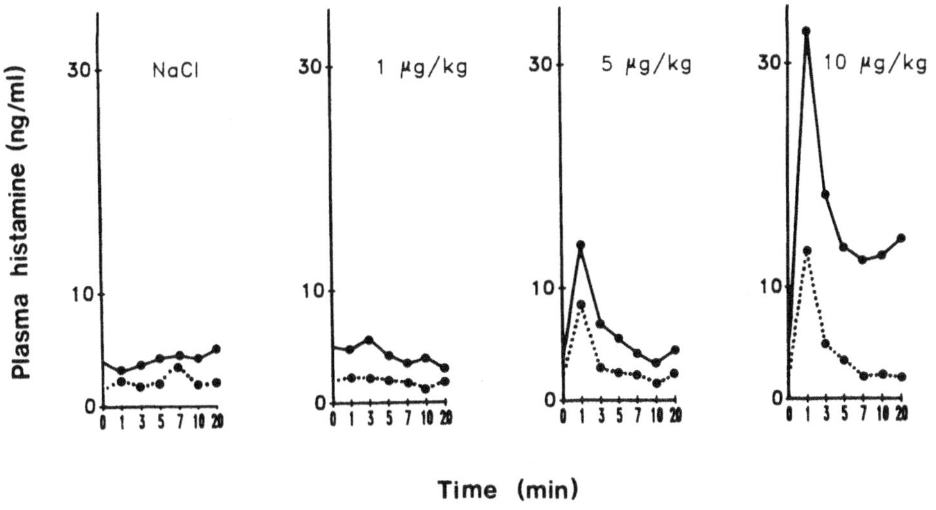

Fig. 8. Illustration of increase and time course of plasma histamine levels in pigs after exogenous administration of histamine up to orders of magnitude which were also observed after food intake. Note the reduced elimination of histamine especially at the highest dose (10 μg/kg b.w.) when a DAO blocker was administered: —— median values of 10 animals pretreated with 100 mg/kg aminoguanidine, ----- median values of animals pretreated with saline. Aminoguanidine – pretreated animals show already higher base-line values

responses such as hypotension and tachycardia, but these increased considerably after 5 μg/kg and 10 μg/kg histamine. Hence, there was no doubt that exogenous histamine can imitate symptoms and reactions of enteral-induced histaminosis.

Table 5. Changes of plasma histamine levels in pigs by different doses of exogenous histamine

Treatment	Histamine bolus (μg/kg b.w.)	Changes in plasma histamine (ng/ml) 1	5	20
		minutes after bolus injection		
with DAO inhibition	0	0.45 (0–1.1)	0.1 (0.–1.2)	0 (0.–1.0)
	1	1.05* (0.7–2.2)	0.45 (0–1.7)	0.2 (0–1.1)
	5	8.0* (6.0–9.4)	1.75 (1.0–2.4)	0.4 (0–1.0)
	10	16.1* (12.3–20.0)	3.9 (0.7–7.9)	0.9 (0.1–2.1)
without DAO inhibition	0	0 (0–0.8)	0 (0–0.1)	0.8 (0–1.3)
	1	0.7** (0–1.4)	0 (0–0)	0 (0–0.4)
	5	4.15** (3.0–5.6)	0.2 (0–0.5)	0 (0–0.3)
	10	10.6** (8.1–13.5)	1.55 (0–2.9)	0 (0–0)

Values for the histamine increase are expressed as medians and (in brackets) as first and third quartiles. 10 of the animals received aminoguanidine (100 mg/kg) for DAO inhibition, 10 saline solution as a placebo control. Basal plasma histamine levels correspond to the injection of saline (0 values). ANOVA with Scheffé test showed significant differences between the 3 doses histamine injected: * ($p < 0.03$); ** ($p < 0.02$)

Discussion

"Enteral-induced histaminosis" is a disease concept which according to Fig. 1 is based on (1) *large amounts of histamine in the gut lumen as the exposition*, (2) *factors which reduce DAO activity in the intestinal wall* as aetiological factors and (3) *free histamine in sufficient amounts in the general circulation* as the main factor in pathogenesis. All these three prognostic variables for the final disease presentation (or manifestation of clinical symptoms) were *separately* demonstrated in man (Senanayake and Vyravanathan, 1981; Kottegoda and Uragoda, 1976; Motil and Scrimshaw, 1979) and animals (Sattler et al., 1988) but never in a *concept of a disease entity* (Wulff, 1981). In a disease taxonomy of our days such a concept must include name of disease, definition, cause (aetiology and pathogenesis), manifestations and prognoses, diagnoses and treatment (Wulff, 1981). Of all these variables three central ones were modelled in the experimental series of this article.

The first variable of the model was the resorption of large amounts of histamine that make an animal ill. The results showed that histamine normally present in large or considerable amounts in certain food (cheese, sauerkraut, wine, fish etc.) can be resorbed with fatal consequences for the organism if DAO is blocked by a classical inhibitor aminoguanidine. Although the animals were of the same breed, weight, age and sex and had the same treatment before the experiment they showed a considerable biological variability both with regards to plasma histamine increase and to clinical symptoms. This observation which is typical for all diseases (Wulff, 1981) reflects the danger of even a moderate histamine load in the gut lumen, if individuals are "sensible" and renders the present recommendations for hazard levels of histamine load with 50 mg published by public health authorities of the Food and Drug Administration (1982) very questionable. However, histamine may also be derived from histidine in haemoglobin after gastrointestinal bleeding or from serum proteins in ileus and septic conditions if bacterial growth permits a large amount of the amino acid be decarboxylated by highly active bacterial histidine decarboxylases (Gale, 1957; Horakova et al., 1971; Riley and Snell, 1979). Hence, enteral-induced histaminosis may be a more common disease than just food-poisoning and is a subject of interest for several medical disciplines including surgery and intensive care.

The second variable of the model is DAO inhibition by commonly used drugs. More than 10% of the drugs tested in this experimental series showed a considerably inhibitory capacity against the intestinal enzyme in vitro in pig and man and have to be considered as potentially dangerous in vivo if it is taken into account that some drugs are given in gram doses over weeks, e.g. the antibiotics or even longer e.g. antihypertensives. More important seems the finding of different types of DAO inhibition. Non competitive inhibition, as demonstrated with pentamidine, does not need necessarily depend on high plasma or tissue concentrations of the drugs. The repetitive administration over several days or weeks may lead to stepwise inactivation of intestinal DAO until all enzymic activity is blocked leading to a progressive resorption of luminal histamine into the circulation. The observation of Loescher et al. (1987) could easily be explained by such a process. A patient with AIDS suffering from Pneumocystis carinii infection and treated with pentamidine developed histamine like symptoms (e.g. hypotension, tachycardia, abdominal cramps, flush, headache) not immediately, but after 8 days of treatment which all disappeared within 2 days after stopping the drug infusion. Pentamidine was also shown in this investigation to be a non competitive DAO inhibitor a finding which confirmed previous observations (Sattler et al., 1985; Karvonen, 1987).

The third variable of the model is the demonstration of increased plasma histamine levels which were shown to make ill by exogenous route (Koch-Dale criterion 3) (Lorenz et al., 1984). Our results with exogenous histamine given as a bolus intravenously demonstrate that the pig is sensitive to histamine elevation in the plasma although basal values and threshold concentrations are somewhat higher than registered in man and dog (Lorenz et al., 1982; Lorenz and Doenicke, 1978). The symptoms and biological reaction occurred immediately (1 min) after histamine injection and plasma histamine increased concomitantly. Thus, the porcine model was shown to be suitable for studying the *pathogenesis* of enteral-induced histaminosis. This is a central criterion in the disease taxonomy according to Wulff (1981).

The diagnosis of the disease is not always easy, especially when skin symptoms are not detected. But the effect of prophylaxis could already be shown in a further series of experiments. The combined use of $H_1 + H_2$ receptor antagonists in pigs did not prevent histamine resorption after DAO blockade, but any occurrence of symptoms although plasma histamine levels reached 100 ng/ml and more (Sattler et al., 1989). This finding fulfills a further criterion in the disease taxonomy.

If the finding in pigs as a model are extrapolated cautiously to man the enormous amount of drugs which is prescribed every year (Schwabe and Paffrath, 1989) has to be considered as a real danger for enteral-induced histaminosis (Table 6). Especially aged people receive drugs and can therefore meet one or more DAO blockers. The first consequences of our findings are therefore to investigate the trias histamine in the gut lumen, DAO-blockers and elevated plasma histamine levels consequently in man in

Table 6. Usage of DAO blocking drugs (packages) in FRG from September 1987–October 1988

INN	Prescription
Chloroquine	71.700
Prajmalium	164.300
Carbocromen	312.000
Pirenzepine	859.300
Promethazine	1.051.400
Amitriptyline	2.597.000
Dihydralazine	2.449.200
Verapamil	4.227.900
Acetylcysteine	6.005.600
Metoclopramide	6.098.600
Ambroxol	11.071.200

Data from: Arzneiverordnungsreport 1989 (Schwabe and Paffrath, 1989)

various conditions (normal food and drinking, patients in intensive care units etc.). However, it is also thinkable then to avoid enteral-induced histaminosis by 3 particular measures:

(1) Avoidance of DAO blocking agents if possible.
(2) Avoidance of histamine-rich food or better food processing which eliminate histamine from the product.
(3) The use of antihistamines in intensive care unit to lower the unpredictable risk of high plasma histamine levels as it was demonstrated for a proportion of patients in septic shock (Dietz et al., 1990) and upper gastrointestinal bleeding (Lindlar et al., 1990).

Acknowledgements

This study was supported by grant of Deutsche Forschungsgemeinschaft (Lo 199/15-2). The great support in producing this manuscript by Barbara Gstrein is very appreciated as well as the help with statistics by H. Sitter (dipl. math.).

References

Bieganski T, Kusche J, Feussner KD, Hesterberg R, Richter H, Lorenz W (1980) Human intestinal diamine oxidase: substrate specificity and comparative inhibitor study. Agents Actions 10:108–110

Bieganski T, Kusche J, Lorenz W, Hesterberg R, Stahlknecht CD, Feussner KD (1983) Distribution and properties of human intestinal diamine oxidase and its relevance for the histamine catabolism. Biochim Biophys Acta 756:196–203

Cortot A, Phillips SF, Malagelada JR (1981) Gastric emptying of lipids after ingestion of a solid-liquid meal in humans. Gastroenterology 80:922–927

Dietz W, Stinner B, Lorenz W, Neugebauer E (1990) Erhöhte Plasmahistaminspiegel in der Frühphase des septischen Schocks: Ergebnisse einer prospektiven Querschnittsstudie. In: Beger HG (Hrsg) Sepsis und septischer Schock. Springer (in press)

Fletcher RH, Fletcher SW, Wagner EH (1982) Clinical epidemiology – the essentials. Williams and Wilkins, Baltimore London, pp 1–223

Food and Drug Administration (1982) Defect action levels for histamine in tuna; availability of guide. Fed Regist 47:40487–40488

Gale EF (1957) Determination of amino acids by use of bacterial amino acid decarboxylase. Methods Biochem Anal 4:285–306

Horàkovà Z, Zierdt CH, Beaven MA (1971) Identification of Lactobacillus as the source of bacterial histidine decarboxylase in rat stomach. Eur J Pharmacol 16:67–77

Karvonen E (1987) Inhibition of diamine oxidase of rat small intestine by pentamidine and berenil (diminazene aceturate). Biochem Pharmacol 36:2863–2864

Kim KS, Backus B, Harris M, Rourke P (1969) Distribution of diamine oxidase and imidazol N-methyltransferase along the gastrointestinal tract. Comp Biochem Physiol 31:137–145

Kottegoda SR, Uragoda CG (1976) Interaction of isoniazid and fish exhibiting high histamine levels. ICRS Med Sci 4:370

Kusche J, Richter H, Hesterberg R, Schmidt J, Lorenz W (1973) Comparison of the ^{14}C-putrescine assay with the NADH test for the determination of diamine oxidase: description of a standard procedure with a high precision and an improved accuracy. Agents Actions 3:148–156

Kusche J, Schmidt J, Schmidt A, Lorenz W (1975) Diamine oxidases in the small intestine of rabbits, dogs and pigs: separation from soluble monoamine oxidases and substrate specificity of the enzymes. Agents Actions 5:440–441

Kusche J, Stahlknecht CD, Lorenz W, Reichert G, Richter H (1977) Diamine oxidase activity and histamine release in dogs following acute mesenteric artery occlusion. Agents Actions 7:81–84

Kusche J, Lorenz W, Stahlknecht CD, Richter H, Hesterberg R, Schmal A, Hinterlang E, Weber D, Ohmann C (1981) Intestinal diamine oxidase and histamine release in rabbit mesenteric ischemia. Gastroenterology 80:980–987

Kusche J, Lorenz W, Hesterberg R (1986) The relevance of the diamine oxidase-histamine system for shock development following intestinal ischemia. In: Pfeiffer EJ (ed) Animal models for intestinal disease. CRC Press, Boca Raton Fl, pp 255–280

Lindlar R, Sattler J, Lorenz W, Schmitt K, Schäfer U, Stinner B (1990) Blut im Darm als pathogener Faktor: Nutzen von Mediatorbestimmungen bei oberer gastrointestinaler Blutung. In: Häring R (Hrsg) Gastrointestinale Blutung. Blackwell Ueberreuter Wissenschaft, Berlin, S44–49

Lineweaver H, Burk D (1934) Determination of enzyme dissociation constants. J Am Chem Soc 56:658–666

Loescher T, Loeschke K, Niebel J (1987) Severe ventricular arrhythmia during pentamidine treatment of AIDS-associated pneumocystis carinii pneumonia. Infections 15:455–455

Lorenz W (1975) Histamine release in man. Agents Actions 5:402–416

Lorenz W, Doenicke A (1978) Anaphylactoid reactions and histamine release by intravenous drugs used in surgery and anaesthesia. In: Watkins J, Ward AM (eds) Adverse response to intravenous drugs. Academic Press, London Grune & Stratton, New York, pp 83–112

Lorenz W, Doenicke A (1985) H1 + H2-blockade: a prophylactic principle in anaesthesia and surgery against histamine-release responses of any degree of severity. N Engl Reg Allergy Proc 6:37–57 (Part I), 6:174–194 (Part II)

Lorenz W, Neugebauer E (1990) Histamine determination: current techniques – fluorometric assays. In: Uvnäs B (ed) Handbook of experimental pharmacology, vol 97. Springer, Berlin Heidelberg New York, pp 9–30

Lorenz W, Reimann HJ, Barth H, Kusche J, Meyer R, Doenicke A, Hutzel M (1972) A sensitive and specific method for the determination of histamine in human whole blood and plasma. Hoppe-Seyler's Z Physiol Chem 353:911–920

Lorenz W, Doenicke A, Schöning B, Neugebauer E (1981) The role of histamine in adverse reactions to intravenous agents. In: Thornton A (ed) Adverse reactions of anaesthetic drugs. Biomedical Press, Elsevier/North-Holland, pp 169–238

Lorenz W, Doenicke A, Schöning B, Ohmann C, Grote B, Neugebauer E (1982) Definition and classification of the histamine-release response to drugs in anaesthesia and surgery: studies in the conscious human subject. Klin Wochenschr 60:896–913

Lorenz W, Thon K, Barth H, Neugebauer E, Reimann HJ, Kusche J (1983) Metabolism and function of gastric histamine in health and disease. J Clin Gastroenterol 5 [Suppl] 1:37–56

Lorenz W, Röher HD, Doenicke A, Ohmann C (1984) Histamine release in anaesthesia and surgery: a new method to evaluate its clinical significance with several types of causal relationship. Clin Anaesth 2:403–426
Lorenz W, Thon K, Neugebauer E, Stöltzing H, Ohmann Ch, Weber D, Schmal A, Hinterlang E, Barth H, Kusche J (1987) Reliability and practicability of the fluorometric-fluoroenzymatic histamine determination in pathogenetic studies on peptic ulcer. Detection limits and problems with specificity. Agents Actions 21:1–25
Lorenz W, Dietz W, Ennis M, Stinner B, Doenicke A (1990) Histamine in anaesthesia and surgery: causality analysis. In: Uvnäs B (ed) Handbook of experimental pharmacology, vol 97. Springer, Berlin Heidelberg New York, pp 385–439
Maslinski C (1975a) Histamine and its metabolism in mammals, part I. Chemistry and formation of histamine. Agents Actions 5:89–107
Maslinski C (1975b) Histamine and its metabolism in mammals, part II. Catabolism of histamine and histamine liberation. Agents Actions 5:183–225
Meyer JH (1987) Motility of the stomach and gastroduodenal junction. In: Johnson LR (ed) Physiology of the gastrointestinal tract, vol 1, 2nd edn. Raven Press, New York, pp 613–629
Meyer JH, Mayer EA, Jehn D, Gu Y, Fink AS, Fried M (1986) Gastric processing and emptying of fat. Gastroenterology 90:1176–1187
Motil KJ, Scrimshaw NS (1979) The role of exogenous histamine in scombroid poisoning. Toxicol Lett 3:219
Neugebauer E, Lorenz W (1981) Histamine in health and disease. Behring Inst Mitt 68:102–133
Neugebauer E, Lorenz W, Maroske D, Barthlen W, Ennis M (1987) The role of mediators in septic/endotoxic shock. A meta-analysis evaluating the current status of histamine. Theor Surg 2:1–28
Neugebauer E, Keyzer J, Oosting E, Lorenz W (1990) Reliability of current techniques for histamine determination in human plasma: the European external quality control study 1988. Agents Actions 30:274–277
Riley WD, Snell EE (1970) Histidine decarboxylase of Lactobacillus 30a. Biochemistry 9:1485–1491
Sattler J, Lorenz W (1987) Nahrungsmittelinduzierte Histaminose: Ein Krankheitsbild mit Diaminoxydasehemmung verschiedener Herkunft. Münch Med Wochenschr 129:551–556
Sattler J, Hesterberg R, Lorenz W, Schmidt U, Crombach M, Stahlknecht CD (1985) Inhibition of human and canine diamine oxidase by drugs used in an intensive care unit: relevance for clinical side effects? Agents Actions 16:91–94
Sattler J, Häfner D, Klotter HJ, Lorenz W, Wagner PK (1988) Food-induced histaminosis as an epidemiological problem: plasma histamine elevation and haemodynamic alterations after oral histamine administration and blockade of diamine oxidase (DAO). Agents Actions 23:361–365
Sattler J, Lorenz W, Kubo K, Schmal A, Sauer S, Lüben L (1989) Food-induced histaminosis under diamine oxidase (DAO) blockade in pigs: further evidence of the key role of elevated plasma histamine levels as demonstrated by successful prophylaxis with antihistamines. Agents Actions 27:212–214
Sattler J, Lorenz W, Lindlar R, Schäfer U (1990) Histamine in gastrointestinal diseases: causality analysis. In: Uvnäs B (ed) Handbook of experimental pharmacology, vol 97. Springer, Berlin Heidelberg New York, pp 441–472
Schayer RW (1959) Catabolism of physiological quantities of histamine in vivo. Physiol Rev 39:116–126

Schayer RW (1966) Enzymatic formation of histamine from histidine. In: Eichler O, Farah A (eds) Handbook of experimental pharmacology. Springer, Berlin Heidelberg New York, pp 688–725

Schwabe U, Paffrath D (1989) Arzneimittelverordnungsreport 89. G Fischer, Stuttgart

Senanayake N, Vyravanathan S (1981) Histamine reactions due to ingestion of tuna fish (Thunnus argentivitattus) in patients on antituberculosis therapy. Toxicon 19:184–185

Shakir KM, Margolis S, Baylin SB (1977) Localization of histaminase (diamine oxidase) in rat small intestinal mucosa: site of release by heparin. Biochem Pharmacol 26:2343–2347

Wollin A, Navert H, Bounous G (1980) Effects of intestinal ischemia on diamine oxidase activity in rat intestinal tissue. Gastroenterology 80:349–355

Wulff HR (1981) Rational diagnosis and treatment, 2nd edn. Blackwell, Oxford

Authors' address: Dr. J. Sattler, Institute of Theoretical Surgery, Centre of Operative Medicine I, University of Marburg, Baldingerstraße, D-3550 Marburg, Federal Republic of Germany

J Neural Transm (1990) [Suppl] 32: 315–318

Methylamine oxidase from *Arthrobacter* P1 as a prototype of eukaryotic plasma amine oxidase and diamine oxidase

W. S. McIntire[1], **D. M. Dooley**[2], **M. A. McGuirl**[2], **C. E. Cote**[2], and **J. L. Bates**[2]

[1] Molecular Biology Division, Department of Veterans Affairs Medical Center, San Francisco; Department of Biochemistry and Biophysics and Department of Anesthesia, University of California, San Francisco, California, and
[2] Department of Chemistry, Amherst College, Amherst, Massachusetts, U.S.A.

Summary. Methylamine oxidase (MAOx) from Gram-positive soil bacterium *Arthrobacter* P1 catalyzes the oxidation of CH_3NH_2 to $H_2C{=}O$ and NH_4^+ via reduction of O_2 to H_2O_2. Past work indicates that MAOx is similar to mammalian plasma amine oxidase (PAO) and diamine oxidase (DAO), plant DAO, and yeast peroxisomal amine oxidase (YAO). All have $M_r \simeq 170{,}000$ and are composed of 2 identical subunits, each of which contains 1 atom of Cu(II) and one molecule of quinonoid cofactor. Herein, we report further evidence as to the striking similarity of these enzymes, and describe properties of MAOx which offer insights into understanding the eukaryotic oxidases. It is our belief that the structure of the quinone cofactor, and the Cu(II) site in MAOx are identical to these sites in PAO and DAO.

Materials and methods

Most of the materials and methods have been enumerated elsewhere (Dooley et al., 1990; McCracken et al., 1987). MAOx was purified as described in the literature (Dooley et al., 1990). The N-terminal amino acid sequencing of pure, undigested MAOx was done at the Protein Sequencing Facility, University of California, Davis on a fee for service basis.

Results

As with eukaryotic DAO and PAO, MAOx reacts with carbonyl reagents. The UV-visible spectra of MAOx and its phenylhydrazine derivative are very much like those of bovine PAO (Morpurgo et al., 1988). More revealing are resonance Raman spectra of the phenylhydrazine derivatives of MAOx, porcine PAO, and bovine PAO, which are essentially the same (Dooley et al., 1990; Knowles et al., 1987). Resonance Raman spectra of p-NO_2-phenylhydrazine, 2,4-diNO_2-phenylhydrazine, and 2-hydrazinopyridine forms of MAOx and bovine PAO are also virtually identical.

EPR and CD spectra of Cu(II) of oxidized MAOx are consistent with the structure of this site in PAO and DAO (Dooley et al., 1990). CH_3NH_2 reduction of MAOx, with Cl^- present, bleaches the λ_{max} at 486 nm to ~ 17% of its original level, indicating that only the quinone cofactor is reduced. This was confirmed by EPR. With Cl^- absent, the resulting UV-visible spectrum is reminiscent of that for reduced plant DAO (Rinaldi et al., 1984), while for porcine DAO only the quinone is substrate reduced (Dooley et al., 1990). Perhaps Cl^- was absent for the plant enzyme, but present for porcine DAO. CN^- reacts with MAOx, plant or porcine DAO at the Cu(II) site. CH_3NH_2 reduction of MAOx-CN^- produces a spectral change like that seen for reduced MAOx minus Cl^-. EPR spectra indicate that Cu(I) and quinone radical exists in reduced MAOx-CN^- and DAO-CN^-, and the spectra of the radicals for the 2 oxidases are identical (Dooley et al., 1990; Finazzi-Argo et al., 1984). Possibly CN^- cause the redox potential of Cu(II) $+ e^- \rightarrow$ Cu(I) to increase, so the dihydroquinone can be oxidize. When MAOx-CN^- is reduced by ^{15}N-CH_3NH_2, the EPR spectrum of the organic radical is significantly different than that seen on ^{14}N-CH_3NH_2 reduction, indicating that substrate-derived nitrogen is bound to reduced cofactor. This confirms that the reaction proceeds through a quinone-imine intermediate, not a carbinolamine intermediate for this class of enzymes (Hartmann and Klinman, 1988).

Fourier transform electron spin echo envelope modulation spectra of the semiquinone radicals of MAOx-CN^- formed by treatment with ^{14}N- or ^{15}N-CH_3NH_2 have been obtained. A peak at 13.48 MHz is due to protons, while a peak at 1.5 MHz is due to residual Cu(II)-imidazole coupling. When the ^{14}N-spectrum is divided by the ^{15}N-spectrum, what remains is a 3.3 MHz feature, that results from ^{14}N-coupling to the unpaired electron. Simulations of a Q-band EPR spectrum of MAOx quinone radical indicate that the observed spectrum is due to two conformations of the radical, which are frozen out at low temperatures. This is similar to EPR spectra observed for some protein-bound tyrosyl radicals.

When bovine PAO reacts with [^{14}C]-benzylamine in the presence of [^3H]-$NaCNBH_3$, one mol of ^{14}C is incorporated into PAO, but ^3H is not incorporated (Hartmann and Klinman, 1987). These finding are consistent with a quinone-type cofactor in PAO. Hartmann and Klinman (1990) have repeated the trapping experiment with MAOx, and the same results are obtained, which provides further evidence for the striking similarity of bacterial MAOx and eukaryotic PAO.

Table 1 presents the N-terminal amino acid sequence of MAOx, which is compared with the sequence of YAO. High homology exists in segment

Table 1. Comparison of amino acid sequences of amine oxidases

	1 10
YAO	Met-Glu-Arg-Leu-Arg-Gln-Ile-Ala-Ser-Gln-Ala-Thr-Ala-Ala-Ser-Ala-Ala-
	20 30
YAO	Pro-Ala-Arg-Pro-Ala-His-Pro-Leu-Asp-Pro-Leu-Ser-Thr-Ala-Glu-Ile-Lys-Ala-Ala-
	◦ • • • • • • • • ◦ • • •
MAOx	Leu-Val-Gly-Val-Ser-His-Pro-Leu-Asp-Pro-Leu-Ser(Ser)Val-Glu-Ile-Ala-???-Ala-
	1 10
	40 50
YAO	Thr-Asn-Thr-Val-Lys-Ser-Tyr-Phe-Ala-Gly-Lys-Lys-Ile-Ser-Phe-Asn-Thr-Val-Thr-
	◦ • • • • ◦ ◦
MAOx	Val-Ala-Ile-Leu-Lys-Glu-Gly-Pro-Ala-Ala-Ala-???-----Ser-Phe(Phe)Phe-Ile-Ser-
	20 30
	60 70
YAO	Leu-Arg-Glu-Pro-Ala-Arg-----Lys-Ala-Tyr-Ile-Gln-Trp-Lys-Glu-Gln → → →
	◦ •
MAOx	Val (Val)Leu-???-???-Pro-???-Lys-???-???-
	40 45

Key: *YAO* = *Hansenula polymorpha* amine oxidase (deduced from cDNA sequence; Bruinenberg et al., 1989); *MAOx* = *Arthrobacter* P1 MAOx; • = identical residues; ◦ = conservative substitution based on properties; (Xxx) = uncertain residue; ---- = inserted space to maintain optimal overlap; ??? = unidentified residue; → → → = continuation of the YAO sequence to the C-terminus

Leu_1-Ala_{19} of MAOx and Pro_{18}-Ala_{36} of YAO (10 identical residues and 6 conservative changes), with fair homology thereafter.

Discussion

All studies indicate that MAOx is nearly identical to eukaryotic DAO, PAO and YAO. Even limited amino acid sequences indicate the similarity of MAOx and YAO. We conclude that MAOx contains the same quinonoid cofactor as DAO, PAO and YAO as the redox center responsible for the oxidation of methylamine, and the electrons so obtained by the quinone are passed to the Cu(II) site, for conversion of O_2 to H_2O_2, which itself is identical to the Cu(II) site in the eukaryotic copper-quinoprotein amine oxidases.

Acknowledgements

We thank Drs. J. McCracken and J. Peisach for help with the spin echo envelope modulation work, and Prof. N. D. Chasteen and Dr. P. Hanna for help with Q-band EPR. Dr. McIntire's research was funded by a UCSF Academic Senate Grant, NIH Program Project Grant HL 16251, and the Department of Veterans Affairs. The work done at Amherst College was supported by NIH grant GM 27659 to Dr. Dooley.

References

Bruinenberg PG, Evers M, Waterham HR, Kuipers J, Arnberg AC, Greet AB (1989) Cloning and sequencing of the peroxisomal amine oxidase from *Hansenulà polymorpha*. Biochim Biophys Acta 1008:157–167

Dooley DM, McIntire WS, McGuirl MA, Cote CE, Bates JL (1990) Characterization of the active site of *Arthrobacter* P1 methylamine oxidase: evidence for cooper-quinone interactions. J Am Chem Soc 112:2782–2789

Finazzi-Argò A, Rinaldi A, Floris G, Rotilio G (1984) A free-radical intermediate in the reduction of plant Cu-amine oxidases. FEBS Lett 176:378–380

Hartmann C, Klinman JP (1988) Pyrroloquinoline quinone: a new redox cofactor in eukaryotic enzymes. BioFactors 1:41–49

Hartmann C, Klinman JP (1987) Reductive trapping of substrate to bovine plasma amine oxidase. J Biol Chem 262:962–965

Hartmann C, Klinman P (1990) Reductive trapping of substrate to methylamine oxidase from *Arthrobacter* P1. FEBS Lett 261:441–444

Knowles PF, Pandeya KB, Rius FX, Spencer CM, Moog RS, McGuirl MA, Dooley DM (1987) The organic cofactor in plasma amine oxidase: evidence for pyrroloquinoline quinone and against pyridoxal phosphate. Biochem J 241: 603–608

McCracken J, Peisach J, Dooley DM (1987) Cu(II) coordination chemistry of amine oxidases: pulsed EPR studies of histidine imidazole, water, and exogenous ligand coordination. J Am Chem Soc 109:4064–4072

Morpurgo L, Befani O, Sabatini S, Mondavi B, Artioc M, Corello F, Massa S, Stefancich G, Avigliano L (1988) Spectroscopic studies of the reaction between bovine serum amine oxidase (copper-containing) and some hydrazides and hydrazines. Biochem J 256:565–570

Rinaldi A, Giartosio A, Floris G, Medda R, Finazzi-Argo A (1984) Lentil seedling amine oxidase: preparation and properties of the copper-free enzyme. Biochem Biophys Res Commun 120:245–294

Authors' address: Dr. W. S. McIntire, Department of Veterans Affairs Medical Center, Molecular Biology Division (151-S), 4150 Clement Street, San Francisco, CA 94121, U.S.A.

J Neural Transm (1990) [Suppl] 32: 319–322

Chronic ethanol feeding and diamine oxidase activity in rat brain and liver*

A. Sessa[1] and **A. Perin**[2]

[1] Centro di Studio sulla Patologia Cellulare, C.N.R., and [2] Istituto di Patologia Generale, Università di Milano, Milan, Italy

Summary. The activity of diamine oxidase, the principal enzyme for diamine catabolism, was evaluated in brain and liver from male rats fed nutritionally complete diets, of which 36% of total calories was given as ethanol or as isocaloric carbohydrates for 4 months. Ethanol caused an increase in diamine oxidase activity in the brain and a decrease in the liver, tissues with a low and high rate of ethanol oxidation, respectively.

Introduction

Diamine oxidase (EC 1.4.3.6, DAO) is involved in the oxidative deamination of short-chain aliphatic amines, like putrescine and cadaverine, as well as of histamine. The physiological role of the enzyme in mammalian tissues is related to the biological function of its natural substrates. Putrescine, the precursor of higher polyamines, has been associated with rapid cell growth and division (Heby, 1981). Histamine, besides its importance in gastric acid secretion and in a wide variety of allergic reactions and inflammation, plays a role as a neurotransmitter or neuromodulator in the central nervous system (Mazurkiewicz-Kwilecki, 1984). Cadaverine, like putrescine, appears to be elevated in tissues stimulated to grow (Rosengren et al., 1981).

An interesting feature of DAO is that it is in vivo induced by diamines or their acetylated derivatives in rat tissues with low enzyme activity (Perin et al., 1986). Moreover, its expression is regulated by the β_2-adrenoceptor/adenyl cyclase system in rat kidney and heart (Desiderio et al., 1985, 1986) as well as by estrogens in rat uterus (Sessa et al., 1990).

Some papers indicate that ethanol influences DAO activity in rodent liver (Bieganski et al., 1984; Sessa et al., 1984), in maternal placenta, embryo

---

* The study was supported in part by the Ministero della Pubblica Istruzione, Rome.

and fetus of rat (Desiderio et al., 1987; Sessa et al., 1987). Here we report data on the effect of chronic ethanol exposure of DAO activity of rat liver and brain, which are target organs of the drug.

Material and methods

Male Sprague-Dawley rats (200–220 g) were pair-fed nutritionally complete liquid diets, of which 36% of total calories was given as ethanol or as isocaloric carbohydrates (controls) for 4 months. DAO activity was measured in liver and brain (telencephalon and diencephalon) homogenates by a radiochemical method as reported in our previous paper (Perin et al., 1986).

Results

The results reported in Fig. 1 show that at the end of 4 months of a liquid diet of ethanol, rat brain exhibited an enzyme activity nearly 2-fold that of controls. In contrast, ethanol exposure decreased DAO activity by about 60% in the liver, expressed as nmol/h/mg protein and as nmol/h/whole liver.

Discussion

The intent of the present study was to determine whether chronic ethanol exposure that roughly paralleled the long-term ethanol intake of alcoholics affected DAO activity and to evaluate a possible correlation with alcohol

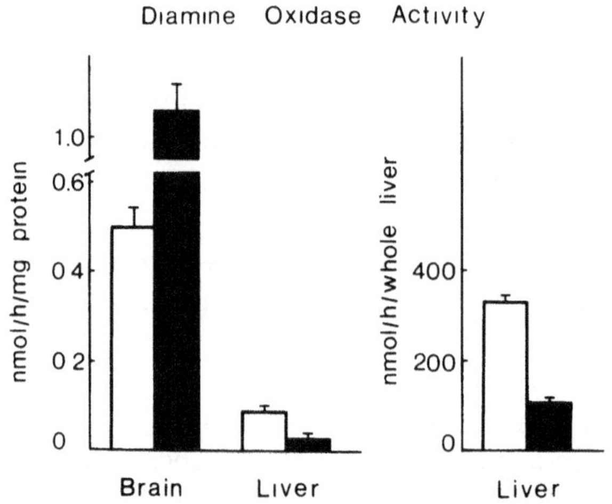

Fig. 1. Effect of chronic ethanol feeding on diamine oxidase activity of rat brain and liver. The results are the mean ± S.E. of six animals. □ Controls; ■ ethanol-treated

metabolism. We therefore studied such enzyme activity in rat liver, the principal organ for alcohol oxidation, compared to that in brain, a tissue with a very low ethanol metabolism and considered virtually devoid of alcohol dehydrogenase.

Our results demonstrated that ethanol given to rats in a nutritionally balanced liquid diet for several months caused a marked increase in cerebral DAO activity, whereas hepatic enzyme activity was significantly diminished. Such a decrease did not depend on an influence of ethanol on total protein content of the liver, since it was to the same extent when measured as specific activity or as activity in the whole liver.

The mechanism(s) by which ethanol differently affects DAO activity in brain and liver is still unclear. As regards the stimulatory effect on cerebral enzyme activity, two observations are important in trying to explain the ethanol action. One concerns the possible implications of the adrenergic receptor system, which has been shown to be involved in the regulation of DAO activity in rat tissues (Desiderio et al., 1985, 1986). Although the data on the effects of ethanol on such a system are scanty and contradictory, chronic ethanol administration to rats has been found to produce an increase in norepinephrine turnover in cerebral cortex (Lynch et al., 1983), which could account for the enhanced DAO activity. The second point considers the stimulation of DAO as a possible consequence of an acquired abnormality of plasma membranes. In this regard, chronic ethanol ingestion has been found to induce changes in the structure, composition and function of cell membranes (Taraschi and Rubin, 1985). As regards the liver, a tissue with a very high rate of ethanol oxidation, the impairment of DAO activity may be due to alcohol metabolism, which has been shown to reduce liver protein synthesis (Rothschild et al., 1987), and/or to the toxic action of the drug on the liver.

References

Bieganski T, Fogel WA, Korsak Z, Maśliński C, Osinska Z, Ulatowska MA (1984) Ethanol involvement in putrescine and histamine oxidation by guinea pig liver. Pol J Pharmacol Pharm 36:633–640

Desiderio MA, Sessa A, Perin A (1985) Regulation of diamine oxidase expression by β_2-adrenoceptors in normal and hypertrophic rat kidney. Biochim Biophys Acta 845:463–468

Desiderio MA, Sessa A, Perin A (1986) Involvement of beta$_2$-adrenoceptors in the regulation of diamine oxidase activity in the heart of spontaneously hypertensive rats. J Hypertens 4 [Suppl] 3:S139–S140

Desiderio MA, Sessa A, Perin A (1987) Polyamines and diamine oxidase activity in maternal, embryonal, and fetal tissues of rat after chronic ethanol consumption. Biochem Biophys Res Commun 142:843–848

Heby O (1981) Role of polyamines in the control of cell proliferation and differentiation. Differentiation 19:1–20

Lynch M, Littleton J, McKernan RM, Durcan MJ, McMillan T, Campbell IC (1983) α-Adrenoceptor number and function in rat cortex after ethanol and immobilisation stress. Brain Res 288:145–149

Mazurkiewicz-Kwilecki JM (1984) Possible role of histamine in brain function: neurochemical, physiological and pharmacological indications. Can J Physiol Pharmacol 62:709–714

Perin A, Sessa A, Desiderio MA (1986) Response of tissue diamine oxidase activity to polyamine administration. Biochem J 234:119–123

Rosengren E, Henningsson A-Ch, Henningsson S, Persson L (1981) Polyamine metabolism as related to growth and hormones. Med Biol 59:320–326

Rothschild MA, Oratz M, Schreiber SS (1987) Effects of ethanol on protein synthesis. In: Rubin E (ed) Alcohol and the cell. Ann NY Acad Sci 492:233–244

Sessa A, Desiderio MA, Perin A (1984) Stimulation of hepatic and renal diamine oxidase activity after acute ethanol administration. Biochim Biophys Acta 801:285–289

Sessa A, Desiderio MA, Perin A (1987) Effect of acute ethanol administration on diamine oxidase activity in maternal, embryonal and fetal tissues. Agents Actions 21:49–53

Sessa A, Desiderio MA, Perin A (1990) Estrogenic regulation of diamine oxidase activity in rat uterus. Agents Actions 29:162–166

Taraschi TF, Rubin E (1985) Biology of disease. Effects of ethanol on the chemical and structural properties of biological membranes. Lab Invest 52:120–131

Authors' address: Dr. A. Sessa, Centro di Studio sulla Patologia Cellulare C.N.R., Via Mangiagalli 31, I-20133 Milan, Italy.

J Neural Transm (1990) [Suppl] 32: 323–326

Amine oxidase activities in chemically-induced mammary cancer in the rat

J. M. Lizcano[1], **E. Escrich**[2], **K. F. Tipton**[1], and **M. Unzeta**[1]

[1] Departament Bioquímica i Biologia Molecular, and [2] Departament Biologia Cellular i Fisiologia, Facultat de Medicina, Universitat Autónoma de Barcelona, Barcelona, Spain

Summary. The activities and distributions of monoamine oxidase (MAO) and semicarbazide-sensitive amine oxidase (SSAO) were studied in solid breast tumours induced in rat by treatment with 7,12-dimethylbenz(α)anthracene (DMBA).

It was observed that increasing degree of malignancy was associated with an increase of MAO-A activity an decrease of MAO-B and SSAO activities. The K_m values did not change significantly with malignancy but V_{max} values for MAO-A increased whereas V_{max} for MAO-B and SSAO diminished with malignancy.

It was detected in the more malignant tumours the presence of endogenous reversible inhibitor of SSAO activity not removed by dialysis.

Introduction

Although alterations in amine metabolism in tumour tissue have been described (Sunkara and Rao, 1981) there have been no detailed studies on the possible involvement of changes in the activity of the semicarbazide-sensitive amine oxidase (SSAO) [amine: oxygen oxidoreductase deaminating) (copper-containing) EC: 1.4.3.6.] in these processes. The present studies were aimed at investing possible correlations between alterations of the activities of the SSAO and monoamine oxidase (MAO) [amine: oxygen oxidoreductase (deaminating) (flavin-containing) Ec: 1.4.3.4.] and the development of mammary tumours induced in rat by treatment with 7,12-dimethylbenz(α)anthracene (DMBA).

Materials and methods

Mammary cancer was induced in 52 day old female Sprague-Dawley rats, of 160–180 g. weight, by a single dose of 1 ml of sesame oil containing 20 mg. of

DMBA per animal, administered by gastric instillation. Rats were fed by AO4 standard diet. Tumours were dissected and were classified according to anatomopathological criteria (Young and Hallowey, 1973) as fibroadenoma (FAD) the benign one and adenocarcinoma (ADC) and infiltrant adenocarcinoma (I-ADC) the more malignant.

Samples were stored in liquid nitrogen, and then disrupted mechanically, homogenized in 50 mM potassium phosphate buffer, pH 7.2, (1:10 w/v) and stored in aliquots at $-80\,^{\circ}C$ until assay.

MAO and SSAO activities were determined radiochemically by a modification (Fowler and Tipton, 1981) of the method of Otsuka and Kobayashi (1964), towards phenylethylamine (PEA), tyramine (Tyr), benzylamine (Bz) and 5-hydroxytryptamine (5-HT) as substrates. Selective inhibition of MAO-A, MAO-B and SSAO activities were obtained by 10^{-7} M clorgyline, 10^{-7} M deprenyl and 10^{-3} M semicarbazide, respectively.

Microsomal pellets from I-ADC tissue, containing 3 mg/ml of protein were added to β-octyl-glucoside (OG) to give a final detergent concentration of 1.25 (w/v) in experiments to test the effects of this detergent on the dependence of initial velocity on enzyme concentration.

Results and discussion

The activities of MAO and SSAO towards 1 μM benzylamine were determined in the presence of the selective inhibitors clorgyline and semicarbazide. SSAO accounted for about 85% of the total activity, whereas MAO-B contributed the remaining 15%. These relative values did not change significantly with the degree of malignancy as assessed histologically. However the development of the tumour was associated with an increase in the specific activity of MAO-A towards 5-HT as substrate and a decrease of specific activities of MAO-B and SSAO towards PEA and Bz as substrates respectively (Fig. 1). No significant differences were observed between the specific activities obtained with single and pooled tumours for the same tumour type. The K_m values of these enzymes towards their specific substrates did not change with the tumour development being the mean K_m values of the different type of tumours were: 110 ± 9 μM for MAO-A, 10.9 ± 2.0 μM for MAO-B and 6.9 ± 1.5 μM for SSAO. The activity changes were solely due to alterations of the maximum velocities, increasing with malignancy in the case of MAO-A (333 ± 3, 551 ± 9 and 636 ± 6 pmol/min mg.) and decreasing in the case of MAO-B (270 ± 9, 123 ± 0.5); all values in the order FAD, ADC, I-ADC. Those results suggest that the changes in activities may result from changes in the enzyme concentrations rather than in their kinetic behaviour.

Although the initial velocities of the reactions catalyzed by MAO-A and MAO-B increased linearly with enzyme concentration, the activity of SSAO showed an hyperbolic dependence, with the velocity-protein concentrations curves only approximating to linearity at homogenate concentrations below

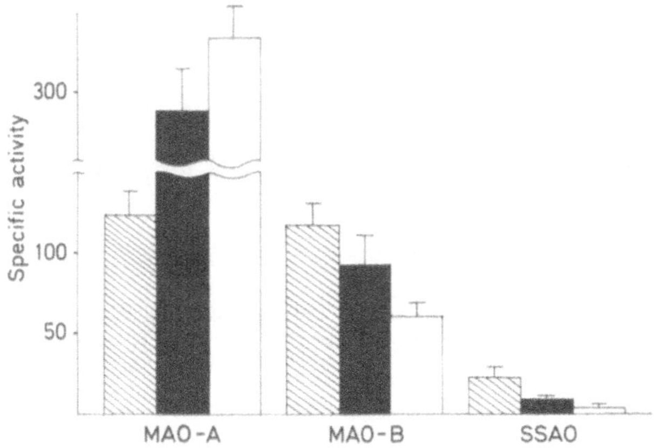

Fig. 1. Amine oxidase activities in mammary tumours of the rat. *Hatched columns*: fibroadenoma tumours. *Black columns*: adenocarcinoma tumours. *Open columns*: infiltrant-adenocarcinoma tumours. Enzyme activities are mean values ± S.E. of 9 individual tumours corresponding to each histopathological type. MAO-A and MAO-B specific activities were determined towards 5-HT (100 μM) and PEA (20 μM), respectively. SSAO specific activity was determined towards 1 μM Bz after inhibition of total MAO activity with 1 mM clorgyline. All values in pmol min^{-1} mg of $protein^{-1}$

about 300 μg per 225 μl assay volume. Such behaviour would be expected if the tumour tissue contained an endogenous reversible inhibitor of SSAO (see Dixon and Webb, 1979). However this non-linear behaviour was not affected by dialysis or treatment of the samples with the detergent β-octyl glucoside (1.25% w/v) suggesting that it may be an intrinsic property of the enzyme itself resulting from subunit association producing an inactive form of the enzyme.

Such non-linearity in the dependence of initial velocity on tissue concentration has not been previously reported for SSAO from other sources. It is possible that it may be a peculiarity of the tumour tissue but, unfortunately, the small mass of rat breast tissue precluded the performance of such studies with the non-carcinogenic material from this source. Further work will be necessary before it can be established whether this behaviour results from the presence of an endogenous high M_r inhibitor or whether the inhibitory process results from subunit interactions of the enzyme itself.

References

Dixon M, Webb EC (1979) Enzymes, 3rd edn. Longman Group, London, pp 48–55

Fowler CJ, Tipton KF (1981) Concentration dependence of the oxidation of tyramine by the two forms of rat liver mitochondrial monoamine oxidase. Biochem Pharmacol 30:3329–3332

Otsuka S, Kobayashi Y (1964) A radioisotopic assay for manoamine oxidase determination in human plasma. Biochem Pharmacol 13:995–1006
Sunkara PS, Rao PN (1981) The transformed cell. Academic Press, New York, pp 267–291
Young S, Hallowey RC (1973) In: Turosov VS (ed) Pathology of tumours in laboratory animals, vol 1. Lyon, pp 31–74

Authors' address: Dr. M. Unzeta, Departament de Bioquímica i Biologia Molecular, Facultat de Medicina, Universitat Autónoma de Barcelona, 08193 Bellaterra (Barcelona), Spain.

J Neural Transm (1990) [Suppl] 32: 327–330

Monoamine oxidase and semicarbazide-sensitive amine oxidase activities in bovine eye

A. Fernandez de Arriba, D. Balsa, K. F. Tipton, and **M. Unzeta**

Departament de Bioquimica i Biologia Molecular, Facultat de Medicina, Universitat Autonoma de Barcelona, Barcelona, Spain

Summary. The activities of MAO-A, MAO-B and the semicarbazide-sensitive amine oxidase (SSAO) were determined in bovine lens, retina, optic nerve, iris and epithelium. No activities were detected in lens but the SSAO activity was found to be rather evenly distributed in the other tissues. The SSAO activities determined with 1 μM benzylamine were about half those determined for MAO-B. MAO-A activity, measured with 5-HT as substrate, was considerably greater than those of MAO-B and SSAO, determined with 2-phenylethylamine and benzylamine, respectively, in all the eye tissues.

Introduction

In addition to the two monoamine oxidase (MAO) [amine: oxygen oxidoreductase (deaminating) (flavin-containing) EC 1.4.3.4.] isoenzymes an amine oxidase activity which is sensitive to inhibition by carbonyl-group reagents, such as semicarbazide, but is not inhibited by the acetylenic monoamine oxidase inhibitors clorgyline and (–)-deprenyl is present in many tissues (for reviews see Callingham and Barrand, 1987). This enzyme has been termed semicarbazide-sensitive amine oxidase (SSAO) or clorgyline-resistent amine oxidase (CRAO) [amine: oxygen oxidoreductase (deaminating) (copper-containing) EC 1.4.3.6.]. It shows high activity towards benzylamine but its physiological functions are obscure. In an attempt to obtain insights into its possible role we have compared the distribution and specificities of SSAO and MAO-A and -B in the different components of bovine eye.

Materials and methods

Bovine eyes were obtained immediately after slaughter of the animals and transported to the laboratory from a local abattoir packed in ice. The optic nerve was cut off and the cornea dissected by means of a circular incision at the corneoscleral junction. The iris, lens and vitrious body were removed, the eyeball was inverted and the retina detached by gently shaking in a 50 mM potassium phosphate buffer, pH 7.2. Finally the epithelium was removed.

All samples were frozen in liquid N_2 and then disrupted mechanically, homogenized in 50 mM potassium phosphate buffer, pH 7.2 (1:10 w/v), filtered through gauze (110 μ) and stored in aliquots at −80 °C. In the case of the retina, the preparations were disaggregated by sonication at low frequency for 15 seconds before each assay.

Monoamine oxidase activity was determined at 37 °C radiochemically by the method of Fowler et al. (1980). 2-phenylethylamine (PEA, [ethyl-1-^{14}C]phenyl ethylamine hydrochloride), benzylamine (Bz, [7-^{14}C]benzylamine hydrochloride), 5-hydroxytryptamine (5-HT, 5-hydroxy[side-chain-2-^{14}C] were used as substrates. The reaction took place in a final volume of 225 μl of 50 mM potassium phosphate buffer, pH 7.2, containing 200–400 μg of protein and was stopped by the addition of 100 μl of 2 M citric acid. The products were extracted into toluene/ethyl acetate 1:1 (v/v) containing 0.6% (w/v) PPO and the radioactivity was measured in a scintillation counter. SSAO activity was determined towards benzylamine (1 μM) as substrate under the same assay conditions.

Results and discussion

Figure 1 shows a representative curve for the inhibition of the amine oxidase activities in bovine optic nerve by clorgyline, (−)-deprenyl and semicarbazide with benzylamine and 5-hydroxytryptamine (5-HT) as substrates. The relative contributions of MAO-B and SSAO activities were 67 ± 3% and 29 ± 8% at a benzylamine concentration of 1 μM respectively and 70 ± 3% and 35 ± 2% at a benzylamine concentration of 100 μM (mean values ± s.e.m., n = 4). MAO-B and SSAO have little activity towards 5-HT (Fowler et al., 1980; Barrand et al., 1982) and the data in Fig. 1 indicate that 5-HT is essentially a substrate only for MAO-A in bovine optic nerve. The specific activities of MAO-B and SSAO towards benzylamine (1 μM) in this tissue were 0.062 ± 0.005 nmol h^{-1} mg protein^{-1} and 0.029 ± 0.002 nmol min^{-1} mg protein^{-1}, respectively, whereas the activity towards 5-HT (100 μM) was 24.5 ± 1.5 nmol h^{-1}mg protein^{-1} (mean values ± s.e.m.; n = 4).

No amine oxidase activities were detectable in bovine lens, but there were significant activities in the other components of the eye tissue examined. The total activities towards 1 μM benzylamine were: 0.063 ± 0.003 nmol h^{-1} mg protein^{-1} in iris, 0.055 ± 0.001 nmol h^{-1} mg protein^{-1} in epithelial tissue and 0.059 ± 0.005 nmol h^{-1} mg protein^{-1} in retina. With 100 μM 5-HT as substrate the specific activities were 9.04 ± 1.62 nmol h^{-1} mg protein^{-1} in iris, 4.98 ± 0.9 nmol h^{-1} mg protein^{-1} in epithelial tissue and 10.2 ± 0.4 nmol h^{-1} mg protein^{-1} in retina. All values are expressed as mean values ± s.e.m. (n = 5). The activities of SSAO towards benzylamine (100 and 1 μM) in the different eye tissues were

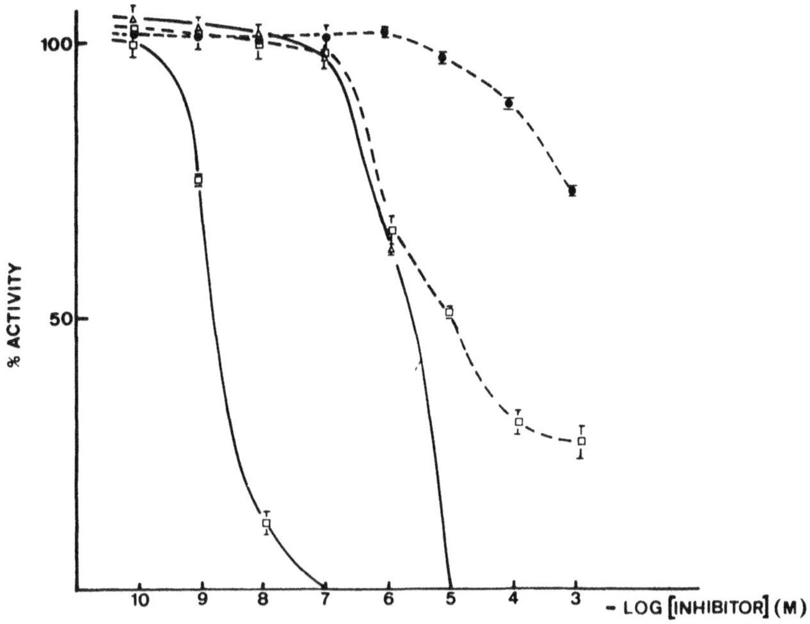

Fig. 1. Inhibition curves by clorgyline (□), semicarbazide (●) and (−)-deprenyl (△) of the amine oxidase activities towards 5-HT (100 μM) (—) and Bz (1 μM) (···) as substrates in optic nerve. The samples were preincubated for 30 min at 37 °C with different inhibitor concentrations (10^{-10}–10^{-3} M). The remaining activity is expressed as percentage of the control activity determined in the absence of inhibitor. Each value is the mean ± S.E.M., (n = 3)

determined by their sensitivities to inhibition by semicarbazide and their insensitivities to inhibition by clorgyline (1 mM). They were found to be similar in retina, iris and epithelium to those in optic nerve indicating a rather even distribution of SSAO activities throughout these eye tissues. In contrast the activities of MAO-A and B, measured with 100 μM 5-HT and 22.2 μM of PEA, respectively, varied between these tissues. The ratios of MAO-A/MAO-B activities were 4.55 ± 0.5 in optic nerve, 2.05 ± 0.4 in epithelial tissue, 2.12 ± 0.62 in iris and 2.35 ± 0.36 in retina.

The rather even distribution of relatively low levels of SSAO in the different eye tissues might be consistent with the enzyme playing a protective role such as in the destruction of methylamine as has been suggested by Precious et al. (1988). MAO-A was the major form of this enzyme present in all the tissues examined being particularly dominant in the optic nerve, perhaps reflecting a specific role in neurotransmitter metabolism in that tissue.

References

Barrand MA, Callingham BA (1982) Monoamine oxidase activities in brown and adipose tissue of the rat: some properties and subcellular distribution. Biochem Pharmacol 31:2177–2184

Callingham BA, Barrand MA (1987) Some properties of semicarbazide-sensitive amine oxidase. J Neural Transm [Suppl] 23:37–54

Fowler CJ, Oreland L, Marcusson J, Winblad B (1980) Titration of human brain monoamine oxidase -A and -B by clorgyline and L-deprenyl. Naunyn-Schmiedebergs Arch Pharmacol 311:263–272

Precious E, Gunn CE, Lyles GA (1988) Deamination of methylamine by semicarbazide-sensitive amine oxidase in human umbilical artery and rat aorta. Biochem Pharmacol 37:707–713

Authors' address: Dr. D. Balsa, Departament de Bioquimica i Biologia Molecular, Facultat de Medicina, Universitat Autonoma de Barcelona, 08193 Bellaterra (Barcelona), Spain.

J Neural Transm (1990) [Suppl] 32: 331–336

Cultured preadipocytes produce a semicarbazide-sensitive amine oxidase (SSAO) activity*

L. Raimondi, R. Pirisino, G. Banchelli, G. Ignesti, L. Conforti, and F. Buffoni

Department of Pharmacology, University of Florence, Italy

Summary. The presence of an amine oxidase activity which is resistant to pargyline, an acetylenic MAO inhibitor, but sensitive to semicarbazide, is described in rat white and brown preadipocytes.

Introduction

High levels of a semicarbazide-sensitive amine oxidase activity (*SSAO*) (EC 1.4.3.6.) in rat white and brown mature adipocytes have already been described by Barrand et al. (1984) and Raimondi et al. (1989).

Mature adipocytes are the final transformation step from undifferentiated precursors (*preadipocytes*). Attempts have been made by many authors to isolate and identify cells which are able to differentiate and express all the features of mature adipocytes when cultured in the presence of adipogenic factors.

Our aim was to investigate whether an SSAO activity is present also in white and brown preadipocytes in order to verify whether this enzymatic activity appears with cell transformation or whether it is already present in the undifferentiated precursor cells.

The presence of SSAO activity in preadipocytes could be a further confirmation that they actually evolve to mature adipocyte-like cells. Moreover this enzyme could be considered a marker of adipose transformation as already known for other enzymatic activities.

For this purpose primary cultures from the white (WAT) and brown (BAT) adipose tissue of young rats were set up.

Materials

Methylene [^{14}C]-benzylamine, 55.7 mCi/mmoles was purchased from Amersham; Dulbecco's Modified Eagle Medium (DMEM, containing pyruvate, 1 g/L of glu-

* This work was supported by a grant from the MPI, Rome.

cose), penicillin, streptomycin, foetal calf serum (FCS) from GIBCO. Biotin, pantothenate, semicarbazide dihydrochloride, catalase, isobuthylmethylxantine (IBMX), dexamethasone, insulin, dihydroxyacetonphosphate were from Boehringer.

Methods

Male Wistar rats weighing between 60–180 gr were used. They were allowed to access to a standard chow, housed at constant temperature and killed by cervical dislocation. Interscapular (BAT) and epididymal WAT were resected under sterile conditions and minced in small pieces.

Stromal vascular cells were prepared according to Björntrop et al. (1978) and Forest et al. (1987). Cells from WAT were seeded in 35 mm multiwell dishes (NUNC) at a density of 10^5 cells/cm^2 whereas cells from BAT in 1.5 mm multiwell dishes (NUNC) at the same density. All cells were grown in a standard medium (DMEM) containing 100 U/ml of penicillin, 100 µg/ml of streptomycin, 33 µM biothin, 17 µM pantothenate, 10% heat inactivated (20 min at 56°) FCS and incubated at 37 °C in air (95%) and CO_2 (5%). Stromal WAT cells reached confluence after 48–72 h from plaiting at a density of 10^5 cells/cm^2, BAT cells after around 92 h at the same density. As soon as they reached confluence (*day 0*) cells were shifted to a standard medium supplemented with IBMX (0.5 mM), dexamethasone, 10^{-6} M, insulin 10^{-9} M for 48 h. (*day 2*), then cultured up to 7–10 days in DMEM plus insulin 10^{-9} M. A series of cells kept in standard medium alone were used controls.

Determination of SSAO activity

At the decided time dishes were thoroughly washed with standard medium without FCS and maintained in this condition for 4–6 h at 37 °C. Cells were then scraped off the dishes and pelleted by centrifugation at 600 × g × 5 min. Cellular pellets were then resuspended in 1 mM potassium phosphate buffer pH 7.8 containing 0.25 M sucrose (sucrose buffer) homogenized (2 × 10^6 cells/ml for white stromal cells and 10^6 cells/ml for the brown ones) and briefly sonicated in a Bronson bath for 2 min. SSAO activity was measured as described by Callingham and Laverty (1973) using 50 µM [^{14}C]-benzylamine as substrate. 70 µl of cell homogenates were incubated in sucrose buffer in the presence of 1 mM pargyline, 2 U/ml of catalase, semicarbazide 10^{-4} M, when used, for 30 min at 37 °C. Reactions were stopped 5 min after substrate addition.

Triglycerides (TG)

The *TG* content of white preadipocytes homogenates was measured according to Eggestein and Kuhlmann (1978).

Protein

The protein content of cell homogenates was assayed according to Lowry et al. (1951).

Glycerol-3-phosphate dehydrogenase activity (GDH)

GDH was measured in total cell homogenates or in a low speed supernatant (600 × g × 5 min) of brown preadipocytes, using dihydroxyacetonphosphate as substrate. Results are expressed as nmoles of reduced nicotinamide adenine dinucleotide/mg of total cell proteins/min according to Weise and Green (1979).

Results

The presence of an SSAO activity was investigated in stromal vascular cells from the WAT and BAT of young rats.

Enzyme levels in white preadipocytes were very low in control cells and rose as much as 5 times in the transformed ones at *day 7*. At that time (*day 7*), TG were already detectable in control (C) cells (Fig. 1) but a significant rise in TG levels was measured in transformed white preadipocytes (T). SSAO activity was detectable concurrently with lipid deposition. No enzymatic activity was measurable at *day 0*.

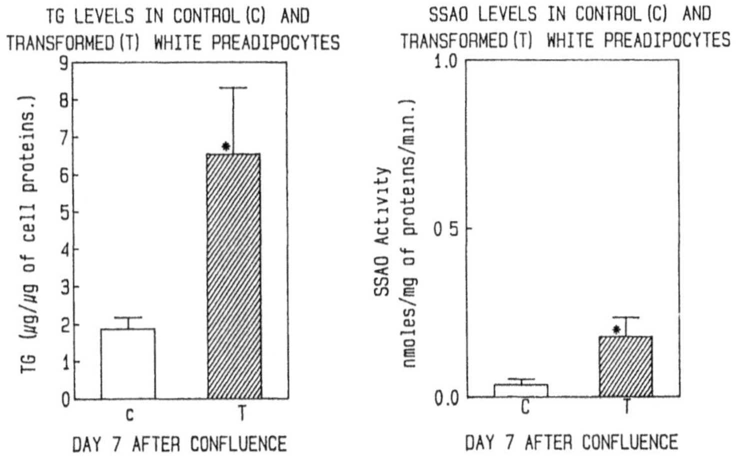

Fig. 1. TG and SSAO levels in control (*C*) and transformed (*T*) white preadipocytes Homogenates of white preadipocytes kept 7 days after confluence in standard medium alone and of transformed cell at day 7 were assayed for TG content. Results are mean ± s.e. of 7 samples coming from different primary cultures. * Student's t test $p < 0.05\%$. SSAO activity was measured as described in Methods. Cell homogenates of control and transformed white cells at day 7 were assayed for enzymatic activity. Results are mean ± s.e. of 4 or 7 different primary cultures. * Student's t test, $p < 0.05\%$. Statistical analysis showed the same significant rise in SSAO activity when it was calculated in grouped data (all the transformed cells vs. the control group) or as paired data (control value vs. the respective transformed sample of the same preparation). Semicarbazide 10^{-4} M completely inhibited this activity

Brown cells did not accumulate any visible lipid droplets, even after 10 days after confluence, in control cells kept in standard medium alone. We therefore concluded that it was highly important to plate brown preadipocytes at the highest density as possible in order they could reach confluence before starting lipid accumulation. The addition of the transforming cocktail (IBMX + Dexamethazone + insulin produced a marked change in cell morphology; they started becoming rounded and accumulating lipids. A time-dependence monitoring of SSAO activity in brown preadipocytes showed that the enzyme was not detectable in control cells cultured up to *day 10*. In same cases, but not all, it started to be detectable at day 2 concurrently with GDH presence.

The SSAO production for cells number rised significatively ($p < 0.05\%$, Student's t test for grouped data $n = 4$–10 samples from different primary cultures) from *day 2* to *day 7* then it henceforth decreased to *day 10* (Fig. 2). The same time-dependent rised was observed for SSAO specific activity (Fig. 2).

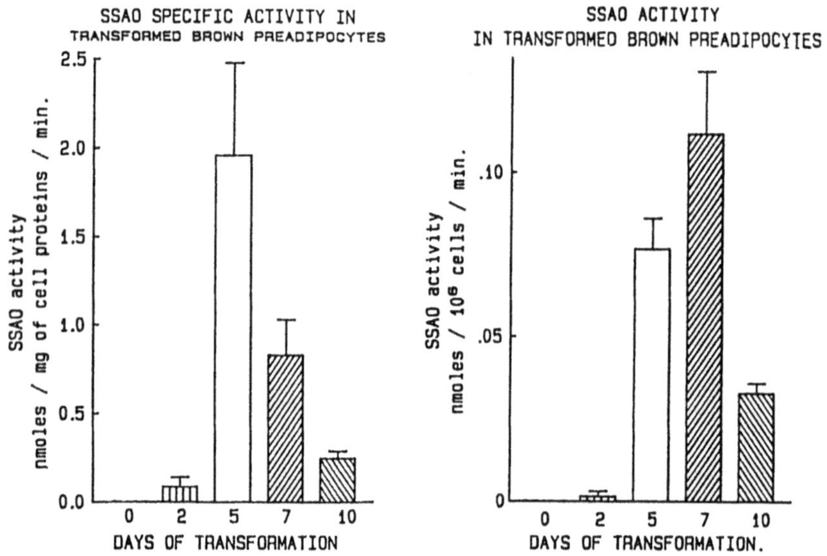

Fig. 2. SSAO levels in transformed brown preadipocytes (a time-course appearance). The time-dependent appearance of SSAO in brown transformed cells and its related specific acrtivity are shown in Fig. 2. Brown cell homogenates were used as described in Methods for determination of SSAO activity. Enzymatic activity was assayed radiochemically, using 50 μM [^{14}C]-benzylamine as substrate. Semicarbazide 10^{-4} M completely abolished this activity. Results are means ± s.e. of 5 or 8 samples from different primary cultures. In control cells no GDH activity was measured up to day 10. GDH started to be detectable 48 h after transformation reaching 166 ± 44 nmoles/mg of cell proteins/min at day 5, 259 ± 69 nmoles at day 7 and then decreased at day 10 (50.5 ± 8.8 nmoles/mg of total cell proteins/min.) Results of GDH activity are mean ± s.e. of 4 samples from different primary cultures

Discussion

The presence of a semicarbazide-sensitive amine oxidase activity in rat white and brown preadipocytes is discussed here. This enzymatic activity appears in both cells with lipid deposition. Lipogenic transforming agents promoted a differentiation of cells that probably occurred spontaneously, at least in the white preadipocytes. The presence of an SSAO activity in control white cells at day 7 might confirm this hypothesis. Brown cells did not show any lipid accumulation before day 2 of transformation, SSAO seemed to appear in a parallel way at the same time. Thenceforth the decrease in enzyme activity went in parallel with GDH behaviour. A time dependent GDH decrease has already been described by Amri et al. (1984) in similar cells. SSAO seems to be expressed early in the first 48 h of transformation, the enzyme production rises significatively from day 0 to day 7 ($p < 0.05\%$, Student's t test for grouped data $n = 4$–10 observations). We also find an increase of cell protein content from day 0 to day 7 that could explain the decrease of SSAO specific activity at day 7 and 10. In conclusion the induction of this SSAO seems to be related to the adipose phenotype expression of stromal vascular cells from adipose tissues. This finding might be an useful tool for monitoring specific adipose conversion of preadipocytes.

References

Amri E, Dani C, Grimaldi P, Doglio A, Ailhaud G (1984) Coupling of growth arrest and expression of early markers during adipose conversion of preadipocyte cells line. Biochem Biophys Res Commun 137:903–910

Barrand MG, Callingham BA, Fox S (1984) Amine oxidases activity in brown adipose tissue of rat: identification of a semicarbazide-sensitive (clorgyline resistant) amine oxidase activity at the fat cell membrane. J Pharm Pharmacol 36:652–568

Björntorp PM, Karlsson M, Pertoft H, Pettersson P, Sjöstorp L, Smith S (1978) Isolation and characterization of cells from rat adipose tissue developing into adipocytes. J Lipid Res 19:316–324

Callingham BA, Laverty R (1973) Studies on the nature of the increased monoamine oxidase activity in the rat heart after adrenalectomy. J Pharm Pharmacol 25:940–947

Eggestein M, Kuhlmann E (1978) Triglycerides and glycerol. Determination after alkaline hydrolysis. In: Bergmeger HU (ed) Methods of enzymatic analysis, 2nd edn. Academic Press, New York, pp 1825–1831

Forest C, Doglio A, Ricquier D, Ailhaud G (1987) A preadipocyte clonal line from mouse brown adipose tissue. Exp Cell Res 168:218–232

Lowry OH, Rosebrough NJ, Farr AL, Randall RJ (1951) Protein measurement with Folin phenol reagent. J Biol Chem 193:265–275

Raimondi L, Ignesti G, Pirisino R, Banchelli G, Buffoni F (1989) Semicarbazide-sensitive amine oxidase activity (SSAO) in rat adipocytes which also acts on histamine. Proc, IXth Meeting of the European Histamine Research Society, Breda, May 18–20

Wise LS, Green H (1979) Participation of one isoenzyme of cytosolic glycerophosphate dehydrogenase in the adipose conversion of 3T3 cells. J Biol Chem 254:273–275

Authors' address: Dr. L. Raimondi, Department of Pharmacology, University of Florence, Viale G. B. Morgagni 65, I-50134 Florence, Italy.

J Neural Transm (1990) [Suppl] 32: 337–339

Semicarbazide-sensitive amine oxidase activity in rat aortic cultured smooth muscle cells

J. R. D. Blicharski and **G. A. Lyles**

Department of Pharmacology and Clinical Pharmacology, University of Dundee, Ninewells Hospital and Medical School, Dundee, United Kingdom

Summary. Metabolism of 5 μM benzylamine (BZ) by rat aortic cultured smooth muscle cells (SMC) is inhibited almost completely by 10^{-3} M semicarbazide and 10^{-6} M propargylamine, but is little affected by 10^{-4} M and 10^{-3} M pargyline and clorgyline, indicating BZ metabolism predominantly by the semicarbazide-sensitive amine oxidase (SSAO) previously characterized in rat aortic homogenates. K_m values of 7–9 μM for BZ metabolism by SSAO in SMC cultures, indicate similar, if not identical properties, to the enzyme in the parent blood vessel.

Introduction

Blood vessels of various species contain a benzylamine (BZ)-metabolising amine oxidase which is inhibited by semicarbazide but is resistant to certain selective inhibitors (clorgyline, pargyline and deprenyl) of monoamine oxidase (MAO) (Lewinsohn, 1984). In the rat aorta, this semicarbazide-sensitive amine oxidase (SSAO) which can also metabolise such amines as β-phenylethylamine, tyramine, tryptamine and methylamine in vitro (Precious et al., 1988), appears from histochemical evidence to be associated with the smooth muscle cell (SMC) layers of the intact vessel (Lyles and Singh, 1985). Here, we have studied BZ metabolism in rat aortic SMC cultures, to determine if SSAO activity towards this substrate is expressed in these cells also.

Materials and methods

SMC were dissociated from thoracic aortae of male rats (around 200 g weight) and were grown to confluence as primary cultures in 80 cm^2 flasks containing Medium

199 with 10% v/v foetal calf serum, antibiotics (penicillin/streptomycin) and amphotericin. Flasks were maintained in a humidified atmosphere (5% CO_2; 95% air). Some primary cultures were subcultured by trypsinization and regrown to confluence as "first-passage" cells in fresh flasks. Cultures exhibited the characteristic "hill-and-valley" morphology of SMC cultures, and also cross-reacted with a fluorescently-labelled anti-SMC antibody (Boehringer-Mannheim). For enzyme assays, cells from 1–2 flasks were detached with a "rubber policeman", collected by centrifugation, and homogenized in 1 mM potassium phosphate buffer pH 7.8. Metabolism of ^{14}C-Bz was determined in 60 min assay incubations, using a previously described radiochemical assay (Precious et al., 1988).

Results and discussion

Deamination of 5 μM ^{14}C-BZ was determined in 4–5 homogenates of "first-passage" cultures which had been preincubated (20 min at 37 °C) with various amine oxidase inhibitors, and activities were expressed as a percentage of corresponding preincubated control samples lacking inhibitors. Remaining activities (mean ± s.e.) after semicarbazide (10^{-3} M) or propargylamine (10^{-6} M) treatment were 10 ± 3% and 6 ± 4%, respectively. These drug concentrations inhibit SSAO activity virtually completely in rat aortic homogenates (Precious et al., 1988), and the results here suggest that BZ deamination is predominantly carried out by SSAO in the cultured cells. Supporting this, acetylenic inhibitors selective for MAO had relatively little effect on BZ metabolism, with remaining activities being 83 ± 9% (10^{-4} M pargyline), 82 ± 6% (10^{-3} M pargyline), 93 ± 3% (10^{-4} M clorgyline), 88 ± 8% (10^{-3} M clorgyline). However, the slight apparent inhibitory effects of these compounds may be indicative of a small MAO contribution to 5 μM BZ metabolism.

Subsequent studies determined apparent kinetic constants for BZ metabolism in homogenates of primary and "first-passage" cultures of SMC. BZ metabolism (2–20 μM) was assayed in preincubated samples lacking or containing 10^{-4} M clorgyline. Some samples were treated thus with clorgyline to inhibit any MAO activity (if present), and to determine if kinetic constants differed from those found in the absence of clorgyline. These constants were estimated by linear regression of data from individual homogenates, plotted by the Lineweaver-Burk method. In *primary* cultures, mean values (± s.e., n = 5 cultures) for K_m (μM) were 12.0 ± 1.8 (no clorgyline) and 8.9 ± 1.3 (with clorgyline); for V_{max} (nmoles product/hr/mg protein) were 4.7 ± 2.2 (no clorgyline) and 4.0 ± 2.3 (with clorgyline). Studies with "first-passage" cultures (n = 5) gave corresponding K_m values of 9.8 ± 2.1 (no clorgyline), 6.7 ± 1.3 (with clorgyline); and V_{max} values of 3.3 ± 0.8 (no clorgyline) and 2.6 ± 0.7 (with clorgyline). Kinetic constants determined in the presence of clorgyline were not significantly different (paired t-test) from those in its absence. Corresponding values of K_m and

V_{max} were also closely similar when comparing primary with "first-passage" cultures, indicating that enzyme properties do not appear to change markedly if, at all, upon subculturing of cells. K_m values for cultured cells here were very similar to those (around 5 μM) usually reported for SSAO in rat aortic homogenates (e.g. Precious et al., 1988).

Cultured SMC from pig (Hysmith and Boor, 1988) and rat aorta (Ramos et al., 1988) have been used recently to investigate the role of SSAO in metabolising allylamine to the cardiovascular toxin, acrolein. The current results showing that the properties of SSAO (towards BZ) in rat aortic SMC cultures are similar to those of SSAO in homogenates of the parent tissue, suggest that studies with cultures may be an additional useful approach for investigating the largely unknown importance of SSAO in blood vessels.

Acknowledgements

This work was supported by funds from the Royal Society, the Smith Kline (1982) Foundation, and the University of Dundee.

References

Hysmith RM, Boor PJ (1988) Role of benzylamine oxidase in the cytotoxicity of allylamine toward aortic smooth muscle cells. Toxicology 51:133–145

Lewinsohn R (1984) Mammalian monoamine-oxidizing enzymes with special reference to benzylamine oxidase in human tissues. Braz J Med Biol Res 17:223–256

Lyles GA, Singh I (1985) Vascular smooth muscle cells: a major source of the semicarbazide-sensitive amine oxidase of the rat aorta. J Pharm Pharmacol 37:637–643

Precious E, Gunn CE, Lyles GA (1988) Deamination of methylamine by semicarbazide-sensitive amine oxidase in human umbilical artery and rat aorta. Biochem Pharmacol 37:707–713

Ramos K, Grossman SL, Cox LR (1988) Allylamine-induced vascular toxicity in vitro: prevention by semicarbazide-sensitive amine oxidase inhibitors. Toxicol Appl Pharmacol 95:61–71

Authors' address: Dr. G. A. Lyles, Department of Pharmacology and Clinical Pharmacology, University of Dundee, Ninewells Hospital and Medical School, Dundee, DD1 9SY, United Kingdom

J Neural Transm (1990) [Suppl] 32: 341–344

Amine oxidase activities in bovine lung

J. M. Lizcano, D. Balsa, K. F. Tipton, and **M. Unzeta**

Departament Bioquímica i Biologia Molecular, Facultat de Medicina, Universitat Autonoma de Barcelona, Barcelona, Spain

Summary. The subcellular distributions of amine oxidase activities MAO-A, MAO-B and SSAO, in bovine lung have been determined. MAO activities are largely confined to the mitochondria whereas SSAO activity is concentrated in the microsomal fraction.

The contribution of different type of amine oxidase present in the microsomal fraction to the metabolism of the benzylamine, tyramine, phenylethylamine, tryptamine and dopamine were determined. The kinetic constants towards benzylamine showed SSAO to have a higher affinity than MAO-B, and the comparison of the V_{max}/K_m relationship for both enzymes indicated that benzylamine would be oxidised more efficiently by SSAO than by MAO-B.

Introduction

Lung has been shown to be potentially important in the metabolism of amines and the activities of monoamine oxidase (MAO) have been characterised in perfused preparations of that tissue (Roth and Gillis, 1975; Bahkle and Youdim, 1979). However little alteration has been paid to any possible contribution of the semicarbazide-sensitive amine oxidase (SSAO) in this tissue. Since this enzyme has recently been shown to be active towards the more volatile short-chain aliphatic amines as well as towards benzylamine (Elliot et al., 1989) it may have important functions in the metabolism of inhaled amines. In order to assess the possible contributions of SSAO we have compared its distribution and specificity in bovine lung with those of MAO-A and MAO-B.

Materials and methods

Bovine lungs were obtained after slaughter of the animals and transported immediately to the laboratory from a local abattoir packed in ice. After weighing, the tissue was homogenized 1:10 (w/v) in 50 mM potassium buffer, pH 7.2 containing sucrose 0.25 M, by the use of a varying-blender. This homogenate was centrifuged at 600 g for 10 min. to remove cell nuclei, debris and unbroken cells (P1). This supernatant was then centrifuged at 600 g for 10 min. to obtain the pos-nuclear supernatant (P2). The resulting supernatant was centrifuged at 10000 g for 10 min, the pellet washed in buffer/sucrose, recentrifuged and finally suspended in buffer to obtain the mitochondrial fraction (MIT). The supernatant from 10000 g was then centrifuged at 10000 g for 10 min. in order to obtain a pellet (P3) which contain small mitochondria, lysosomes and possibly peroxisomes. Finally this supernatant was centrifuged at 200000 g for 30 min. and the pellet from this suspended in buffer as the microsomal fraction (MIC). The supernatant resulting from this microsomal sedimentation was saved and use as the cytosolic fraction (CYT). All samples were stored in aliquots at − 20 °C until assay.

MAO and SSAO activities were determined radiochemically (Fowler and Tipton, 1981) towards phenylethylamine (PEA), 5-hydroxytryptamine (5-HT), benzylamine (Bz), tyramine (Tyr), dopamine (Dop) and tryptamine (Ta) as substrates. Selective inhibition of MAO and SSAO activities were obtained by incubation with 1 mM clorgyline and 1 mM semicarbazide, respectively.

Results and discussion

The subcellular distribution of MAO-A, MAO-B and SSAO activities are shown in Fig. 1. The activities of MAO are largely confined to the mitochondrial fraction, as has been well-documented in other tissues (Tipton, 1986 for review). The SSAO activity was apparently concentrated in the microsomal

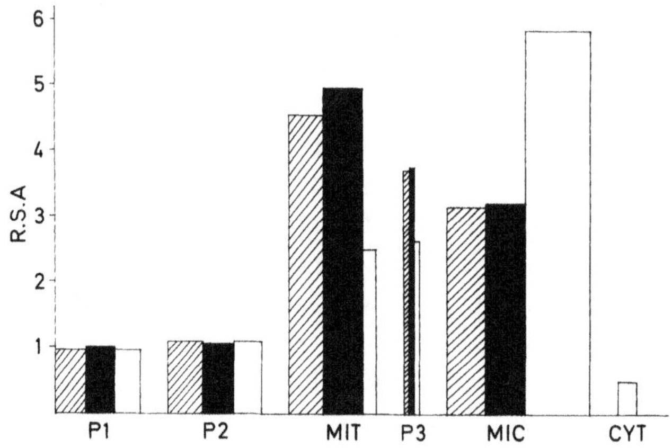

Fig. 1. Relative specific activity (R.S.A.) values for amine oxidase activities in subcellular fractionations obtained from bovine lung. Values are relative to the activities in the total homogenate (R.S.A. = 1.0). The width of each column corresponds to the percentage of the total activity of that enzyme present in the fraction. *Hatched columns*: MAO-A activity, measured towards 100 μM 5-HT as substrate. *Black columns*: MAO-B activity measured towards 20 μM PEA after inhibition of SSAO activity by treatment with 1 mM semicarbazide. *Open columns*: SSAO activity towards 20 μM Bz, determined after inhibition of the MAO-B activity by treatment with 1 mM clorgyline. Other details are given in the text

fraction and this was confirmed by the use of specific markers, of known subcellular distribution, and by electron microscopy of the isolated fractions (results not shown). It has been suggested that SSAO may be associated with the plasma membrane in brown adipose tissue (Barrand and Callingham, 1982). However the present results are unable to distinguish between a plasma membrane or endoplasmatic reticulum location of this enzyme in bovine lung.

The inhibition of activities towards several substrates given by clorgyline and semicarbazide allowed the contributions of different amine oxidases to the metabolism of these amines to be determined in the microsomal fraction of bovine lung. Benzylamine (Bz) at a concentration of 20 μM is metabolized largely by SSAO (90%) whereas MAO-B contributes only about 10% of the total activity. Higher concentrations of Bz (100 μM) had little effect on this distributions of activities (94% SSAO and 6% MAO-B). Phenylethylamine at a concentration of 20 μM is mainly metabolised by MAO-B (91%) with the remaining 9% of the activity being due to SSAO. 100 μM dopamine is deaminated by MAO-A (80%), MAO-B (12%) and SSAO (8%). Tyramine at a concentration of 100 μM is metabolised 29% by MAO-A, 71% by MAO-B with no detectable SSAO activity. In case of tryptamine at a concentration of 100 μM the behaviour was similar, with only MAO-A (87%) and MAO-B (13%) contributing to the oxidation of this amine and with no SSAO activity being detectable. 5-hydroxytryptamine at a concentration of 100 μM was oxidised exclusively by MAO-A, thus this amine is not substrate for SSAO activity in bovine lung.

The specific activities, in pmol min^{-1} mg of $protein^{-1}$, towards 20 μM Bz were 979 for SSAO and 98 for MAO-B. At 100 μM Bz these values were 3101 for SSAO and 201 for MAO-B. The specific activities towards 20 μM PEA were 181 for MAO-B and 18 for SSAO. In case of 5-HT (100 μM) the specific activity was 2220 (MAO-A). The specific activities towards 100 μM Tyr were 687 for MAO-A and 1682 for MAO-B. In case of Dop (100 μM) the specific activities were 1690 for MAO-A, 249 for MAO-B and 170 for SSAO. Towards 100 μM Ta as substrate, MAO-A had a specific activity of 1851, whereas that of MAO-B was 276.

The K_m (μM) and V_{max} (pmol min^{-1} mg of $protein^{-1}$) values towards Bz as substrate were 86.72 ± 11.71 and 5326.1 ± 199.1 for SSAO and 139.65 ± 14.17 and 457.5 ± 24.5 for MAO-B, respectively. The values of V_{max}/K_m for SSAO (61.4) and MAO-B (3.3) suggest that in bovine lung both activities would contribute to the oxidation of benzylamine. In total homogenates from lung the specific activities of MAO-B and SSAO towards 100 μM benzylamine were 515 and 91 pmol min^{-1} mg of $protein^{-1}$, respectively. These data and the kinetics parameters presented above indicate that SSAO may make an important contribution to the metabolism of benzylamine in the lung. Thus it is important to reassess the role of the lung in amine metabolism by taking account of the contributions of SSAO.

References

Bakhle YS, Youdim MBH (1979) The metabolism of 5-hydroxytryptamine and β-phenyl-ethylamine in perfused rat lung and in vitro. Br J Pharmacol 65:147–154

Barrand MA, Callingham BA (1984) Monoamine oxidase activities in brown adipose tissue of the rat: some properties and subcellular distribution. Biochem Pharmacol 31:2177–2184

Elliot J, Callingham BA, Sharman DF (1989) Semicarbazide-sensitive amine oxidase (SSAO) of the rat aorta. Biochem Pharmacol 38:1507–1515

Fowler CJ, Tipton KF (1981) Concentration dependence of the oxidation of tyramine by the two forms of rat liver mitochondrial monoamine oxidase. Biochem Pharmacol 30:3329–3332

Roth JA, Gillis CN (1975) Multiple forms of amine oxidases in perfused rabbit lung. J Pharmacol Exp Ther 194:537–544

Tipton KF (1986) Enzymology of monoamine oxidase. Cell Biol Funt 4:79–87

Authors' address: Dr. M. Unzeta, Departament de Bioquímica i Biologia Molecular, Facultat de Medicina, Universitat Autonoma de Barcelona, 08193 Bellaterra (Barcelona), Spain

J Neural Transm (1990) [Suppl] 32: 345–349

Mucosal mono- and polyamine oxidase activities in digestive tract are distributed complementary to diamine oxidase*

W. A. Fogel

Department of Biogenic Amines, Polish Academy of Sciences, Lodz, Poland

Summary. The distribution of mucosal monoamine and polyamine oxidases along rat gastrointestinal tract presents a complementary pattern to that of diamine oxidase. The maximal diamine oxidase values are found in the last two thirds of small intestine, whereas the highest activities of monoamine and polyamine oxidase occur in the proximal duodenum and in the large intestine.

Introduction

The mucosa of gastrointestinal tract is permanently exposed to mono-, di-, and polyamines which are either ingested with food (Rice et al., 1976) or formed and released locally (Beaven, 1970; Baylin et al., 1978).

Some of them e.g. histamine, serotonin, tyramine are potent vasoactive agents, some e.g. histamine, serotonin influence gastric and intestinal secretion (Beaven, 1970; Hardcastle and Hardcastle, 1987) as well as have or may have, e.g. spermine, a role in intestinal motility (Beaven, 1970; Martin and Tansy, 1986); some (putrescine, histamine) exert trophic effect on the mucosa (Tutton, 1976; Majumdar and Johnson, 1987). Normally, free access of these compounds to the general circulation or to the sites of their action is largely prevented due to mucosal amine oxidases. There are numerous studies concerning diamine oxidase (DAO EC 1.4.3.6) in mammalian alimentary tract as reviewed recently by Bieganski (1983); less attention has been paid to monoamine oxidase (MAO EC 1.4.3.4) (Kusche et al., 1975), and even less to polyamine oxidase (PAO) (Suzuki et al., 1984).

The aim of the present study was to examine in parallel the distribution of these three enzymes along the digestive tract of rat.

* Supported by Polish Academy of Sciences grant WPR 1/90.

Materials and methods

Chemicals

Radioisotopes were obtained from Amersham International, UK: (1,4-^{14}C) putrescine dihydrochloride, sp. act 118 mCi/mmol; 5-hydroxy-(side chain 2-^{14}C) tryptamine creatinine sulphate, sp. act 57 mCi/mmol or from New England Nuclear, Boston USA: tryptamine bisuccinate (side chain 2-^{14}C) sp. act 60 mCi/mmol. Putrescine, N^1acetylspermine and aminoguanidine were from Sigma; 5-hydroxytryptamine, tryptamine, pargyline, histamine, homovanillic acid and horse radish peroxidase were from Serva. The other compounds were obtained from commercial sources in the purest available form.

Animals

Male Wistar rats with an average weight of 242 g (235–250 g) were used. Animals were sacrificed by decapitation, stomach and whole intestine were removed and sectioned into the following parts: A– pars prepylorica; B– proximal and C– distal halves of duodenum; D– proximal, E– middle and F– distal thirds of small intestine; G– proximal and H– distal halves of large intestine.

Each segment was flushed with a cold saline, split longitudinally, mucosa gently scraped and immediately frozen in liquid nitrogen. The samples were kept at – 50 °C and assays were performed within 3–4 days.

Assay of enzyme activities and protein

The mucosal scrapings were thawed, immediately homogenized in 35 volumes of 0.1 M Na^+K^+-phosphate buffer pH 7.4 using Ultra Turrax homogenizer followed by low energy sonication and resulting homogenates used as the enzymes source.

MAO activity was determined using as substrates ^{14}C serotonin (0.5 mM) and ^{14}C tryptamine (0.1 mM). The basic procedure employed was that described by Wurtman and Axelrod (1963) for tryptamine. The oxidative deamination products of tryptamine were extracted into toluene, these of ^{14}C serotonin into toluene-ethylacetate (1 : 1 v/v). DAO activity was measured with putrescine – by isotopic method exactly as it was previously reported (Fogel et al., 1985) and with histamine – by fluorometric method according to Matsumoto et al. (1982). The last procedure was also used to determine the activity of PAO with N^1 acetylspermine (0.1 mM) as substrate. Enzyme activities were expressed as picomoles of oxidized amine per min of incubation at 37 °C per mg protein.

Protein concentrations of the homogenates were estimated by the method of Lowry et al. (1951).

Results

As it can be seen in the upper panel in Fig. 1, characteristic pattern of DAO distribution showing a progressive rise of the enzyme activity from traces in prepyloric area through duodenum until attaining maximum in the last two thirds of the small intestine, and followed by the decline in the large bowel, is

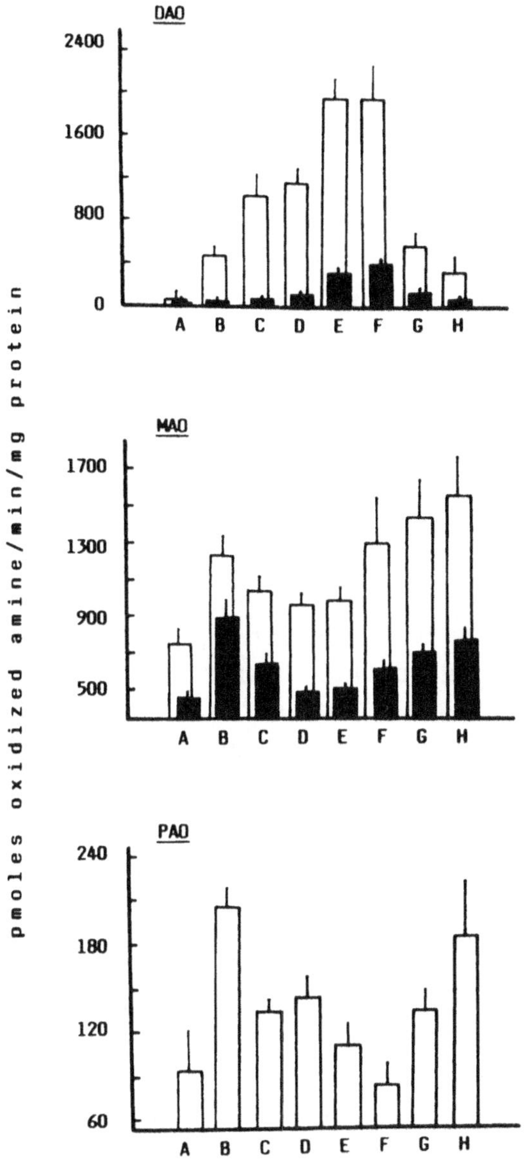

Fig. 1. The distribution of mucosal amine oxidase activities along rat alimentary canal. Following amine substrates were used – upper panel: putrescine (open columns), histamine (filled columns); middle panel: serotonin (open columns), tryptamine (filled columns); bottom panel: N^1acetylspermine. Ordinate: *A* corresponds to prepyloric area; *B*, *C* proximal and distal halves of duodenum; *D*, *E*, *F* proximal, middle and distal thirds of small intestine; *G*, *H* proximal and distal halves of large intestine. Each value is a mean ± SEM for n = 8–10 rats

manifested with either of the DAO substrates – putrescine and histamine (correlation coefficient $r = .922\, p < 0.001$), the difference between the two being only quantitative. This pattern is, however, clearly distinct from these obtained while using serotonin and tryptamine (middle panel) or N^1 acetylspermine (bottom panel) as substrates. The monoamine and polyamine oxidizing activities are distributed in a similar way along the gastrointestinal tract (MAO 5HT vs MAO tryptamine $r = .74$ $p < 0.04$; MAO tryptamine vs PAO $r = .806$ $p < 0.02$). The highest activities are present in proximal duodenum and in the large intestine, the lowest – in the small intestine.

Discussion

The present study indicates that distribution of the enzymic activity oxidizing mono- and polyamines is complementary to that of diamine oxidase. Peak of DAO activity in the small intestine covers the area of the low, in relation to the other segments, MAO and PAO activities. In the stomach, proximal duodenum and in the large intestine where DAO is absent or low, MAO and PAO activities are high. This would suggest that gastrointestinal mucosa all along is fortified with an efficient barrier against biologically active amines which may exert a wide diversity of effects. The pattern of DAO distribution shown in Fig. 1 is in accord with the previously published data on the subject (for ref. see Bieganski, 1983). The very high correlation between the profiles of activities transforming putrescine and histamine gives additional support to the identity of diamine oxidase and histaminase (Beaven and Shaff, 1975). Lower reaction velocity obtained with histamine, e.g. 8–20% of that with putrescine, also agrees well with the earlier studies (Beaven and Shaff, 1975; Kusche et al., 1975).

If one assumes that serotonin is MAO-A and tryptamine MAO-B substrate (Youdim and Finberg, 1982) the data in Fig. 1 are compatible with the notion that in mucosa of GI tract in the rat both MAO forms are present, their distribution is alike but MAO-A activity prevails. One cannot exclude however, that conversion of serotonin and tryptamine was accomplished by the same enzyme form and the differences found were caused by different affinity of substrates.

The high correlation with MAO distribution shows PAO. It has been claimed that N^1acetylspermine is not oxidized by rat liver MAO (Suzuki et al., 1984). If this holds true also for rat intestinal MAO one can ascribe this high correlation to the parallel distribution of two distinct enzymes.

The presence of PAO in all examined segments of gastrointestinal tract indicates that interconversion of poliamines may be important, alternative to the de novo synthesis from ornithine, source of putrescine, the amine necessary for mucosal proliferation.

References

Baylin SB, Stevens SA, Shakir KMM (1978) Association of diamine oxidase and ornithine decarboxylase with maturing cells in rapidly proliferating epithelia. Biochem Biophys Acta 541:415–419

Beaven MA (1970) Selected topics on the function of biogenic amines in exocrine organs with special reference to histamine in stomach. In: Blum JJ (ed) Biogenic amines as physiological regulators. Prentice-Hall, Englewood Cliffs NJ, pp 207–222

Beaven MA, Shaff RE (1975) Study of the relationship of histaminase and diamine oxidase activities in various rat tissues and plasma by sensitive isotopic assay procedures. Biochem Pharmacol 24:979–984

Bieganski T (1983) Biochemical, physiological and pathophysiological aspects of intestinal diamine oxidase. Acta Physiol Pol 34:139–154

Fogel WA, Ulatowska M, Adach K, Osinska Z (1985) A sum of ^{14}C-putrescine metabolites as a measure of DAO activity. Column chromatography assay. Agents Actions 16:99–101

Hardcastle J, Hardcastle PT (1987) The secretory actions of histamine in rat small intestine. J Physiol 388:521–532

Kusche J, Lorenz W, Schmidt J (1975) Oxidative deamination of biogenic amines by intestinal amine oxidases: histamine is specifically inactivated by diamine oxidase. Hoppe-Seyler's Z Physiol Chem 356:1485–1496

Majumdar APN, Johnson LR (1987) Effect of putrescine on oxyntic gland and colonic mucosal growth in rats. Life Sci 41:961–966

Martin JS, Tansy MF (1986) Comparison of the inhibitory effects of spermine, papaverine and adrenaline upon isolated segments of the rat small intestine. Clin Exp Pharmacol Physiol 13:87–90

Matsumoto T, Furuta T, Nimura Y, Suzuki O (1982) Increased sensitivity of the fluorometric method of Snyder and Hendley for oxidase assays. Biochem Pharmacol 31:2207–2209

Rice SL, Eitenmiller RR, Koehler PE (1976) Biologically active amines in food: a review. J Milk Food Technol 39:353–358

Suzuki O, Matsumoto T, Katsumata Y (1984) Determination of polyamine oxidase activities in human tissues. Experientia 40:838–839

Tutton PJM (1976) The influence of histamine on epithelial cell proliferation in the jejunum of the rat. Clin Exp Pharmacol Physiol 3:369–373

Wurtman RJ, Axelrod J (1963) A sensitive and specific assay for the estimation of monoamine oxidase activity. Biochem Pharmacol 12:1439–1441

Youdim MBH, Finberg JPM (1982) Monoamine oxidase inhibitor antidepressants. In: Grahame-Smith DG, Hippius H, Winokery G (eds) Psychopharmacology. Excerpta Medica, Amsterdam, p 38

Author's address: Prof. Dr. W. A. Fogel, Department of Biogenic Amines, Polish Academy of Sciences P.O. Box 225, Tylna 3, PL-90-950 Lodz, Poland.

J Neural Transm (1990) [Suppl] 32: 351–356

Does FAD-dependent polyamine oxidase contribute to the metabolism of milacemide?

M. Strolin Benedetti, G. Cocchiara, M. Colombo, and **P. Dostert**

Research and Development – Erbamont Group, Farmitalia Carlo Erba, Milan, Italy

Summary. Milacemide, a secondary amine derivative, was previously demonstrated to be a substrate of MAO-B and to be insensitive to the action of copper-dependent amine oxidases. In the present study, it was investigated whether the FAD-dependent secondary amine metabolizing enzyme polyamine oxidase (PAO), could participate in the metabolism of milacemide. For this purpose, the urinary metabolic pattern of oral ^{14}C-milacemide was determined in rats with and without pretreatment with the irreversible PAO inhibitor MDL 72527 and, for comparison, after inhibition of MAO-B by 1-deprenyl.

While 1-deprenyl was shown to significantly decrease the urinary excretion of glycinamide and of an unknown metabolite (UK_1), pretreatment with MDL 72527 did not modify the elimination of milacemide as glycinamide but produced a decrease in UK_1 equal to that induced by 1-deprenyl. The percent of the dose of milacemide eliminated as unchanged drug was slightly but significantly increased after PAO inhibition, though considerably less than after 1-deprenyl. These data suggest that milacemide might be a substrate of PAO. If confirmed, this result would constitute the first example of the involvement of the FAD-dependent PAO in drug metabolism.

Introduction

Milacemide [2-(pentylamino)acetamide] is an atypical potential neuropsychotropic agent with anticonvulsant properties. By analogy with pentylamine, a known substrate of MAO-B (Strolin Benedetti et al., 1981), milacemide was shown to be a selective MAO-B substrate, with apparent K_m values of 30–90 µM and approximately 1,300 µM for MAO-B and MAO-A, respectively (Janssens de Varebeke et al., 1988) and also to behave as a mechanism-based inhibitor of MAO-B, producing irreversible and partially reversible inhibition in vitro and in vivo, respectively (Janssens de Varebeke et al., 1989). The mechanism by which the inhibition of MAO-B occurs in the presence of milacemide is still unclear.

Although pentylamine was shown to be also a substrate of semicarbazide-sensitive amine oxidase (SSAO) (Guffroy et al., 1983), we have recently established that milacemide is not a substrate of rat copper-containing amine oxidases (Strolin Benedetti et al., 1988), a result in line with the fact that only primary amines are thought to be substrates of SSAO (Barrand and Callingham, 1982). Besides several copper-containing amine oxidases which are capable of oxidatively deaminating the terminal amino function of diamines or polyamines and their natural monoacetyl derivatives, a FAD-dependent polyamine oxidase (PAO) was shown to be present virtually in all rat tissues (Seiler et al., 1980). This enzyme is involved in the interconversion of polyamines, its role being of oxidizing a secondary amine group in monoacetyl spermidine and- spermine to form putrescine and spermidine, respectively (Seiler, 1987).

In the present study, in an attempt to examine whether PAO could be involved in the metabolism of the secondary amine derivative milacemide, the urinary metabolic profile of ^{14}C-milacemide was determined in rats with and without pretreatment with the irreversible PAO inhibitor MDL 72527 (N^1,N^2-bis(2,3-butadienyl)-1,4-butanediamine) (Bolkenius et al., 1985). For comparison, the urinary metabolism of milacemide was also studied after pretreatment with the MAO-B inhibitor 1-deprenyl. While PAO inhibition by 1-deprenyl cannot be ruled out by analogy with the weak PAO inhibitory activity of the propargyl MAO inhibitor pargyline (Hölltä, 1977), liver MAO-B activity was assayed after MDL 72527 given at a dose (20 mg/kg, i.p.) known to produce complete inhibition of PAO in all organs of the rat (Bolkenius et al., 1985).

Materials and methods

Male Sprague Dawley rats (Charles River) weighing 200–220 g were used. Milacemide HCl was synthesized at Farmitalia Carlo Erba and ^{14}C-milacemide HCl (^{14}C-M), labelled in the CH_2 of the acetamide residue, specific activity 15.8 μCi/mmol, radiochemical purity $\geq 98\%$, was synthesized by Prof. Luu-Duc at the Laboratoire de Chimie Pharmaceutique, Grenoble, France. 2-Phenyl[1-^{14}C]ethylamine hydrochloride (PEA) was purchased from Amersham, U.K., MDL 72527 was a generous gift of Dr. N. Seiler, Merrell Dow Research Institute, Strasbourg, France. Doses are expressed in terms of the free base.

Urinary metabolism of ^{14}C-milacemide

Rats were individually housed in glass metabolism cages. Three groups of 4 rats each were used. Rats received either 16 mg/kg ^{14}C-M (final specific activity 6.5 μCi/mg) or the same dose of ^{14}C-M 1 h after the oral administration of l-deprenyl at a dose (10 mg/kg) known to inhibit rat MAO-B for at least 24 h (Strolin Benedetti et al., 1983), or the same dose of ^{14}C-M 1 h after MDL 72527 (20 mg/kg, i.p.). ^{14}C-M was administered by gastric gavage. Urine collection and metabolic pattern analysis were performed as described by Colombo et al. (this issue). Since radioactivity excretion was almost complete after 24 h (Colombo et al., this issue), the metabolic pattern of milacemide in the 3 groups of rats was compared in the 24 h-urines only.

Ex-vivo inhibition of liver MAO-B by MDL 72527

Three rats were administered with MDL 72527 (20 mg/kg, i.p.) and three rats received the vehicle. Animals were killed by decapitation 4 h after administration; livers were removed, rinsed in saline, blotted on filter paper, frozen in liquid nitrogen and stored at $-20\,^{\circ}C$ until use. The tissues were homogenized in phosphate buffer 0.1 M pH 7.4, using Ultra-Turrax (1 g tissue/16 ml buffer). MAO-B activity was measured as described by Colombo et al. (this issue).

Results

The percent values of the dose of milacemide excreted in the 3 groups of rats in the 24 h-urines as total radioactivity, glycinamide, unchanged milacemide and as non-identified radioactive peaks UK_1, UK_2 and UK_3 are presented in Table 1.

The percent of the dose eliminated as total radioactivity increased significantly in rats pretreated with l-deprenyl and did not vary when rats received MDL 72527. The percent amount of milacemide eliminated as unchanged drug was considerably increased after l-deprenyl ($55.9 \pm 6.2\%$) and also significantly increased, though markedly less ($21.3 \pm 4.6\%$), after MDL 72527, compared to the non-pretreated group ($12.1 \pm 2.5\%$).

Excretion of the metabolites UK_2 and UK_3 was not influenced by either pretreatment. The percent of the dose of milacemide eliminated as glycinamide was significantly lower after l-deprenyl, from $14.3 \pm 2.5\%$ (non-pretreated group) to $4.6 \pm 1.2\%$ (l-deprenyl group), whereas it remained unchanged after MDL 72527. Pretreatment with l-deprenyl or MDL 72527 significantly lowered to the same extent the percent of the dose of milacemide excreted as UK_1, from 9.6% (non-pretreated group) to 4.4% (both pretreated groups). It is to note that homogeneity of radioactive peaks UK_1, UK_2 and UK_3 has not been established.

Rat liver MAO-B was found to be inhibited by $38 \pm 4\%$ 4 h after oral administration of MDL 72527.

Discussion

Metabolism of milacemide was found to be strongly affected after inhibition of MAO-B by l-deprenyl, which, however, did not totally reduce the

Table 1. Major radioactive components in rat 24 h-urines after oral administration of ^{14}C-milacemide (16 mg/kg) given alone or 1 h following pretreatment with 1-deprenyl (10 mg/kg, p.o.) or with MDL 72527 (20 mg/kg, i.p.). Results are expressed as percent of the administered dose

	% of the dose (mean ± SD)					
Treatment	UK_1	Glycinamide	UK_2	UK_3	Milacemide	Total radioactivity
Milacemide (M)	9.6 ± 0.5	14.3 ± 2.5	10.5 ± 1.0	1.3 ± 0.3	12.1 ± 2.5	47.8 ± 4.5
M after 1-deprenyl	4.4 ± 0.6**	4.6 ± 1.2**	6.7 ± 4.6	8.2 ± 8.7	55.9 ± 6.2**	79.8 ± 3.7**
M after MDL 72527	4.4 ± 1.0**	14.6 ± 1.9	7.8 ± 2.7	1.1 ± 0.3	21.3 ± 4.6*	49.2 ± 4.7

UK_1, UK_2 and UK_3 unknown metabolites 1, 2 and 3; pretreated groups compared to the group treated with milacemide only: * $p < 0.05$, ** $p < 0.01$, Dunnett's test

formation of glycinamide. It might be that glycinamide formation after 1-deprenyl reflects the contribution of MAO-A to milacemide metabolism, or that a residual MAO-B activity persisting after a 10 mg/kg dose of 1-deprenyl (Strolin Benedetti et al., 1983) is enough to somewhat deaminate milacemide in 24 h. The large involvement of MAO-B in the metabolism of milacemide is clearly reflected by the statistically significant increase in the percent of the dose excreted as unchanged milacemide after 1-deprenyl. Participation of glycine in a number of metabolic steps should explain why the percent of the dose of milacemide excreted in 24 h as total radioactivity was significantly lower in the absence of pretreatment than after 1-deprenyl.

Though administration of MDL 72527 was found to somewhat inhibit liver MAO-B activity (38% after 4 h), the residual MAO activity appears to be enough to contribute to the metabolism of milacemide to the same extent as in the absence of the PAO inhibitor. In fact, glycinamide was found to account for the same percent of the dose of milacemide (about 14%) in the absence of pretreatment and after MDL 72527. The percent of the dose excreted as unchanged milacemide was significantly increased after the PAO inhibitor although markedly less than after 1-deprenyl, indicating that PAO should contribute to the metabolism of milacemide. That, of the three unknown metabolites, only the percent of the dose excreted as UK_1 was significantly decreased after MDL 72527, suggests that UK_1 might result from the deamination of milacemide by PAO. If this is true, a 10 mg/kg dose of 1-deprenyl would also inhibit PAO, as UK_1 formation was found to be equally reduced after 1-deprenyl and MDL 72527. Determination of the chemical structure of UK_1 is in progress and should allow to elucidate whether this metabolite is formed by the action of PAO. If so, this would constitute the first example of the involvement of PAO in the metabolism of drugs (Strolin Benedetti, 1989).

Acknowledgements

The authors wish to thank Dr. C. Efthymiopoulos for the statistical analysis and G. Garattini for typing the manuscript.

References

Barrand MA, Callingham BA (1982) Monoamine oxidase activities in brown adipose tissue of the rat: some properties and subcellular distribution. Biochem Pharmacol 31:2177–2184

Bolkenius FN, Bey P, Seiler N (1985) Specific inhibition of polyamine oxidase in vivo is a method for the elucidation of its physiological role. Biochim Biophys Acta 838:69–76

Guffroy C, Fowler CJ, Strolin Benedetti M (1983) The deamination of n-pentylamine by monoamine oxidase and a semicarbazide-sensitive amine oxidase of rat heart. J Pharm Pharmacol 35:416–420

Höllta E (1977) Oxidation of spermidine and spermine in rat liver: purification and properties of polyamine oxidase. Biochemistry 16:91–100

Janssens de Varebeke P, Cavalier R, David-Remacle M, Youdim MBH (1988) Formation of the neurotransmitter glycine from the anticonvulsant milacemide is mediated by brain monoamine oxidase. J Neurochem 50:1011–1016

Janssens de Varebeke P, Pauwels G, Buyse C, David-Remacle M, De Mey J, Roba J, Youdim MBH (1989) The novel neuropsychotropic agent milacemide is a specific enzyme-activated inhibitor of brain monamine oxidase B. J Neurochem 53:1109–1116

Seiler N (1987) Inhibition of enzymes oxidizing polyamines. In: McCann PP, Pegg AE, Sjoerdsma A (eds) Inhibition of polyamine metabolism. Academic Press, Orlando, pp 49–77

Seiler N, Bolkenius FN, Knödgen B, Mamont P (1980) Polyamine oxidase in rat tissues. Biochim Biophys Acta 615:480–488

Strolin Benedetti M (1989) Oxidative drug metabolism not cytochrom P-450 dependent. Actual Chim Thér 16:337–356

Strolin Benedetti M, Sontag N, Boucher T, Kan JP (1981) Aliphatic amines as MAO substrates in the rat: the effect of selective inhibitors on the deamination of n-pentylamine. In: Usdin E, Weiner N, Youdim MBH (eds) Function and regulation of monoamine enzymes. Macmillan, London, pp 527–538

Strolin Benedetti M, Dostert P, Guffroy C, Tipton KF (1983) Partial or total protection from long-acting monoamine oxidase inhibitors by new short-acting $MAOI_s$ of type A MD 780515 and type B MD 780236. Mod Probl Pharmacopsychiatry 19:82–104

Strolin Benedetti M, Marrari P, Moro E, Dostert P, Roncucci R (1988) Do amine oxidases contribute to the metabolism of milacemide? Pharmacol Res Commun 20 [Suppl] 4:135–136

Authors' address: Dr. P. Dostert, Farmitalia Carlo Erba, R&D, Via C. Imbonati 24, I-20159 Milan, Italy.

Catechol-O-methyl transferase

J Neural Transm (1990) [Suppl] 32: 359–368

Behaviour and properties of catechol-O-methyltransferase from human placenta

N. Nic a' Bháird, and **K. F. Tipton**

Department of Biochemistry, Trinity College, Dublin, Ireland

Summary. A procedure is reported for the purification of human placental catechol-O-methyltransferase. The preparation is apparently homogeneous and behaves as a monomer with an approximate M_r of 23,000. The sequence of the first 21 amino acid residues from the N-terminal end of the protein is reported. The activity of the enzyme is strongly influenced by the nature of the buffer in which it is assayed.

Introduction

Catechol-O-methyltransferase (EC 2.1.1.6; COMT) catalyses the methylation of a range of catechol derivatives including the catecholamines adrenaline, noradrenaline and dopamine and their deaminated derivatives (see Nic a' Bháird et al., 1990 for review). In experimental animals it has been shown to play an important role in the extraneuronal transport and metabolism of the catecholamine neurotransmitters (Trendelenburg et al., 1987). Studies on the properties and function of the enzyme in human tissues have been hampered by the relatively low activities present and the instability of the enzyme. Jeffrey and Roth (1985) obtained a relatively high degree of purification (about 4,000 fold) of the enzyme from human brain, although the preparation was not homogeneous at that stage and further studies were made difficult by the instability of the preparation.

The possibility that the instability of the enzyme might, at least in part, result from proteolysis occuring in the tissue preparation may make it unreliable to work with post-mortem human tissues. As an alternative we have studied the enzyme from human placenta. This paper describes its purification, procedures for stabilizing the purified enzyme and some of its properties.

Materials and methods

Assays

Catechol-O-methyltransferase was assayed radiochemically by a slight modification of the method of Zürcher and Da Prada (1982) at 37 °C in the presence of 100 mM potassium phosphate buffer, pH 7.4, and 500 μM catechol.

Protein concentrations above 1 mg/ml were determined by the method of Markwell et al. (1978) and at lower concentrations by the method of Mejbaum-Katzenellenbogen and Dobryszcka (1959).

Electrophoresis

Slab polyacrylamide gel electrophoresis was carried by the general method of Weber and Osborn (1969) using a 5% stacking gel and a resolving gel of 20% in the absence and presence of 0.1% SDS. Gels were stained with Coomassie Blue and destained as described by Weber and Osborn (1969). Standard proteins were supplied by Sigma Chemical Company.

N-terminal amino acid sequence analysis

A fraction of purified COMT (100 pmol) was exposed to trichloroacetic acid (at a final concentration of 10%) for 1 hour at 4 °C. The protein precipitate was sedimented by centrifugation at 20,000 × g for 10 min and was washed with acetone and centrifuged as before. The resulting pellet was resuspended in 100 μl 10% trifluoroacetic acid. The amino acid sequence was determined with an Applied Biosystems model 477A protein sequencer, connected to an Applied Biosystems model 120A PTH analyzer.

The effects of pH on COMT activity

The activity and stability of COMT in a number of different buffers was studied over a range of pH values from 5.0 to 8.5. The pH-activity profile of COMT was determined by substituting buffers of varying pH in the radiochemical assay.

The stability profile of COMT was investigated by incubating aliquots of the enzyme at 37 °C for 30 min with buffers of different pH values. Following this incubation the enzyme was assayed at pH 7.4 as described above. The activity was found to be stable under these conditions over the pH range 5.5–8.5. However inactivation was observed at pH values below 5.5 with some 20% of the activity being lost on preincubation in potassium acetate buffer at pH 5.0.

Purification of COMT

All operations were carried out at 4 °C or on ice. 400 g of ice-cold fresh placenta were homogenised in 450 ml 10 mM potassium phosphate buffer, pH 8.0, containing 130 mM KCl, 1 mM EDTA, 1 mM $MgCl_2$, 5 mM dithiothreitol, 10 mM 2-mercaptoethanol and 0.2 mM phenylmethanesulphonyl chloride in a Waring Blender. The homogenate was centrifuged at 17,700 × g for 40 min.

Acid treatment

The supernatant was decanted and titrated over a period of 10 min to pH 5.0 with 1 M acetic acid and centrifuged at 17,700 × g for 30 min.

Ammonium sulphate fractionation

The supernatant was adjusted to pH 8.0 with 2 M ammonia solution and solid ammonium sulphate (33.8 g/100 ml) was added slowly with stirring. The pH was maintained at 8.0 and the solution was allowed to stand with stirring for 20 min before centrifuging at 17,700 × g for 20 min. The precipitate was resuspended in the minimum volume of 10 mM potassium phosphate buffer, pH 8.0, containing 1 mM EDTA, 1 mM $MgCl_2$, 10 mM 2-mercaptoethanol, 5 mM dithiothreitol and 0.2 mM phenylmethanesulphonyl chloride, by homogenising the solution in a hand-held Thomas homogeniser. Undissolved material was removed by centrifuging the solution at 48,000 × g for 10 min.

Gel-filtration

The supernatant was applied to a column of Sephadex G-75 (6 × 60 cm) which was equilibrated and developed in 10 mM potassium phosphate buffer, pH 8.0, containing 10 mM 2-mercaptoethanol, 1 mM EDTA and 1 mM $MgCl_2$. Fractions containing COMT activity were pooled and solid dithiothreitol was added to a concentration of 5 mM.

Cation exchange chromatography

The pH was adjusted to 6.0 using 1 M acetic acid. The solution was applied to a column of carboxymethylcellulose (CM-52) (6.5 × 3 cm) which was equilibrated and developed in 10 mM potassium phosphate buffer, pH 6.0, containing 1 mM EDTA, 1 mM $MgCl_2$ and 10 mM 2-mercaptoethanol. The eluate, which contained the COMT activity, was monitored for the appearance of protein by measuring the absorbance of the eluate at 280 nm. All of the protein which eluted from the CM-52 column was collected and pooled and the pH was adjusted to pH 8.0 using a 2 M ammonia solution.

Anion exchange chromatography

The post CM-52 fraction was applied to a column of diethylaminoethyl cellulose (DE-52) (2 × 20) which was equilibrated in 10 mM potassium phosphate buffer, pH 8.0, containing 1 mM EDTA, 1 mM $MgCl_2$ and 10 mM 2-mercaptoethanol. The column was then extensively washed with this buffer until the absorbance of the eluate at 280 nm was below 0.02 absorbance units. A linear salt gradient (2 × 250 ml from 0–500 mM KCl in 10 mM potassium phosphate buffer, pH 8.0, containing 1 mM EDTA, 1 mM $MgCl_2$, 10 mM 2-mercaptoethanol and 5 mM dithiothreitol) was applied to the column. Fractions containing COMT activity were pooled and solid ammonium sulphate was added with stirring to 70% saturation. This solution was centrifuged at 48,000 × g for 10 min and the resulting pellet was resuspended in 1–2 ml of 100 mM potassium phosphate buffer, pH 8.0, containing 1 mM EDTA, 1 mM $MgCl_2$, 10 mM 2-mercaptoethanol and 2 mM dithiothreitol.

Gel-filtration

The resuspended pellet was applied to a column of Sephadex G-75 Superfine (2 × 80 cm) which was equilibrated and developed in 100 mM potassium phosphate buffer, pH 8.0, containing 1 mM EDTA, 1 mM $MgCl_2$, 10 mM 2-mercaptoethanol and 2 mM dithiothreitol. Fractions containing COMT activity were pooled, dithiothreitol was added to a concentration of 5 mM, BSA was added to a concentration of 1 mg/ml, glycerol was added to a final concentration of 40% (v/v) and the enzyme was stored at − 20 °C until required.

Results and discussion

Table 1 summarises the purification of the enzyme. Placenta is a tissue richly supplied with blood and since COMT is present in erythrocytes (see e.g. Assicot and Bohuon, 1971) it might be expected that some of the purified enzyme might arise from that source. However, the activity of COMT in erythrocytes is low (0.32 nmol/min/ml whole blood; Schultz and Nissinen, 1989) compared to that in placental homogenate (6.0 nmol/min/ml) and thus any contribution of the erythrocyte COMT to the final preparation of the enzyme is not likely to be significant. The procedure adopted involved gel-filtration, ion-exchange chromatography and ammonium sulphate fractionation. The affinity chromatographic procedure which was used for the purification of the enzyme from pig liver and brain (Gulliver and Tipton, 1978 and 1979) was found not to be successful with the human placental enzyme. Neither was hydrophobic chromatography on phenyl-Sepharose; which had been used with the human brain enzyme (Jeffrey and Roth, 1985), successful since neither the enzyme nor any substantial amounts of contaminating proteins were retarded on passage through a column of this material, either at low or high concentrations of ammonium sulphate. The nature of the interactions of the human brain enzyme with this material in obscure, since Jeffrey and Roth (1985) found binding of the enzyme to occur only at low ionic strengths under which conditions purely hydrophobic interactions would be at their weakest.

The enzyme prepared in this way was apparently homogeneous by the criteria of electrophoresis in the presence and absence of sodium dodecyl sulphate. The former procedure resulted in a mobility corresponding to a relative molecular mass of 22,800. Gel filtration of the enzyme on Sephadex G-75 gave a M_r value of 23,000. Thus this enzyme is monomeric with a relative molecular mass similar to that reported for the enzyme from human liver (25,000 Ball et al., 1971) rat liver (29,000 Anderson and D'Iorio, 1968; White and Wu, 1975) and pig liver and brain (23,000, Gulliver and Tipton,

Table 1. Purification of catechol-O-methyltransferase from human placenta

Purification step	Volume (ml)	Activity (nmol/min/ml)	Total activity (nmol/min)	Protein concentration (mg/ml)	Total protein (mg)	Specific activity (nmol/min/mg)	Yield (%)	Purification fold
Homogenate	940	6.0	5640	37.5	35250	0.16	100	–
Supernatant	585	9.2	5376	31.25	18281	0.29	95	1.8
Redissolved Ammonium Sulphate pellet	98	22.9	2241	50.0	4900	0.46	40	2.9
Pooled G-75 Fractions	200	12.9	2586	1.25	250	10	46	65
Post CM-52 Fractions	220	11.3	2490	0.70	154	16	44	101
Pooled DE-52 Fractions	43	39.5	1698	0.425	18.3	93	30	581
Pooled G-75 Superfine Fractions	11	92.8	1020	0.125	1.4	742	18	4639

Fractions were assayed by the method of Zürcher and Da Prada (1982). All assays were carried out at a final concentration of S-adenosyl methionine of 900 µM. A final concentration of catechol of 500 µM was used except for the assay of the redissolved ammonium sulphate pellet which was assayed at a final concentration of 5 mM

1978, 1979). This value for human placental COMT, however, differs markedly from the values of 52,000 (Gugler et al., 1970) and 49,000 (Darmenton et al., 1976) previously reported for the enzyme from this source.

The dependence of the activity of the enzyme upon the assay pH is shown in Fig. 1. There was a broad optimum at about pH 7.0–7.6 when the activity was determined in 100 mM phosphate buffer. However, the behaviour in 100 mM Tris-HCl or triethanolamine-HCl buffers appeared to be significantly different (Fig. 1). The activity of the enzyme was also dependent on the buffer concentration. When assayed in the presence of varying concentrations of either sodium or phosphate buffers, pH 7.4, the activity increased with increasing buffer concentration up to a maximum at about 200 mM, after which a decrease was observed (Fig. 2). These effects were not due to ionic strength changes since assay in 100 mM phosphate buffer containing various concentrations of KCl, in the range 0–1.5 M, had no

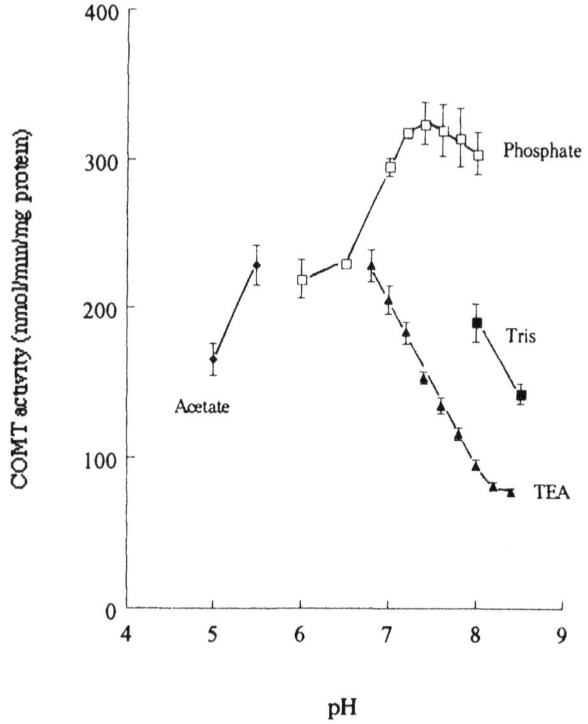

Fig. 1. pH activity profile of catechol-O-methyltransferase. COMT was assayed by the method of Zürcher and Da Prada (1982) at various pH values in the range 5.0 to 8.5. A number of different buffers appropriate to each pH value were used

□ 100 mM potassium phosphate
◆ 100 mM potassium acetate
■ 100 mM Tris/HCl
▲ 100 mM triethanolamine/HCl

Points shown are the mean ± SD of three separate determinations

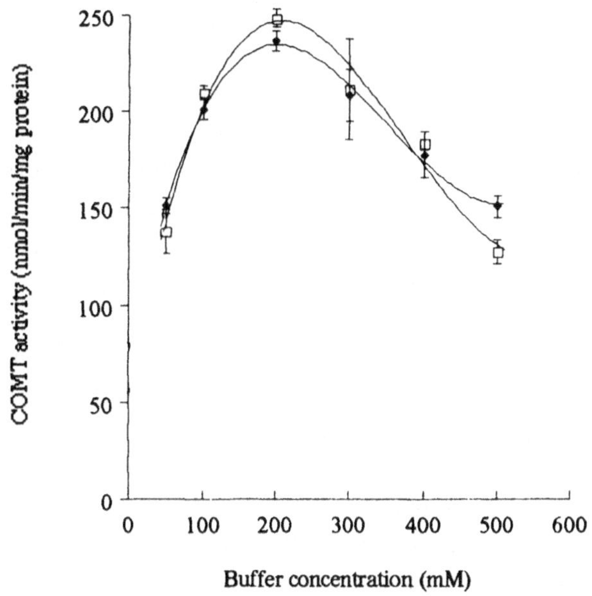

Fig. 2. The effect of buffer concentration on the activity of catechol-O-methyltransferase. COMT was assayed by the method of Zürcher and Da Prada (1982) in the presence of increasing concentrations of either potassium phosphate buffer pH 7.4 (□) or sodium phosphate buffer pH 7.4 (■). Points shown are the mean ± SD of three determinations

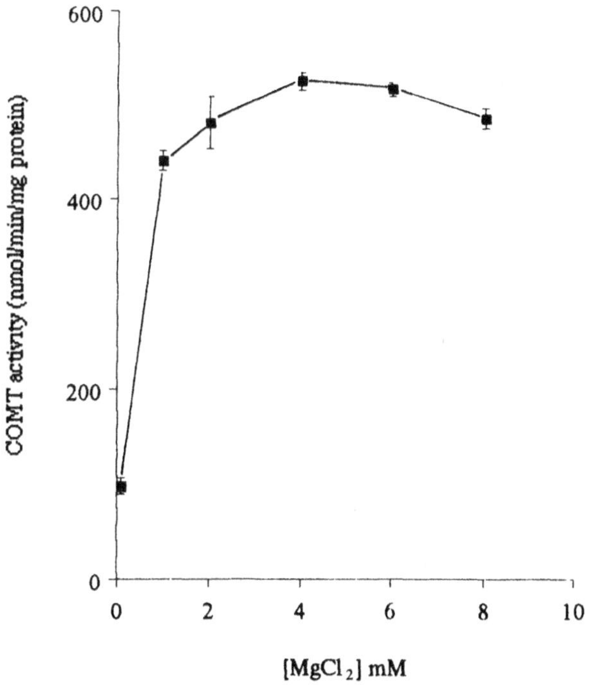

Fig. 3. The effect of magnesium chloride concentrations on the activity of catechol-O-methyltransferase. COMT was assayed by the method of Zürcher and Da Prada (1982) in the presence of increasing concentrations of magnesium chloride. Points shown are the mean ± SD of three determinations

significant effects on the activity of the enzyme. There was a slight inhibitory effect (10%) of KCl when the concentration was increased to 2.0 M.

The inhibition of the enzyme by higher concentrations of phosphate may be due to chelation of the magnesium ions which are essential for activity (see e.g. Axelrod and Tomchick, 1958). Under the assay conditions used in the present work a total concentration of 4 mM $MgCl_2$ was found to be optimal for enzyme activity (Fig. 3). The Michaelis constants towards catechol was determined to be about 39 μM when the concentration of AdoMet was held constant at 450 μM. The corresponding value for AdoMet, determined at 500 μM catechol was 47 μM. This latter value refers to total AdoMet concentrations without correction for the different enantiomers that may be present (see Hoffman, 1986).

Amino terminal analysis gave the partial sequence shown in Table 2.

As has been reported for purified or partially purified COMT from several other sources (e.g. Jeffrey and Roth, 1985; White and Wu, 1975; Anderson and D'lorio, 1968) the enzyme purified by our procedure was unstable. When stored at 4 °C there was a 50% loss of activity in 7 days, whereas complete loss of activity was observed if the enzyme was stored at

Table 2. The N-terminal amino acid sequence of Catechol-O-methyltransferase

Position	Amino acid
1	Valine
2	Glutamine
3	Threonine
4	Valine
5	Proline
6	Leucine
7	Serine
8	Arginine
9	Leucine
10	Phenylalanine
11	Aspartate
12	Histidine
13	Alanine
14	Methionine
15	Leucine
16	Glutamine
17	Alanine
18	Histidine
19	Arginine
20	Alanine
21	Histidine

– 20 °C overnight. The addition of glycerol to a final concentration of 40% (v/v) increased the stability of the enzyme such that 50% of the activity was retained after 3 weeks storage at – 20 °C. The inclusion of 1% (v/v) BSA increased the stability further such that only 10% of its initial activity was lost after 6 months of storage at – 20 °C. Thus 40% glycerol and 1% BSA were routinely added to the enzyme preparations for storage purposes. It is to be hoped that the apparently homogeneous preparations of COMT obtained by this procedure will facilitate further studies on the kinetic behaviour, immunochemistry and molecular biology of the enzyme.

References

Anderson PJ, D'Iorio A (1968) Purification and properties of catechol-O-methyltransferase. Biochem Pharmacol 17:1943–1949

Assicot M, Bohuon C (1971) Presence of two distinct catechol-O-methyltransferase activities in red blood cells. Biochemie 53:871–874

Axelrod J, Tomchick R (1958) Enzymatic O-methylation of epinephrine and other catechols. J Biol Chem 233:702–705

Ball P, Knuppen R, Breuer H (1971) Purification and properties of a catechol-O-methyltransferase of human liver. Eur J Biochem 21:517–525.

Darmenton P, Cronenberger L, Pacheco H (1976) Purification de la catéchol-O-méthyltransférase (EC 2.1.1.6) du placenta humain et études de quelques propriétés. Biochemie 58:1401–1403

Gugler R, Knuppen R, Breuer H (1970) Reinigung und Charakterisierung einer S-adenosyl methionin: catechol-O-methyltransferase der menschlichen Placenta. Biochim Biophys Acta 220:10–21

Gulliver PA, Tipton KF (1978) The purification and properties of pig-liver catechol-O-methyltransferase. Eur J Biochem 88:439–444.

Gulliver PA, Tipton KF (1979) The purification and properties of pig-brain catechol-O-methyltransferase. J Neurochem 21:517–525

Hoffman JL (1986) Chromatographic analysis of the chiral and covalent instability of S-adenosyl-L-methionine. Biochemistry 25:4444–4449

Jeffrey DR, Roth JA (1985) Purification and kinetic mechanism of human brain soluble catechol-O-methyltransferase. J Neurochem 44:881–885

Markwell MAK, Hass SM, Bicker LL (1978) A modification of the Lowry procedure to simplify protein determination in membrane and lipoprotein samples. Anal Biochem 87:206–210

Mejbaum-Katzenellenbogen W, Dobryszcka W (1959) New method for quantitative determination of serum proteins separated by paper electrophoresis. Clin Chem Acta 4:515–522

Nic a' Bháird N, Goldberg R, Tipton KF (1990) Catechol-O-methyltransferase and its role in metabolism. Adv Neurol 53 (in press)

Schultz E, Nissinen E (1989) Determination of catechol-O-methyltransferase activity in erythrocytes by high performance liquid chromatography with electrochemical detection. Biomed Chromatogr 3:64–67

Trendelenburg U, Cassis L, Grohmann M, Langloh A (1987) The functional coupling of neuronal and extraneuronal transport with intracellular monoamine oxidase. J Neural Transm [Suppl] 23:91–101

Weber K, Osborn M (1969) The reliability of molecular weight determinations by dodecyl sulphate/polyacrylamide gel electrophoresis. J Biol Chem 244: 4406–4412
White HL, Wu CJ (1975) Properties of catechol-O-methyltransferase from brain and liver of rat and human. Biochem J 45:135–143
Zürcher G, Da Prada M (1982) Rapid and sensitive single-step radiochemical assay for catechol-O-methyltransferase. J Neurochem 38:191–195

Authors' address: Prof. Dr. K. F. Tipton, Department of Biochemistry, Trinity College, Dublin 2, Ireland.

J Neural Transm (1990) [Suppl] 32: 369–374

Monoclonal antibodies recognizing both soluble and membrane bound catechol-O-methyltransferase

B. Bertocci[1], **G. Garotta**[2], **G. Zürcher**[3], **V. Miggiano**[2], and **M. Da Prada**[3]

[1] Pharmacological Department, University of Florence, Italy
[2] Central and [3] Pharmaceutical Research Departments
F. Hoffmann-La Roche Ltd, Basel, Switzerland

Summary. Both cytosolic, soluble and membrane-bound catechol-O-methyltransferase (COMT) from pig and rat liver or kidney were recognized by mouse monoclonal antibodies (MAbs) raised against soluble COMT isolated from pig liver. In ELISA, the MAbs Co 16 and Co 54 reacted better with the pig than with the rat enzyme. The MAb Co 60 showed good reactivity with both pig and rat COMT. In addition, all three MAbs recognize the soluble (23 kDa) as well as the membrane-bound (26 kDa) forms of the COMT enzyme.

Introduction

Catechol-O-methyltransferase (COMT, EC 2.1.1.6.), is widely distributed in human and animal tissues and catalyzes the transfer of a methyl group from S-adenosyl-L-methionine (SAM) to endogenous and exogenous substrates containing a catechol moiety. This enzyme plays an important role in the inactivation by O-methylation of neurotransmitter catecholamines (CA) such as dopamine (DA) and noradrenaline (NA) (Axelrod, 1966).

The occurrence of at least two distinct COMT forms, a soluble (S-COMT) and a membrane-bound (MB-COMT), was reported for brain (Borchardt and Cheng, 1978; Roth, 1980), heart (Roth, 1980), liver (Tong and D'Iorio, 1977) and erythrocytes (Assicot and Bohuon, 1971). The MB-COMT has an apparent molecular mass of about 26 kDa. It is different from the cytosolic S-COMT which shows a molecular mass of 23 kDa (Grossman et al., 1985) and which seems to be an integral membrane protein, bound in tiny amounts ($<$ 10% of the total COMT activity) to microsomes (Roth, 1980).

MB-COMT exhibits a K_m value for catechol substrates about 100-fold lower than that of S-COMT and might, therefore, substantially contribute to the O-methylation of endogenous CA (Jeffery and Roth, 1984).

To date, the primary structure and the mechanisms by which MB-COMT and S-COMT O-methylate CA are still largely unknown (Nissinen, 1984). In the present investigations we studied the reactivity of both MB- and S-COMT with three monoclonal antibodies (MAbs) raised against S-COMT partially purified from pig liver.

Materials and methods

Preparation of monoclonal antibodies

Mice were immunized with pig liver S-COMT partially purified by ammonium sulfate precipitation and gel filtration (Zürcher and Da Prada, 1979) and their spleen fused with azaguanin-resistant PAI-O myeloma (Galfre et al., 1977; Stocker et al., 1982). Out of several positive hybridomas three Ig G-producing clones were selected (Bertocci et al., 1989) and different MAbs purified from cell culture supernatants on Sepharose beads coupled with rabbit anti-mouse Ig (Stähli et al., 1983).

Preparation of subcellular fractions

After homogenization of tissues in potassium phosphate buffer (10 mM, pH 7.5) containing 2-mercaptoethanol (5 mM), subcellular fractions were prepared from pig liver as well as from rat liver and kidney. Nuclei were separated from the crude homogenate by slow-speed centrifugation (600 × g, 20 minutes), then the mitochondrial pellet was obtained by centrifugation of the supernatant at 11,000 × g for 20 min. Finally, the microsomal pellet was separated from the remaining supernatant by high-speed centrifugation (100,000 × g, 60 min) (Jeffery and Roth, 1984).

To reduce contamination with S-COMT loosely bound to the sedimenting fractions, mitochondrial and microsomal pellets were washed twice with phosphate buffer.

Gel electrophoresis and immunoblotting

SDS-PAGE was performed on slab gel with 15% acrylamide according to Laemmli (1970). Proteins were then electroblotted to nitrocellulose (Towbin et al., 1979) and incubated with different MAbs anti S-COMT (10 µg MAb/ml). Bound MAbs were detected using ^{125}I-sheep anti-mouse Ig (~ 10^6 cpm/ml) and radioactivity visualized by exposure to Kodak X-Ray film.

Results

The capacity of MAbs, Co 16-2H/2 (Co 16), Co 60-1B/7 (Co 60) and Co 54-5F/8 (Co 54), to react with S-COMT partially purified from pig and rat liver was studied by a solid-phase antibody binding assay (ELISA). The different

MAbs had different capacities to recognize pig and rat S-COMT. Co 16 and Co 54 showed the best reactivity against pig S-COMT whereas Co 60 showed the same capacity to react with pig and rat COMT (Fig. 1).

Western blot analysis (Fig. 2) performed on pig liver crude homogenates (lane 1), on mitochondrial (lane 2) and on microsomal fractions (lane 3) as well on the high-speed supernatant (lane 4), showed that a protein band of 23 kDa was recognized by the MAb Co 16. The MAbs Co 60 and Co 54 were less efficient in detecting S-COMT. In the microsomal fraction Co 16 detected a second protein band at 26 kDa likely representing MB-COMT (Fig. 2, lane 3).

The MAb Co 60, the only which showed good reactivity against rat S-COMT, was used to investigate the distribution of the enzyme in microsomes and in microsome-free high-speed supernatant prepared from rat liver and kidney. As shown in Fig. 3, this MAb detected in the high-speed supernatant from both tissues a protein with the apparent molecular mass of 23 kDa possibly representing S-COMT (liver = 1A; kidney = 2A). When microsomal fractions were analyzed Co 60 detected similar amounts of the 23 and 26 kDa forms of COMT (liver = 1B; kidney = 2B).

Discussion

Both S- and MB-COMT were recognized by MAbs produced against pig S-COMT. The Western blot analysis performed on the enzyme extracted from pig and rat liver showed that these MAbs detected 2 proteins whose

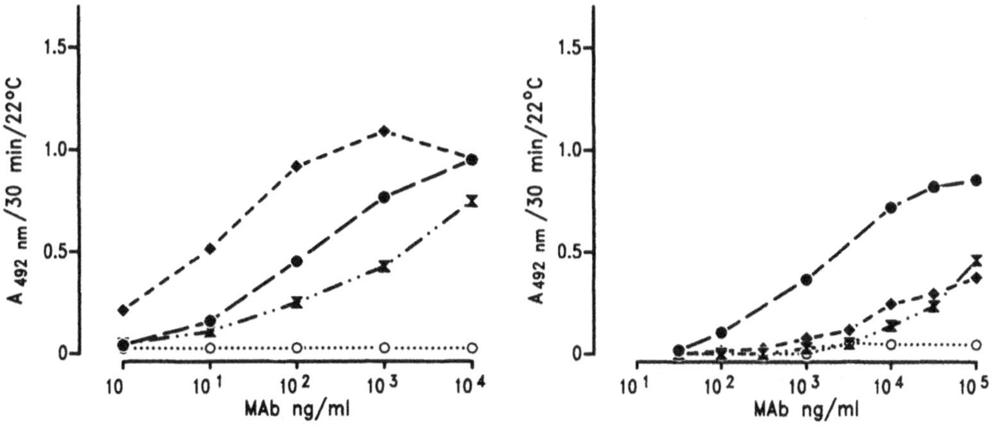

Fig. 1. Detection of S-COMT from pig (left) and rat (right) liver by MAbs. Purified MAbs were serially diluted (0.5 ng–5 μg) in microtiter wells coated with 500 ng of partially purified COMT from pig and rat liver. The antibodies bound to the enzyme were detected by peroxidase-conjugate sheep Ig G anti-mouse immunoglobulins and quantitated at 492 nm by a Titertek Multiscan reader. Co 16-2H/2 (■), Co 60-1B/7 (●), Co 54-5F/8 (⧗), not related MAb (○)

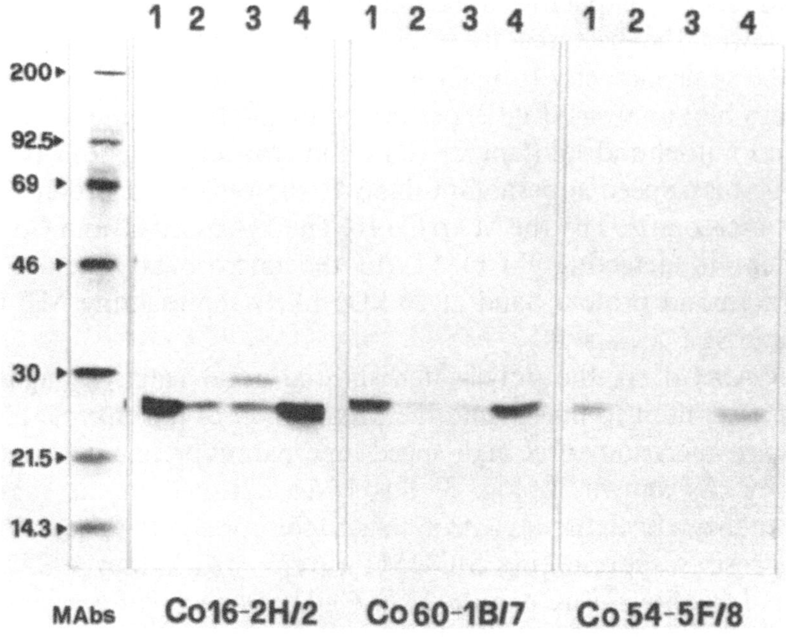

Fig. 2. Western blot analysis of different subcellular fractions prepared by differential centrifugation from pig liver homogenates. Proteins (~ 120 µg per sample) of *1* crude homogenate, *2* mitochondria, *3* microsomes and *4* supernatant (100,000 × g) were resolved by SDS-PAGE and immunoblotted with three anti-S-COMT MAbs

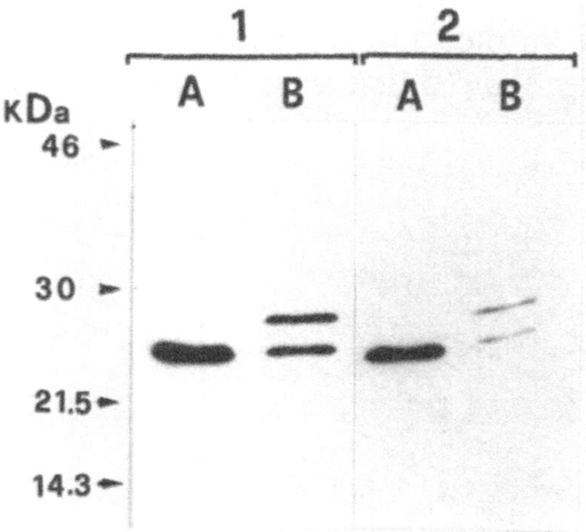

Fig. 3. Western blot analysis of COMT from rat liver and kidney subcellular fractions. Rat liver *1* and kidney *2* COMT present in supernatant (*A*) and in microsomes (*B*) sedimented at 100,000 × g, 60 min were exposed to the MAb Co 60-1B/7 and bound immunoglobulins detected by ^{125}I-labeled sheep anti-mouse Ig

apparent molecular masses of 23 and 26 kDa are compatible with those previously reported for rat S- and MB-COMT (Grossman et al., 1985).

Moreover, our results show that the cytosolic form of COMT is the only detectable in crude homogenates or high-speed supernatants (Fig. 2). However, detectable amounts of MB-COMT became evident, in traces, in mitochondrial fractions and, in larger amounts, in microsomal membranes (Fig. 2). In our conditions, the 66 kDa COMT described by Grossman et al. (1985) was never detected.

In spite of extensive washing with phosphate buffer (two or more times), the microsomal membranes reveal the concomitant presence of S- and MB-COMT forms.

The biological significance of the two COMT forms is, up to now, obscure. In any case, our experiments show that S- and MB-COMT share identical epitopes.

References

Assicot M, Bohuon C (1971) Presence of two distinct catechol-O-methyltransferase activities in red blood cells. Biochimie 53:871–874

Axelrod J (1966) Methylation reactions in the formation and metabolism of catecholamines and other biogenic amines. Pharmacol Rev 18:95–113

Bertocci B, Garotta G, Zürcher G, Miggiano V, Da Prada M (1989) Immunoaffinity chromatography and characterization of pig liver COMT with monoclonal antibodies. J Neurochem 52 [Suppl]: S83D

Borchardt RT, Cheng CF (1978) Purification and characterization of rat heart and brain catechol-O-methyltransferase. Biochem Biophys Acta 522:49–62

Galfre G, Howe SC, Milstein C, Butcher GW, Howard JC (1977) Antibodies to major histocompatibility antigens produced by hybrid cell lines. Nature 266:550–552

Grossman MH, Creveling CR, Rybczynski R, Brauerman M, Isersky C, Breakefield XO (1985) Soluble and particulate forms of rat catechol-O-methyltransferase distinguished by gel electrophoresis and immune fixation. J Neurochem 44:421–432

Jeffery DR, Roth JA (1984) Characterization of membrane-bound and soluble catechol-O-methyltransferase from human frontal cortex. J Neurochem 42:826–832

Laemmli UK (1970) Cleavage of structural proteins during the assembly of the head of bacteriophage T4. Nature 227:680–685

Nissinen E (1984) The site of O-methylation by membrane-bound catechol-O-methyltransferase. Biochem Pharmacol 33:3105–3108

Roth JA (1980) Presence of membrane-bound catechol-O-methyltransferase in human brain. Biochem Pharmacol 29:3119–3122

Stähli C, Miggiano V, Stocker J, Staehelin T, Häring P, Takacs B (1983) Distinction of epitopes by monoclonal antibodies. Methods Enzymol 92:242–253

Stocker JW, Forster HK, Miggiano V, Stähli C, Staiger C, Takacs B, Staehelin Th (1982) Generation of 2 new mouse myeloma cell lines 'PAI' and 'PAI-O' for hybridoma production. Res Discl 217:155–157

Tong JH, D'Iorio A (1977) Solubilization and partial purification of particulate catechol-O-methyltransferase from rat liver. Can J Biochem 55:1108–1113

Towbin H, Staehelin T, Gordon J (1979) Electrophoretic transfer of proteins from polyacrylamide gels to nitrocellulose sheets: procedure and same application. Proc Natl Acad Sci 76:4350–4354

Zürcher G, Da Prada M (1979) Radioenzymatic assay of fentomole concentration of dopa in tissues and body fluids. J Neurochem 33:631–639

Authors' address: Prof. Dr. M. Da Prada, F. Hoffmann-La Roche Ltd, CH-4002 Basel, Switzerland.

J Neural Transm (1990) [Suppl] 32: 375–380

Ro 40-7592: inhibition of COMT in rat brain and extracerebral tissues

G. Zürcher, A. Colzi, and M. Da Prada

Pharmaceutical Research Department, F. Hoffmann-La Roche Ltd, Basel, Switzerland

Summary. A random biochemical screening followed by lead optimization culminated in the discovery of Ro 40-7592 (3,4-dihydroxy-4′-methyl-5-nitrobenzophenone). Ro 40-7592 was found to inhibit COMT in a competitive fashion both in the CNS and in the periphery (K_i for the liver enzyme = 30 nM). Ro 40-7592 (30 mg/kg p.o.) combined with benserazide (15 mg/kg p.o.) and a low dose of L-DOPA (10 mg/kg p.o.) almost completely blocked (for about 6 h) the formation of 3-O-methyldopa (3-OMD) in brain and plasma, producing a long-lasting increase of L-DOPA in plasma and a parallel marked increase of L-DOPA and dopamine in the brain. Ro 40-7592, combined with peripheral decarboxylase inhibitors and L-DOPA (as in Madopar® and Sinemet®), will offer substantial advantages in the therapy of Parkinson's disease, i.e. enhanced bioavailability and prolonged plasma half-life of L-DOPA, pronounced DOPA sparing effect and blockade of 3-OMD formation.

Introduction

The amino acid precursor of dopamine (DA) 3,4-dihydroxy-L-phenylalanine (L-DOPA, levodopa) is administered with the aim to restore normal levels of DA in the brain of parkinsonian patients where neuronal DA synthesis and storage are largely impaired due to nigro-striatal degeneration of the dopaminergic pathways. However, an optimal and steady effect of this DA substitution therapy is hampered by pharmacokinetic and pharmacodynamic factors (Dubuis, 1987; Luquin et al., 1989) which prevent the bulk of the orally administered L-DOPA from reaching the brain as intact amino acid. In order to alleviate the resulting motor fluctuations in Parkinson's disease (PD), the two enzymes primarily involved in the peripheral metabolism of L-DOPA, i.e. the aromatic L-amino acid decarboxylase (AADC) and the catechol-O-methyltransferase (COMT), have to be inhibited.

As depicted in Fig. 1, L-DOPA in the extracerebral tissues is decarboxylated to a large extent to DA and, to a minor extent, O-methylated to 3-O-methyldopa (3-OMD). However, when coadministered together with the peripheral AADC inhibitors benserazide (Madopar®) or carbidopa (Sinemet®), L-DOPA is largely converted into 3-OMD in the gastrointestinal tract, in the liver and in the kidney, organs displaying high COMT activity (Zürcher et al., 1990).

Fluctuations in response, erratic pharmacokinetics and frequent dosing due to the short half-life of L-DOPA (approx. 1 h) all cause problems in the long-term therapy of parkinsonian patients with standard as well as with controlled-release Madopar® and Sinemet® formulations (Da Prada et al., 1987). Since these problems might in part be related to the conversion of relevant amounts of L-DOPA into 3-OMD, we started in 1982 an intensive search for a selective and reversible compound able to inhibit, after oral administration, both cerebral and extracerebral COMT activity. This effort resulted in the synthesis and biological characterization of the novel, potent and orally active COMT inhibitor Ro 40-7592 (3,4-dihydroxy-4-methyl-5-nitrobenzophenone) (Borgulya et al., 1989).

The present study in rats describes some biochemical characteristics of Ro 40-7592, a compound which in healthy volunteers has already shown good tolerability (up to 800 mg) and produced long-lasting inhibition of erythrocyte COMT.

Materials and methods

In male albino rats (Fü-albino, 150–200 g body weight) anaesthetized with Vetanarcol® (0.4 ml/kg i.p.), blood from the abdominal artery was collected into a plastic syringe containing an anticoagulant (1 vol 1% EDTA in saline for 9 vol blood)

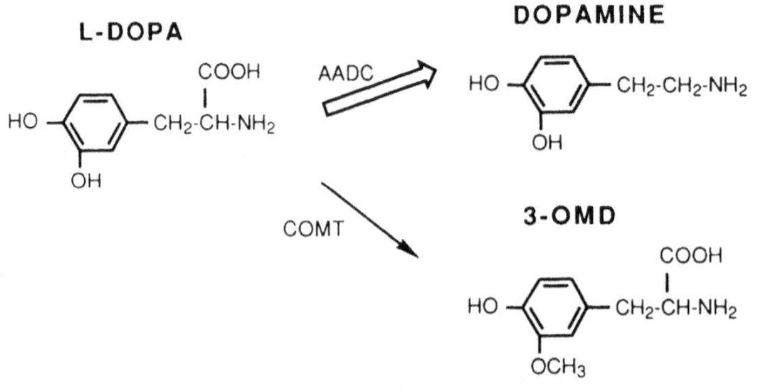

Fig. 1. Extracerebral biotransformation of L-DOPA by aromatic L-amino acid decarboxylase (*AADC*) and by catechol-O-methyltransferase (*COMT*)

and then centrifuged to yield plasma samples, stored at − 80°C until analysis. L-DOPA and 3-OMD were measured in deproteinized plasma and brain samples by direct injection of the supernatants into an automated column switching system allowing on-line prepurification of the samples by cation exchange chromatography followed by reverse-phase high-performance liquid chromatography (HPLC) with electrochemical detection (Zürcher and Da Prada, 1990).

In whole brain of rats sacrificed by decapitation, L-DOPA 3-OMD, DA and noradrenaline (NA) were also measured by HPLC with electrochemical detection (Da Prada et al., 1989). L-DOPA (10 mg/kg) plus benserazide (15 mg/kg), as well as this combination plus the COMT inhibitor Ro 40-7592 (30 mg/kg) were suspended using all-glass Potter-Elvehjem homogenizer in 0.9% NaCl. The volumes for p.o. administration were 10 ml/kg body weight.

Results

After oral administration of L-DOPA in combination with benserazide the plasma concentration of DOPA increased rapidly and peak levels were reached already 30 min after dosing (Fig. 2, left). In these conditions the large amount of 3-OMD formed reached its maximum 3 h after drug administration. The results in Fig. 2 (left) also show that at the doses used the concentration of DOPA in plasma reached control values about 5 h after dosing, at a time at which plasma 3-OMD levels were still very high (about 3 µg 3-OMD/ml plasma). As shown in Fig. 2 (left) Ro 40-7592 (30 mg/kg p.o.) coadministered with L-DOPA and benserazide virtually suppressed the formation of 3-OMD and, at the same time, drastically increased L-DOPA bioavailability. Ro 40-7592 produced long-lasting elevations of L-DOPA plasma levels (increase of the peak concentration of L-DOPA in plasma from 1.5 µg to 3 µg/ml). The value of the area under the curve (AUC) was about 4 times greater after L-DOPA combined with benserazide and Ro 40-7592 than after L-DOPA plus benserazide (Fig. 2, left). As expected (Fig. 2, right) the levels of L-DOPA and 3-OMD measured in the whole rat brain closely paralelled the levels and the time-courses of these amino acids in the blood. Thus, the 3-OMD level was very low also in the brain after Ro 40-7592, whereas the concentration of L-DOPA substantially increased for several hours after dosing when Ro 40-7592 was administered with L-DOPA and benserazide. As expected (Fig. 3), DA formed by the cerebral decarboxylation of L-DOPA was markedly increased in the rat brain for at least 6 h after dosing. In the same experiment, brain NA levels remained unchanged after L-DOPA plus benserazide and were slightly increased 6 and 8 h after Ro 40-7592 combined with L-DOPA and benserazide.

Discussion

The present experiments in rats as well as previous findings (Da Prada et al., 1990; Zürcher et al., 1990) provide ample evidence that the novel nitrocatechol derivative Ro 40-7592 is a potent and orally active inhibitor of both

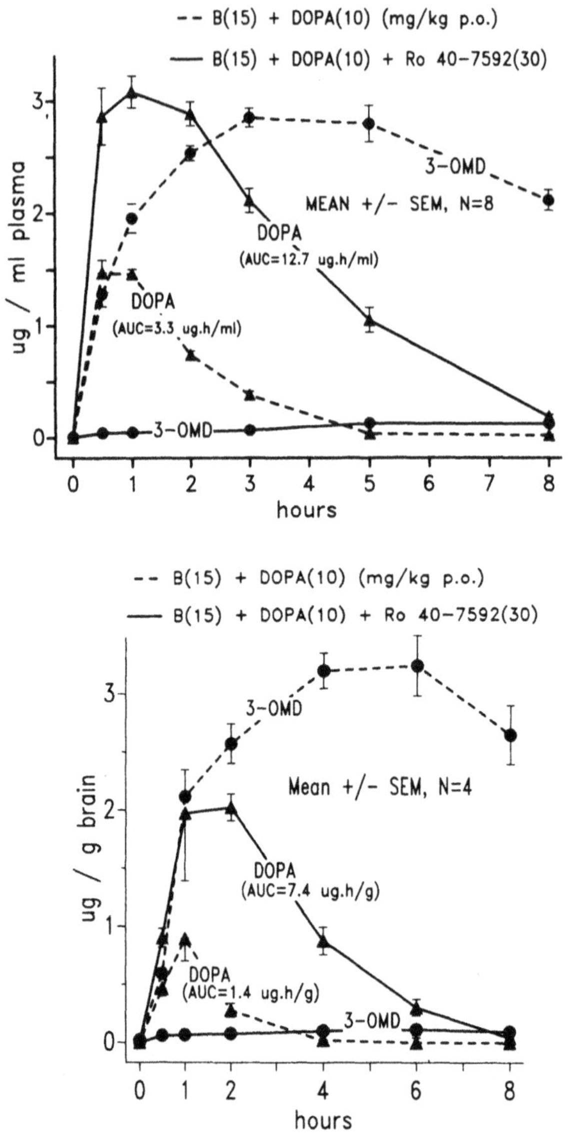

Fig. 2. Time-course of DOPA and 3-OMD changes in plasma (upper) and brain (lower) of rats administered DOPA together with benserazide or DOPA, benserazide and the COMT inhibitor Ro 40-7592

peripheral and central COMT activity. As published elsewhere (Da Prada et al., 1990; Zürcher et al., 1990) COMT inhibition in the CNS can be assessed by the decrease of the O-methylated metabolites of catecholamines, e.g. homovanillic acid and 3-methoxytyramine. Therefore, these metabolites can be considered as good markers of brain COMT inhibition. Concerning peripheral COMT activity, its suppression can be revealed by the lack of formation and accumulation of 3-OMD in plasma (Fig. 2).

An almost complete blockade of the conversion of L-DOPA into 3-OMD is the prerequisite for a markedly enhanced oral bioavailability of

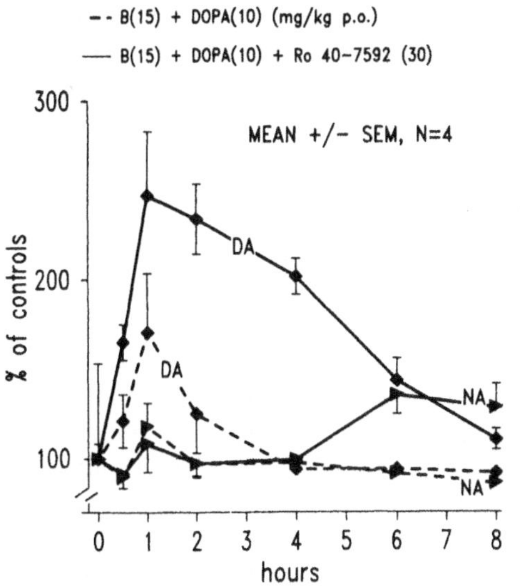

Fig. 3. Time-course of changes of dopamine (*DA*) and noradrenaline (*NA*) in whole rat brain. The animals were administered DOPA together with benserazide or DOPA, benserazide and the COMT inhibitor Ro 40-7592. Control values (nanograms per gram of wet tissue, range) were: *DA*: 550 to 650; *NA*: 350 to 450

L-DOPA when Ro 40-7592 is coadministered with Madopar® or Sinemet®. Since 3-OMD and L-DOPA are actively transported into the brain by the same carrier system, the maintenance of low 3-OMD plasma concentrations should result in a more efficient and constant transport of L-DOPA through the blood-brain barrier. Therefore, in the therapy of PD a first advantage obtained by the marked COMT inhibition is the substantial reduction of the daily amounts of exogenous L-DOPA to be administered.

In our opinion, in the DA substitution therapy of PD, maximum benefit will be achieved by the concomitant administration of Ro 40-7592 with a controlled-release formulation of L-DOPA e.g. Madopar® HBS. In fact, Madopar® HBS by itself permits to avoid abrupt variations of L-DOPA plasma concentrations and thus to reduce the frequency of motor performance fluctuations. However, although controlled-release formulations of L-DOPA can maintain DOPA in plasma at relatively constant levels, they cannot assure a satisfactory L-DOPA bioavailability. However, when Ro 40-7592 is given in combination with Madopar® HBS, L-DOPA bioavailability is substantially increased allowing a reduction of the frequency and of the daily dose of DOPA.

Current clinical trials indicate that in healthy subjects Ro 40-7592 is well tolerated up to the relatively high dose of 800 mg p.o. and produces a dose-dependent inhibition of the erythrocyte COMT activity. At 400 mg maximum COMT inhibition (> 90%) in red blood cells was already obtained 30 min after dosing and complete recovery of the COMT activity was achieved 24 h after dosing.

The present experiments strongly support the concept that the dose of Ro 40-7592 needed for optimum therapy with Madopar® standard or Madopar® HBS can be easily assessed by monitoring in volunteers the plasma level of L-DOPA and 3-OMD.

References

Borgulya J, Bruderer H, Bernauer K, Zürcher G, Da Prada M (1989) Catechol-O-methyltransferase – inhibiting pyrocatechol derivatives: synthesis and structure-activity studies. Helv Chim Acta 72:952–968

Da Prada M, Kettler R, Zürcher G, Schaffner R, Haefely WE (1987) Inhibition of decarboxylase and levels of DOPA and 3-O-methyldopa: a comparative study of benserazide versus carbidopa in rodents and of Madopar standard versus Madopar HBS in volunteers. Eur Neurol 27 [Suppl] 1:9–20

Da Prada M, Kettler R, Keller HH, Burkard WP, Muggli-Maniglio D, Haefely WE (1989) Neurochemical profile of moclobemide, a short-acting and reversible inhibitor of monoamine oxidase type A. J Pharmacol Exp Ther 248:400–414

Da Prada M, Zürcher G, Kettler R, Colzi A (1990) New therapeutic strategies in Parkinson's disease: inhibition of MAO-B by Ro 19-6327 and of COMT by Ro 40-7592. In: Bernardi G, Carpenter M, Di Chiara G, Stanzione P (eds) Proc, 3rd Triennal Meeting "Basal Ganglia 89". Plenum Press, New York (in press)

Dubuis R (1987) "On-off" phenomenon: description, incidence and management. Eur Neurol 27 [Suppl] 1:3–8

Luquin MR, Vaamonde J, Obeso JA (1989) Levodopa and 3-O-methyldopa plasma levels in parkinsonian patients with stable and fluctuating motor response. Clin Neuropharmacol 12:46–54

Zürcher G, Da Prada M (1990) A single automated HPLC column-switching technique for the measurement of DOPA and 3-O-methyldopa in plasma. J Chromatogr (Bioch Appl) (in press)

Zürcher G, Keller HH, Kettler R, Borgulya J, Bonetti EP, Eigenmann R, Da Prada M (1990) Ro 40-7592, a novel, very potent, and orally active inhibitor of catechol-O-methyltransferase: a pharmacological study in rats. Adv Neurol 53:497–503

Authors' address: G. Zürcher, B.Sc., Pharmaceutical Research Department, F. Hoffmann-La Roche Ltd, CH-4002 Basel, Switzerland.

J Neural Transm (1990) [Suppl] 32: 381–386

Effects of the COMT inhibitor, CGP 28014, on plasma homovanillic acid and O-methylation of exogenous L-DOPA in the rat

P. C. Waldmeier, P. De Herdt, and **L. Maitre**

Research Department, Pharmaceuticals Division, Ciba-Geigy Ltd., Basel, Switzerland

Summary. Methods potentially suitable for the determination of the COMT inhibitory properties of CGP 28014 (N-(2-pyridone-6-yl)-N′,N′-di-n-propylformamidine) in humans were tested in rats: the measurement of effects on HVA levels and on the conversion of exogenously administered L-DOPA into its O-methylated derivative, in serum and striatum. The compound decreased HVA as well as the conversion of L-DOPA to O-methyl-DOPA similarly in serum and striatum. The latter parameter was considered to be more useful for the assessment of COMT inhibition in humans, because it is less affected by diurnal changes, dietary effects, physical activity etc.

Introduction

CGP 28014 (N-(2-pyridone-6-yl)-N′,N′-di-n-propylformamidine) is a potent inhibitor of catechol-O-methyltransferase (COMT) in vivo, with little effect on the enzyme activity in vitro. In this and other respects (but not in its chemical structure), it resembles tropolone (Waldmeier et al., 1990). COMT inhibitors are not likely to cause overt symptoms in humans, which could be taken as an index of pharmacological action. To establish the appropriate dose range for clinical studies, a biochemical method to estimate the COMT inhibitory effect of the compound in volunteers is needed. Due to a lack of suitable compounds, such methods have not yet been developed. Among a number of theoretically possible approaches, the measurement of the effects of CGP 28014 on serum homovanillic acid (HVA) levels, with and without minimization of the contribution of peripheral dopamine metabolism by peripheral monoamine oxidase inhibition by debrisoquin (Sternberg et al., 1983), and on serum levels of 3-O-methyl-DOPA (3-OMD) formed from exogenously administered L-DOPA, were considered feasible for volunteer studies. To see whether and to what extent these parameters could reflect central COMT inhibition, the effects of CGP 28014 were investigated in the striatum and serum of rats in parallel, as a feasibility test for studies in humans.

Methods

HVA and 3-OMD in the striata of female Tif:RAlf(SPF) rats weighing 160–180 g were determined by HPLC with electrochemical detection as described before (Waldmeier and Feldtrauer, 1987).

Determination of HVA in rat serum

Serum prepared from trunk blood of decapitated rats was stored at − 80 °C. One ml 0.15 M chloroacetic acid pH 2.9 containing 1000 ng vanillic acid (VA) was added to 1 ml thawed serum. After centrifugation, HVA was isolated on 1 ml Bond Elut® SAX anion exchange disposable columns (Analytichem Int., Harbor City, CA 90710, U.S.A.) and quantified by HPLC with coulometric detection (model 5100 A, ESA, Bedford, Mass., U.S.A.; detector 1: + 0.25 V, detector 2 switched off). The mobile phase was 0.15 M chloroacetate buffer of pH 2.9 containing 0.03 mM sodium octylsulfate and 0.2 mM EDTA (1 ml/min), and the column a μBondapak C_{18} 300 ×3.9 mm (Waters, Milford, Mass., U.S.A.). Typical retention times were 8.2 min for VA and 9.5 min for HVA.

Determination of the accumulation of O-methyl-DOPA formed from exogenously administered L-DOPA, and DOPA itself, in rat serum

Serum was prepared from the trunk blood obtained at decapitation and frozen at − 80 °C. To 1 ml of thawed serum, 1 ml of 0.15 M chloroacetic acid buffer pH 2.9, 0.125 ml 2 M $HClO_4$ and 100 ng 3,4-dihydroxyphenylpropionic acid (DOPPA) or α-methyldopamine (α-MDA) were added. L-DOPA and 3-OMD were determined by HPLC with coulometric detection (as above; detector 1: + 0.43 V, detector 2 switched off), using a μBondapak C_{18} column. The mobile phase (0.15 M chloroacetic acid buffer pH 2.9 containing 0.32 mM sodium octylsulphate and 0.2 mM EDTA) was pumped at 0.9 ml/min. Retention times were 7.3 min (DOPA), 18.3 min (3-OMD), 27.2 min (DOPPA) and 34.1 min (α-MDA).

Results

Serum vs. striatal HVA

No change in serum HVA levels vs. controls nor in the effect of CGP 28014 was observed after application of the debrisoquin treatment schedule recommended by Sternberg et al. (1983). Debrisoquin doses were therefore

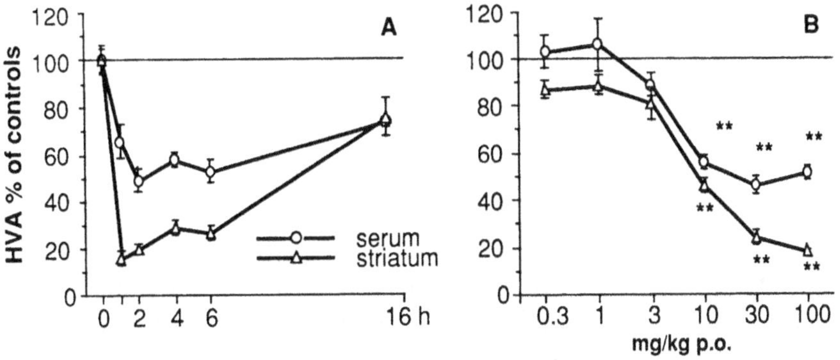

Fig. 1. Comparison of the effects of CGP 28014 on serum and striatal HVA in the same rats: Groups of 8 rats were decapitated at various intervals (**A**) or 2 h after graded oral doses of CGP 28014 (pooled from 2 experiments; **B**). Serum was obtained from trunk blood. Striata were dissected from the brains of 6 of the 8 rats per group. Data are means ± SEM in percent of controls. The means of the absolute control values were between 4.99 and 7.48 ng/ml in serum and 501 and 535 ng/g in striatum

Table 1. Time-course of serum DOPA levels after administration of DOPA-methylester HCl and effect of CGP 28014

Time after DOPA methylester HCl	DOPA ng/ml	
	– CGP 28014	+ CGP 28014
30 min	1228 ± 57	1682 ± 73
60 min	1555 ± 30	1861 ± 19
90 min	1349 ± 18	1848 ± 20
2 hours	1294 ± 13	1832 ± 15
2 hours	2196 ± 228	3546 ± 162
3 hours	520 ± 25	932 ± 73
4.5 hours	85 ± 8	138 ± 10
6 hours	36 ± 1	72 ± 6

Two experiments were carried out, each with 5 rats per group. In both, 30 mg/kg p.o. CGP 28014 were administered 15 min before 50 mg/kg i.p. DOPA methylester hydrochloride. In the first, the animals were killed 0.5, 1, 1.5 and 2 h, in the second 2, 3, 4, 5 and 6 h thereafter. Data are means ± SEM

increased, using the same time schedule. However, even doses as high as fourfold those recommended by Sternberg et al. did not lead to a decrease in serum HVA (not shown). Therefore, this approach was dropped.

CGP 28014 at 100 mg/kg p.o. lowered HVA levels in serum and striatum with a similar time-course. The maximal decrease in serum was somewhat smaller than in the striatum (Fig. 1A). While doses up to 10 mg/kg p.o. caused very similar decreases in serum and striatum, no further decrease was obtained in the former by higher doses in contrast to the striatum (Fig. 1B).

Serum vs. striatal 3-OMD

CGP 28014 dose-dependently decreased the formation of 3-OMD from exogenous L-DOPA, to similar extents in serum and in the striatum. The ED_{50} values were approximately 3 mg/kg p.o. or i.p. in both (Fig. 2). The time-course of the levels of 3-OMD after administration of 50 mg/kg i.p. L-DOPA methylester were rather similar in serum and in striatum. Maxima were reached after ~2–3 h, and 3-OMD levels decreased relatively slowly thereafter; control levels, normally below the detection limit of this method, were clearly not reached after 6 h, neither in serum nor in striatum. CGP 28014 at 30 mg/kg p.o. markedly reduced the production of 3-OMD over the entire period (Fig. 3). On the other hand, the compound increased DOPA levels in serum in relative terms more effectively at later than at earlier times after administration of DOPA methylester. This is expected, because, by inhibiting COMT, CGP 28014 prolongs the half-life of DOPA.

Discussion

The comparison of time-courses and dose-response relationships of the effect of CGP 28014 on the levels of HVA in rat striatum and serum in rats suggests that, in humans, measurement of plasma HVA levels may provide a

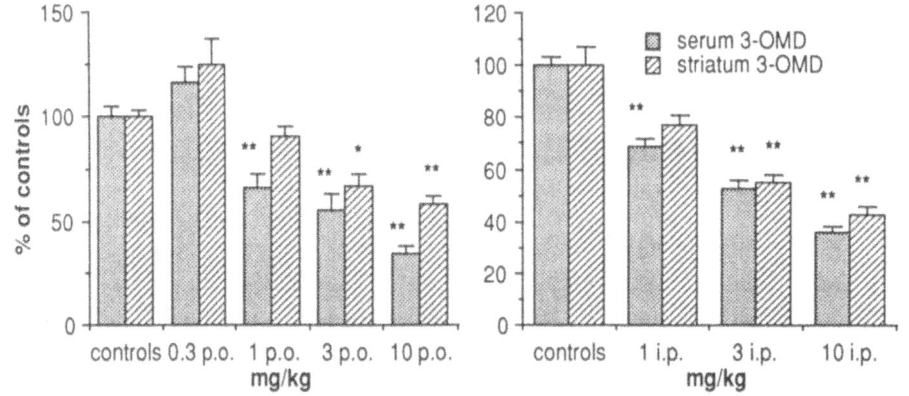

Fig. 2. Effects of CGP 28014 on 3-OMD in rat serum and striatum after treatment with DOPA methylester HCl: Groups of 5 rats were treated p.o. or i.p. with CGP 28014 15 min before 50 mg/kg i.p. DOPA methylester HCl, and decapitated 1 h later. Trunk blood was collected and striata were dissected. Data are means ± SEM in percent of controls. Absolute control values (i.e. those from animals treated with DOPA methylester HCl only) were 1621 ± 78 ng/ml and 1314 ± 44 ng/ml serum, and 1547 ± 47 ng/g and 1348 ± 96 ng/g striatum

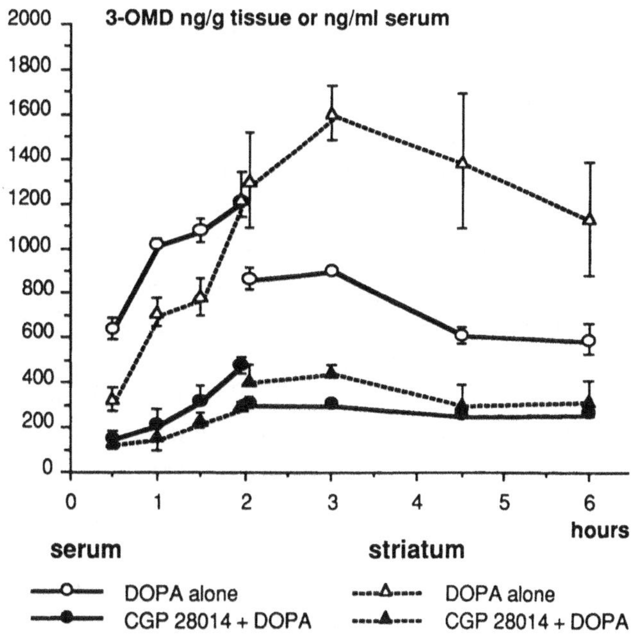

Fig. 3. Effect of the time elapsing after administration of DOPA methylester HCl on the inhibition of the formation of 3-OMD in serum and striatum: Two experiments were carried out, each with 5 rats per group. In both, 30 mg/kg p.o. CGP 28014 were administered 15 min before 50 mg/kg i.p. DOPA methylester HCl. In the first, the animals were killed 0.5, 1, 1.5 and 2 h, in the second 2, 3, 4, 5 and 6 h thereafter. All data are means ± SEM

means of estimating the effect of the drug on central catechol-O-methylation, by analogy. In the rat, the time-courses of the inhibitory effects of the drug in serum and striatum were parallel; the extent of the effect in the former was somewhat smaller. This was confirmed in the comparison of the effects of graded doses; however, it should be noted that a clearcut difference was only obvious at doses exceeding the ED_{50}. In human studies, attention should be paid to the diurnal rhythm of serum HVA levels, which in the rat was not very marked (not shown).

No comment can be given with respect to a potential usefulness of attempting to suppress the contribution to serum HVA of the periphery, since in our rats and under our conditions, debrisoquin was not able to produce such an effect even at considerably higher doses than those recommended in the literature.

The experiments in which the formation of 3-OMD from DOPA methylester was measured lead to two different practical conclusions. First, they suggest the practicability of similar tests in humans to assess the COMT inhibitory effects of the drug. It is likely that this test will provide more reliable data than the measurement of serum HVA, because the latter is prone to be affected by physical activity, nutritional factors etc., and may therefore show a larger intra- and interindividual variability. Second, they demonstrate the potential usefulness of CGP 28014 in the treatment of parkinsonism, in conjunction with L-DOPA, because the compound reduced 3-OMD levels both in the serum and the striatum, and increased serum DOPA concentrations. Although a beneficial effect of the ability to spare DOPA might be expected, the prevention of 3-OMD formation is probably more important. This metabolite of DOPA, which has a much higher elimination half-life than DOPA, competes with transport of DOPA both in the intestinal mucosa and at the blood-brain barrier. It is also a substrate of aromatic amino acid decarboxylase and may compete with DOPA in dopamine synthesis. Increased plasma levels of 3-OMD and high 3-OMD/DOPA ratios have been correlated with poor response of patients to levodopa therapy (for review of these issues see Männistö and Kaakkola, 1989).

References

Männistö PT, Kaakkola S (1989) New selective COMT inhibitors: useful adjuncts for Parkinson's disease? TIPS 10:54–56

Sternberg DE, Heninger GR, Roth RH (1983) Plasma homovanillic acid as an index of brain dopamine metabolism: enhancement with debrisoquin. Life Sci 32:2447–2452

Waldmeier PC, Feldtrauer JJ (1987) On the role of O-methylation in the metabolism of S-adenosylmethionine in rat brain. Biochem Pharmacol 36:2855–2861

Waldmeier PC, Feldtrauer JJ, Hauser K, Bittiger H, Bischoff S, von Sprecher G (1990) CGP 28014, a new non-catecholic COMT inhibitor. Naunyn Schmiedebergs Arch Pharmacol 341 [Suppl] R 97. (abstract no 386)

Authors' address: Dr. P. C. Waldmeier, Research Department, Pharmaceuticals Division, Ciba-Geigy Ltd., CH-4002 Basel, Switzerland.

J Neural Transm (1990) [Suppl] 32: 387–391

Effect of the new selective COMT inhibitor CGP 28014 A on the formation of 3-O-methyldopa (3OMD) in plasma of healthy subjects

P. R. Bieck, E. Nilsson, and **K. H. Antonin**

Human Pharmacology Institute, Ciba-Geigy GmbH, Tübingen, Federal Republic of Germany

Summary. CGP 28014 A is a specific inhibitor of catechol-O-methyl transferase (COMT). The effect on COMT was assessed with the *levodopa test* in 5 unmedicated subjects and after pretreatment with 200–600 mg CGP 28014 p.o. Plasma concentrations of DOPA, 3-O-methyldopa (3OMD), 3,4-dihydroxyphenylacetic acid (DOPAC) and homovanillic acid were measured. CGP 28014 A decreased 3OMD not dose-related by 67% ($p < 0.05$). This, and an increase of DOPAC shows COMT inhibition by this compound in humans.

Introduction

Parkinson's disease which is characterized by spasticity, tremors and an inability to initiate movements, is caused by a degeneration of the nigrostriatal pathway, resulting in a deficiency of dopamine (DA) in the striatum. The symptoms are alleviated by treatment with the DA precursor levodopa (DOPA), which can cross the blood brain barrier, whereas DA cannot. After entering the brain, DOPA is decarboxylated to DA (Cotzias et al., 1975). Because of the high activity of decarboxylase within the peripheral blood vessels, a large portion of the dose is prevented from entering the CNS. Therefore, high doses of DOPA causing unwanted effects have to be administered in order to achieve adequate brain levels. It is possible to compensate for the loss by coadministration of only peripherally acting decarboxylase inhibitors, such as carbidopa or benserazide. After such combination therapy, about 5–10% of oral DOPA reaches the brain (Maennistoe and Kaakkola, 1989). Compensatorily enhanced for the inhibited peripheral decarboxylation of DOPA is its catabolism by catechol-O-methyl transferase (COMT) in the gastrointestinal tract and kidneys to

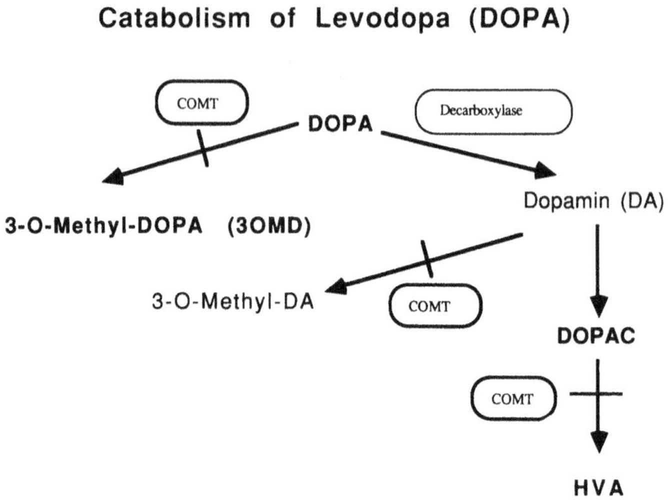

Fig. 1. Catabolism of levodopa (DOPA)

3-O-methyldopa (3OMD). It has been suggested that 3OMD by inhibiting the active transport of DOPA into the CNS and by competing with DOPA in DA synthesis, decreases the efficacy of this treatment (Maennistoe and Kaakkola, 1989). It can be postulated that a selective inhibitor of COMT might counteract these negative effects. Waldmeier et al. (1990) have shown that the formation of 3OMD in rat brain and plasma is inhibited by the new *COMT inhibitor CGP 28014*, (N-(6-hydroxy-2-pyridil-)-N′-di-n-propyl formamidine methane sulfonate). The effect of this compound on COMT of *humans* was studied after oral levodopa by measuring plasma concentrations of DOPA, 3OMD and of the two DA metabolites 3,4-dihydroxyphenylacetic acid (DOPAC) and homovanillic acid (HVA) (Fig. 1).

Methods

Five healthy male volunteers participated. 250 mg levodopa (Brocadopa®) was administered p.o. in the morning on 3 days, separated by one week. On the 2nd and 3rd study day, single rising oral doses of CGP 28 014 (200–600 mg) were given 0.5 h prior to levodopa. Blood samples were collected for 10 h. DOPA, 3OMD, DOPAC and HVA were measured with a specific HPLC method. To 1 ml plasma were added: 100 µl (1 µg) of the internal standard homovanillyl alcohol (4-hydroxy-3-methoxyphenethanol), 1 ml of 0.15 mol/L chloracetic acid buffer pH 2.9 (CAB) and 0.125 ml 2.0 mol/L perchloric acid. The mixture was kept on ice for 1 h before centrifugation at 40 000*g for 20 min at 4 °C. From a mixture of 700 µl supernatant and 200 µl CAB, 100 µl were injected into the HPLC. Separation was carried out using a Nucleosil C-18 column. The mobile phase consisted of CAB:acetonitrile (96:4, v/v), containing 0.32 mmol/L n-octyl sodium sulfate and 0.2 mmol/L EDTA. The flow was 0.9 ml/min. The compounds were detected coulometrically at 0.43 V.

The linear range for all 4 substances was 0.15–5 umol/L. In plasma samples stored at − 80 °C, no deterioration of the substances was found within 2 months.

The area under the plasma level-time curves (AUC_{0-10}) for DOPA and the metabolites before and after treatment with CGP 28014 A was calculated by the trapezoidal rule and expressed as ((μmol/L)*h). Mean ± SD or SEM are given. The paired t-test of Student was applied to test for statistical differences.

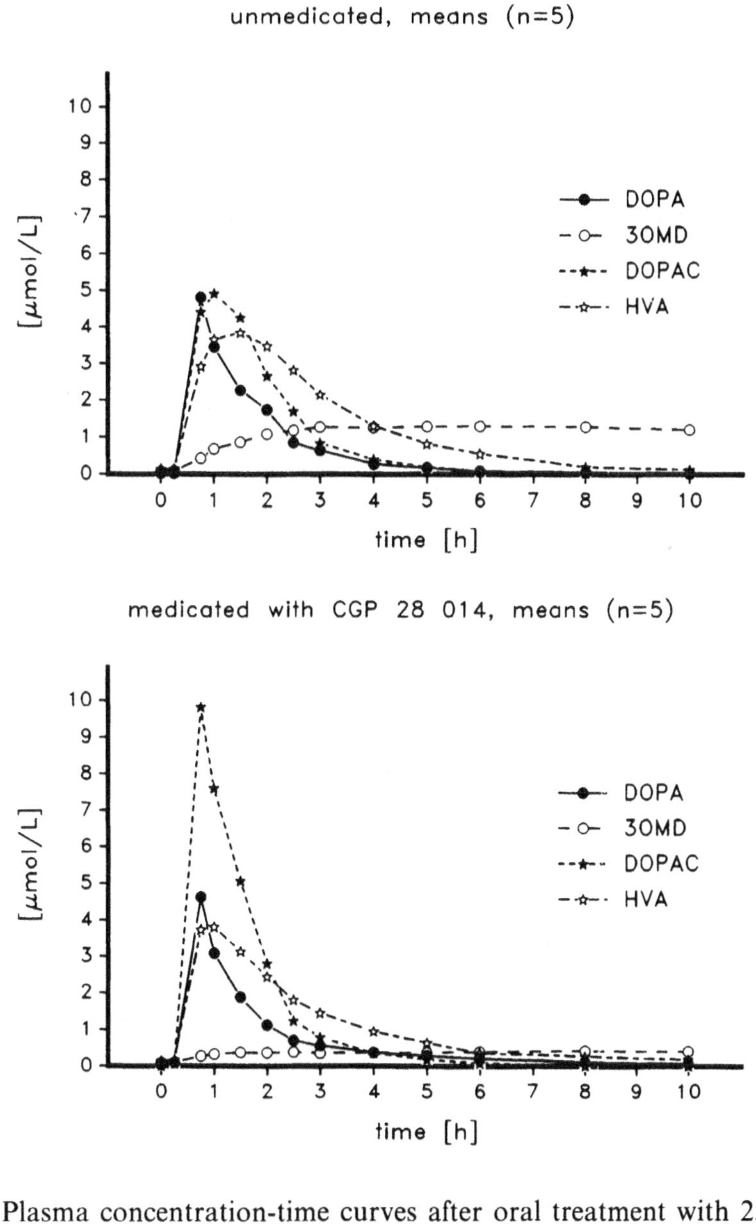

Fig. 2. Plasma concentration-time curves after oral treatment with 250 mg levodopa in 5 subjects. *Upper panel*: unmedicated. *Lower panel*: pretreatment with single oral doses of 200–600 mg CGP 28014 A

Results and discussion

Plasma concentrations of DOPA and metabolites in unmedicated subjects

Mean plasma concentration-time curves of DOPA, 3OMD, DOPAC and HVA after oral administration of 250 mg levodopa to 5 untreated subjects are shown in Fig. 2. Maximal concentrations of 4.6, 1.3, 4.9 and 3.9 µmol/L were reached within 0.75 h (DOPA), 0.75 h (DOPAC), 1.0 (HVA) and 2.5 h (3OMD). Only the concentrations of 3OMD were elevated for 10 h after intake of DOPA. Expressed as % of the total AUC of all 4 substances, the contributions of DOPA, 3OMD, DOPAC and HVA are 17%, 27%, 24% and 32%, respectively.

Table 1. Ratio of 3OMD/DOPA after administration of 250 mg levodopa alone and after pretreatment with single oral doses of the COMT inhibitor CGP 28014 (COMT-I). Values from 2 kinetic studies in 5 subjects

CGP 28014	Unmedicated + DOPA	COMT-I + DOPA	Change
200	1.60	.72	− 55%
300	2.18	.69	− 68%
400	2.26	.72	− 68%
500	1.51	.35	− 77%
600	1.73	.60	− 65%

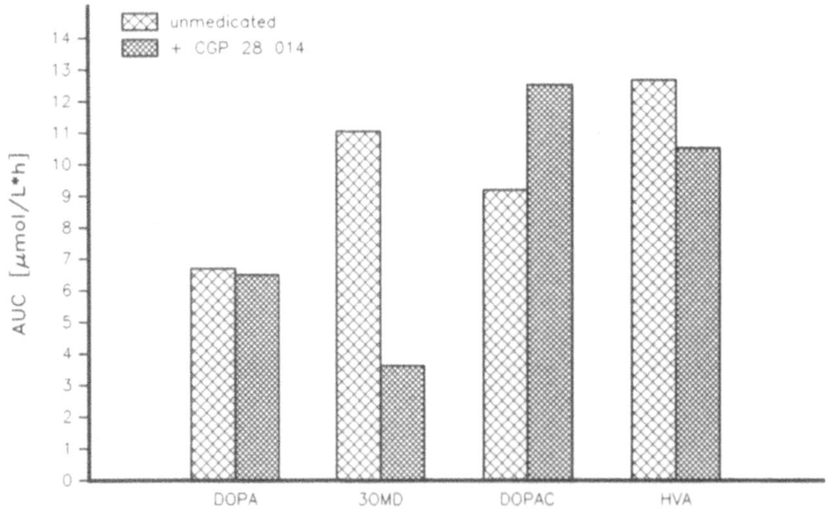

Fig. 3. Plasma concentrations of DOPA and metabolites, expressed as AUC (0–10 h) after 250 mg levodopa alone and after pretreatment with single oral doses of 200–600 mg CGP 28014 A

Plasma concentrations of DOPA and metabolites in subjects treated with CGP 28014A

After treatment with CGP 28014 A, 3OMD concentrations were about 1/3 lower, but again measurable for 10 h. The dopamine metabolite DOPAC was found to be twice as high after COMT inhibition (Fig. 2).

Figure 3 shows the effect of acute treatment with 200–600 mg of the COMT inhibitor CGP 28014 A on the amounts of DOPA and metabolites in plasma. Whereas the AUC of DOPA is not changed, 3OMD in plasma is reduced by 67% ($p < 0.05$). DOPAC was non significantly increased by 36% and HVA reduced by 17%. The values for the *ratio 3OMD/DOPA* before and during treatment with increasing single doses of the COMT inhibitor are shown in Table 1. The changes were not dose-related. However, in all 5 subjectes the ratio was lowered by more than 55%. In analogy to animal experiments, it could be shown that – by measuring plasma concentrations of DOPA and its metabolic products – acute administration of the new selective compound CGP 28014 A in doses from 200 to 600 mg inhibits COMT activity in humans.

Acknowledgement

We thank Ms. G. Pflugrad for preparing the illustrations.

References

Cotzias GC, Papvasiliou PS, Ginos JZ, Tolosa ES (1975) Treatment of Parkinson's disease and allied conditions. In: Tower DB (ed) The nervous system, vol 2. The clinical neurosciences. Raven Press, New York, pp 323–329

Maennistoe PT, Kaakkola S (1989) New selective COMT inhibitors: useful adjuncts for Parkinson's disease? TIPS 10:54–56

Waldmeier PC, Feldrauer J-J, Hauser K, Bittiger H, Bischoff S, von Sprecher G (1990) CGP 28014 A, a new non-catecholic COMT inhibitor. Naunyn-Schmiedebergs Arch Pharmacol 386 [Suppl]: R 97

Authors' address: Prof. Dr. P. Bieck, Humanpharmakologisches Institut, Ciba-Geigy, Waldhörnlestrasse 22, D-7400 Tübingen, Federal Republic of Germany.

Uptake processes

J Neural Transm (1990) [Suppl] 32: 395–404

Oxidative deamination of noradrenaline in human blood vessels

W. Osswald

Laboratório de Farmacologia, Faculdade de Medicina, Porto, Portugal

Summary. Human vascular tissue (saphenous vein and uterine artery) was incubated with tritiated noradrenaline. In both vessels, oxidative deamination predominated over O-methylation. Deamination was due to the action of MAO type A and B, as well as of a semicarbazide-sensitive oxidase. Whereas DOPEG was a good index of intraneuronal deamination, NMN, DOMA and OMDA appeared to be entirely extraneuronal in origin. Extraneuronal deamination plays an important role in human vascular tissue and may contribute to the inactivation of circulating catecholamines.

Introduction

In a large number of animal species blood vessels have been shown to be endowed with the transport and enzymatic mechanisms needed to metabolize catecholamines and to them the role of major sites of inactivation of circulating catecholamines has been ascribed (Osswald and Branco, 1973; Boenisch et al., 1974; Trendelenburg, 1980). In fact, they are well equipped for the metabolic handling of both the transmitter noradrenaline being released from varicosities (situated at the medioadvential region in most blood vessels) and the blood borne catecholamines reaching the vessel wall through an endothelium provided with uptake and metabolizing mechanisms (for reviews, see Osswald and Guimarães, 1983; Kopin, 1985; Osswald, 1986, 1988).

However, it is evident that the very large amount of information obtained in the past concerning the metabolic handling of catecholamines by blood vessels relates to various animal species, with the conspicuous exception of the human one. Although it is clear that ethical limitations, as well as logistic reasons may explain why the subject has not received the attention of research workers in the past, it remains nevertheless a fact that undue extrapolations have been made from results obtained in experimental animals to humans. This has specially been the case in what concerns receptor activation and smooth muscle contraction, as aptly pointed out by Docherty (1987). We decided, therefore, to investigate the fate of noradrenaline in human vascular tissue and used for this purpose two different blood vessels: the saphenous vein and the uterine artery. The present report deals with the data so far obtained, of published and unpublished source (Branco and Osswald, 1988; Branco et al., 1990).

Methods

Blood vessels

The saphenous vein segments were obtained from patients (mostly male, between 45 and 60 years of age) who were subjected to coronary bypass operations. Segments from the distal part of the vein were removed, immediately placed in oxygenated, ice-cold Krebs-Henseleit solution and transported to the laboratory; less than 1 h elapsed between surgical removal of the vein and the beginning of the experiment.

The uterine arteries were obtained from premenopausal women subject to hysterectomy due to nonmalignant pathological changes of the uterus. The uterus and its arteries were removed en bloc and transported to the laboratory in ice-cold and oxygenated Krebs-Henseleit solution. The arteries were dissected free from surrounding tissue and segments prepared under binocular magnifying. Also in this case interval between surgery and actual experiment was less than 1 h.

Determination of endogenous catechols

A small ring of the vessel was cut, blotted, weighed and placed in a tube containing 2 ml perchloric acid 0.1 mol/l. The tube was kept during 24 h at $-20\,^{\circ}C$ and the acid extract used for determination of noradrenaline, dopamine, dihydroxyphenylglycol (DOPEG) and dihydroxymandelic acid (DOMA) by high pressure liquid chromatography with electrochemical detection (HPLC-ED) as described elsewhere (Branco et al., 1990).

Incubation experiments

Segments of the remaining tissue, weighing 30–50 (uterine artery) or 80–100 mg (saphenous vein) were incubated in 2 ml of Krebs-Henseleit solution at $37\,^{\circ}C$, gassed with 95% O_2 and 5% CO_2 and subjected to constant agitation, during 30 min (pre-incubation period). ^3H-noradrenaline 0.1 μmol/l was present during the incubation period which followed, also of 30 min duration. Whenever the influence of drugs was studied, they were present during the preincubation and incubation periods in the concentrations indicated in the Results. At the end of the incubation period, the tissue was removed, rapidly washed in ice-cold Krebs-Henseleit solution, blotted, weighed and placed in 2 ml of cold perchloric acid (see above). The incubation medium was acidified by addition of 0.2 ml perchloric acid 2 mol/l. After 24 h at $-20\,^{\circ}C$ aliquots of the tissue extract and of the acidified medium were passed by chromatography columns; ^{3}H-noradrenaline and its tritiated metabolites were determined by liquid scintillation counting as described by Graefe et al. (1973), the fractions thus obtained corresponding to noradrenaline, DOPEG, DOMA, normetanephrine – NMN and O-methylated and deaminated metabolites – OMDA.

The Krebs-Henseleit solution had the following composition (in mmol/l): NaCl 118.1; KCl 4.7; $CaCl_2$ 1.9; KH_2PO_4 1.18; $MgSO_4$ 1.18; $NaHCO_3$ 25.8; D-glucose 5.0; Na_2 EDTA 0.027; Na fumarate 2.0; Na pyruvate 1.0; 1-glutamine 1.0; 1-methionine 0.1. After it was found, in comparative assays, that omission of the last four ingredients did not significantly change the metabolic pattern of noradrenaline, the Krebs-Henseleit solution was prepared without these compounds. Most of the results obtained in the experiments in which the uterine artery was used were done with the "simple" Krebs-Henseleit solution.

In vitro denervation of the uterine artery

In a series of experiments, fragments of uterine artery were incubated with 6-hydroxydopamine containing Krebs solution (6-OHDA 1.2 mmol/l). After a series of tentatives the conclusion was reached that exposure to 6-OHDA during three periods of 20 min separated by washout periods of 20 min and followed by preincubation and incubation periods as described above resulted in a loss of 85–95% of endogenous noradrenaline; this procedure was then adopted. Control fragments were incubated for the same periods of time in Krebs-Henseleit solution without 6-OHDA.

Drugs used

Amezinium hydrochloride (BASF, Ludwigshafen, Federal Republic of Germany), cocaine hydrochloride (Uquipa, Lisboa, Portugal), cortisol phosphate (Sigma, St. Louis, MO, U.S.A); clorgyline hydrochloride (May and Baker, Dagenham, U.K.), semicarbazide hydrochloride (E. Merck, Darmstadt, Federal Republic of Germany), 6-hydroxydopamine hydrobromide (Sigma, St. Louis, MO, U.S.A.), selegyline hydrochloride (deprenyl; Prof. J. Knoll, Budapest, Hungary), $(-)$-^{3}H-2,5,6-noradrenaline spec. activity 40.8 Ci/mmol (NEN, Dreieich, Federal Republic of Germany); Ro 01-2812 (3,5-dinitropyrocatechol; F. Hoffmann-La Roche, Basel, Switzerland).

Statistics

Independent Student's t test and Tuckey-Krammer analysis was used for the comparison of means of the different groups. The accepted level of significance was 95% ($P < 0.05$).

Results

The saphenous vein and the uterine artery had similar contents in noradrenaline (namely 0.33 and 0.45 μg/g) but differed in the contents in dopamine (3 and 17 ng/g, respectively), DOPEG (2.8 and 4.1 μg/g) and DOMA (6.2

and 13.0 µg/g). Thus, in both vessels the deaminated metabolites DOPEG and DOMA largely exceeded the amount of noradrenaline.

In what concerns the metabolism of ^{3}H-noradrenaline, again there were more similaritie than differences in the respective patterns. In fact, both blood vessels formed small amounts of NMN and large amounts of DOPEG and DOMA, the predominance of deamination over O-methylation being more marked in the uterine artery than in the saphenous vein (Fig. 1).

Cocaine (30 µmol/l) markedly reduced the accumulation of noradrenaline and practically abolished the formation of DOPEG, while increasing the formation of NMN and slightly increasing that of DOMA. These results were uniform in the saphenous vein and the artery; the production of OMDA, however, was increased in the saphenous vein and decreased in the uterine artery (Fig. 2).

Amezinium (10 µmol/l) was tested in the saphenous vein and had effects which were entirely superposable to those of cocaine.

The selective inhibitors of MAO-A and type B, clorgyline (0.1 µmol/l) and selegyline (10 µmol/l) reduced strongly the formation of DOPEG and DOMA and less markedly that of OMDA. Clorgyline was slightly more potent in reducing DOPEG formation; selegyline was more effective in reducing DOMA and OMDA formation. Both drugs markedly enhanced NMN production. There were no striking differences between the two vessels under study (Fig. 3).

Semicarbazide (100 µmol/l) had effects which were very similar to those of chlorgyline and selegyline, although less marked.

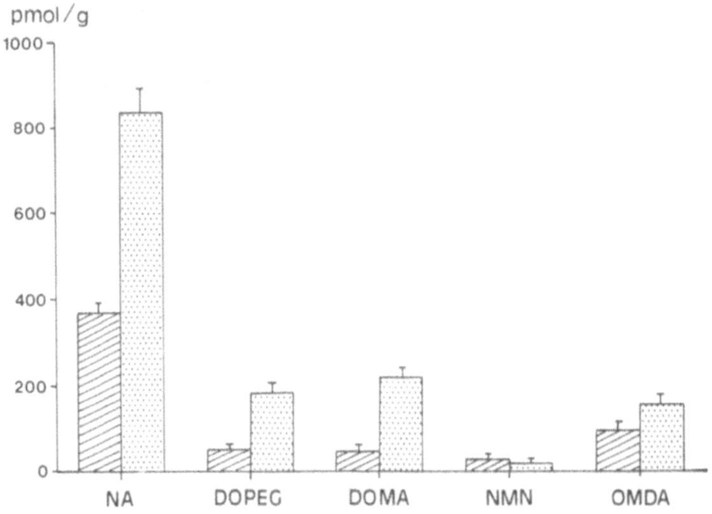

Fig. 1. Accumulation and metabolism of ^{3}H-noradrenaline by human isolated blood vessels (▨ saphenous vein and ⬚ uterine artery) incubated during 30 min with ^{3}H-noradrenaline 0.1 µmol/l. Results are expressed in pmol/g/30 min (mean ± SEM, n = 11 for the saphenous vein and n = 13 for the uterine artery)

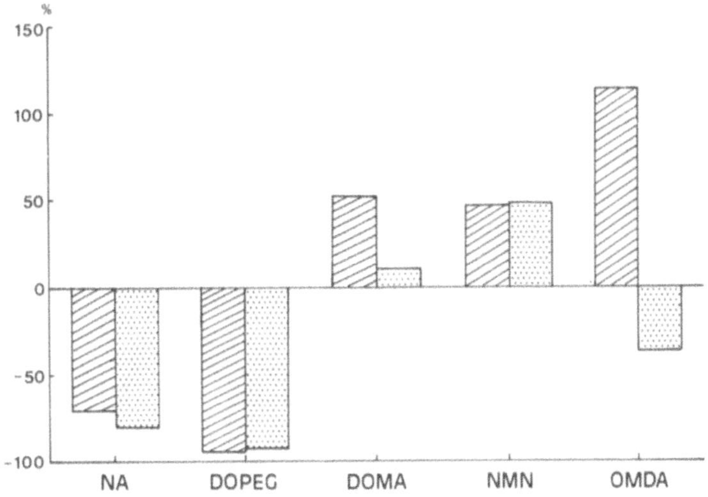

Fig. 2. Effects of cocaine on accumulation and metabolism of ^{3}H-noradrenaline in the saphenous vein ▨ and uterine artery ▢. Results expressed as Δ % (n = 5)

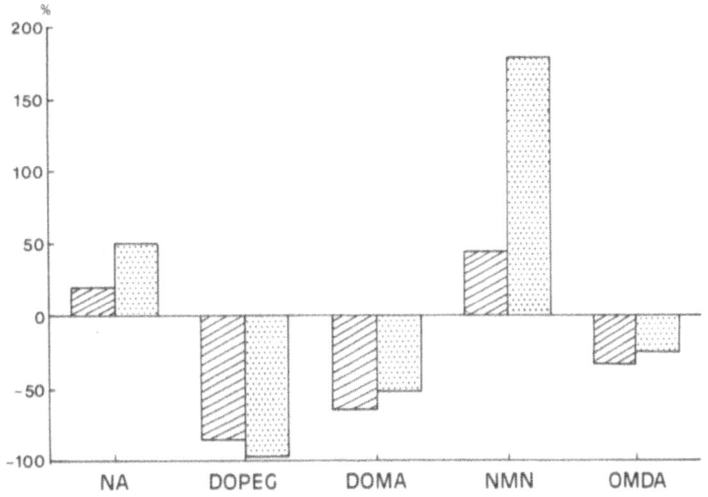

Fig. 3. Effects of clorgyline on accumulation and metabolism of ^{3}H-noradrenaline in the saphenous vein ▨ and uterine artery ▢. Results expressed as Δ % (n = 5)

As shown in Fig. 4 cortisol (40 μmol/l) caused the saphenous vein to produce more DOPEG and to halve the formation of NMN, DOMA and OMDA.

The catechol-O-methyltransferase inhibitor Ro 01-2812 (1 μmol/l), tested in the same vessel, favoured the formation of DOPEG and DOMA, almost abolished that of NMN and markedly reduced that of OMDA (Fig. 4).

Denervation effected by incubation of the uterine artery with 6-OHDA (see Methods) resulted in a very reduced accumulation of noradrenaline and

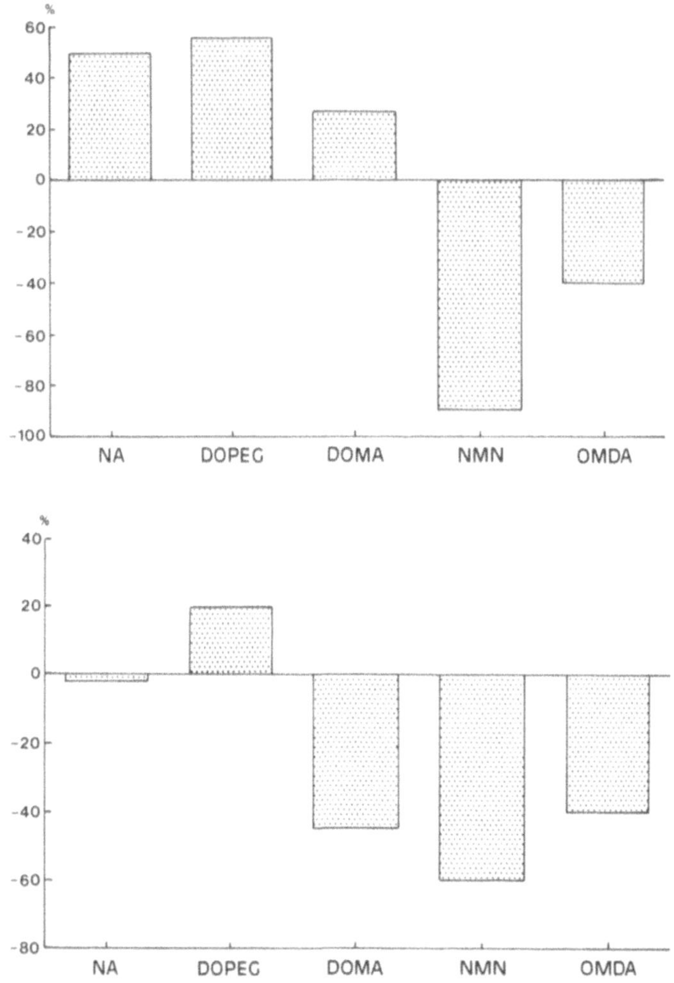

Fig. 4. Effects of cortisol (lower panel) and of Ro 12812 (upper panel) on the accumulation and metabolism of ^{3}H-noradrenaline in the saphenous vein. Results expressed as Δ % (n = 7 and n = 5, respectively)

a marked reduction in the amount of DOPEG formed; the other metabolites were only slightly affected by chemical sympathectomy.

Discussion

As it is self-evident, ethical rules and logistic circumstances may limit quite severely the resource to human blood vessels for experimental uses. For example, the change of attitude in cardiac surgery by shifting from saphenous vein to internal mammary artery bypass makes the former vessels increasingly more difficult to obtain. We were fortunate enough to begin our studies at an occasion when saphenous veins were more readily available and our laboratory lies in close vicinity to the surgical rooms, so that experiments can be made on very fresh material.

We have chosen to concentrate on the saphenous vein and the uterine artery due to the availability of these vessels but also because they represent two very different types of human vessels: a large subcutaneous vein with varicosities distributed throughout the media and a medium-sized artery with varicosities located at the adventitio-medial region. However, although there were some differences between the two blood vessels, the similarities were much more striking, as well as the differences regarding vessels of other animal species.

First of all, the two human vessels had a relatively low noradrenaline content, lower than that observed in most animal vessels. Although the noradrenaline content can not be equated with density of adrenergic innervation (Bevan et al., 1980), in this case there are reasons to believe that there is a scarcity of adrenergic innervation, since the content of dopamine and the proportion to that of noradrenaline does not hint at an unusually high turnover rate and ultramicroscopical studies confirm a relatively poor adrenergic innervation (I. Azevedo, personal communication). Contrasting with the low endogenous noradrenaline levels are the remarkably high amounts of the deaminated metabolites DOPEG and DOMA found in the vessels. Since these amounts are disproportionate to the noradrenaline content and, on the other hand, DOMA appears to be almost exclusively formed outside the nerve, it is most probable that they originate from circulating catecholamines, elevated due to the surgical stress. This would constitute a further proof for the important role played by blood vessels as sites of loss for circulating catecholamines, as discussed by Trendelenburg (1980) and Osswald and Guimarães (1983).

The experiments in which blood vessel segments were incubated with ^{3}H-noradrenaline showed some differences between the saphenous vein and the uterine artery but again the similarities were more impressive than the differences. In fact, the uterine artery had a larger capacity to accumulate and to metabolize noradrenaline, but the general pattern of metabolism did not differ substantially from one vessel to the other. Unusual features were the predominance of deaminated metabolites over O-methylated ones, the large amounts of DOMA produced by the tissues and the modest amounts of NMN being formed. Since this general pattern was observed whether the incubation medium contained or not additional sources of energy and a methyl donor, it is clear that no artefact is to blame for this picture.

The use of a number of drugs as experimental tools helped to better define the types of transmembranous transports and of intracellular enzymic machinery responsible for the metabolic pattern described.

The use of cocaine (and amezinium) and of cortisol clearly demonstrated the existence of the two classic neuronal and extraneuronal uptake mechanisms in the human blood vessels used. Cocaine (and amezinium) reduced the accumulation of noradrenaline and abolished the formation of DOPEG, as has been repeatedly observed in a large variety of tissues; whereas cortisol did not change the accumulation of noradrenaline and reduced the formation of its metabolites, with the exception of DOPEG. Thus, the results pinpoint the DOPEG formed after exposure of the tissue to a low concentration of noradrenaline as an excellent index of neuronal uptake of noradrenaline followed by intraneuronal deamination, in good agreement with earlier results (Starke et al., 1981; Majewski et al., 1982; Graefe and Bönisch, 1988).

On the other hand, it is evident that DOMA is almost if not exclusively formed at extraneuronal deaminating sites. In fact, exclusion of neuronal uptake by cocaine or amezinium caused a larger production of DOMA, whereas cortisol reduced its formation. Segments of the artery denervated in vitro by 6-OHDA had a very low DOPEG production but the formation of DOMA was scarcely affected. All these data are consistent with an extraneuronal origin of DOMA.

Although the deamination pathway predominates over O-methylation, it is clear that in both blood vessels there is a competition between the two enzymatic systems. Inhibition of catechol-O-methyl transferase by the novel and potent inhibitor Ro 01-12812 resulted in markedly reduced formation of NMN and OMDA and in increases in deaminated metabolites and accumulated noradrenaline. Conversely, blockade of deamination (by clorgyline) resulted in large increases in NMN formation and noradrenaline accumulation. The two major enzymatic pathways thus appear to be in a state of dynamic balance.

If the role of oxidative deamination and the sites where it takes place are well understood, the nature of the enzyme or enzymes responsible for deamination is far from being defined. Our results clearly show that in concentrations considered to be selective for MAO type A and MAO type B (Johnston, 1968; Knoll and Magyar, 1972; Fowler et al., 1978) clorgyline and selegyline were almost indistinguishable in their effects, clorgyline being usually the more potent of the two drugs. This is in sharp contrast with what happens in other tissues, where noradrenaline metabolism is not affected by inhibition of MAO type B. It has been shown that noradrenaline in a substrate for both types of MAO, in homogenates of the canine saphenous vein, but that MAO type B does not play any role in the metabolism of the amine in the intact tissue (Caramona and Osswald, 1985). In the human vascular tissue, however, and presuming that selegyline selectively inhibits MAO type B, we have a very different situation, with this type of the enzyme intervening actively in the process of deamination of noradrenaline.

In the same line of reasoning, and since semicarbazide exerted evident inhibitory effects of deamination, we should conclude that human vascular tissue also contains semicarbazide sensitive amine oxidase or SSAO (Coquil et al., 1973; Clarke et al., 1982; Lyles, 1984).

Thus, one could conclude that the two blood vessels owe their deamination capacity to the presence of MAO type A, MAO type B and SSAO. However, the results can also be explained by the presence of an admittedly hypothetical amine oxidase of a different type.

Taken together, the data reported herein show that the human saphenous vein and uterine artery are able to take up, accumulate and metabolize noradrenaline in a pattern which differs from that reported for blood vessels of other animal species. An important role played by extraneuronal deamination appears to be characteristic for human blood vessels. Thus, undue extrapolations from other vessels should be avoided.

Acknowledgements

The author is indebted to his colleagues D. Branco, M. Caramona, D. Araújo and J. Ferreira de Almeida for permission to quote unpublished results. Thanks are also due to Dr. M. da Prada (Basel), Prof. J. Knoll (Budapest), BASF (Ludwigshafen) and Uquipa (Lisbon) for generous gifts of Ro 01-2812, selegyline, amezinium and cocaine, respectively. The technical assistance of Misses Manuela Moura and Prazeres Cleto is gratefully acknowledged.

References

Bevan JA, Bevan RD, Duckles SP (1980) Adrenergic regulation of vascular smooth muscle. In: Bohr DF, Somlyo AP, Sparks Jr HV (eds) Vascular smooth muscle. Am Physiol Soc, Bethesda (Handbook of physiology, vol 2, section 2, pp 515–566)

Bönisch H, Uhlig W, Trendelenburg U (1974) Analysis of the compartments involved in the extraneuronal storage and metabolism of isoprenaline in the perfused heart. Naunyn Schmiedebergs Arch Pharmacol 283:233–244

Branco D, Caramona MM, Araujo D, Osswald W (1990) The role of monoamine oxidase in the metabolism of exogenous noradrenaline by the human saphenous vein. (Submitted for publication)

Branco D, Osswald W (1988) The influence of *Ruscus* extract on the uptake and metabolism of noradrenaline in the normal and varicose human saphenous vein. Phlebology 3 [Suppl] 1:33–39

Caramona MM, Osswald W (1985) Effects of clorgyline and (−) deprenyl on the deamination of normetanephrine and noradrenaline in strips and homogenates of the canine saphenous vein. Naunyn Schmiedebergs Arch Pharmacol 328: 396–400

Clarke DE, Lyles GA, Callingham BA (1982) A comparison of cardiac and vascular clorgyline-resistant amine oxidase and monoamine oxidase. Inhibition by amphetamine, mexiletine and other drugs. Biochem Pharmacol 31:27–35

Coquil JF, Goridis C, Mark G, Neff NH (1973) Monoamine oxidase in rat arteries: evidence for different forms and selective localization. Br J Pharmacol 48: 590–599

Docherty JR (1987) The use of the human saphenous vein in pharmacology. Trends Pharmacol Sci 8:358–361

Fowler CJ, Callingham BA, Manle TJ, Tipton KF (1978) Monoamine oxidase A and B: a useful concept? Biochem Pharmacol 27:97–101

Graefe KH, Bönisch H (1988) The transport of amines across the axonal membranes of noradrenergic and dopaminergic neurons. In: Trendelenburg U, Weiner N (eds) Catecholamines I. Springer, Berlin Heidelberg New York Tokyo, pp 193–245

Johnston JP (1968) Some observations upon a new inhibitor of monoamine oxidase in brain tissue. Biochem Pharmacol 17:1285–1297

Knoll J, Magyar K (1972) Some puzzling pharmacological effects of monoamine oxidase inhibitors. In: Costa E, Sander M (eds) Monoamine oxidase – new vistas. Raven Press, New York (Adv biochem psychopharmacol, pp 393–408)

Kopin IJ (1985) Catecholamine metabolism: basic aspects and clinical significance. Pharmacol Rev 37:333–364

Lyles GA (1984) The interaction of semicarbazide-sensitive amine oxidase with MAO inhibitors. In: Tipton KF, Dostert D, Strolin Benedetti M (eds) Monoamine oxidase and disease. Academic Press, London, pp 547–556

Majewski H, Hedler L, Steppeler A, Starke K (1982) Metabolism of endogenous and exogenous noradrenaline in the rabbit perfused heart. Naunyn Schmiedebergs Arch Pharmacol 319:125–129

Osswald W, Branco D (1973) The effects of drugs and denervation on removal and accumulation of noradrenaline in the perfused hind-limb of the dog. Naunyn Schmiedebergs Arch Pharmacol 277:175–190

Osswald W, Guimarães S (1983) Adrenergic mechanisms in blood vessels: morphological and pharmacological aspects. Rev Physiol Biochem Pharmacol 96: 53–122

Osswald W (1986) Inactivation of catecholamines in the blood vessel wall. In: Magro A, Osswald W, Reis D, Vanhoutte PM (eds) Central and peripheral mechanisms of cardiovascular regulation. Plenum Press, New York London, pp 1–31

Osswald W (1988) Disposition of catecholamines in the blood vessel wall. In: Bevan JA, Majewski H, Maxwell RA, Story DF (eds) Vascular neuroeffector mechanisms. IRL Press, Oxford, pp 321–329

Trendelenburg U (1980) A kinetic analysis of the extraneuronal uptake and metabolism of catecholamines. Rev Physiol Biochem Pharmacol 87:33–115

Author's address: Dr. W. Osswald, Laboratório de Farmacologia, Faculdade de Medicina, 4200 Porto, Portugal.

J Neural Transm (1990) [Suppl] 32: 405–412

Modification of alpha-2 presynaptic receptor activity and catecholamine release following chronic MAO inhibition

J. P. M. Finberg, G. Ari, G. Lavian, and **D. Hovevey-Sion**

Rappaport Family Institute for Research in the Medical Sciences and Faculty of Medicine, Technion, Haifa, Israel

Summary. Neuronal release of noradrenaline from the isolated rat vas deferens has been studied following long-term MAO inhibition. Fractional release of ^{3}H-noradrenaline in response to electrical field stimulation was reduced following chronic clorgyline treatment, but yohimbine equalised the fractional release to that of control animals. Endogenous release of noradrenaline was increased 1.8 fold in yohimbine-treated tissues from rats treated chronically with clorgyline, while tissue noradrenaline content increased 2.3 fold.

Electrophysiological activity measured from the renal nerve of anaesthetised rats decreased following acute treatment with clorgyline. This effect was absent following pretreatment with yohimbine.

From isolated tissue studies, there was no conclusive evidence for down-regulation of α_2-presynaptic receptors by long-term MAO inhibition. Acute MAO inhibition with clorgyline produces distinct physiological effects in the CNS, but not in peripheral sympathetic nerves.

Introduction

Acute inhibition of MAO activity has little or no effect on the physiological consequences of sympathetic nerve activation, but markedly potentiates the pharmacological actions of tyramine, and other indirectly-acting sympathetic amines. On the other hand, administration of MAO inhibitors over the course of several weeks increases neuronal noradrenaline (NA) release (Finberg and Kopin, 1986), is effective in the alleviation of depressive symptomatology in man, and reduces the density of β and α_2-adrenergic receptors in the cerebral cortex of rats (Sugrue, 1981; Cohen et al., 1982). Various studies have also shown a reduction in the biochemical, electrophysiological and behavioural effects of clonidine in the CNS following chronic MAO inhibition (see Youdim and Finberg, 1983).

One hypothesis which has been proposed to explain these events, is that chronic MAO inhibition reduces the effective number of α_2-presynaptic inhibitory autoreceptors, and as a result, neuronal release of NA is gradually increased, resulting in down-regulation of post-synaptic beta receptors (Siever et al., 1982). However, the reduction of α_2-adrenoceptor density in the CNS as measured by ligand binding technique cannot directly be related to reduction in presynaptic receptors, because most receptors measured by the binding technique are post-synaptic (U'Pritchard et al., 1980). Alpha$_2$-receptors in the CNS may be situated on cell bodies, dendrites or axon terminals, and only the last category correspond to presynaptic receptors, with the same local inhibitory function on NA release as those receptors situated on axon terminals in the sympathetic nervous system. It is therefore of interest to determine whether chronic MAO inhibition can cause a down-regulation of α_2-presynaptic inhibitory sites.

Inhibition of MAO results in increase in both cytoplasmic (Stefano and Trendelenburg, 1984) and vesicular (Gomez and Fillenz, 1982; Buu, 1989) NA contents, and chronic MAO inhibition could increase neuronal NA release by an action unrelated to alteration in α_2-receptor activity, by increasing the amount of NA released per nerve pulse, as a result of an increased content of NA per storage vesicle. In animal and human studies, MAO inhibition has actually been found to decrease the concentration of dopamine beta-hydroxylase, and increase free NA levels, in the CSF (Lerner et al., 1979, 1980), which would indicate an increase in the amount of NA released per pulse, with a decrease in the number of storage vesicles released, perhaps as a result of a decrease in neuronal firing rate. Direct electrophysiological evidence has shown a decrease in firing rate of the locus coeruleus following acute administration of an MAO inhibitor (Campbell et al., 1985). This in turn may be the result of an increased activation of NA input to the locus coeruleus, which would reduce activity of these cells by an action at somatodendritic α_2-receptors, or an increased activation of these receptors by NA released from the coerulean cell itself. Down-regulation of post-synaptic α_2-adrenoceptors situated on cell bodies in the locus coeruleus would increase NA release from the whole neurone, since activation of these receptors normally reduces neuronal activity. These cells, however, maintained normal inhibitory activity to clonidine over the course of a long-term MAO inhibitor treatment (Blier and DeMontigny, 1985).

When whole body turnover of NA is considered, MAO inhibition may increase release from a primary neuronal site in the CNS, but this primary action may lead to a reduction in general sympathetic activity, and a reduction in whole body NA turnover. Total body excretion of NA and metabolites was reduced in patients treated with clorgyline (Linnoila et al., 1982).

In an attempt to understand the effects of MAO inhibition at the level of the primary neuronal site, we have elected to study the effects of MAO inhibition on sympathetic nerves, because similar processes of amine synthesis, release and metabolism of NA subsequent to release occur in the CNS and peripheral sympathetic nerves, and because definite presynaptic effects can be studied in the periphery. Most of our studies have been carried out with clorgyline, a selective inhibitor of MAO-A, since MAO-A is the subtype found in sympathetic nerves (Jarrott, 1971), and because clorgyline is devoid of amphetamine-like and other side effects, but possesses clinical antidepressant efficacy.

Using the pithed rat model, Finberg and Kopin (1986) determined release of endogenous NA subsequent to stimulation of the entire sympathetic outflow from the spinal cord. Although an acute treatment with the MAO-A inhibitor clorgyline produced no change in the plasma concentration of NA attained following sympathetic stimulation, chronic treatment resulted in an increased release, both before and after α_2-adrenoceptor blockade with yohimbine. Yohimbine caused the same degree of enhancement of NA release before and after chronic clorgyline treatment. The increased neurotransmitter release, therefore, could not be related to a decrease in functional presynaptic inhibitory receptor activity.

Using the isolated rat vas deferens as a model system to study sympathetic function, chronic treatment with clorgyline but not deprenyl was found to reduce the presynaptic effect of clonidine to inhibit twitch response of the organ to electrical field stimulation at 0.1 Hz (Finberg and Tal, 1985). Both acute and chronic treatments with clorgyline and nialamide reduced the fractional release of ^{3}H-NA from isolated vas deferens and anococcygeus muscle in response to electrical field stimulation (Hovevey-Sion and Finberg, 1987), although release induced by veratridine or 50 mM KCl was enhanced. The finding of a reduction in fractional release of ^{3}H-NA does not mean that endogenous NA release is also reduced, since tissue NA stores may be substantially increased following MAO inhibition. Further studies were therefore carried out in an attempt to assess endogenous NA release in the isolated vas deferens following chronic clorgyline treatment.

The effect of acute MAO-A inhibition by clorgyline on the level of sympathetic outflow from the CNS was studied by electrophysiological monitoring of renal nerve activity in the anaesthetised rat.

Materials and methods

Tissue preparation

Vasa deferentia from adult Sprague-Dawley rats (220 + 20 g body weight) were removed from close to junction with bladder to start of epididymis. Tissues were suspended in Krebs' solution contained in a 3 ml capacity organ bath, at 37 °C, gassed with 95% O_2, 5% CO_2, under a resting tension of 0.5 g. Electrical field stimulation was applied between lower (hook) and upper (ring) platinum electrodes using a Grass S44 stimulator. Tissue tension was measured with a Statham isometric transducer (UC2), and displayed on a Gould chart recorder.

Catecholamine determination

Catecholamines were determined in Krebs' solution following extraction on acid-washed alumina, as described by Eisenhofer et al. (1987). Tissue NA content was determined following homogenisation of the tissues in perchloric acid, and centrifugation at 4 °C. Aliquots of the perchloric acid eluent from the alumina, or of supernatant of tissue homogenates, were injected onto a Spherisorb S30DS2 column of a Varian HPLC apparatus, and NA in the eluate detected using an ECA Coulochem detector. Output of the detector was recorded using an integrating recorder, and quantitation of peaks carried out by comparing area under the curve with that of standard solutions.

Release of ^{3}H-NA

Tissue NA stores were labelled by preincubation for 30 min with ^{3}H-NA (0.5 μM). Subsequently, the tissues were washed over a period of 1.5 h before collection of eluant from the organ bath. Tritium content in aliquots of Krebs' solution was determined by scintillation counting. At the end of the experiment, tissues were solubilised in Soluene 100, for determination of total tritium content. Fractional release of tritium (i.e. the amount released by electrical stimulation divided by the total amount in the tissue) was calculated as described previously (Hovevey-Sion and Finberg, 1987).

Renal nerve activity

Rats were anesthetised with a pentobarbital/chloral hydrate mixture, and the left renal nerve exposed via a flank approach. A pair of fine Pt electrodes was placed around the renal nerve, and the amplified signal was filtered, integrated, and the calibrated rate signal displayed on a Graphtec chart recorder.

Results

Endogenous NA release

In these experiments, the Krebs' solution contained desipramine (1 μM), corticosterone (28 μM) and yohimbine (0.5 μM). Release of endogenous NA was elicited by electrical field stimulation with 1 ms pulses at various frequencies for 100 s. Only the results obtained at 1 Hz are reported here. Stimulation-induced release was calculated as the difference between NA content in the Krebs' solution in the period immediately following stimulation, and the basal release prior to stimulation. As can be seen (Table 1), stimulation-induced release was increased 1.8 fold as a result of the chronic treatment with clorgyline, while the tissue content of NA was increased 2.3

Table 1. Release of endogenous and tritiated noradrenaline from the isolated rat vas deferens by electrical field stimulation (1 Hz X 100 s)

		Chronic saline	Chronic clorgyline
a) Fractional release of ^{3}H-noradrenaline (%)[a]			
Control	mean	0.26 (12)	0.18 (6)
	SEM	0.018	0.052
Yohimbine	mean	0.41	0.41
(0.5 μM)	SEM	0.03	0.04
b) Release of endogenous noradrenaline (pg)[c]			
	mean	310 (8)	543 (9)[b]
	SEM	48	61
c) Tissue noradrenaline content (ng/vas deferens)			
	mean	131 (8)	304 (10)[b]
	SEM	17	21
d) Fractional release of endogenous noradrenaline (%)[c]			
		0.24	0.18

[a] 1 mg kg^{-1} daily i.p. for 21 days. Animals killed 24 h after last injection.
[b] Significant difference from saline group.
[c] Measurements of endogenous release done in presence of yohimbine (0.5 μM), desipramine (1 μM) and corticosterone (28 μM)

fold. Data obtained separately on the fractional release of ^{3}H-NA are shown in the same table. In the absence of any antagonist drugs, fractional release of ^{3}H-NA was 30% lower in tissues from animals treated chronically with clorgyline, but in the presence of a maximally effective concentration of yohimbine, the fractional release was equivalent in the saline and clorgyline-treated groups.

Sympathetic nerve activity

The intraperitoneal administration of clorgyline (2 mg/kg) was followed by a gradual reduction in nerve activity together with a fall in arterial pressure, which reached a peak after 30 min. At this time nerve activity was 29 $\pm$ 6.6% lower than in the control period, and mean blood pressure had fallen to 83 $\pm$ 4 from a control value of 114 $\pm$ 8 mmHg (SEM, N = 6). This situation can be compared with the effect of vasodilator drugs, which cause a fall in blood pressure with an (reflex) increase in renal nerve activity. Prior treatment of the animals with yohimbine (0.3 mg/kg), resulted in a rise in the level of resting renal nerve activity, but the decrease in both renal nerve activity and blood pressure by the subsequent administration of clorgyline was blocked.

Discussion

The data from experiments on release of tritiated NA show that in the presence of yohimbine, electrical field stimulation results in the same fractional release of NA following chronic MAO inhibition as in control animals. Since tissue NA content is increased by chronic MAO inhibition, this should lead to an increased release of endogenous NA, if release of tritiated noradrenaline reflects release of the endogenous neurotransmitter. In the presence of yohimbine, we were able to detect an enhanced release of endogenous NA following chronic clorgyline treatment (+75%).

These findings are in accord with results obtained in the whole animal cardiovascular system, showing an increase in endogenous NA release in plasma of the pithed rat, following chronic clorgyline treatment (Finberg and Kopin, 1986).

We can find no proof for down-regulation of presynaptic α_2-receptors following chronic MAO inhibition. Such a down-regulation, if it occurred, should result in a reduction in the effectiveness of yohimbine to enhance NA release, but in fact, yohimbine produced a greater enhancement of ^{3}H-NA fractional release after chronic clorgyline treatment. This is because of continued α_2-receptor mediated inhibition of release during the chronic clorgyline treatment. We were also unable to elicit a reduction in response to yohimbine by chronic treatment with clonidine in the pithed rat (Finberg and Kopin, 1987). Down-regulation of presynaptic α_2-receptors by prolonged elevation in synaptic NA levels has not been demonstrated to occur, and if it did, would be contrary to the putative functional role of these receptors (Finberg, 1987).

Observations of a reduced responsiveness to clonidine following chronic MAO inhibition, which have been explained by a reduction in presynaptic receptor density, may in fact be explained by competition with an increased synaptic NA concentration. In a similar way, presynaptic responses to clonidine were diminished by acute in vitro blockade of neuronal amine reuptake (Goncalves et al., 1989). Some of these observations are also consistent with a down-regulation of postsynaptic, as opposed to pre-synaptic, receptors.

Acute MAO inhibition normally produces no change in the effects of sympathetic nerve stimulation, but decreases the level of spontaneous locus coeruleus activity (Campbell et al., 1985). Our finding of a decrease in central sympathetic tone following acute clorgyline administration also shows that functional, immediate changes may result from MAO inhibition in the CNS, and that these effects may be mediated, at least in part, by activation of

α_2-receptors. This finding is in direct support of the observation by Ross et al. (1985) of a decrease in plasma NA concentration following administration of clorgyline to depressed patients, and offers an explanation for the antihypertensive effect of MAO inhibitors.

At present, it is not clear whether the effect at the central sites mediating decreased sympathetic outflow is the result of increased exocytotic NA release. The difference between CNS and sympathetic nerves could be the result of different rates of neurotransmitter release. The rate of increase in tissue NA content following MAO inhibition has been shown to be related to the level of neuronal activity (Clarke and Sampath, 1973). A higher NA turnover rate in the CNS neurons would lead to a greater rate of increase in vesicular NA, provided that feedback inhibition of tyrosine hydroxylase is not limiting.

In conclusion, the acute inhibition of neuronal MAO activates α_2-receptors in the CNS, leading to a reduction in central sympathetic tone. Chronic inhibition of neuronal MAO increases neuronal exocytotic NA release without producing a down-regulation of α_2-presynaptic autoreceptors. This enhanced neuronal NA release may result in down-regulation of post-synaptic receptors, and may be involved in the antidepressant action of these drugs.

References

Blier P, De Montigny C (1985) Serotonergic but not noradrenergic neurons in rat CNS adapt to long-term treatment with monoamine oxidase inhibitors. Neuroscience 16:949–955

Buu NT (1989) Modification of vesicular dopamine and norepinephrine by monoamine oxidase inhibitors. Biochem Pharmacol 38:1685–1692

Campbell IC, Gallager DW, Hamburg MA, Tallman JF, Murphy DL (1985) Electrophysiological and receptor studies in rat brain: effects of clorgyline. Eur J Pharmacol 111:355–365

Clarke DE, Sampath SS (1973) Studies on the functional role of intraneuronal monoamine oxidase. J Pharmacol Exp Ther 187:539–549

Cohen RM, Campbell IC, Dauphin M, Tallman JF, Murphy DL (1982) Changes in α- and β-receptor densities in rat brain as a result of treatment with monoamine oxidase inhibiting antidepressants. Neuropharmacology 21:293–298

Eisenhofer G, Ropchak TG, Stull RW, Goldstein DS, Keiser HR, Kopin IJ (1987) Dihydroxyphenylglycol and intraneuronal metabolism of endogenous and exogenous norepinephrine in the rat vas deferens. J Pharmacol Exp Ther 241:547–553

Finberg JPM (1987) Antidepressant drugs and down-regulation of presynaptic receptors. Biochem Pharmacol 36:3557–3562

Finberg JPM, Kopin IJ (1986) Chronic clorgyline treatment enhances release of norepinephrine following sympathetic stimulation in the rat. Naunyn-Schmiedebergs Arch Pharmacol 332:236–242

Finberg JPM, Kopin IJ (1987) Chronic clonidine treatment produces desensitisation of post- but not presynaptic α_2-adrenoceptors. Eur J Pharmacol 138:95–100

Finberg JPM, Tal A (1985) Reduced peripheral presynaptic adrenoceptor sensitivity following chronic antidepressant treatment in rats. Br J Pharmacol 84:609–617

Gomez J, Fillenz M (1982) The effect of tranylcypromine on noradrenergic nerve terminals in different brain regions. Neurochem Int 4:135–141

Goncalves J, Carvalho F, Guimares S (1989) Uptake inhibitors do not change the effect of imidazoline α_2-adrenoceptor agonists on transmitter release evoked by single pulse stimulation in mouse vas deferens. Naunyn-Schmiedebergs Arch Pharmacol 339:288–292

Hovevey-Sion D, Finberg JPM (1987) Chronic inhibition of monoamine oxidase reduces noradrenaline release in rat vas deferens and anococcygeus muscle. Naunyn-Schmiedebergs Arch Pharmacol 336:70–76

Jarrott B (1971) Occurrence and properties of monoamine oxidase in adrenergic neurons. J Neurochem 18:7–16

Lerner P, Major LF, Murphy DL, Lipper S, Lake CR, Lovenberg W (1979) Dopamine-β-hydroxylase and norepinephrine in human cerebrospinal fluid: effects of monoamine oxidase inhibitors. Neuropharmacology 18:423–426

Lerner P, Major LF, Dendel PS, Campbell IC, Murphy DL (1980) Central and peripheral dopamine beta hydroxylase: responses to long term treatment with monoamine oxidase inhibitors. Neuropharmacology 19:877–881

Linnoila M, Karoun F, Potter WZ (1982) Effect of low dose clorgyline on 24-hour binary monoamine excretion in patients with rapidly cycling bipolar affective disorder. Arch Gen Psychiatry 39:513–516

Ross RJ, Scheinin M, Lesieur P, Rudorfer MV, Hauger RL, Siever LJ, Linnoila M, Potter WZ (1985) The effect of clorgyline on noradrenergic function. Psychopharmacology 85:227–230

Siever LJ, Uhde TW, Murphy DL (1982) Possible subsensitization of alpha$_2$-adrenergic receptors by chronic monoamine oxidase inhibitor treatment in psychiatric patients. Psychiatry Res 6:293–302

Stefano FJE, Trendelenburg U (1984) Saturation of monoamine oxidase by intraneuronal noradrenaline accumulation. Naunyn Schmiedebergs Arch Pharmacol 328:135–141

Sugrue MF (1981) Effect of chronic antidepressant administration on rat frontal cortex α_2 and β adrenoceptor binding. Br J Pharmacol 74:760P–761P

U'Prichard DC, Reisine TD, Mason ST, Fibiger HC, Yamamura HI (1980) Modulation of rat brain α- and β-adrenergic receptor populations by lesion of the dorsal noradrenergic bundle. Brain Res 187:143–154

Youdim MBH, Finberg JPM (1983) Monoamine oxidase inhibitor antidepressants. In: Psychopharmacology, vol 1. Excerpta Medica, Amsterdam, p 38

Authors' address: Dr. J. P. M. Finberg, Rappaport Family Institute for Research in the Medical Sciences, Technion, Haifa, Israel.

J Neural Transm (1990) [Suppl] 32: 413–419

Biochemical characterization and purification of the neuronal sodium-dependent noradrenaline transporter*

H. Bönisch[1], **G. Martiny-Baron**[1], **B. Blum**[2], and **J. Michael-Hepp**[2]

[1] Department of Pharmacology, University of Bonn, and
[2] Department of Pharmacology, University of Würzburg, Federal Republic of Germany

Summary. The protein properties of the neuronal sodium-dependent noradrenaline (NA) transporter of PC12 (rat pheochromocytoma) cells and of bovine adreno-medullary cells were studied by means of binding of ^{3}H-desipramine (^{3}H-DMI). ^{3}H-DMI binding was decreased by proteases, phospholipase A_2, by disulfide reducing agents and by the sulfhydryl-group alkylating agent N-ethylmaleimide. The NA transporter was partially purified by anion exchange and affinity chromatography. Tritiated desmethylxylamine (^{3}H-DMX) bound irreversibly and in a DMI-sensitive manner to two PC12 membrane proteins (32 kd and 53 kd) which may represent components of the NA transporter.

Introduction

After its release the neurotransmitter noradrenaline (NA) is removed from the synaptic gap by reuptake into the noradrenergic nerve terminals; this transport system ("uptake$_1$", Iversen, 1967) possesses the following pharmacological properties (for review see Graefe and Bönisch, 1988): 1) beside the catecholamines NA, dopamine and adrenaline (but not isoprenaline), indirectly acting phenethylamines (e.g. amphetamine) are transported substrates; 2) substrate transport by uptake$_1$ is absolutely dependent on Na^+ and Cl^-; both ions are obviously co-substrates, and the direction of the Na^+-gradient determines the transport direction; 3) uptake$_1$ is selectively and competitively inhibited by nisoxetine and the tricyclic antidepressant desipramine (DMI; K_i about 4 nmol/l).

* Supported by the Deutsche Forschungsgemeinschaft.

Tritiated desipramine (^{3}H-DMI) has been used to pharmacologically characterize its binding site at the neuronal NA transporter (for references see Graefe and Bönisch, 1988); ^{3}H-DMI binding is absolutely dependent on Na^+ and Cl^- and is stereoselectively inhibited by the enantiomers of cocaine and oxaprotiline. Using membranes from PC12 (rat pheochromocytoma) cells, which are endowed with uptake$_1$ (Greene and Rein, 1977), we have obtained evidence that 3H-DMI binds to the substrate recognition site of the neuronal NA transporter (Bönisch and Harder, 1986; Schömig and Bönisch, 1986; Schömig et al., 1988a, b). In preliminary reports (Bönisch and Michael-Hepp, 1989; Michael-Hepp and Bönisch, 1990), we have shown that ^{3}H-DMI labels the NA transporter of bovine adreno-medullary plasma membranes and that a derivative of xylamine may be useful to label the NA transporter. The present study was undertaken to extend these investigations and to label and purify the transporter protein.

Materials and methods

Experiments with intact PC12 cells and preparations of purified plasma membranes were carried out as described earlier (Friedrich and Bönisch, 1986; and Harder and Bönisch, 1984, respectively). Crude PC12 membranes were isolated after hypotonic lysis of the cells essentially as described by Albus et al. (1984).

For the preparation of bovine adreno-medullary membranes, adrenal medullae (30 g) were homogenized (4 × 30 s; polytron setting 6) in 0.27 mol/l sucrose containing 57 mmol/l ascorbic acid; the homogenate was freed of large fragments by centrifugation at 600 g (6 min) and then centrifuged at 15,000 g (15 min). The resulting supernatant was diluted with 140 mmol/l KCl (containing 20 mmol/l HEPES/NaOH pH 7.0, 2 mmol/l $MgCl_2$ and 1 mmol/l DTT) and thereafter centrifuged at 165,000 g (30 min). The pellet was resuspended in the same buffer (plus 2 mmol/l EDTA), layered on a 32% sucrose cushion and centrifuged at 81,000 g (60 min). The plasma membranes on top of the sucrose cushion were diluted (1:10) with buffer A (composition in mmol/l: NaCl 135, KCl 5, $MgSO_4$ 1, Tris/HCl pH 7.4 10, DTT 1) and centrifuged at 165,000 g (30 min). The pelleted membranes were resuspended in buffer A to obtain a protein content of 3 mg/ml and stored at −70 °C.

The binding of ^{3}H-DMI to either membrane fractions or to solubilized NA transporter fractions were performed by a filtration assay using GF/F filters as outlined previously (Bönisch and Harder, 1986).

Results

Effects of thiol and disulfide reagents on ^{3}H-DMI binding

PC12 and bovine adreno-medullary membranes were first preincubated (for 20 min) with the disulfide reducing agents 2-mercapto ethanol (2-ME), dithiothreitol (DTT) and dithioerythritol (DTE) or the thiol group alkylating agent N-ethyl maleimide (NEM) and then incubated (for 40 min) in the additional presence of ^{3}H-DMI (1 nmol/l). Increasing concentrations of these agents caused decrease of ^{3}H-DMI binding with the relative order of potency NEM = DTE > DTT > 2-ME (Table 1).

Loss of ^{3}H-DMI binding activity by enzymes

PC12 membranes (intact and digitonin solubilized) as well as bovine adreno-medullary membranes were exposed to subtilopeptidase A or phospholipase A_2, and ^{3}H-DMI binding was determined as described above. Both enzymes caused concentration-dependent decrease of ^{3}H-DMI binding; in solubilized PC12 membranes, the decrease of binding by phospholipase A_2 was strongly reduced (Table 1).

^{3}H-DMI binding was also decreased by trypsin (Table 1; examined only in intact PC12 membranes). The nearly complete loss of ^{3}H-DMI binding activity by trypsin (250 mg/l) was abolished in the presence of 25 mg/ml soybean trypsin inhibitor.

Solubilization and purification of the ^{3}H-DMI binding site

The anionic detergents sodium cholate (0.1%) and sodium deoxycholate (0.1%), the zwitter-ionic detergents CHAPS (0.5%) and CHAPSO (0.5%) as well as the non-ionic detergents octylglucoside (0.2%), sucrose laurate (0.05%), dodecylmaltoside (1%), digitonin (0.5%), triton X-100 and nonidet

Table 1. Inhibition (inactivation) of ^{3}H-DMI binding by thiol reagents (I) and by enzymes (II)

	IC_{50} (I: mmol/l; II: mg/l) A	B	C
(I) Thiol reagents:			
dithioerythritol (DTE)	1.8	1.5	n.d.
dithiothreitol (DTT)	50	25	n.d.
2-mercato ethanol (2-ME)	196	55	n.d.
N-ethylmaleimide (NEM)	5.5	1.4*	n.d.
(II) Enzymes:			
subtilopeptidase A	25.4	6.1	16.1
phospholipase A_2	0.013	0.010	4.8
trypsin	n.d.	75	n.d.

A bovine adreno-medullary membranes; *B* PC12 membranes; *C* digitonin solubilized PC12 membranes; *n.d.* not determined; mean values (SEM < 15%) of 3–5 experiments; * Taken from Schömig et al. (1988)

P40 were examined for solubilization of the ^{3}H-DMI binding site in an active state. With the exception of the latter two detergents all surfactants revealed solubilization without pronounced loss of ^{3}H-DMI binding activity. Membranes solubilized with CHAPS or digitonin lost only about 25% of ^{3}H-DMI binding activity after storage for two weeks at 4 °C.

To examine the micellar size of the solubilized ^{3}H-DMI binding site, digitonin-solubilized PC12 membranes were centrifuged (at 200,000 g for 25 h) in a linear sucrose gradient (5–40%); the gradient was thereafter fractionated (25 fractions) and the fractions were tested for ^{3}H-DMI binding activity. A single peak of binding activity was found in a fraction corresponding to a globular protein with a molecular size of about 110 kilodaltons (kd).

First experiments to purify the digitonin-solubilized ^{3}H-DMI binding site (of PC12 membranes) were carried out with a weak anion exchange resin (DEAE-Tris-acryl) and by affinity chromatography on an imipramine-based agarose support. About 90% of the solubilized ^{3}H-DMI binding protein were retained by the anion exchange resin and about 60% could be eluted from the resin with 40 mmol/l NaCl; the eluate contained a 10-fold purified ^{3}H-DMI binding protein. Analysis of the protein composition of the eluate by SDS-polyacrylamide gel electrophoresis (SDS-PAGE) revealed an enrichment of a 32 kd (31–33 kd) polypeptide.

By means of the affinity support about 50% of the ^{3}H-DMI binding protein was retained by the affinity column. The binding protein was eluted with either a sodium-free solution (KCl 1 mol/l) or a competing ligand (e.g amezinium 1 mmol/l). Both procedures revealed an about 80-fold purification of ^{3}H-DMI binding (specific binding activity: 400 pmol/mg protein). SDS-PAGE showed a major protein of about 32 kd beside some additional proteins (e.g. a 40 kd and 50 kd protein).

Affinity labeling of the NA transporter

We have synthetized a derivative of xylamine lacking the ortho methyl group (desmethyl-xylamine, DMX) and have tested DMX for irreversible inhibition of ^{3}H-NA uptake into intact PC12 cells and for irreversible inhibition of ^{3}H-DMI binding to PC12 membranes. DMX was as active as xylamine in irreversibly inhibiting both parameters (IC_{50} about 3–8 μmol/l). To study the incorporation of DMX into PC12 membranes, PC12 cells were incubated (for 5–120 min at 37 °C) with tritiated DMX (0.2 μmol/l, 5 μCi/ml), and in the absence or presence of Na^+, Cl^-, nisoxetine or DMI (10 μmol/l). Irreversible labeling of PC12 membrane proteins by ^{3}H-DMX was time-dependent and reached a maximum after about 60 min. Incorporation of ^{3}H-DMX (after 60 min of incubation) into crude membranes was inhibited by nisoxetine and DMI (by about 90–95%) and by the absence of Na^+

(90%) or Cl^- (70%). Analysis of these membranes by SDS-PAGE revealed a sodium and chloride-dependent and nisoxetine and DMI-sensitive labeling of two major proteins with a relative mass of about 32 kd (31–33 kd) and 53 kd (52–55 kd).

Discussion

The data presented support the protein nature of the ^{3}H-DMI binding site of PC12 and bovine adreno-medullary membranes since it was completely inactivated by subtilopeptidase A and trypsin. In this connection it is interesting to note, that the "specific" binding of ^{3}H-imipramine to rat cortical membranes is only partly sensitive to inactivation by subtilopeptidase A (Marcusson et al., 1985); furthermore it has been suggested that the protease-sensitive binding site is the recognition site of the 5-HT transporter whereas the protease-resistant site is presumably a lipid recognition site. For binding of ^{3}H-DMI to the binding site of the neuronal NA transporter, both intact disulfide bonds and free sulfhydryl (thiol) groups are essential. This conclusion is based on the finding that the disulfide bond reducing agents 2-ME, DTT and DTE and the thiol alkylating agent NEM "inactivated" ^{3}H-DMI binding. The "inactivation" by NEM obviously takes place at the substrate recognition site of the NA transporter (Schömig et al., 1988b).

Many membrane proteins require lipids for their function (for review see McElhaney, 1982); this has for example been shown for the binding of ^{3}H-imipramine to the 5-HT transporter of platelets (Wennogle et al., 1981). The inactivation by phospholipase A_2 of ^{3}H-DMI binding indicates that lipids are also crucial to the function of the ^{3}H-DMI binding of the NA transporter.

To label components of the neuronal NA transporter a tritiated derivative of xylamine (^{3}H-DMX) has been shown to be useful. ^{3}H-DMX was covalently incorporated into a 32 kd and a 53 kd protein of PC12 membranes; the 53 kd protein was labeled in a less pronounced manner than the 32 kd protein. Since the incorporation of ^{3}H-DMX was inhibited under condition which inhibit substrate binding to the recognition site of the NA transporter, both labeled proteins may represent components of the NA transporter. The 32 kd protein might either be a functional subunit or a fragment of the NA transporter. The 10- and 80-fold purification of the 32 kd protein by anion exchange and affinity chromatography, respectively, argues against an accidental singular finding. A molecular size of 110 kd was found by determining ^{3}H-DMI binding activity in fractions obtained after equilibrium centrifugation of digitonin-solubilized PC12 membranes; this is compatible with the idea that the 32 kd protein might represent a functional unit, since digitonin (which has a micellar size of about 80 kd) probably largely contributed to the measured size of 110 kd. Interestingly, a similar small peptide of 30 kd, labeled by the photoactive 5-HT uptake inhibitor ^{3}H-2-nitroimipramine, has been proposed to represent a macromolecular component of the 5-HT transporter of human platelets (Wennogle et al., 1985).

Further evidence for the identity of the 32 kd protein and the NA transporter protein may be obtained by further purification of the protein. Bovine adreno-medullary membranes contain the NA transporter, thus making a large scale purification of the protein possible. First experiments using this material and the methods described here revealed promising results.

References

Albus U, Habermann E, Ferry DR, Glossmann H (1984) Novel 1,4-dihydropyridine (Bay K 8644) facilitates calcium-dependent [^{3}H]noradrenaline release from PC12 cells. J Neurochem 42:1186–1189

Bönisch H, Harder R (1986) Binding of ^{3}H-desipramine to the neuronal noradrenaline carrier of rat phaeochromocytoma cells (PC12 cells). Naunyn-Schmiedebergs Arch Pharmacol 334:403–411

Bönisch H, Michael-Hepp J (1989) Binding of ^{3}H-desipramine to the noradrenaline carrier of the plasma membrane of bovine adrenal medulla. Naunyn-Schmiedebergs Arch Pharmacol 340 [Suppl]: R 38

Friedrich U, Bönisch H (1986) The neuronal noradrenaline transport system of PC12 cells: kinetic analysis of the interaction between noradrenaline, Na^{+} and Cl^{-} in transport. Naunyn-Schmiedebergs Arch Pharmacol 333:246–252

Graefe K-H, Bönisch H (1988) The transport of amines across the axonal membranes of noradrenergic and dopaminergic neurones. In: Trendelenburg U, Weiner N (eds) Handb exp pharmacol, vol 90/I; Catecholamines I, pp 193–245

Greene LA, Rein G (1977) Release, storage and uptake of catecholamines by a clonal cell line of nerve growth factor (NGF) responsive pheochromocytoma cells. Brain Res 129:257–263

Harder R, Bönisch H (1984) Large-scale preparation of plasma membrane vesicles from PC-12 pheochromocytoma cells and their use in noradrenaline transport studies. Biochim Biophys Acta 775:95–104

Iversen LL (1967) The uptake and storage of noradrenaline in sympathetic nerves. Cambridge University Press, Cambridge

Marcusson J, Fowler CJ, Hall H, Ross SB, Winblad B (1985) "Specific" binding of [^{3}H]imipramine to protease-sensitive and protease-resistant sites. J Neurochem 44:705–711

McElhaney RN (1982) Effects of membrane lipids on transport and enzymic activities. Current Topics in Membranes and Transport 17:317–379

Michael-Hepp J, Bönisch H (1990) Labelling of the neuronal noradrenaline-carrier with a xylamine derivative. Naunyn-Schmiedebergs Arch Pharmacol 341 [Suppl]: R 82

Schömig E, Bönisch H (1986) Solubilization and characterization of the ^{3}H-desipramine binding site of rat phaeochromocytoma cells (PC12 cells). Naunyn-Schmiedebergs Arch Pharmacol 37:412–417

Schömig E, Körber M, Bönisch H (1988a) Kinetic evidence for a common binding site for substrates and inhibitors of the neuronal noradrenaline carrier. Naunyn-Schmiedebergs Arch Pharmacol 337:626–632

Schömig E, Michael-Hepp J, Bönisch H (1988b) Inhibition of neuronal noradrenaline uptake (uptake$_1$) and desipramine binding by N-ethylmaleimide (NEM). Naunyn-Schmiedebergs Arch Pharmacol 337:633–636

Wennogle LP, Beer B, Meyerson LR (1981) Human platelet imipramine recognition sites: biochemical and pharmacological characterization. Pharmacol Biochem Behav 15:975–982

Wennogle LP, Ashton RA, Schuster DI, Murphy RB, Meyerson LR (1985) 2-Nitroimipramine: a photoaffinity probe for the serotonin uptake/tricyclic binding site complex. EMBO J 4:971–977

Authors' address: Prof. Dr. H Bönisch, Institut für Pharmakologie und Toxikologie, Universität Bonn, Reuterstrasse 2b, D-5300 Bonn 1, Federal Republic of Germany.

J Neural Transm (1990) [Suppl] 32: 421–429

The importance of plasma 3,4-dihydroxyphenylglycol (DOPEG) in analyses of the sympathetic nervous system in vivo

K.-H. Graefe, Th. Halbrügge, M. Gerlich, and **J. Ludwig**

Institut für Pharmakologie and Medizinische Poliklinik, University of Würzburg, Federal Republic of Germany

Summary. In the anaesthetized rabbit, plasma DOPEG was used as a tool to estimate both, the total-body rate of neuronal re-uptake of noradrenaline (NA) and the factor F by which the NA concentration in the synaptic cleft exceeds that in plasma. The NA re-uptake rate was 3.5 times higher than the rate of NA net release into plasma and amounted to 863 pmol kg^{-1} min^{-1}. F was 3.4.

In a study in humans it was found that essential hypertension appears to be associated with an enhanced formation of that part of DOPEG which originates from NA leaking out of the transmitter storage vesicles.

Introduction

DOPEG is a primary noradrenaline (NA) metabolite which is formed through the action of monoamine oxidase and aldehyde reductase (Kopin, 1985). In-vitro experiments with isolated tissues have shown that DOPEG originates mainly from NA metabolism within the noradrenergic neurone (Langer, 1974; Graefe and Henseling, 1983; Kopin, 1985). According to these results, there are two sources of presynaptic DOPEG formation: 1. DOPEG formed from NA spontaneously leaking out of the storage vesicles into the axoplasm ("leakage source") and 2. DOPEG formed from synaptic NA after its uptake or re-uptake into the neurone ("uptake source"). While the latter source of DOPEG formation runs dry when neuronal uptake is blocked (e.g., by desipramine), the former does not. On the other hand, the "leakage source" of DOPEG formation is stimulated at first and eventually abolished by reserpine. Results obtained in the rat vas deferens also indicate that 70–80% of the released NA that is neuronally recaptured is sequestered in storage vesicles, with the remainder being converted to DOPEG (Eisenhofer et al., 1987). All these results from in-vitro studies suggest that measurements of DOPEG give valuable information on sympathetic nervous function. Therefore, in the work presented here, the DOPEG concentration in plasma was exploited as a tool to study the sympathetic nervous system in vivo.

Methods

Studies in rabbits

The details of the experiments have been described elsewhere (Halbrügge et al., 1989). Rabbits were anaesthetized with urethane and α-chloralose, and the right carotid artery, the right atrium and the left femoral vein were cannulated with polyethylene tubing. After a control period of 30 min, constant-rate infusions of NA or DOPEG via the left femoral vein were commenced. The right carotid artery served to record blood pressure and heart rate, and mixed central-venous blood was sampled from the right atrium. Drugs were administered into an ear vein. Plasma NA and DOPEG concentrations were measured simultaneously by reversed-phase HPLC and electro-chemical detection (Halbrügge et al., 1988).

In six animals, DOPEG was infused for 75 min at a rate of 2.9 nmol kg^{-1} min^{-1}. Blood was sampled immediately before, at 30, 60 and 75 min during and at various times after the infusion. The plasma clearance of DOPEG (Cl_{DOPEG}) was obtained from the infusion-induced steady-state increase in plasma DOPEG (ΔDOPEG) and the rate of DOPEG infusion (I_0; $Cl_{DOPEG} = I_0/\Delta DOPEG$). The "peeling technique" (Henseling et al., 1976) was used to analyse the time-dependent fall in log ΔDOPEG during the postinfusion period, and half-lives for the disappearance of infused DOPEG from plasma were obtained from slopes (half-life = log 0.5/slope).

In twenty-nine control animals and in 13 desipramine-treated animals, NA (1.2 to 5.9 nmol kg^{-1} min^{-1}) was infused for 75 min, and blood was sampled immediately before and at 60 and 75 min during the infusion. Desipramine was administered in two doses of 1 mg/kg each, 60 and 30 min before the start of infusion. Steady-state increases in plasma NA (ΔNA) and DOPEG (ΔDOPEG) were obtained from the plasma levels observed during NA infusion after subtraction of the baseline levels of these catechols. The plasma clearance of infused NA (Cl_{NA}) was calculated as described above ($Cl_{NA} = I_0/\Delta NA$). The net release of *endogenous* NA into plasma was estimated from the baseline plasma NA concentration times Cl_{NA}. The same principle was used to determine the rate of appearance in plasma of *endogenous* DOPEG (baseline plasma DOPEG times Cl_{DOPEG}) or of DOPEG formed from infused NA (ΔDOPEG times Cl_{DOPEG}).

Eight animals were given clonidine or yohimbine to reduce or enhance sympathetic activity. After the baseline control period, both the clonidine (n = 4) and the yohimbine (n = 4) group of animals underwent three 30-min treatment periods. Clonidine (10 and 30 μg/kg) or yohimbine (0.5 and 1.0 mg/kg) was given at the beginning of the 1st and 2nd period, respectively, and at the beginning of the 3rd period, all animals received 2 mg/kg desipramine. DOPEG and NA concentrations were determined before the 1st and at the end of each of the three treatment periods.

Results are given as arithmetic means ± SEM, and linear correlations were analysed according to established procedures.

Studies in humans

Fourty-seven normotensive subjects (20 women and 27 men, aged 21–65) and 58 out-patients diagnosed to have essential hypertension (26 women and 32 men, aged 18–74) were studied. The subjects with essential hypertension were either previously untreated for hypertension or else had discontinued any antihypertensive medication 2 weeks prior to testing. All subjects gave their informed consent to the study protocol, which was approved by the local Ethics Committee.

After 30 min of supine rest, subjects were asked to sit quietly for 30 min and, thereafter, to stand quietly in place for further 30 min. At the end of each of the three 30-min periods, blood pressure and heart rate were determined and venous blood was collected from a cannula inserted into a forearm vein. Eleven normotensive controls and 12 hypertensives volunteered to repeat the test procedure 1 to 2 days later, but now 3 h after oral administration of 1.5 mg/kg desipramine. The test under this condition consisted of 30 min of recumbency followed by 30 min of quiet standing, with observations being made and blood samples being taken at the end of both 30-min periods. Plasma concentrations of NA and DOPEG were simultaneously assayed by means of the HPLC method described by Halbrügge et al. (1988).

Results are given as geometric means together with their 95% confidence limits. Slopes of regression lines were calculated by the method of least squares. The nonparametric U-test according to Wilcoxon, Mann and Whitney was used to assess group differences.

Results and discussion

The total-body rate of neuronal re-uptake of endogenous NA

When rabbits (n = 6) were infused with DOPEG, the plasma DOPEG concentration reached a steady level within less than 30 min. The steady-state plasma concentration of total DOPEG during the infusion was 13.4 times above the level of *endogenous* DOPEG. These results gave a plasma clearance of DOPEG (Cl_{DOPEG}) of 84.0 ± 4.0 ml kg^{-1} min^{-1}. The terminal half-live for the disappearance of infused DOPEG from plasma was 2.3 ± 0.2 min. Hence, DOPEG has a high plasma clearance and is rapidly eliminated from the body. Similar results were obtained by Eisenhofer et al. (1989) in the conscious rat.

Rabbits (n = 29) infused with NA showed steady-state increases in plasma NA (ΔNA) and DOPEG (ΔDOPEG) which were linearly related to the rate of NA infusion. Hence, the plasma clearance of NA (Cl_{NA}) and the ratio of ΔDOPEG/ΔNA (expressed in %) were independent of the NA infusion rate and amounted to 75.8 ± 4.2 ml kg^{-1} min^{-1} and 9.4 ± 1.5%, respectively. In animals treated with desipramine, Cl_{NA} and the ratio ΔDOPEG/ΔNA were significantly reduced (Fig. 1). The results indicate that 35.4% of the total NA removal and 75.3% of the increment in plasma DOPEG were due to neuronal uptake of the infused NA.

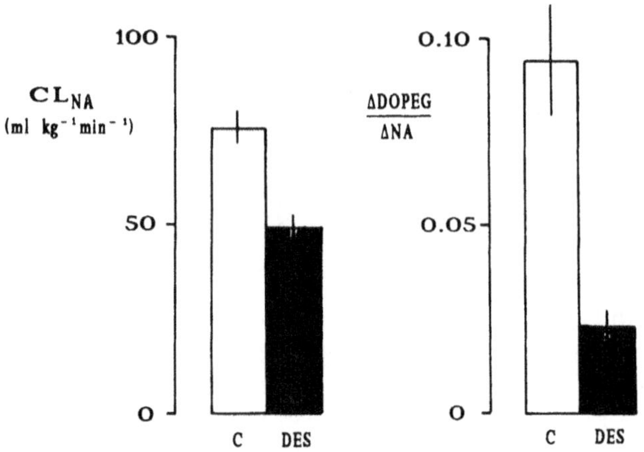

Fig. 1. Effects of desipramine on the plasma clearance (*Cl*) of infused noradrenaline (*NA*; left panel) and the increment in plasma DOPEG relative to that in plasma NA (ΔDOPEG/ΔNA; right panel) induced by NA infusion; anaesthetized rabbits. Shown are means ± SEM from 29 control animals (*C*) and 13 desipramine (2 mg/kg)-treated animals (*DES*). The difference between neighboured columns in both panels was statistically significant ($P < 0.005$)

The rate of net release into plasma of *endogenous* NA and the rate of appearance in plasma of *endogenous* DOPEG (determined from the baseline levels of plasma NA and DOPEG and the corresponding clearance values; see Methods) were 246 and 402 pmol kg^{-1} min^{-1}, respectively. Desipramine reduced ($P < 0.01$) these rates by 39.5 and 47.9%, respectively. The drug-induced decrease in NA net release is probably due to desipramine reducing sympathetic tone (Schultheiss and Szabo, 1990), whereas the decrease in DOPEG appearance is a consequence of the drug blocking the neuronal re-uptake of NA (see below).

From the results of Fig. 1 and Cl_{DOPEG}, the appearance of plasma DOPEG resulting from neuronal uptake of the infused NA can be calculated and expressed as a fraction of neuronal NA uptake. This fraction ($DOPEG_{uptake}$) may be obtained from

$$DOPEG_{uptake} = \frac{(Cl_{DOPEG}/Cl_{NA})(\Delta DOPEG/\Delta NA - \Delta DOPEG_{DES}/\Delta NA_{DES})}{(1/Cl_{NA})(Cl_{NA} - Cl_{NA,DES})} \quad (1)$$

(where the subscript DES denotes the presence of desipramine). The numerator of the fraction in equation (1) gives the appearance rate of the neuronally formed DOPEG ($Cl_{DOPEG} \times \Delta DOPEG/\Delta NA$ of neuronal origin) as a proportion of total NA removal (Cl_{NA}), whereas the denominator represents the neuronal NA uptake as a proportion of total NA removal. $DOPEG_{uptake}$ obtained from equation (1) was $7.9\%/35.4\% = 0.223$, indicating that 22.3% of the infused NA that was taken up by sympathetic nerves appear in plasma as DOPEG. Eisenhofer et al. (1989) were the first to point out that $DOPEG_{uptake}$ should also be given by

$$DOPEG_{uptake} = \frac{(DOPEG - DOPEG_{DES})Cl_{DOPEG}}{\text{Rate of neuronal re-uptake of NA}} \quad (2)$$

(where DOPEG and $DOPEG_{DES}$ are the plasma concentrations of *endogenous* DOPEG in the presence and absence of desipramine, respectively). The significance of equation (2) is obvious, because the numerator of its fraction term is the appearance rate of DOPEG derived from the neuronal re-uptake of *endogenous* NA (i.e., the desipramine-induced reduction of the appearance rate of *endogenous* DOPEG = 47.9% of 402 pmol kg^{-1} min^{-1}; see above). Thus, with $DOPEG_{uptake}$ being known from NA infusion experiments, the rate of neuronal NA re-uptake can be calculated; it was 863 pmol kg^{-1} min^{-1} and, hence, exceeded the NA net release into plasma (246 pmol kg^{-1} min^{-1}) by a factor of 3.5. Since the part played by extraneuronal uptake and metabolism is unknown, our results show that in the anaesthetized rabbit less than 22% of the released NA entered plasma and less than 78% was recaptured by sympathetic nerves. Similar results were obtained in conscious rabbits by Eisenhofer et al. (1990) who used infusions of trace amounts of ^{3}H-NA in their experiments. Moreover, these authors demonstrated that the rate of NA re-uptake increased in proportion to the rate of NA net release into plasma when the sympathetic activity in these animals was increased by a factor of 3 to 4. Therefore, it was concluded that, under the condition of a 3- to 4-fold increase in NA release, the neuronal uptake process operates below saturation.

As shown by Eisenhofer et al. (1989), Cl_{DOPEG} cancels out when equations (1) and (2) are combined. This gives a fairly simple equation which is highly useful in studies aiming at estimating the total-body rate of neuronal NA re-uptake in vivo:

$$\text{NA re-uptake rate} = \frac{(DOPEG - DOPEG_{DES})(Cl_{NA} - Cl_{NA,\,DES})}{\Delta DOPEG/\Delta NA - \Delta DOPEG_{DES}/\Delta NA_{DES}} \quad (3)$$

(where the meaning of abbreviations is as described above).

The synaptic NA concentration

As already mentioned above, the values of ΔDOPEG and ΔNA produced in the rabbit by the infusion of NA were linearly correlated with each other. The slope of the regression line relating ΔDOPEG to ΔNA (i.e., the mean ratio of ΔDOPEG/ΔNA) was 0.094 ± 0.015 ($n = 29$) in the absence and 0.023 ± 0.004 ($n = 13$) in the presence of desipramine. Therefore, the neuronal component of the slope was 0.071.

Changes in the plasma concentration of *endogenous* DOPEG and NA induced by clonidine or yohimbine (see Methods) were also linearly correlated with each other, but the regression line relating ΔDOPEG to ΔNA under this condition had a slope (i.e., a mean ratio of ΔDOPEG/ΔNA) of 0.842 ± 0.140 ($n = 8$). In animals treated with desipramine, the plasma concentration of *endogenous* DOPEG was reduced (see above) and remained virtually unchanged when the level of sympathetic activity was varied. This observation confirms results obtained in humans (Goldstein et al., 1988; Ludwig et al., 1988; see also below) and indicates that changes in plasma DOPEG associated with changes in NA release result from NA recaptured by, and subsequently metabolised within, the sympathetic nerves.

A schematic representation of the regression lines relating ΔDOPEG (of neuronal origin) to ΔNA is shown in Fig. 2. It is obvious that the plasma DOPEG response to any given change in plasma NA was much more pronounced during NA release ($slope_R = 0.842$) than during NA infusion ($slope_I = 0.071$). This difference in slope suggests a marked difference between the NA concentration in the synaptic cleft and that in plasma, because it is the synaptic NA concentration that is pertinent to DOPEG formation. Hence, a given ΔDOPEG in plasma will, under conditions of both NA release and NA infusion, reflect the same ΔNA in the synaptic cleft. The slope of the regression line relating ΔDOPEG in plasma to ΔNA in the synaptic cleft ($slope_S$; broken line in Fig. 2) will be given by

$$slope_S = slope_R/F = slope_I \times F \quad (4)$$

(where F is the factor quantifying the difference between synaptic and plasma NA concentrations). Rearranging equation (4) gives $slope_S = \sqrt{slope_R \times slope_I}$ and $F = \sqrt{slope_R/slope_I}$. Consequently, $slope_S$ was 0.244, and F was 3.4. According to these results, changes in plasma DOPEG amount to 24.4% of the changes in the synaptic NA concentration, and the NA concentration in the synaptic cleft normally exceeds that in plasma by a factor of 3.4. It is worthy of note that a completely different experimental approach arrived at virtually identical values of F in the pithed rat (Kopin et al., 1984) and in humans (Goldstein et al., 1986).

Plasma DOPEG in essential hypertension

Forearm venous plasma DOPEG and NA were determined at rest and during graded orthostatic stress (see Methods) in normotensive controls and in outpatients diagnosed to have essential hypertension; there was no statistically significant group difference in age. The results are summarized in Table 1. At rest as well as during sitting and standing, normotensives and hypertensives had similar plasma NA concentrations. However, plasma

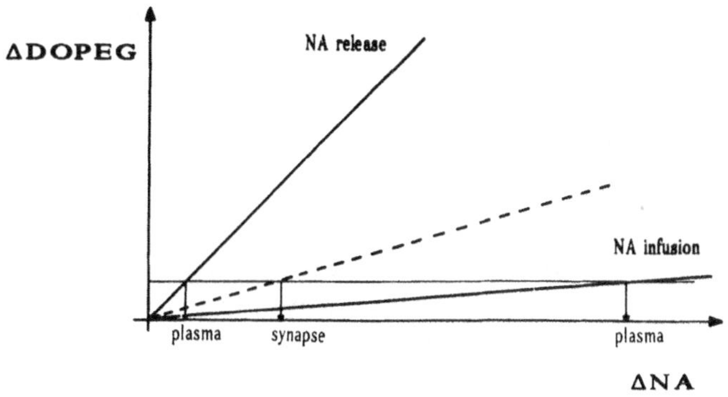

$$slope_S = slope_R/F = slope_I \cdot F$$

$$slope_S = \sqrt{slope_R \cdot slope_I}$$

$$F = \sqrt{slope_R/slope_I}$$

Fig. 2. Schematic representation of the regression lines relating increases in plasma DOPEG of neuronal origin (ΔDOPEG) to increases in plasma noradrenaline (ΔNA) observed under conditions of NA release ($slope_R = 0.842$) and NA infusion ($slope_I = 0.071$). The broken line (with slopes) relates ΔDOPEG in plasma to ΔNA in the synaptic cleft. Since the NA concentration pertinent to the presynaptic DOPEG formation is that in the synaptic cleft, a given ΔNA there will, irrespective of whether induced by NA release or NA infusion, be reflected by the same ΔDOPEG in plasma. Hence, the ΔNA in the synapse observed at this ΔDOPEG in plasma (horizontal line) is 3.4 times higher than that in plasma during NA release and 3.4 times lower than that in plasma during NA infusion. For further details see text

DOPEG concentrations were higher in hypertensives than in normotensives ($P < 0.01$ for the three conditions). Both groups of subjects showed a linear relationship between plasma DOPEG (ordinate) and plasma NA (abscissa). This relation had very much the same slope in the two groups [normotensives: 0.99 (0.84; 1.16) and hypertensives: 1.19 (1.01; 1.41)], but an ordinate

Table 1. Plasma concentrations of noradrenaline (NA) and DOPEG in normotensive control subjects (n = 47) and in patients with essential hypertension (n = 58)

		Normotensives	Hypertensives
NA (nmol/l)	supine	1.05 (0.91; 1.22)	1.19 (1.05; 1.35)
	sitting	1.68 (1.49; 1.91)	1.92 (1.74; 2.11)
	standing	2.66 (2.38; 2.97)	2.90 (2.62; 3.22)
DOPEG (nmol/l)	supine	3.89 (3.58; 4.21)	5.63 (5.24; 6.05)
	sitting	4.46 (4.11; 4.85)	6.40 (5.97; 6.87)
	standing	5.51 (5.06; 6.01)	7.93 (7.36; 8.57)

Given are geometric means (with 95% confidence limits in parentheses) of forearm venous plasma concentrations measured after 30 min of supine rest, 30 min of sitting, and 30 min of standing

intercept that was higher in hypertensives than in normotensives [3.75 (3.37; 4.18) vs. 2.58 (2.34; 2.96) nmol/l; $P < 0.01$]. Eleven normotensive controls and 12 hypertensives were also tested 3 h after oral administration of desipramine (1.5 mg/kg) to block neuronal re-uptake. The drug did not alter plasma NA concentrations either at rest or during standing, but reduced plasma DOPEG at rest [by 25.1 (19.0; 33.3) % in normotensives and by 31.2 (24.9; 39.3) % in hypertensives] and abolished the plasma DOPEG response to upright posture in both groups of individuals (data not shown). Moreover, the mean plasma DOPEG level observed at rest and during standing in the presence of desipramine was, in both groups, virtually identical with the ordinate intercept of the regression line described above for the control condition. In other words, the group difference with respect to plasma DOPEG concentrations remained unchanged after administration of desipramine.

These results confirm earlier observations (Goldstein et al., 1988; Ludwig et al., 1988; including those described here for the rabbit) and show that the plasma DOPEG level observed in the presence of desipramine is independent of the level of sympathetic tone. Since it also corresponds to the ordinate intercept of the regression line relating plasma DOPEG to plasma NA under control conditions, it probably reflects that part of plasma DOPEG which originates from NA leaking out of the transmitter storage vesicles. Hence, the elevated plasma DOPEG concentrations found here in essential hypertension indicate an enhanced "leakage source" of DOPEG formation in this disease. The reason for this is not clear. It could be due to an enhanced NA leakage out of any given vesicle, to an enhanced number of vesicles per varicosity, or to an enhanced density of sympathetic innervation. The available evidence obtained in spontaneously hypertensive rats points toward an hypernoradrenergic innervation of vascular tissues in this animal type of primary hypertension (Head, 1989). Therefore, our results may be interpreted to mean that essential hypertension, like the spontaneous hypertension in the rat, is associated with an increased density of noradrenergic innervation.

Acknowledgements

The studies were supported by the Deutsche Forschungsgemeinschaft (Gr 490/5) and by the Ernst und Hedda Wollheim-Stiftung.

References

Eisenhofer G, Ropchak TG, Stull RW, Goldstein DS, Keiser HR, Kopin IJ (1987) Dihydroxyphenylglycol and intraneuronal metabolism of endogenous and exogenous norepinephrine in the rat vas deferens. J Pharmacol Exp Ther 241:547–553

Eisenhofer G, Goldstein DS, Kopin IJ (1989) Plasma dihydroxyphenylglycol for estimation of noradrenaline neuronal re-uptake in the sympathetic nervous system in vivo. Clin Sci 76:171–182

Eisenhofer G, Cox HS, Esler MD (1990) Parallel increases in noradrenaline reuptake and release into plasma during activation of the sympathetic nervous system in rabbits. Naunyn-Schmiedebergs Arch Pharmacol 342:328–335

Goldstein DS, Zimlichman R, Stull R, Keiser HR, Kopin IJ (1986) Estimation of intrasynaptic norepinephrine concentrations in humans. Hypertension 8:471–475

Goldstein DS, Eisenhofer G, Stull R, Folio CJ, Keiser HR, Kopin IJ (1988) Plasma dihydroxyphenylglycol and the intraneuronal disposition of norepinephrine in humans. J Clin Invest 81:213–220

Graefe K-H, Henseling M (1983) Neuronal and extraneuronal uptake and metabolism of catecholamines. Gen Pharmacol 14:27–33

Halbrügge T, Gerhardt T, Ludwig J, Heidbreder E, Graefe K-H (1988) Assay of catecholamines and dihydroxyphenylethyleneglycol in human plasma and its application in orthostasis and mental stress. Life Sci 43:19–26

Halbrügge T, Wölfel R, Graefe K-H (1989) Plasma 3,4-dihydroxyphenylglycol as a tool to assess the role of neuronal uptake in the anaesthetized rabbit. Naunyn-Schmiedebergs Arch Pharmacol 340:726–732

Head RJ (1989) Hypernoradrenergic innervation: its relationship to functional and hyperplastic changes in the vasculature of the spontaneously hypertensive rat. Blood Vessels 26:1–20

Henseling M, Eckert E, Trendelenburg U (1976) The distribution of ^{3}H-($\pm$)-noradrenaline in rabbit aortic strips after inhibition of the noradrenaline-metabolizing enzymes. Naunyn-Schmiedebergs Arch Pharmacol 292:205–217

Kopin IJ (1985) Catecholamine metabolism: basic aspects and clinical significance. Pharmacol Rev 37:333–364

Kopin IJ, Zukowska-Grojec Z, Bayorh MA, Goldstein DS (1984) Estimation of intrasynaptic norepinephrine concentrations at vascular neuroeffector junctions in vivo. Naunyn-Schmiedebergs Arch Pharmacol 325:298–305

Langer SZ (1974) Selective metabolic pathways for noradrenaline in the peripheral and in the central nervous system. Med Biol 52:372–383

Ludwig J, Gerhardt T, Halbrügge T, Walter J, Graefe K-H (1988) Plasma concentrations of noradrenaline and 3,4-dihydroxyphenylethyleneglycol under conditions of enhanced sympathetic activity. Eur J Clin Pharmacol 35:261–267

Schultheiss A, Szabo B (1990) Desipramine inhibits sympathetic nerve activity in anaesthetized rabbits. Naunyn-Schmiedebergs Arch Pharmacol 341 [Suppl]: R87

Authors' address: Dr. K.-H. Graefe, Institut für Pharmakologie und Toxikologie, Universität Würzburg, Versbacher Strasse 9, D-8700 Würzburg, Federal Republic of Germany

J Neural Transm (1990) [Suppl] 32: 431–435

The involvement of desipramine-sensitive processes in the extraction of various catecholamines from plasma in the anaesthetized rabbit*

T. Halbrügge, B. Friedgen, A. Thyen, and **K.-H. Graefe**

Department of Pharmacology, University of Würzburg, Würzburg, Federal Republic of Germany

Summary. The effects of desipramine on the total body plasma clearance (Cl) as well as the systemic fractional extraction (ER_s) of noradrenaline, dopamine, adrenaline, and isoprenaline, and on the pulmonary fractional extraction (ER_p) of dopamine and adrenaline were determined in the anaesthetized rabbit. Under control conditions, noradrenaline, dopamine, adrenaline, and isoprenaline had Cl values of 59.5, 116.2, 68.0, 78.4 (ml kg^{-1} min^{-1}) and ER_s values of 0.709, 0.789, 0.694, 0.669, respectively. The ER_p of dopamine was 0.235, while that of adrenaline did not differ from zero. Desipramine reduced the Cl of noradrenaline (by 20%), but did not affect that of the other amines. It also reduced the ER_s of noradrenaline (by 31%), dopamine (by 10%) and adrenaline (by 10%); the ER_p of dopamine was diminished by 84%.

Introduction

Inhibition of neuronal uptake by desipramine has been demonstrated to reduce the plasma clearance of noradrenaline in humans and in the anaesthetized rabbit (Esler et al., 1981; Halbrügge et al., 1989). Comparisons of the effect of desipramine with respect to clearance and extraction from plasma of various catecholamines have not yet been reported in the literature. Therefore, the present study was undertaken to compare the clearance and extraction of noradrenaline, dopamine and adrenaline in the anaesthetized rabbit and to determine the effect of desipramine on these parameters. Isoprenaline, which is known to be a poor substrate of neuronal uptake, was also included.

* This study was supported by the Deutsche Forschungsgemeinschaft (Gr 490/5-2)

Material and methods

Experimental procedure

Rabbits (2.0–2.5 kg; n = 16) of either sex were anaesthetized by i.v. injection of Saffan® (1:3 mixture of alfadolone and alfaxolone) and artificially respired as described by Szabo et al. (1989). The left femoral artery and vein, the ascending aorta (via the right carotid artery) and the right ventricle (via the right jugular vein) were cannulated with polyethylene tubing. Blood pressure and heart rate were recorded via the right carotid artery. Arterial and mixed central-venous blood (2 ml each) was collected (in ice cold tubes) from the femoral artery and the right ventricle, respectively. A solution containing ^{3}H-(−)-noradrenaline, ^{3}H-(−)-adrenaline, ^{3}H-dopamine, and ^{3}H-(±)-isoprenaline was infused via the ascending aorta (83 nCi kg^{-1} min^{-1} for each). Unlabelled dopamine (3.26 nmol kg^{-1} min^{-1}) was infused simultaneously via the left femoral vein. Arterial and mixed central-venous blood samples were taken simultaneously prior to the start of infusion and at 60 (P_1), and 90 min (P_2) during the infusion. They were centrifuged immediately. In 8 out of these experiments, desipramine (0.5 mg kg^{-1} followed by 2 mg kg^{-1} h^{-1}) was given i.v. immediately after P_1.

Analytical procedure

After adsorption onto Al_2O_3 and desorption with 0.1 mol l^{-1} $HClO_4$, plasma levels of noradrenaline, adrenaline and dopamine were determined by reversed phase HPLC and electrochemical detection, with the plasma levels being corrected for the analytical recovery of the individual catecholamines determined in each assay (Halbrügge et al., 1988a). Timed collections of the eluant leaving the HPLC system also allowed fractionation and quantification of ^{3}H-labelled noradrenaline, adrenaline, dopamine, and isoprenaline by liquid scintillation counting. In each experiment "blank" plasma, which was spiked with known amounts (10 µl) of the infusate containing the four ^{3}H-catecholamines, was subjected to HPLC and liquid scintillation counting, so as to determine the individual infusion rate of each ^{3}H-labelled catecholamine (see below).

Calculations

The clearance of ^{3}H-labelled catecholamines from plasma was determined from the ratio of "^{3}H-catecholamine infusion rate/steady-state arterial plasma level of ^{3}H-catecholamine". Both parameters of the above ratio were based on the amount of radioactivity co-eluting with the corresponding unlabelled catecholamine from the HPLC system after Al_2O_3 extraction from plasma. This procedure circumvented the need to determine analytical recoveries of the ^{3}H-catecholamines.

Extraction ratios of the ^{3}H-catecholamines across the systemic circulation (ER_s) were obtained from the ratio of central-venous (C_v) to arterial (C_a) plasma concentrations [i.e., $ER_s = 1 - (C_v/C_a)$]. Pulmonary extraction ratios (ER_p) for i.v. infused unlabelled dopamine and endogenous adrenaline were calculated from the ratio C_a/C_v [i.e., $ER_p = 1 - (C_a/C_v)$] as described by Halbrügge et al. (1988b).

Statistics

Results are presented as arithmetic means ± SEM or together with their 95% confidence limits in parentheses. Multiple comparisons were carried out by analysis of variance modified for repeated measures followed by Scheffe's test. To test the significance of desipramine-induced changes, the U-test (according to Wilcoxon, Mann, and Whitney) was used. Differences were taken to be statistically significant when $P < 0.05$.

Substances used in this study

Saffan® ampoules (Opopharma, Zürich, Switzerland); desipramine HCl (Sigma, Deisenhofen, FRG); (−)-noradrenaline bitartrate, (−)-adrenaline, dopamine HCl, and (±)-isoprenaline HCl (Serva, Heidelberg, FRG); ^{3}H-7-(−)-noradrenaline (NET-377, 13 Ci mmol^{-1}), ^{3}H-(N-Methyl)-(−)-adrenaline (NET-623, 78 Ci mmol^{-1}), ^{3}H-7-dopamine (NET-131, 37 Ci mmol^{-1}) (NEN, Dreieich, FRG); ^{3}H-7-(±)-isoprenaline (TRK 295, 7 Ci mmol^{-1}) (Amersham, Braunschweig, FRG).

Results and discussion

In the sixteen rabbits used in this study, baseline (preinfusion) heart rate and mean arterial pressure were 267 ± 6 beats min^{-1} and 64 ± 3 mm Hg, respectively. No time-dependent change in either parameter was observed in separate control experiments (n = 6). Neither the low-dose infusion of unlabelled dopamine nor desipramine caused changes in heart rate or mean arterial pressure.

The results concerning the control values of total body plasma clearance (Cl) and the systemic extraction ratio (ER_s) of the four catecholamines are shown in Table 1. Dopamine exhibited the highest value of Cl and ER_s, with both values being significantly different from those obtained for noradrenaline, adrenaline, and isoprenaline. Cl and ER_s values did not show further differences between the four catecholamines. The Cl values reported here for noradrenaline and adrenaline are similar to those obtained by Medvedev et al. (1990). The present study also confirms our previous results concerning the extent of noradrenaline extraction across the systemic circulation of the rabbit (Halbrügge et al., 1988b), where ER_s was found to be 0.650.

Inhibition of neuronal uptake by desipramine caused a reduction in noradrenaline clearance by about 20%, but did not affect the clearances of dopamine, adrenaline, and isoprenaline (Table 2). On the other hand, desipramine reduced the ER_s not only of noradrenaline, but also that of dopamine and adrenaline, which, like noradrenaline, are both known to be substrates of neuronal uptake (for review, see Graefe and Bönisch, 1988). It is likely, that the ER_s-diminishing effect of desipramine was large enough to reduce the Cl of noradrenaline, but did not suffice in the case of dopamine and adrenaline to reduce their Cl values. Neither the Cl nor ER_s of isoprena-

Table 1. Steady-state total body clearance (Cl) and systemic extraction ratio (ER_s) of intraarterially infused noradrenaline (NA), dopamine (DA), adrenaline (A), and isoprenaline (ISO) in the anaesthetized rabbit (n = 16)

	NA	DA	A	ISO
Cl ($ml\ kg^{-1}\ min^{-1}$)	59.5 ± 4.8**	116.2 ± 11.4	68.0 ± 6.0**	78.4 ± 7.6**
ER_s	0.709 ± 0.030*	0.789 ± 0.027	0.694 ± 0.035**	0.669 ± 0.046**

Given are arithmetic means ± SEM. Asterisks indicate statistically significant differences when compared with DA: * $P < 0.05$; ** $P < 0.01$

Table 2. Effect of desipramine on the total body clearance (Cl) and the systemic extraction ratio (ER_s) of noradrenaline (NA), dopamine (DA), adrenaline (A), and isoprenaline (ISO) in the anaesthetized rabbit (n = 8)

	NA	DA	A	ISO
Cl	0.819* (0.660; 0.978)	1.163 (0.883; 1.442)	1.114 (0.899; 1.328)	1.259 (0.962; 1.556)
ER_s	0.691** (0.572; 0.810)	0.906* (0.814; 0.998)	0.902* (0.829; 0.975)	0.952 (0.827; 1.077)

Shown are results in the presence of desipramine expressed as a fraction of baseline control values (arithmetic means with 95% confidence limits in parentheses). Asterisks indicate statistically significant effects of desipramine: * $P < 0.05$; ** $P < 0.01$

line was affected by desipramine. This is in agreement with the results of Ludwig et al. (1989) who determined the Cl and ER_s of unlabelled isoprenaline in humans. The present study and the results by Ludwig et al. (1989) confirm earlier observations, that isoprenaline is handled by non-neuronal processes only (Goldstein et al., 1983).

Pulmonary extraction ratios (ER_p) were measured for intravenously infused unlabelled dopamine and endogenous adrenaline. The central-venous baseline plasma levels of *endogenous* dopamine and adrenaline were 0.14 ± 0.02 and 0.93 ± 0.18 $nmol\,l^{-1}$, respectively. Under control conditions a ER_p value for dopamine of 0.235 (0.149; 0.321; $P < 0.01$) was obtained, whereas no pulmonary extraction of adrenaline was observed [$ER_p = -0.015$ (−0.104; 0.073)]. Desipramine significantly reduced (by 84%) the ER_p of dopamine to 0.038 (−0.012; 0.087). These results on the pulmonary extraction of dopamine closely resemble results reported by Bryan et al. (1989) who showed an 83% inhibition by cocaine (10 $\mu mol\,l^{-1}$) of dopamine uptake by the isolated perfused rat lung. This desipramine-sensitive extraction process has all the properties of neuronal uptake and is believed to be non-neuronal in nature (Bryan et al., 1988; Bakhle and Vane, 1974).

References

Bakhle YS, Vane JR (1974) Pharmacokinetic function of the pulmonary circulation. Physiol Rev 54:1007–1045

Bryan LJ, O'Donnell SR, Westwood NN (1988) The uptake process for catecholamines in endothelial cells in rat perfused lungs is the same as uptake$_1$ in noradrenergic neurones. Br J Pharmacol 95:539P

Bryan LJ, O'Donnell SR, Westwood NN (1989) Further evidence that the uptake process for catecholamines in pulmonary endothelial cells is the same as uptake$_1$ in noradrenergic neurones. Clin Exp Pharmacol Physiol [Suppl] 14:27

Esler M, Jackman G, Leonhard P, Skews H, Bobik A, Korner P (1981) Effect of norepinephrine uptake blockers on norepinephrine kinetics. Clin Pharmacol Ther 29:12–20

Goldstein DS, Horwitz D, Keiser HR, Polinsky RJ, Kopin IJ (1983) Plasma l-^{3}H-norepinephrine, d-^{14}C-norepinephrine, and d,l-^{3}H-isoproterenol kinetics in essential hypertension. J Clin Invest 72:1748–1758

Graefe K-H, Bönisch H (1988) Neuronal transport of amines across the axonal membrane of noradrenergic and dopaminergic neurones. In: Trendelenburg U, Weiner N (eds) Handbook of experimental pharmacology, vol 90 I, Catecholamines I. Springer, Berlin Heidelberg New York Tokyo, pp 193–245

Halbrügge T, Gerhardt T, Ludwig J, Heidbreder E, Graefe K-H (1988a) Assay of catecholamines and dihydroxyphenylethyleneglycol in human plasma and its application in orthostasis and mental stress. Life Sci 43:19–26

Halbrügge T, Ungell A-L, Wölfel R, Graefe K-H (1988b) Total body, systemic and pulmonary clearance and fractional extraction of unlabelled and differently ^{3}H-labelled noradrenaline in the anaesthetized rabbit. Naunyn-Schmiedebergs Arch Pharmacol 338:361–367

Halbrügge T, Wölfel R, Graefe K-H (1989) Plasma 3,4-dihydroxyphenylglycol as a tool to assess the role of neuronal uptake in the anaesthetized rabbit. Naunyn-Schmiedebergs Arch Pharmacol 340:726–732

Ludwig J, Halbrügge T, Vey G, Walter J, Graefe K-H (1989) Haemodynamics as a determinant of the pharmacokinetics of and the plasma catecholamine responses to isoprenaline. Eur J Clin Pharmacol 37:493–500

Medvedev OS, Esler MD, Angus JA, Cox HS, Eisenhofer G (1990) Simultaneous determination of the plasma noradrenaline and adrenaline kinetics. Naunyn-Schmiedebergs Arch Pharmacol 341:192–199

Szabo B, Hedler L, Starke K (1989) Peripheral presynaptic and central effects of clonidine, yohimbine and rauwolscine on the sympathetic nervous system in rabbits. Naunyn-Schmiedebergs Arch Pharmacol 340:648–657

Authors' address: Dr. T. Halbrügge, Institut für Pharmakologie der Universität Würzburg, Versbacher Strasse 9, D-8700 Würzburg, Federal Republic of Germany

J Neural Transm (1990) [Suppl] 32: 437–440

Human Caki-1 cells are the first model for extraneuronal transport of noradrenaline (uptake$_2$) which is based on a clonal cell line*

E. Schömig, J. Babin-Ebell, C.-L. Schönfeld, H. Russ, and **U. Trendelenburg**

Department of Pharmacology, University of Würzburg, Federal Republic of Germany

Summary. The neurotransmitter noradrenaline is inactivated by active transport out of the synaptic cleft – either back into the adrenergic neuron or into extraneuronal cells. Transport studies on isolated cells provide many advantages. However, an experimental model for the extraneuronal uptake of noradrenaline which is based on a clonal cell line was not known until now. The human renal carcinoma cell line Caki-1 is the first clonal cell line known to express the extraneuronal transport system for noradrenaline.

Introduction

Inactivation of the neurotransmitter noradrenaline which has been released into the synaptic cleft results from active transport into adjacent cells. Two clearly distinct uptake mechanisms for noradrenaline have been identified. (1) the desipramine-sensitive neuronal transport system (uptake$_1$) and (2) the corticosterone-sensitive extraneuronal transport system (uptake$_2$). In the past, studies on uptake$_2$ were based on isolated organs such as the isolated incubated rabbit aorta, the rat salivery gland, and the isolated perfused rat heart (for review see Gillespie, 1976; Trendelenburg, 1988). A clonal cell line with uptake$_2$ was not known until now. However, a model for uptake$_2$ which is based on a clonal cell line would facilitate work in this field significantly. Transport experiments on isolated cells are not complicated by heterogeneous cell material, by distribution of substrate in the extracellular space, and by concentration gradients within the tissue. We therefore tested various cell lines for the occurrence of uptake$_2$.

* Supported by the Deutsche Forschungsgemeinschaft (SFB176, Scho373) and the Dr. Robert Pfleger Stiftung.

Material and methods

Several clonal cell lines (A10, rat vascular smooth muscle; H9c2, rat myoblasts; S49 mouse lymphocytes; OK, opossum kidney; Caki-1, human kidney) were grown in surface culture on plastic flasks (Falcon 175 cm^2, Becton Dickinson, Heidelberg, FRG) in a humified atmosphere (5% CO_2). These cell lines were chosen because they are derived from extraneuronal tissues which are innervated by adrenergic neurones (except S49 lymphocytes). The culture medium was composed either of Dulbecco's Modified Eagle Medium, Minimum Essential Medium, or RPMI and contained 10% fetal calf serum (all from Gibco, Eggenstein, FRG). For the experiments, the cells were seeded on plastic culture dishes (Ø 60 mm, Nunc, Roskilde, Denmark). The metabolizing enzymes, monoamine oxidase and catechol-O-methyl transferase were inhibited by 10 μmol/l pargyline and 10 μmol/l U-0521 to prevent intracellular metabolism of noradrenaline. After incubation with ^{3}H-noradrenaline, intracellular ^{3}H-noradrenaline was measured by scintillation counting. For more detail see Schömig and Schönfeld (1990).

Results and discussion

The cells were incubated with 10 nmol/l ^{3}H-noradrenaline at 37 °C for 15 min in the absence of any inhibitor or in the presence of either 10 μmol/l corticosterone or 100 μmol/l O-methylisoprenaline. Corticosterone- and O-methylisoprenaline are known inhibitors of uptake$_2$. We were not able to demonstrate either corticosterone-sensitive or O-methylisoprenaline-sensitive uptake of ^{3}H-noradrenaline in A10, H9c2, S49, and OK cells.

However, in Caki-1 cells there was marked corticosterone and O-methylisoprenaline-sensitive uptake of ^{3}H-noradrenaline. After an incubation period of 15 min the accumulation of ^{3}H-noradrenaline amounted to 0.89 ± 0.03 pmol/(mg protein) ($n = 47$) in the absence and to 0.22 ± 0.09 pmol/(mg protein) in the presence of 100 μmol/l O-methylisoprenaline. The fraction of total uptake which was inhibited by 100 μmol/l O-methylisoprenaline was defined as specific. Specific accumulation of ^{3}H-noradrenaline increases linearly with time for at least 20 min (Schömig and Schönfeld, 1990). Hence, initial rates of uptake were determined after 15 min of incubation.

Several lines of evidence support the hypothesis that Caki-1 cells express uptake$_2$ (Schömig and Schöfeld, 1990). (1) Initial rates of specific ^{3}H-noradrenaline uptake were markedly inhibited by the uptake$_2$ inhibitors corticosterone (1 μmol/l) and O-methylisoprenaline (10 μmol/l) but not by the uptake$_1$ inhibitors desipramine and cocaine. (2) Uptake$_1$ is known to depend absolutely on extracellular sodium and chloride (Graefe et al., 1984; Sanchez Armass and Orrego, 1977). On the other hand, uptake$_2$ depends only modestly on the concentration of extracellular sodium and not at all on the concentration of extracellular chloride (Bönisch et al., 1985). Indeed, replacement of 75 mmol/l sodium by lithium during 20 min of preincubation and 15 min of incubation diminished ^{3}H-noradrenaline uptake in Caki-1 cells by about 25%. Replacement of 75 mmol/l chloride by isethionate failed to affect ^{3}H-noradrenaline uptake. (3) Depolarization is known to inhibit uptake$_2$ (Bönisch et al., 1985). Depolarization of the Caki-1 cells by 50 mmol/l potassium (which replaced 50 mmol/l sodium) reduced ^{3}H-noradrenaline uptake in Caki-1 cells by about 50%. (4) There is a highly significant positive correlation between the potency of various compounds to inhibit ^{3}H-noradrenaline uptake in Caki-1 cells and to inhibit uptake$_2$ in a model system for uptake$_2$, namely the isolated incubated rabbit aorta (Fig. 1). However, on an average, the compounds inhibited ^{3}H-noradrenaline uptake about seven times more potently in Caki-1 cells than in rabbit aorta. This deviation is probably due to a steady-state concentration gradient within the isolated organ. Near to the surface, the concentration of ^{3}H-noradrenaline may be higher than in the center of the tissue because of limited diffusion combined with an active uptake$_2$ which removes ^{3}H-noradrenaline from the extracellular space. A concentration gradient of the substrate within a tissue may decrease the apparent potency for inhibitors of enzymatic processes as has been shown by Greene (1976) with elegant model calculations.

Interestingly enough, uptake$_2$ in Caki-1 cells and rabbit aorta is inhibited by cimetidine, quinidine and procainamide which are substrates of the renal transport mechanism for organic cations (Fig. 1). These results indicate

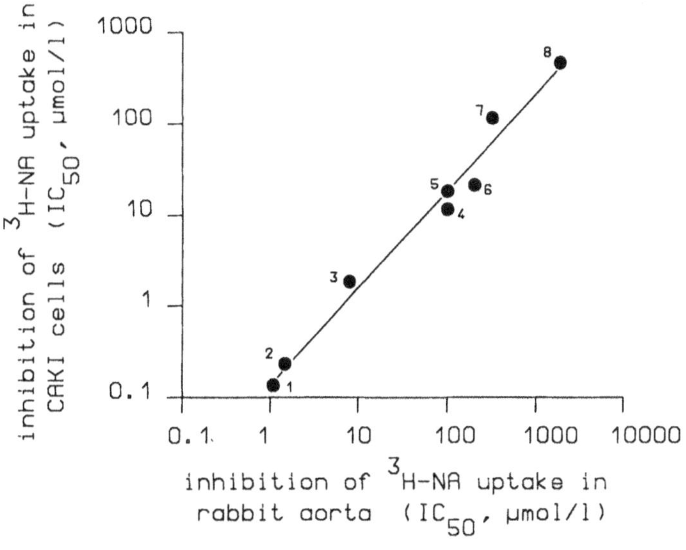

Fig. 1. Correlation between the IC_{50}'s of various compounds for the inhibition of ^{3}H-noradrenaline uptake in Caki-1 cells and the IC_{50}'s for the inhibition of uptake$_2$ in the isolated incubated rabbit aorta. Each symbol represents a compound: corticosterone (*1*), papaverine (*2*), O-methylisoprenaline (*3*), cimetidine (*4*), clonidine (*5*), quinidine (*6*), procainamide (*7*), (−)noradrenaline (*8*). The positive correlation is highly significant ($r = 0.9914$; $n = 8$; $p < 0.001$). The slope of the regression line is 1.06. Data are taken from Schömig and Schönfeld (1990)

at least a striking similarity between uptake$_2$ and the renal transport mechanism for organic cations.

Caki-1 cells are the first model for uptake$_2$ which is based on a clonal cell line. Caki-1 cells can be expected to facilitate functional studies on uptake$_2$ significantly and, moreover, may serve as primary material for molecular characterization of the uptake$_2$ carrier.

References

Bönisch H, Bryan LJ, Henseling M, O'Donnell SR, Stockmann P, Trendelenburg U (1985) The effect of various ions on uptake$_2$ of catecholamines. Naunyn-Schmiedebergs Arch Pharmacol 328:407–416

Gillespie JS (1976) Extraneuronal uptake of catecholamines in smooth muscle and connective tissue. In: Paton DM (ed) The mechanism of neuronal and extraneuronal transport of catecholamines. Raven Press, New York, pp 325–354

Graefe K-H, Zeitner C-J, Fuchs G, Keller B (1984) Role played by sodium in the membrane transport of ^{3}H-noradrenaline across the axonal membrane of adrenergic neurones. In: Fleming WW, Graefe K-H, Langer SZ, Weiner N (eds) Neuronal and extraneuronal events in autonomic pharmacology. Raven Press, New York, pp 51–62

Greene (1976) The kinetics of enzyme action and inhibition in intact tissues and tissue slices, with special reference to cholinesterase. J Pharm Pharmacol 28:265–274

Sanchez Armass S, Orrego F (1977) A major role for chloride in ^{3}H-noradrenaline transport by rat heart adrenergic nerves. Life Sci 20:1829–1838

Schömig E, Schönfeld CL (1990) Extraneuronal noradrenaline transport (uptake$_2$) in a clonal cell line (Caki-1 cells). Naunyn-Schmiedebergs Arch Pharmacol 341:404–410

Trendelenburg U (1988) The extraneuronal uptake and metabolism of catecholamines. In: Trendelenburg U, Weiner N (eds) Handb exp pharmacol 90/I:279–319

Authors' address: Dr. E. Schömig, Department of Pharmacology, University of Würzburg, Versbacher Strasse 9, D-8700 Würzburg, Federal Republic of Germany.

J Neural Transm (1990) [Suppl] 32: 441–445

The synaptic noradrenaline concentration in humans as estimated from simultaneous measurements of plasma noradrenaline and dihydroxyphenylglycol (DOPEG)

J. Ludwig, M. Gerlich, T. Halbrügge, and **K.-H. Graefe**

Medizinische Poliklinik and Institut für Pharmakologie, University of Würzburg, Würzburg, Federal Republic of Germany

Summary. A linear relationship between plasma DOPEG (ordinate) and plasma noradrenaline (NA; abscissa) was found during orthostasis as well as during NA infusion. The slope of the former ($slope_R$) was 13 times steeper than that of the latter ($slope_I$). This difference in slopes suggests that the NA concentration in the synaptic cleft is markedly higher than in plasma. The factor (F) characterizing this concentration difference was obtained from $F^2 = slope_R/slope_I$. It amounted to 3.6.

Introduction

In many isolated tissues DOPEG, a primary deaminated metabolite of NA, is formed mainly within sympathetic nerves (Kopin, 1985). There are two sources of neuronal DOPEG formation. One source originates from NA leaking out of the storage vesicles, the other from NA being recaptured subsequent to its release (Graefe and Henseling, 1983). While the former source of DOPEG formation is abolished after blockade of neuronal uptake (e.g., by desipramine), the latter is not. In-vivo studies have shown that the DOPEG content in plasma is predominantly presynaptic in origin (Eisenhofer et al., 1988; Goldstein et al., 1988; Ludwig et al., 1988).

Similar absolute increases in plasma NA and DOPEG were observed in response to sympathetic activation (Goldstein et al., 1988; Ludwig et al., 1988), whereas during NA infusion the increase in plasma DOPEG relative to that in plasma NA was much smaller (Goldstein et al., 1988). Halbrügge et al. (1989) analysed this finding in more detail in the rabbit. They calculated a more than 3-fold NA concentration difference between synaptic cleft and plasma from the difference between the plasma DOPEG responses to NA release and NA infusion. Their experimental approach can best be

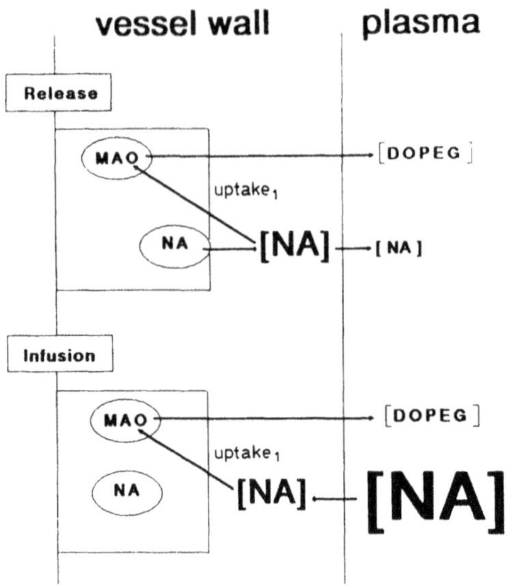

Fig. 1. Schematic representation showing the neuronal source of increments in plasma DOPEG during NA release (upper panel) and during NA infusion (lower panel). It is concluded that the NA concentration gradient between plasma and synaptic cleft during NA release is the reciprocal of that during NA infusion

understood by considering Fig. 1. The primary aim of the present study was to determine as to whether there is a similar NA concentration difference in humans.

Materials and methods

Thirteen healthy volunteers (10 males and 3 females) gave their written informed consent to the study procedure which was approved by the Faculty Ethics Committee.

On study day 1, subjects rested first for 30 min in supine position, were then asked to sit quietly for 30 min and, thereafter, to stand quietly for further 30 min. Forearm venous blood was sampled at the end of each 30-min period. On study day 2, after 30 min of rest, NA was infused for two consecutive 30-min periods. The infusion rates were 0.43 and 0.86 nmol/(kg min) in the 1st and 2nd infusion period, respectively. Mean arterial blood pressure and heart rate were recorded at 5-min intervals. Femoral arterial and forearm venous blood was sampled at the end of each 30-min period. In addition, venous blood was sampled 30 min after the end of the 2nd infusion period (washout period). Plasma levels of NA and DOPEG were determined by HPLC and electrochemical detection. Results are given as arithmetic means ($\pm$ SEM) or as geometric means with 95% confidence limits. Slopes of regression lines were calculated by the method of least squares.

Results

Graded orthostasis induced increases in plasma NA and DOPEG (data not shown). The slope of the regression line ($slope_R$) relating plasma DOPEG (ordinate) to plasma NA (abscissa) was 1.02 (0.81; 1.295).

The plasma concentrations of NA and DOPEG obtained on study day 2 are given in Table 1. There was no difference between the venous (C_v) and arterial (C_a) plasma concentration of NA at the end of the resting period, indicating that there was no NA net extraction in the forearm under this condition. In contrast, during NA infusion the forearm fractional extraction of NA ($ER = 1 - C_v/C_a$) was 0.542 (± 0.046) and 0.544 (± 0.045) during the 1st and 2nd infusion period, respectively. The NA clearance from arterial plasma (CL_a) was 37.1 (± 2.84) ml/(kg min) during the 1st and 30.9 (± 2.06) ml/(kg min) during the 2nd period ($P < 0.02$). Table 1 shows that there was no significant difference between C_v and C_a for DOPEG. The venous plasma level of DOPEG at the end of the washout period was above baseline ($P < 0.05$). Since, during NA infusion, the arterial plasma NA concentration is the relevant substrate concentration for DOPEG formation, the arterial data were used to relate plasma DOPEG to plasma NA. The mean slope ($slope_I$) of this relation was 0.078 (0.067; 0.091).

The concentration difference for NA between plasma and synaptic cleft was calculated from the following equation: $slope_R/F = slope_I\ F$ (Halbrügge et al., 1989), where F is the factor by which the NA concentration in the synaptic cleft exceeds that in plasma. F was found to be 3.6 (3.1; 4.18).

Table 1. Plasma concentrations of noradrenaline (NA) and dihydroxyphenylglycol (DOPEG) before, during and after NA infusion in 13 healthy subjects

	NA (nmol/l)	DOPEG (nmol/l)
$C_a(0)$	1.16 (0.90; 1.50)	4.45 (3.93; 5.03)
$C_a(1)$	13.36 (11.73; 15.20)	5.10 (4.55; 5.71)
$C_a(2)$	29.94 (26.40; 33.96)	6.76 (6.12; 7.47)
$C_v(0)$	1.19 (0.98; 1.45)	4.68 (4.00; 5.48)
$C_v(1)$	5.73 (4.53; 7.24)	5.10 (4.50; 5.78)
$C_v(2)$	12.64 (9.86; 16.20)	6.20 (5.59; 6.87)
$C_v(w)$	1.19 (1.01; 1.39)	5.28 (4.73; 5.89)

Given are geometric means with their 95% confidence limits. C_a arterial plasma concentration, C_v venous plasma concentration. (0) = end of resting period; (1) = at 30 min during the 1st NA infusion (0.43 nmol kg^{-1} min^{-1}); (2) = at 30 min during the 2nd NA infusion (0.86 nmol kg^{-1} min^{-1}); (w) = 30 min after the end of the 2nd NA infusion (washout period)

Discussion

Since DOPEG is formed presynaptically, it was assumed that, irrespective of whether NA is infused or released, any given increase in the synaptic NA concentration causes identical increases in plasma DOPEG. On the basis of this assumption, the factor can be estimated by which the NA concentration at neuroeffector junctions exceeds that in plasma. The results indicate a factor of 3.6. Hence, plasma concentrations of endogenous NA have to be multiplied by 3.6 to give average cleft NA concentrations. Interestingly, our result matches exactly that obtained by Goldstein et al. (1986) who estimated synaptic NA levels in humans from measurements of blood pressure and arterial plasma NA responses to NA infusion and yohimbine-induced increases in NA release. Moreover, our results are in close agreement with those obtained in the anaesthetized rabbit (Halbrügge et al., 1989) and in the pithed rat (Kopin et al., 1984).

However, it is open to discussion whether the method presented here permits the precise determination of the synaptic NA levels. As heart rate decreases during NA infusion, sympathetic activity is probably reduced via the baroreceptor. As a consequence, baseline plasma DOPEG is expected to decrease. On the other hand, part of the infused NA is likely to enter storage vesicles after its neuronal uptake and, hence, augment their NA content. Thus, the part of plasma DOPEG originating from vesicular NA leakage inevitably will increase. However, both sources of error may cancel each other.

References

Eisenhofer G, Goldstein DS, Ropchak TG, Nguyen HQ, Keiser HR, Kopin IJ (1988) Source and physiological significance of plasma 3,4-dihydroxyphenylglycol and 3-methoxy-4-hydroxyphenylglycol alanine in the rat. J Auton Nerv Syst 24:1–14

Goldstein DS, Zimlichmann R, Stull R, Keiser HR, Kopin IJ (1986) Estimation of intrasynaptic norepinephrine concentrations in humans. Hypertension 8:471–475

Goldstein DS, Eisenhofer G, Stull R, Folio CJ, Keiser HR, Kopin IJ (1988) Plasma dihydroxyphenylglycol and the intraneuronal disposition of norepinephrine in humans. J Clin Invest 81:213–220

Graefe K-H, Henseling M (1983) Neuronal and extraneuronal uptake and metabolism of catecholamines. Gen Pharmacol 14:27–33

Halbrügge T, Wölfel R, Graefe K-H (1989) Plasma 3,4-dihydroxyphenylglycol as a tool to assess the role of neuronal uptake in the anaesthetized rabbit. Naunyn-Schmiedebergs Arch Pharmacol 340:726–732

Kopin IJ, Zukowska-Grojec Z, Bayorh MA, Goldstein DS (1984) Estimation of intrasynaptic norepinephrine concentrations at vascular neuroeffector junctions in vivo. Naunyn-Schmiedebergs Arch Pharmacol 325:298–305

Kopin IJ (1985) Catecholamine metabolism: basic aspects and clinical significance. Pharmacol Rev 37:333–364

Ludwig J, Gerhardt T, Halbrügge T, Walter J, Graefe K-H (1988) Plasma concentrations of noradrenaline and 3,4-dihydroxyphenylethyleneglycol under conditions of enhanced sympathetic activity. Eur J Clin Pharmacol 35:261–267

Authors' address: Dr. J. Ludwig, Medizinische Poliklinik der Universität, Klinikstrasse 6-8, D-8700 Würzburg, Federal Republic of Germany.

J Neural Transm (1990) [Suppl] 32: 447–456

Inhibition of MAO activity, ^{3}H-imipramine binding, ^{3}H-paroxetine binding and ^{3}H-5-HT uptake by human cerebrospinal fluid

T. Egashira, S. Goto, F. Murayama, and **Y. Yamanaka**

Department of Pharmacology, Medical College of Oita, Japan

Summary. Addition of small amount of human cerebrospinal fluid (CSF) inhibited both types of MAO in monkey brain mitochondria. The specific binding of ^{3}H-paroxetine decreased remarkably with increasing CSF volumes, while ^{3}H-imipramine binding was slightly inhibited. Scatchard analysis of ^{3}H-paroxetine binding in the presence and absence of CSF indicated that the inhibitory effect was associated with a decreased B_{max} without an appreciable change in Kd. Addition of CSF induced an inhibition of uncompetitive ^{3}H-5-HT uptake to monkey cerebral membranes. These results indicate that the materials in human CSF inhibit ^{3}H-paroxetine binding, and modulate the uptake system for 5-HT.

Introduction

Since the possible physiological relationship between platelet monoamine oxidase (MAO) activity and some psychiatric diseases was reported (Miron et al., 1980; Davidson et al., 1980; Oreland et al., 1981), several endogenous MAO modulators have been detected in the human plasma (Berrttini and Vogel, 1978; Yu and Boulton, 1979; Giambalvo, 1984). In addition a further putative endogenous MAO inhibitor has been isolated from human cerebrospinal fluid (CSF) (Becker et al., 1983). The materials in dog CSF also possess the action to activate the ^{3}H-imipramine binding and ^{14}C-5-HT uptake as well as MAO inhibition (Egashira et al., 1986, 1987). While, there are many reports on the possible existence of the endogenous inhibitors of ^{3}H-imipramine binding and ^{3}H-5-HT uptake (Barbaccia et al., 1983; Langer et al., 1984; Rehavi et al., 1985; Barkai et al., 1986; Abraham et al., 1987).

These observations have prompted us to search for an endogenous psychotic drug-like substances in CSF. The present report shows that human CSF which inhibit the MAO activity, also inhibit the uptake of

^{3}H-5-HT and the binding of ^{3}H-imipramine as well as ^{3}H-paroxetine to monkey brain. We now report the existence of endogenous MAO inhibiting factors in human CSF and demonstrate that these factors are associated with the neuronal 5-HT uptake transporter.

Materials and methods

The human cerebrospinal fluid (CSF) samples obtained from normal and patients in the course of routine diagnostic lumbar punctures. Monkey was anesthetized with ketalar, given 30 mg/kg s.c., and the brain was quickly removed after withdrawing blood. The monkey brain and human CSF samples were stored at − 30 °C until used.

MAO activity

Monkey brain mitochondrial fractions were prepared by differential centrifugation as described earlier (Egashira et al., 1984), the mitochondria suspended in 0.32 M sucrose were used as the enzyme preparations for determination of MAO activity. MAO activity was estimated by radioisotopic assay with ^{14}C-5-HT (substrate for type A MAO, final concentration at 200 μM) and ^{14}C-β-phenylethylamine (β-PEA) (substrate for type B MAO, final concentration at 50 μM) as described earlier (Egashira et al., 1987).

Receptor binding assay

The monkey cerebral cortex was homogenized in 25 vol. of ice-cold buffer (50 mM Tris-HCl containing 100 mM NaCl and 5 mM KCl, pH 7.4). P_2 fractions obtained by centrifugation of this homogenate were used as the crude membrane preparations for the receptors binding. ^{3}H-imipramine binding to monkey cortical membrane preparations was performed according to the method of Kinnier et al. (1981). Desipramine at 10 μM final concentration was used to determine non-specific binding. ^{3}H-paroxetine binding to monkey cortical membrane preparations using fluoxetine at 10 μM final concentration to assess non-specific binding was performed according to the method of Habert et al. (1985). Aliquots of crude membrane suspension were incubated at 0 °C for 60 min (imipramine binding) or at 22 °C for 180 min (paroxetine binding). The incubation was terminated by rapid filtration of the membrane suspensions under reduced pressure through Whatman GF/B fiber glass filters and each filter was rapidly washed three times with 5 ml of ice-cold 50 mM Tris-HCl buffer. Then, the filters were dried and the radioactivities were determined in Triton X-100-toluene scintillation fluid in a liquid scintillation spectrometer.

5-HT uptake

The uptake procedure of ^{3}H-5HT into the P_2 fraction was carried out by the method of Barbaccia et al. (1983). The monkey cerebral cortex was homogenized in 9 vol. of 0.32 M sucrose and the P_2 fraction was obtained by differential centrifugation. The P_2 fractions were incubated for 2 min at 37 °C (for total uptake) or 0 °C (for non-specific uptake) with ^{3}H-5-HT and Krebs bicarbonate buffer. The incubation was terminated by rapid filtration of the samples under reduced pressure through GF/B glass fiber filters and each filter was washed three times with 5 ml of ice-cold saline. Then, the filters were dried and the radioactivities present were counted by liquid scintillation spectrometry after adding Triton X-100-toluene scintillation fluid.

Protein determination

Protein concentrations were determined by the method of Lowry et al. (1951) with bovine serum albumin as the standard.

Results

1. Effects of human CSF on MAO activity in monkey brain mitochondria

The effect of adding varying amounts of human CSF to monkey brain mitochondrial MAO in vitro was studied using 5-HT and β-PEA as substrates. When the brain mitochondria were incubated with increasing amounts of CSF, the increases in the inhibition of MAO activity with 5-HT and β-PEA were observed (Fig. 1A). Addition of 50 µl CSF inhibited MAO activity about 45% and 20% with 5-HT and β-PEA as substrates, respectively. A kinetic study of the inhibition of MAO by human CSF addition was investigated with Lineweaver-Burk double reciprocal plots. The inhibition was competitive at least toward 5-HT, while with β-PEA as substrate, the inhibition was non-competitive (Fig. 2).

2. Effect of human CSF on high affinity ^{3}H-imipramine binding to monkey cortical membranes

The effects of adding varying amounts of human CSF on high affinity ^{3}H-imipramine binding were studied. When P_2 fractions were incubated with increasing amounts of CSF, the slight inhibition in the binding of ^{3}H-imipramine was observed (Fig. 1B); the high affinity binding of ^{3}H-imipramine decreased about 30% in the presence of 50 µl CSF. The specific binding of ^{3}H-imipramine to monkey cortical membranes was saturable in the presence and absence of CSF. Scatchard analysis of specific high affinity ^{3}H-imipramine binding indicated a single population of binding sites with an apparent dissociation constant, K_d, of 4.16 ± 1.15 nM with 50 µl CSF and 4.46 ± 2.70 nM without CSF. The maximal concentration of binding sites (B_{max}) was estimated to be 279 ± 111 fmol/mg with 50 µl CSF and 305 ± 99 fmol/mg without CSF (Table 1).

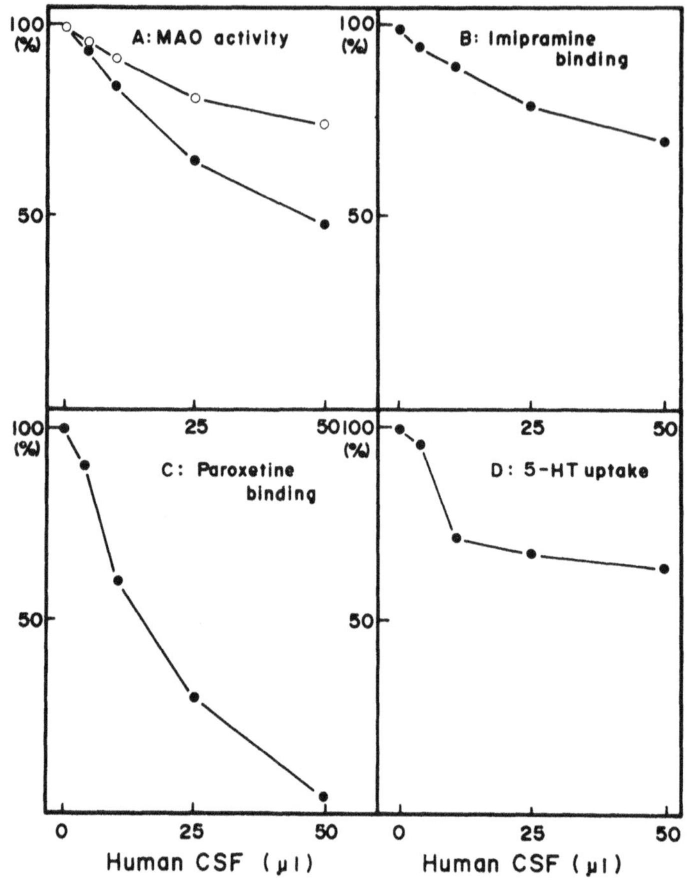

Fig. 1. Effects of adding of human CSF on monkey brain MAO activity, ^{3}H-imipramine binding, ^{3}H-paroxetine binding and ^{3}H-5-HT uptake. **A** MAO activities to 200 μM 5-HT(●——●) and 50 μM β-PEA (○——○) as substrates were measured with addition of human CSF at 37 °C for 20 min. The control values for MAO activity were 0.98 nmole/min/mg protein for 5-HT and 1.12 nmole/min/mg protein for β-PEA, respectively. **B** Specific binding was assayed at a final ^{3}H-imipramine concentration of 0.77 nM in the presence of different volumes of human CSF. Membrane suspensions were incubated at 0 °C for 60 min. Control specific binding was 0.045 pmol/mg of monkey cortical membranes. **C** Specific binding was assayed at a final ^{3}H-paroxetine concentration of 0.05 nM in the presence of different volumes of human CSF. Membrane suspensions were incubated at 22 °C for 180 min. Control specific binding was 0.024 pmol/mg of monkey cortical membranes. **D** 5-HT uptake was assayed with 0.25 μM ^{3}H-5-HT in the presence of different volumes of human CSF and the incubation was carried out at 37 °C for 2 min. Control 5-HT uptake activity was 4.64 pmol/min/mg of monkey cerebral cortex. Each value represents the mean percentages of the control values in triplicate experiments

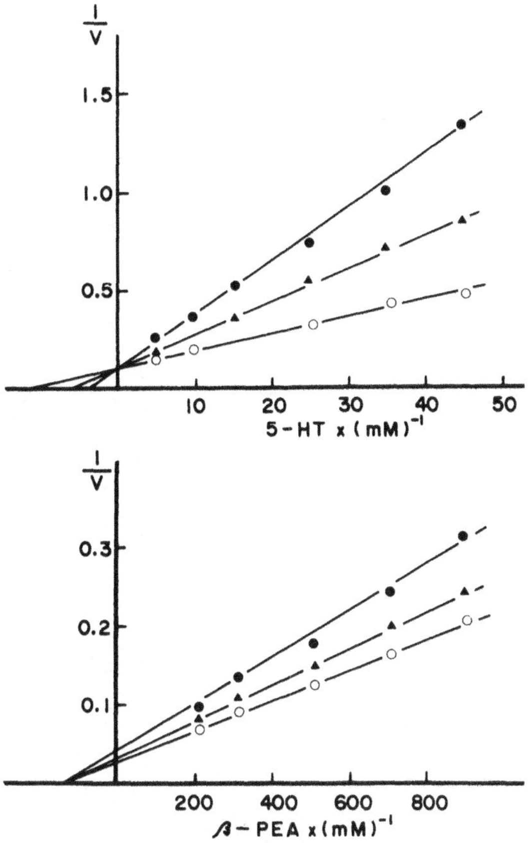

Fig. 2. Effect of human CSF on MAO activity in monkey brain mitochondria. Lineweaver-Burk plots of the reciprocal of the initial velocity of 5-HT and β-PEA oxidation against the reciprocal of the substrate concentration in the presence of human CSF. Abscissa: 1/substrate concentration in mM, Ordinate: 1/initial velocity in nmole/min/mg of protein. Substrate used were 5-HT (upper) and β-PEA (bottom) assay in the absence (○——○) and presence (▲—▲, ●—●) of 25 and 50 μl human CSF. Each point represent the mean MAO activity assayed in triplicate experiments

3. Effect of human CSF on ^{3}H-paroxetine binding to monkey cortical membranes

When different volumes of human CSF were added to the incubation mixtures containing 50 pM of ^{3}H-paroxetine, ^{3}H-paroxetine binding was strongly displaced with increasing amounts of CSF; the binding of ^{3}H-paroxetine decreased about 90% of the control in the presence of 50 μl CSF (Fig. 1C). Specific ^{3}H-paroxetine binding was assayed at different ligand concentrations (13 to 570 pM) in the presence and absence of CSF. Scatchard analysis of specific ^{3}H-paroxetine binding indicated a single population of binding site with an apparent K_d of 327 ± 274 pM with CSF

Table 1. Effects of human CSF on ^{3}H-imipramine binding, ^{3}H-paroxetine binding and ^{3}H-5-HT uptake

	Imipramine binding B_{max}	K_d	Paroxetine binding B_{max}	K_d	5-HT uptake V_{max}	K_m
Control	305 ± 99	4.46 ± 2.7	220 ± 71	625 ± 338	5.84 ± 0.24	0.42 ± 0.04
With CSF	279 ± 111	4.16 ± 1.1	109 ± 33*	327 ± 274	4.00 ± 0.76*	0.17 ± 0.01**

Values are means ± SE for three separate preparations. 50 µl CSF was used in this experiment. ^{3}H-imipramine binding: B_{max} = fmol/mg protein, K_d = nM. ^{3}H-paroxetine binding: B_{max} = fmol/mg protein, K_d = pM. ^{3}H-5-HT uptake: V_{max} = pmol/min/mg protein, K_m = µM. * $p < 0.05$ and ** $p < 0.01$ compared to control value (Student's t test)

Specific binding represents total binding minus binding in the presence of 10 µM desipramine or 10 µM fluoxetine, at free imipramine or paroxetine concentrations ranging from 0.4 to 13.2 nM or 13 to 570 pM, in the absence and presence of human CSF, respectively. Each B_{max} and K_d was estimated by Scatchard analysis of specific binding.

Kinetic analysis of ^{3}H-5-HT uptake into monkey cerebral cortex and the effect of human CSF were determined over the concentration range of 50 to 750 nM at 37 °C

and 625 ± 338 pM without CSF. B_{max} was estimated to be 109 ± 33 fmol/mg with CSF and 220 ± 71 fmol/mg without CSF (Table 1).

4. *Effect of human CSF on ^{3}H-5-HT uptake into monkey cerebral cortex*

Addition of human CSF induced an inhibition of ^{3}H-5-HT uptake (Fig. 1D). Kinetic analysis of ^{3}H-5-HT uptake into monkey cerebral cortex and the effects of human CSF were determined. The uptake was concentration dependent and exhibited apparent saturation ($K_m = 0.42 \pm 0.04\,\mu M$ and $V_{max} = 5.84 \pm 0.24$ pmol/min/mg). Human CSF 50 µl linhibited the uptake uncompetitively with an apparent K_m of $0.17 \pm 0.01\,\mu M$ and a V_{max} of 1.00 ± 0.76 pmol/min/mg.

Discussion

There have been many reports on the possible presence of MAO modulators in human plasma (Yu and Boulton, 1979; Giambalvo, 1984), urine (Glover et al., 1980) and CSF (Becker et al., 1983). We also have recently demonstrated that addition of dog CSF to the dog brain mitochondria inhibited A-form MAO competitively at least toward 5-HT, but was non-competitive with B-form MAO (Egashira et al., 1986). In this study, human CSF also possessed the inhibitory action of MAO in monkey brain mitochondria. The present results agree with the finding of Becker et al. (1983), although they have used the rat brain mitochondria as the enzyme preparation. While some endogenous MAO modulators have been found in CSF, but there have been few studies on their possible role as physiological regulators of the neurotransmission in CNS.

We reported that the materials in dog CSF induced a marked activation of ^{3}H-imipramine binding and ^{14}C-5-HT uptake and indicated that the materials in dog CSF may represent endogenous lithium-like substances (Egashira et al., 1987). In contrast, the materials in the human CSF inhibited the ^{3}H-imipramine and ^{3}H-paroxetine binding and ^{3}H-5-HT uptake. It may be concluded that human CSF contain structurally different types of the materials which act like MAO inhibitor drugs.

It is wellknown that antidepressive agents are potent inhibitor of serotonin uptake (Carlsson et al., 1968; Lindbrink et al., 1971; Ross et al., 1975). Using a radiolabelled form of the antidepresant imipramine, a specific high affinity binding sites in the neuronal membranes has been identified (Raisman et al., 1979). This recognition site labeled by ^{3}H-imipramine has been shown to be distinct but allostericall coupled to the 5-HT transporter (Briley et al., 1982; Barbaccia et al., 1983; Wennogle and Meyerson, 1985).

Recently, it has been reported that paroxetine is a more specific and more potent inhibitor of 5-HT uptake than imipramine, which renders it a better candidate for labelling of the 5-HT transport system (Magnussen et al., 1982; Mellerup et al., 1984; Habert et al., 1985). Moreover, Mellerup et al. (1985) have found that the binding sites for ^{3}H-imipramine and ^{3}H-paroxetine are located on different parts of the macromolecular complex.

Using the specific 5-HT uptake inhibitor, fluoxetine or desipramine at a final concentration of 10 µM to define non-specific binding, the present study demonstrates specific, saturable binding of ^{3}H-paroxetine or ^{3}H-imipramine to monkey cortical membranes in a concentration dependent manner. The comparison of ^{3}H-imipramine binding and ^{3}H-paroxetine binding to monkey cortical membranes showed that the values of B_{max} and K_d for ^{3}H-imipramine binding are higher than these for ^{3}H-paroxetine binding. These results indicate that the number of binding sites for ^{3}H-paroxetine and ^{3}H-imipramine are also located in monkey cortical membranes and are closely associated with the 5-HT transport system.

Assuming the ^{3}H-imipramine or ^{3}H-paroxetine binding site may function as a regulator of the 5-HT uptake system, the presence of endogenous effectors of this site is feasible. Recently, several laboratories have searched for endogenous 5-HT transport modulators. A few low molecular weight materials have been detected or partially purified from either human plasma (Angel and Paul, 1984; Brusov et al., 1985; Abraham et al., 1987) or from rat brain (Barbaccia et al., 1983; Rehavi et al., 1985). The present results appear to be in good agreement with the finding of Barkai et al. (1986) that direct addition of plasma to ^{3}H-imipramine binding receptors assay produced a significant inhibition of binding. ^{3}H-paroxetine binding to monkey cortical membranes was also potently inhibited by the materials in human CSF, although it showed a weak inhibition on ^{3}H-imipramine binding. It was revealed that the antagonism by human CSF to ^{3}H-paroxetine binding was due to a decreased V_{max} with no significant change in K_d. Moreover, the presence of the materials in human CSF resulted in decreased total ^{3}H-5-HT uptake uncompetitively. These results indicate that the materials in human CSF most probably act directly at the recognition site labeled by ^{3}H-paroxetine, but not ^{3}H-imipramine recognition sites. Occupation of the site by these materials is likely to modulate the 5-HT uptake process allosterically. Further studies are clearly needed to investigate the role of the materials in human CSF on the inhibitions of ^{3}H-imipramine or ^{3}H-paroxetine binding, ^{3}H-5-HT uptake and MAO activity.

References

Abraham KI, Ieni JR, Meyerson LR (1987) Purification and properties of a human plasma endogenous modulator for the platelet tricyclic binding/serotonin transport complex. Biochim Biophys Acta 923:8–21

Angel I, Paul SM (1984) Inhibition of synaptosomal 5-^{3}H-hydroxytryptamine uptake by endogenous factor(s) in human blood. FEBS Lett 171:280–284

Barbaccia ML, Gandolfi O, Chuang DH, Costa E (1983) Modulation of neuronal serotonin uptake by a putative endogenous ligand of imipramine recognition sites. Proc Natl Acad Sci USA 80:5134–5138

Barkai AI, Baron M, Kowalik S, Cooper B (1986) Inhibition of ^{3}H-imipramine binding by plasma from depressed and normal subjects. Psychiatry Res 17: 261–267

Becker RE, Giambalvo CT, Fox RA, Macho M (1983) Endogenous inhibitors of monoamine oxidase present in human cerebrospinal fluid. Science 221:476–478

Berrettini WH, Vogel WH (1978) Evidence for an endogenous inhibitor of platelet MAO in chronic schizophrenia. Am J Psychiatry 135:605–607

Briley M, Langer SZ, Sette M (1982) Allosteric interaction between the ^{3}H-imipramine binding site and the serotonin uptake mechanism. Br J Pharmacol 74:817–818

Brusov OS, Fomenko AM, Katasonov AB (1985) Human plasma inhibitors of platelet serotonin uptake and imipramine receptor binding: extraction and heterogeneity. Biol Psychiatry 20:235–244

Carlsson A, Fuxe K, Ungerstedt U (1968) The effect of imipramine on central 5-hydroxytryptamine neurons. J Pharm Pharmacol 20:150–151

Davidson JRT, Mcleod MN, Turnbull CD, White HL, Feuer EJ (1980) Platelet monoamine oxidase activity and the classification of depression. Arch Gen Psychiatry 37:771–773

Egashira T, Yamamoto T, Yamanaka Y (1984) Some interrelated properties of A and B form monoamine oxidase in monkey brain mitochondria. Jpn J Pharmacol 34:327–334

Egashira T, Takano R, Yamanaka Y (1986) Demonstration of endogenous inhibitors of monoamine oxidase in dog cerebrospinal fluid. Jpn J Pharmacol 42:583–586

Egashira T, Takano R, Yamanaka Y (1987) Modulation of neuronal MAO activity, 5-HT uptake and imipramine binding by endogenous substances in dog cerebrospinal fluid. Biochem Pharmacol 36:1781–1785

Giambalvo CT (1984) Purification of endogenous modulators of monoamine oxidase from plasma. Biochem Pharmacol 33:3929–3932

Glover V, Reveley MA, Sandler M (1980) A monoamine oxidase inhibitor in human urine. Biochem Pharmacol 29:467–470

Habert E, Graham D, Tahraoui L, Claustre Y, Langer SZ (1985) Characterization of ^{3}H-paroxetine binding to rat cortical membranes. Eur J Pharmacol 118: 107–114

Kinnier WJ, Chuang DM, Gwynn G, Cost E (1981) Characteristics and regulation of high affinity ^{3}H-imipramine binding to rat hippocampal membranes. Neuropharmacology 20:411–419

Langer SZ, Raisman R, Tahraoui L, Scatton B, Niddam R, Lee CR, Claustre Y (1984) Substituted tetrahydro-β-carbolines are possible candidates as endogenous ligand of the ^{3}H-imipramine recognition site. Eur J Pharmacol 98: 153–154

Lindbrink P, Jonsson G, Fuxe K (1971) The effect of imipramine like drugs and antihistamine drugs on uptake mechanisms in the central noradrenaline and 5-hydroxytryptamine neurons. Neuropharmacology 10:521–536

Lowry OH, Rosebrough NJ, Farr AL, Randal RJ (1951) Protein measurement with folin phenol reagent. J Biol Chem 193:265–275

Magnussen I, Tonderk K, Engback F (1982) Paroxetine, a potent selective long-acting inhibitor of synaptosomal 5-HT uptake in mice. J Neural Transm 55:217–226

Mellerup ET, Plenge P, Engelstoft M (1984) High affinity binding of ^{3}H-paroxetine and ^{3}H-imipramine to human platelet membranes. Eur J Pharmacol 96:303–309

Mellerup ET, Plenge P, Nielsen M (1985) Size determination of binding polymers for ^{3}H-imipramine and ^{3}H-paroxetine in human platelet membranes. Eur J Pharmacol 106:411–413

Miron B, Levitt M, Perlman R (1980) Low platelet monoamine oxidase activity: a possible biochemical correlate of borderline schizophrenia. Psychiatry Res 3:329–336

Oreland L, Wiberg A, Askerg A, Traskman L, Sjostand L, Thoren P, Bertilsson L, Tybring G (1981) Platelet MAO activity and monoamine metabolites in cerebrospinal fluid in depressed and succidal patients and healthy controls. Psychiatry Res 4:21–30

Raisman R, Briley M, Langer SZ (1979) Specific tricyclic antidepressant binding sites in rat brain. Nature 281:148–150

Rehavi M, Ventura I, Sarne Y (1985) Demonstration of endogenous imipramine like material in rat brain. Life Sci 36:687–693

Ross SB, Renyi AL (1975) Tricyclic antidepressant agents I. Comparison of the inhibition of the uptake of ^{3}H-noradrenaline and ^{14}C-5-hydroxytryptamine in slices and crude synaptosome preparations of the midbrain, hypothalamus regions of the rat brain. Acta Pharmacol Toxicol 36:382–394

Wennogle LP, Meyerson LR (1985) Serotonin uptake inhibitors differentially modulate high affinity imipramine dissociation in human platelet membranes. Life Sci 36:1541–1550

Yu PH, Boulton AA (1977) Activation of platelet monoamine oxidase by plasma in the human. Life Sci 25:31–36

Authors' address: Dr. T. Egashira, Department of Pharmacology, Medical College of Oita, 1-1, Idaigaoka, Hazama-cho, Oita 879-56, Japan.

J Neural Transm (1990) [Suppl] 32: 457–461

α_2-Adrenoceptor responsivity in depression: effect of chronic treatment with moclobemide, a selective MAO-A-inhibitor, versus maprotiline

K. P. Lesch, G. Laux, and **T. Mueller**

Department of Psychiatry, University of Würzburg, Federal Republic of Germany

Summary. The effect of chronic treatment with the selective and reversible MAO-A-inhibitor moclobemide (MOC) vs. the norepinephrine reuptake inhibitor maprotiline (MAP) on α_2-adrenoceptor responsivity was studied by clonidine (CLON)-evoked growth hormone (GH) release in major depressive disorder. Compared to matched controls the depressed patients showed attenuated CLON-induced GH responses before treatment with MOC or MAP. Chronic treatment with both MOC and MAP significantly improved the depressive symptomatology. Although a trend toward increased GH responses to CLON was demonstrated after treatment in both groups, neither MOC nor MAP had a complete effect on restoration of α_2-adrenoceptor responsivity. No difference in insulin-like growth factor I (IGF-I) plasma concentrations before and after treatment with MOC or MAP was found. Our results support the view that antidepressants with different mechanisms of action may be capable of restoring α_2-adrenoceptor function during recovery from a major depressive episode.

Introduction

In 1980, Matussek et al. reported that growth hormone (GH) responses to the α_2-adrenoceptor agonist clonidine (CLON) are impaired in patients with endogenous depression. Furthermore, the effect of chronic antidepressant drug treatment was evaluated by several investigators. While Charney et al. (1982) reported that treatment with the norepinephrine (NE) reuptake inhibitor desipramine and the nonselective NE/serotonin reuptake inhibitor amitriptyline did not alter GH responses to CLON in either treatment responders or nonresponders, Checkley et al. (1981) found an enhanced CLON-induced GH response (although not significant) in 5 out of 6 depressed patients. Also no significant difference in α_2-adrenoceptor responsivity in 10 patients with depression was reported between before and during chronic treatment with the MAO-inhibitor clorgyline (Siever et al., 1982). Finally, the GH response to CLON in drug-free recovered patients with a major depressive disorder was attenuated although a trend toward normalization was seen.

Recently, our group suggested that peripheral mechanisms (e.g. glucocorticoid-dependent modulation of the receptor-G-protein-effector system complex, insulin-like growth factor I [IGF-I]-dependent feedback regulation) may be involved in the attenuation of neuroendocrine secretion following pharmacologic challenges (Lesch et al., 1988, 1990; Lesch and Beckmann, 1990). In order to further investigate the effect of four weeks of antidepressant treatment with the reversible and selective MAO-A-inhibitor moclobemide (MOC) versus the tetracyclic NE reuptake inhibitor maprotiline (MAP) on α_2-adrenoceptor sensitivity, the CLON-evoked GH release and plasma IGF-I concentrations were studied in patients with a major depressive disorder before and after chronic treatment (Erb et al., 1988).

Subjects and methods

CLON challenge tests were conducted in ten patients (8 women, 2 men; age, 45 ± 13.3 [± SD] years) with a major depressive disorder (DSM-III-R and RDC criteria; Hamilton Depression Rating Scale [HDRS] score, 26 ± 2.2) before and after treatment, with 300 mg/day MOC for four weeks, in 13 depressed patients (9 women, 4 men; age, 48.4 ± 14.7 years; HDRS, 28.9 ± 1.4) before and treatment with 150 mg/day MAP for 4 weeks, and in 15 healthy subjects matched for age, gender, body mass index and ovarian status.

The CLON challenge tests were conducted on all subjects at rest in bed at 8 a.m. after an overnight fast. Thirty minutes after insertion of an i.v. line in an antecubital vein, 2 µg/kg of the selective α_2-adrenoceptor agonist CLON, diluted in 10 ml saline (0.9%), was injected slowly over 10 min. For measurement of GH, blood was collected at −15, 0, 15, 30, 45, 60, 90 and 120 min into heparinized tubes, centrifuged, and stored at −80 °C until analysis. IGF-I concentrations were determined in −15 min EDTA samples. GH and IGF-I plasma concentrations were measured using standard RIA techniques. The GH responses following CLON of patients and controls were calculated as the net area under the curve (AUC), using trapezoidal integration. The data were analyzed using nonparametric statistical methods; group comparisons by Mann-Whitney U-test, and correlations by Spearman's rank order correlation. All significance levels are two-tailed.

Results

Compared with the controls, both groups of patients with depression exhibited attenuated GH hormone responses to CLON either before treatment with MOC (392 ± 111 vs 95 ± 39 ng·min/ml, $p < 0.01$) or before treatment with MAP (534 ± 190 vs 124 ± 72 ng·min/ml, $p < 0.01$). After treatment the mean HRSD scores were significantly improved in both groups (MOC, 26.4 ± 2.2 vs 10.3 ± 1.9, $p < 0.001$; MAP, 28.9 ± 1.4 vs 11.3 ± 2.0, $p < 0.001$). Although a trend toward increased GH responses to CLON was demonstrated, neither treatment with MOC (288 ± 132 ng·min/ml; $p < 0.1$) nor treatment with MAP (277 ± 106 ng·min/ml; $p < 0.1$) were capable of completely restoring α_2-adrenoceptor responsivity despite clinical remission of depressive symptomatology. No correlation was found between CLON-induced GH release and severity of depression. While plasma IGF-I concentrations were increased in both groups of depressed patients as compared to controls (MOC, 1.39 ± 0.40 vs 0.76 ± 0.31 U/ml $p < 0.05$; MAP, 131 ± 0.30 vs 0.69 ± 0.26 U/ml, $p < 0.05$), no difference was seen between pre- and posttreatment levels (MOC, 1.37 ± 0.17 U/ml, ns; MAP, 1.73 ± 0.47 U/ml, ns).

Discussion

Despite clinical recovery, neither MOC nor MAP had a complete effect on restoration of α_2-adrenoceptor responsivity, although a tendency towards increased GH responses to CLON was demonstrated after both antidepressant treatment regimen. Since α_2-adrenoceptors are located both pre- and postsynaptically on noradrenergic terminals and target cells, their activation should exert complex effects on noradrenergic neurotransmission. Hence, the stimulation of postsynaptic binding sites would facilitate noradrenergic

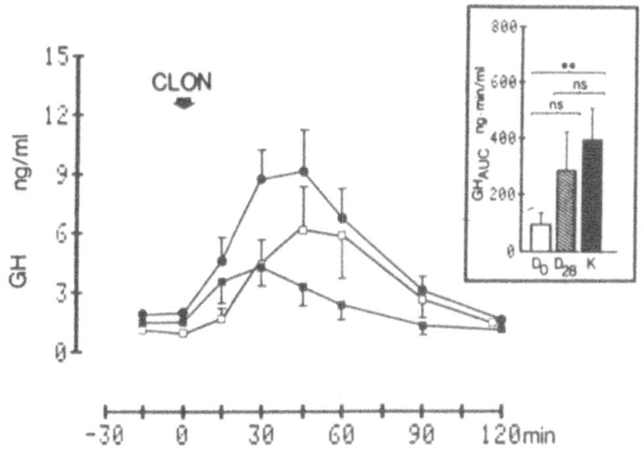

Fig. 1. Growth hormone (GH) responses to clonidine (CLON) in patients with depression ($n = 10$) before (■, D_0) and after treatment with moclobemide for four weeks (□, D_{28}) and in matched controls (●, K). **$p < 0.01$ (Mann-Whitney U-test, two-tailed)

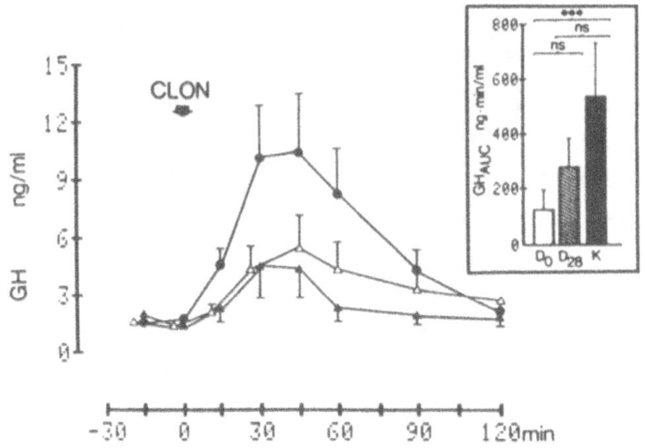

Fig. 2. GH responses to CLON in depressed patients (n = 13) before (▲, D_0) and after treatment with maprotiline for 4 weeks (△, D_{28}) and in matched controls (●, K). *** $p < 0.001$

transmission, while activation of presynaptic receptors should suppress the release of NE. Indeed, data in the literature support this view, since CLON has been shown to induce complex effects involving pre- and/or postsynaptic α_2-adrenergic receptors (Lesch et al., 1989). Therefore, the interpretation of attenuated CLON-induced GH release as evidence of a defective central postsynaptic α_2-adrenoceptor-related function remains limited. Consequently, our study fails to answer the questions regarding the significance or blunted GH responses to CLON in patients with depression before treatment, the relevance of peripheral feedback mechanisms and the role of α_2-adrenoceptor responsivity in the mechanism of action of different antidepressant drug treatment strategies. In conclusion, our findings support the hypothesis of a α_2-adrenoceptor/postreceptor effector system-related dysfunction in depression but the pathophysiological significance of the interrelationship of receptor responsivity and neuroendocrine abnormalities which are frequently associated with depression remains to be elucidated in detail.

References

Charney DS, Heninger GR, Sternberg DE (1982) Failure of chronic antidepressant treatment to alter growth hormone response to clonidine. Psychiatry Res 7:135–138

Checkley SA, Stade AP, Shur E, Dawling S (1981) A pilot study of the mechanism of action of desipramine. Br J Psychiatry 138:248–251

Erb A, Laux G, Lesch KP (1988) Effects of moclobemide and maprotiline on hypothalamic-pituitary-somatotropic (HPS) axis function in depression. Psychopharmacology 96 [Suppl]: 279

Lesch KP, Laux G, Erb A, Pfueller H, Beckmann H (1988) Growth hormone (GH) response to GH-releasing hormone in depression: correlation with GH release following clonidine. Psychiatry Res 25:301–310

Lesch KP, Rupprecht R (1989) Psychoneuroendocrine research in depression. II. Hormonal responses to releasing hormones as a probe for hypothalamic-pituitary-endorgan dysfunction. J Neural Transm 75:179–194

Lesch KP, Mayer S, Disselkamp-Tietze J, Hoh A, Wiesmann M, Osterheider M, Schulte HM (1990) 5-HT_{1A} receptor responsivity in unipolar depression: evaluation of ipsapirone-induced ACTH and cortisol secretion in patients and controls. Biol Psychiatry (in press)

Lesch KP, Beckmann H (1990) Zur Serotonin-Hypothese der Depression. Fortschr Neurol Psychiat (in press)

Matussek N, Ackenheil M, Hippius H, Mueller F, Schroeder HS, Schultes A, Wasilewski B (1980) Effect of clonidine on growth hormone release in psychiatric patients and controls. Psychiatry Res 2:25–36

Siever LJ, Uhde TW, Insel TR, Roy BF, Murphy DL (1982) Growth hormone response to clonidine unchanged by chronic clorgyline treatment. Psychiatry Res 7:139–144

Authors' address: Dr. K. P. Lesch, Laboratory of Clinical Science NIMH, NIH Clinical Center 10/3D41, Bethesda, MD 20892, USA

J Neural Transm (1990) [Suppl] 32: 463–467

Supersensitivity to catecholamines after inhibition of extraneuronal uptake (uptake$_2$) or O-methylation

J. Proença[1], **M. Q. Paiva**[2], and **S. Guimarães**[2]

Department of Pharmacology, Faculties of [1] Pharmacy and [2] Medicine, University of Porto, Portugal

Summary. The influence of inhibition of uptake$_2$ and of O-methylation on β-adrenoceptor-mediated responses of guinea-pig trachea and rat uterus to some catecholamines was compared. The responses were more markedly enhanced by inhibition of O-methylation (by U-0521) than by inhibition of uptake$_2$ (by hydrocortisone). The ranking order for the degree of supersensitivity caused by either U-0521 or hydrocortisone was different in the two tissues, suggesting differences between the O-methylating systems of those tissues. In experiments with ^{3}H-isoprenaline metabolism, hydrocortisone had a weak inhibitory effect on O-methylation in both tissues.

Introduction

Whenever a "site of loss" for an agonist is able to generate a concentration gradient from the incubation medium to the extracellular space, inhibition of the "site of loss" increases the concentration of this agonist in the biophase, causing supersensitivity to it (for reviews see Trendelenburg, 1972, 1988).

Uptake$_2$ followed by O-methylation by catechol-O-methyl transferase (COMT) is an important factor determining the active concentration of sympathomimetic agonists, particularly at β-adrenoceptor level (Guimarães and Paiva, 1981a and b; for a review see Guimarães, 1982). Inhibition of this system increases the amine concentration at the biophase, thus increasing the magnitude and the duration of the response.

As shown by several authors, in the guinea-pig trachea the O-methylating system is more effective in modulating β_2- than β_1-adrenoceptor-mediated responses to isoprenaline (Paiva and Guimarães, 1976; Bryan et al., 1981; Proença et al., 1988). On the other hand, inhibition of this non-neuronal "site of loss" can be achieved by inhibition of either the first (uptake) or the second step (enzyme) (Guimarães et al., 1985; Guimarães and Trendelenburg, 1985).

The present study was undertaken to compare the influence of inhibition of uptake$_2$ and of O-methylation on β-adrenoceptor-mediated responses to several catecholamines of guinea-pig trachea and rat uterus, two tissues where β_2-adrenoceptors predominate.

Materials and methods

Spirally cut *guinea-pig tracheas* prepared according to Constantine (1965) were contracted by carbachol (0.3 μmol/l) to allow determinations of concentration-response curves for the relaxation caused by sympathomimetic amines. Two concentration-response curves were determined per preparation: the first in the absence and the second in the presence of U-0521 (an inhibitor of COMT) or hydrocortisone (an inhibitor of uptake$_2$). The *rat uterus* were obtained from animals 3 months age (250–300 g) during estrus determined by the microscopic examination of vaginal smears. After having been dissected out and prepared, the uterus was contracted by 48 mmol/l KCl. Then, two concentration-response curves for the relaxant action of either catecholamine were determined per preparation: the first in the absence and the second in the presence of U-0521 (or hydrocortisone). To block α-adrenoceptors, 7 μmol/l phentolamine was present in the medium.

Results

Experiments on the guinea-pig trachea

Both inhibition of uptake$_2$ and O-methylation caused supersensitivity to the catecholamines used, although the supersensitivity caused by inhibition of O-methylation exceeded that caused by inhibition of uptake$_2$ (Table 1). As Table 1 shows, the degree of supersensitivity caused by inhibition of both uptake$_2$ and O-methylation varied according to the amine.

The influence of inhibition of uptake$_2$ and of O-methylation on the metabolism of ^{3}H-(−)-isoprenaline was also compared. While inhibition of uptake$_2$ by hydrocortisone reduced by $25.2 \pm 7.9\%$ ($n = 4$) the amount of ^{3}H-O-methylated isoprenaline (^{3}H-OMI) formed during 30 min from 0.1 μmol/l ^{3}H-(−)-isoprenaline, inhibition of O-methylation by U-0521 reduced the formation of ^{3}H-OMI by $95.2 \pm 3.7\%$ ($n = 4$).

Experiments on the rat uterus

As for the guinea-pig trachea, the responses to all the agonists were enhanced by either U-0521 or hydrocortisone (Table 2). Again, for all the agonists, the supersensitivity caused by U-0521 was higher than that caused by hydrocortisone. The

Table 1. Influence of U-0521 (60 μmol/l) or hydrocortisone (200 μmol/l) on the sensitivity of guinea-pig trachea to sympathomimetic amines. The tracheal muscle was previously contracted by 0.3 μmol/l carbachol in the presence of 12 μmol/l cocaine (to inhibit neuronal uptake)

Amine	EC_{50} (control)	F_1	EC_{50} (control)	F_2
(−)-Isoprenaline (n = 6)	10.5 nmol/l (8.5–13.0)	3.9 (2.5–5.0)	9.1 nmol/l (7.5–11.0)	1.9* (1.7–2.1)
(−)-Adrenaline (n = 6)	36.3 nmol/l (28.2–46.9)	2.2 (1.9–2.5)	39.1 nmol/l (24.6–58.2)	1.8 (1.7–2.1)
Dopamine (n = 6)	30.1 μmol/l (21.7–41.6)	2.2 (1.7–2.7)	25.4 μmol/l (20.1–32.1)	1.5* (1.3–1.7)
(±)-Dobutamine (n = 6)	2.3 μmol/l (1.7–3.1)	3.9 (3.6–4.2)	2.3 μmol/l (1.8–2.9)	1.4* (1.3–1.5)

F_1 ratio EC_{50} control/EC_{50} after U-0521; F_2 ratio EC_{50} control/EC_{50} after hydrocortisone. The values represent geometric means and confidence limits.
* Significant different from the corresponding F_1 value

Table 2. Influence of U-0521 (60 μmol/l) or hydrocortisone (200 μmol/l) on the sensitivity of rat uterus to sympathomimetic amines. The uterine muscle was previously contracted by 48 mmol/l KC1 in the presence of 12 μmol/l cocaine (to inhibit neuronal uptake) and of 7 μmol/l phentolamine (to inhibit α-adrenoceptors)

Amine	EC_{50} (control)	F_1	EC_{50} (control)	F_2
(−)-Isoprenaline (n = 6)	(2.2 nmol/l (1.4–3.7)	2.0 (1.4–3.3)	2.4 nmol/l (1.1–4.5)	1.3* (1.0–1.5)
(−)-Adrenaline (n = 6)	12.0 nmol/l (8.8–16.2)	2.4 (1.8–3.0)	14.6 nmol/l	1.4* (1.2–1.6)
Dopamine (n = 6)	12.0 μmol/l (8.5–17.2)	2.1 (1.6–2.7)	15.0 μmol/l (10.8–22.9)	1.7 (1.5–1.9)
(±)-Dobutamine (n = 6)	1.1 μmol/l (1.0–1.2)	4.2 (3.4–4.9)	1.0 μmol/l (0.9–1.1)	2.4* (1.9–3.1)

F_1 ratio EC_{50} control/EC_{50} after U-0521; F_2 ratio EC_{50} control/EC_{50} after hydrocortisone. The values represent geometric means and confidence limits.
* Significant different from the corresponding F_1 value

ranking order for the degree of supersensitivity caused by U-0521 was different from that observed in the guinea-pig trachea.

The influence of inhibition of uptake$_2$ and of O-methylation on the metabolism of ^{3}H-(−)-isoprenaline was also compared. While inhibition of uptake$_2$ by hydrocortisone reduced by 15.2+2.3% (n = 4) the amount of ^{3}H-OMI formed during

30 min from 0.1 μmol/l ^{3}H-(−)-isoprenaline, inhibition of O-methylation by U-0521 reduced by 85.2±9.1% (n = 4) the formation of ^{3}H-OMI.

Discussion

In the present study it was observed that, both in the guinea-pig trachea and rat uterus, the inhibition of COMT by U-0521 practically abolished the formation of ^{3}H-OMI (full inhibition of COMT), while inhibition of uptake$_2$ by hydrocortisone only reduced ^{3}H-OMI formation (partial inhibition of uptake$_2$). These results obtained with isoprenaline explain why β-adrenoceptor-mediated effects caused by this amine (and probably those caused by the other amines used on those tissues) were enhanced more markedly by inhibition of COMT than by inhibition of uptake$_2$.

When one compares the ranking order for the degree of supersensitivity by either U-0521 or hydrocortisone observed in these tissues, it becomes apparent that it is different: in the guinea-pig trachea, U-0521-induced supersensitivity is isoprenaline = dobutamine > dopamine = adrenaline, whereas in the rat uterus the order is: dobutamine > adrenaline > dopamineeisoprenaline; on the other hand, hydrocortisone-induced supersensitivity is: isoprenaline > adrenaline > dopamine > dobutamine in the guinea-pig trachea and dobutamine > dopamine > adrenaline > isoprenaline (a reverse order) in the rat uterus. How to explain this varying ranking order from structure to structure?

There is evidence that suggests that in some tissues there is a membrane-bound form of COMT (Roth, 1980; Raxworthy et al., 1982). In the rabbit aorta the existence of two kinds of COMT, a soluble and a membrane-bound species have been described which preferentially inactivate oestrogens and isoprenaline, respectively (Head et al., 1985; Reid et al., 1986). It is possible that some differences in the distribution of the enzyme species in the two tissues and some differences in the affinity of the catecholamines to them might account for the above referred differences in the ranking order for the degree of supersensitivity.

In the rat heart, OMI formation is totally inhibited by corticosterone (Uhlig et al., 1974). However, in the canine saphenous vein (Azevedo and Osswald, 1976) and in the guinea-pig trachea and rat uterus (present experiments), hydrocortisone only reduced the formation of OMI by 38, 25 and 15% respectively. This interesting variability is probably due to the heterogeneity of extraneuronal system in the different tissues.

References

Azevedo I, Osswald W (1976) Uptake, distribution and metabolism of isoprenaline in the dog saphenous vein. Naunyn-Schmiedebergs Arch Pharmacol 295: 141–147

Bryan LJ, Cole JJ, O'Donnell SR, Wanstall JC (1981) A study designed to explore the hypothesis that beta-1 adrenoceptors are "innervated" receptors and beta-2 adrenoceptors are "hormonal" receptors. J Pharmacol Exp Ther 216:395–400

Constantine JW (1965) "Spirally cut tracheal strip preparation". J Pharm Pharmacol 17:384

Guimarães S (1982) Two adrenergic biophases in blood vessels. Trends Pharmacol Sci 3:159–161

Guimarães S, Paiva MQ (1981a) Two distinct adrenoceptor-biophases in the vasculature: one for α- and the other for β-agonists. Naunyn-Schmiedebergs Arch Pharmacol 316:195–199

Guimarães S, Paiva MQ (1981b) Two distinct biophases for adrenaline released by electrical stimulation or tyramine from the sympathetic nerve endings of the dog saphenous vein. Naunyn-Schmiedebergs Arch Pharmacol 316:200–204

Guimarães S, Trendelenburg U (1985) Deviation supersensitivity and inhibition of saturables sites of loss. Trends Pharmacol Sci 6:371–374

Guimarães S, Paiva MQ, Moura D, Proença J (1985) The saturability of a site of loss and the degree of supersensitivity to agonists which are substrates of this site of loss. Naunyn-Schmiedebergs Arch Pharmacol 329:30–35

Head RJ, Irvine RJ, Barone S, Stitzel RE, De la Lande IS (1985) Non intracellular, cell-associated O-methylation of isoproterenol in the isolated rabbit thoracic aorta. J Pharmacol Exp Ther 324:184–189

Paiva MQ, Guimarães S (1976) Influence of the blockade of extraneuronal uptake of the affinity of some amines for the β-adrenergic receptors of guinea-pig trachea and atria. Arch Pharmacol Toxicol 2:185–193

Proença J, Paiva MQ, Guimarães S (1988) The guinea-pig trachea O-methylating system is more effective in modulating β_2- than β_1-adrenoceptor-mediated responses to isoprenaline. J Pharm Pharmacol 40:76–77

Raxworthy MJ, Gulliver PA, Hughes PJ (1982) The cellular location of catechol-O-methyltransferase in rat liver. Naunyn-Schmiedebergs Arch Pharmacol 320:182–188

Reid JJ, Stitzel RE, Head RJ (1986) Characterization of the O-methylation of catechol oestrogens by intact rabbit thoracic aorta and subcellular fractions thereof. Naunyn-Schmiedebergs Arch Pharmacol 334:17–28

Roth SA (1980) Presence of membrane-bound catechol-O-methyltransferase in human brain. Biochem Pharmacol 29:3119–3122

Trendelenburg U (1972) Factors influencing the concentration of catecholamines. In: Blaschko E, Muscholl E (eds) Catecholamines. Springer, Berlin Heidelberg New York (Handbook of experimental pharmacology, vol 33, pp 726–761)

Trendelenburg U (1988) The extraneuronal uptake and metabolism of catecholamines. In: Trendelenburg U, Weiner N (eds) Catecholamines I. Springer, Berlin Heidelberg New York (Handbook of experimental pharmacology vol 90, pp 279–319)

Uhlig W, Bönisch H, Trendelenburg U (1974) The O-methylation of extraneuronally stored isoprenaline in the perfused rat. Naunyn-Schmiedebergs Arch Pharmacol 283:245–261

Authors' address: Dr. S. Guimarães, Department of Pharmacology, Faculty of Medicine, 4200 Porto, Portugal.

Sulfation

J Neural Transm (1990) [Suppl] 32: 471–479

Urinary dopamine sulfate: regulations and significance in neurological disorders

E. Kienzl[1], **K. Eichinger**[2], **E. Sofic**[3], **K. Jellinger**[1], **P. Riederer**[3], **W. Kuhn**[3], **G. Fuchs**[4], and **G. Laux**[3]

[1] Ludwig Boltzmann Institute of Clinical Neurobiology, Lainz-Hospital, Vienna, and
[2] Institute of Organic Chemistry, Technical University, Vienna, Austria
[3] Clinical Neurochemistry Section, Department of Psychiatry, University of Würzburg School of Medicine, Würzburg, and
[4] Parkinson Clinic, Wohlfahrt, Federal Republic of Germany

Summary. Dopamine-3-O-sulfate (DA-3-O-S) and dopamine-4-O-sulfate (DA-4-O-S) are important end products of L-dopa metabolism. Therefore they may give indications of disturbances in the peripheral metabolism of catecholamines, when measured in urine samples of patients with Parkinson's disease (PD). In addition, information about the reliability of DA sulfatation after L-dopa therapy may be of significance for its role in the elimination of DA from the peripheral nervous system.

Although DA-3-O-S appears to be the predominant sulfo-conjugate in urine, there are no changes in PD nor in depression syndrome compared to controls with or without other neurological disorders. By contrast, DA-4-O-S is significantly decreased in de novo PD subjects. However, a similar reduction is notable in patients with other neurological disorders. In depressed persons the loss of this compound was less pronounced as compared to de novo PD. Treatment with combined L-dopa therapy caused increased excretion of DA-3-O-S, while changes in DA-4-O-S were only marginal. It is concluded that urinary DA-3-O-S cannot be used as marker for PD, while DA-4-O-S is significantly reduced in a variety of neurological disorders and in particular in de novo PD. Further studies are necessary to elucidate its role as possible peripheral marker to distinguish preclinical PD and depression syndrome.

Introduction

The role of brain dopamine (DA) in the pathophysiology of PD is well established. Main products of the dopaminergic pathways are detectable in both blood and urine: via MAO 3,4-dihydroxy-phenylacetic acid (DOPAC) and via Catechol-O-methyl-transferase (COMT) homovanillic acid (HVA). However, the chemical mechanisms of the overall process of conjugation with sulfate and to a lesser extend with glucuronid received relatively little interest. Sulfatation represents a major pathway in the metabolism of catecholamines but its physiological significance remains unresolved. DA sulfate was first considered to be an end metabolite destined for excretion, since sulfatation has long been associated with detoxification mechanisms. The possibility that sulfate esters of DA play a role as metabolic intermediates may be of interest, since, in contrast to free DA, they are present in considerable amounts in plasma and urine (Van Loon, 1983; Yamamoto et al., 1985; Scott and Elchisak, 1983). In humans, 95–99 % of circulating DA are sulfate conjugated, and considerable amounts of DA-sulfates are excreted in urine (Johnson et al., 1988). The ratio of free and total DA in urine has been found to be approximately the same as in plasma (Rutledge and Hoehn, 1973; Westerink and ten Kate, 1986). DA undergoes sulfoconjugation to produce two sulfate esters: DA-3-O-S and DA-4-O-S (Johnson, 1980). Since urinary excretions directly reflect the plasma levels of DA-3-O-S and DA-4-O-S, urinary assays could provide in vivo informations on central nervous system (CNS) functions and dysfunctions, e.g. in neurological disorders (Jenner and Rose, 1973; Bronaugh et al., 1975; Elchisak, 1987; Kienzl and Eichinger, 1988).

Patients and methods

24 h urine samples were obtained from controls and the following groups of patients:

1. Controls, free of neurological, psychiatric or other organic diseases (9 females and 6 males aged 57 ± 2.9 years) (mean ± sem).
2. 16 drug naive patients with PD (11 females, 5 males, age 63.6 ± 2.9 years, showing of early and late onset (range 36 to 77 years) of stage I and II (Hoehn and Yahr, 1967).
3. 19 patients with PD (12 females, 7 males; age 69.9 ± 2.5 years) treated with amantadine sulfate and anticholinergics or the partial dopamine agonist "terguride". The duration of the disease was 5.9 ± 4.4 years. Most were at stage II–III (Hoehn and Yahr).
4. 21 patients with PD, 16 females and 5 males; age 73.1 ± 1.5 years, duration of disease 7.7 ± 4.4 years; Hoehn-Yahr stage II–IV) treated with L-Dopa plus peripherally acting decarboxylase inhibitor benserazide (Madopar® 62, 5 mg/d up to 6 times 125 mg/day) and amantadine sulfate.
5. 11 patients, 4 males, 7 females, aged 60.7 ± 3.3 years, suffering from a variety of other neurological disorders, including multiple sclerosis (n = 3), polyneuropathy (n = 2), cerebral infarction (n = 2), myasthenia gravis, arthritis, cerebral atrophy and Wilsons disease (one case each), treated for their respective disease according to usual strategies.

6. 8 patients (6 females, 2 males, aged 45.1 ± 4.2 years) with major depression diagnosed according to DSM III.

Sample preparation

The specimens were in each case 24 h urines, kept frozen at −30 °C after being acidified to about pH 2 with 25 % hydrochloric acid. For HPLC-analysis 2 ml urine were extracted twice with 1 ml hexane to remove lipids and 100 µl glacial acid was added. Prior to dilution (1:10) with the mobile phase the sample was pressure filtered through a 0.45 µm filter ACRO LC 25 (Gelman Sciences, Ann Arbor, Mi, USA). Prior to injection the samples were diluted once more 1:10 with the mobile phase. No further sample preparation was necessary. All procedures were carried out at room temperature. Because of the stability of the DA-sulfate-isomers time for storage, extraction and analysis could be neglected.

Standard preparation and HPLC-system

Reaction of anhydrous dopamine hydrochloride with sulfuric acid (conc.) resulted to give DA-4-O-S. A mixture of chlorsulfonic acid and pyridine in chloroform was found to be an excellent reagent for the synthesis if a nearly equimolar mixture of DA-3-O-S and DA-4-O-S. The reversed phase HPLC determinations were carried out using a model 5000 solvent delivery module and a detector Coulochem 5100A with a 5010 analytical cell (ESA, Bedford, MA., USA) according the method of Kienzl and Eichinger (1988) (Fig. 1, 2). The 24 h urine excretion of the principal DA metabolites DOPAC and HVA was measured using an established HPLC method (Sofic, 1986).

Statistics

Arithmetic means and SDs were calculated. The results were analyzed with the Mann-Whitney-U-Test.

Results

DA-3-O-sulfate and Da-4-O-sulfate

In control subjects without neurologic disorders the mean amount of each DA-isomer in the 24 h urinary excretion was approximately 7 µM/24 h (n = 13) (Table 1 and Fig. 3). The average excretion of DA-3-O-S in subjects with various neurological disorders or depression was not considerably decreased as compared to controls. Drug naive PD patients showed a markedly decreased excretion of DA-3-O-S (4.4 µM/24 h, n = 16) and DA-4-O-S (1.3 µM/24 h, n = 16). However, the extreme low levels in de novo PD patients did not differ significantly from the decreased DA-4-O-S levels in subjects with other neurological disorders or depression. In PD patients treated with L-DOPA (Madopar®), the urinary DA-3-O-S and DA-4-O-S

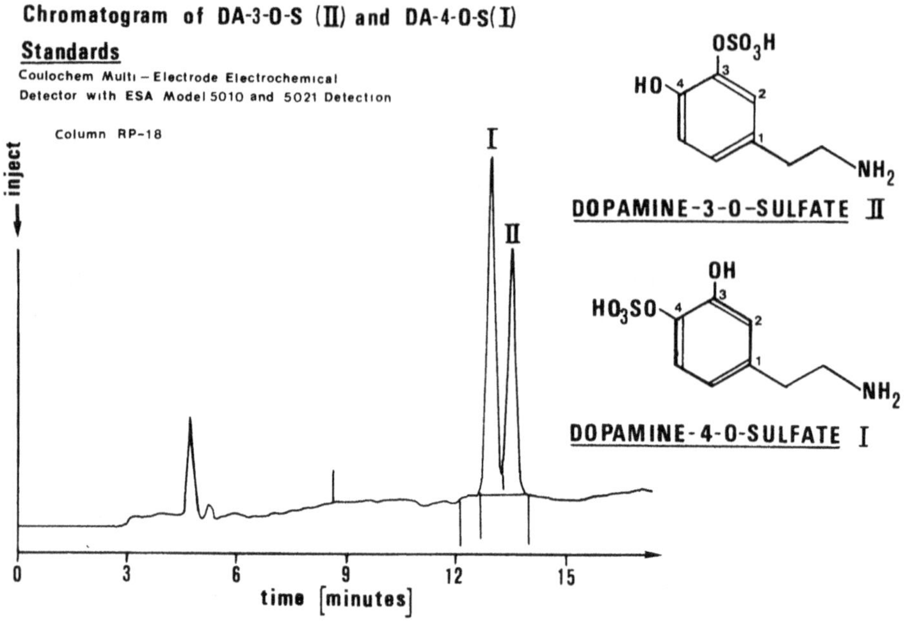

Fig. 1. Chromatogram of the standards of DA-3-O-S and DA-4-O-S. Potentials were +0.50V (conditioning cell), 0.35V (detector 1) and 0.45V (detector 2). Stationary phase was a 250 × 4.6 mm S5 ODS (5 μm) column (Spherisorb, Norwalk, CT, USA). The mobile phase consisted of a solution of 0.8 ml (6 mmol) anhydrous acetic acid, 250 mg Na-1-heptane sulfonate and 80 mg EDTA in 992 ml water (LiChrosolv, Merck)

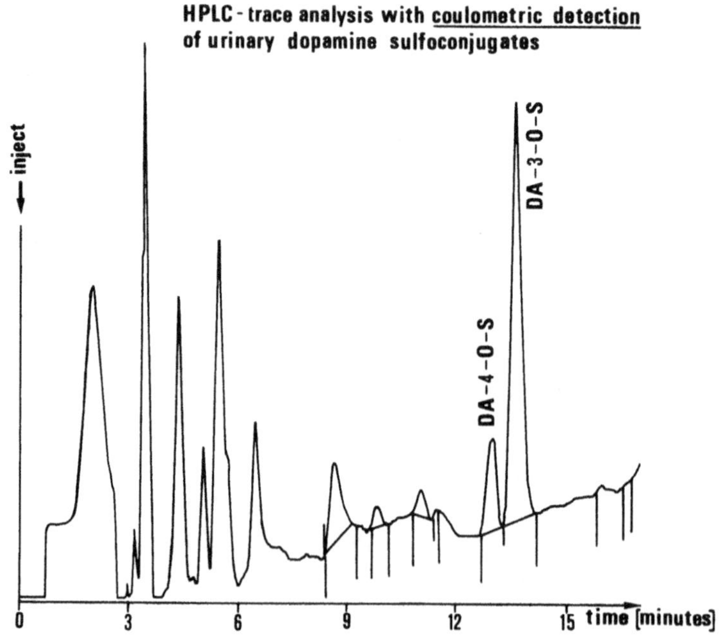

Fig. 2. Typical pattern of DA-3-O-S and DA-4-O-S in the urine of a patient suffering from Parkinson's Disease (dilution 1:100)

Table 1.

Group	n	DA4-O-S µmol/24 h	DA-3-O-S µmol/24 h	DOPAC µmol/24 h	HVA µmol/24 h
CONTROLS	13	7.8 ± 1.6	7.3 ± 1.8	13.6 ± 8.0	27.1 ± 8.0
PD de novo	16	1.3 ± 0.2*	4.4 ± 0.6	4.5 ± 0.8	17.2 ± 1.3
PD no L-DOPA	19	4.6 ± 0.8	18.1 ± 4.0*	47.8 ± 16.0*	52.5 ± 11.6
PD + L-DOPA	21	10.7 ± 1.2	40.5 ± 3.3*	56.7 ± 6.3*	98.9 ± 24.4*
Depression	8	2.7 ± 1.3*	7.2 ± 1.2	4.6 ± 0.8	14.6 ± 2.4*
Div. Neurol. Disorders	11	2.1 ± 0.5*	6.0 ± 1.4	3.03 ± 0.2*	19.4 ± 2.9

Mean ± SD. Statistics: * **p** ≤ 0.01 versus corresponding values in controls

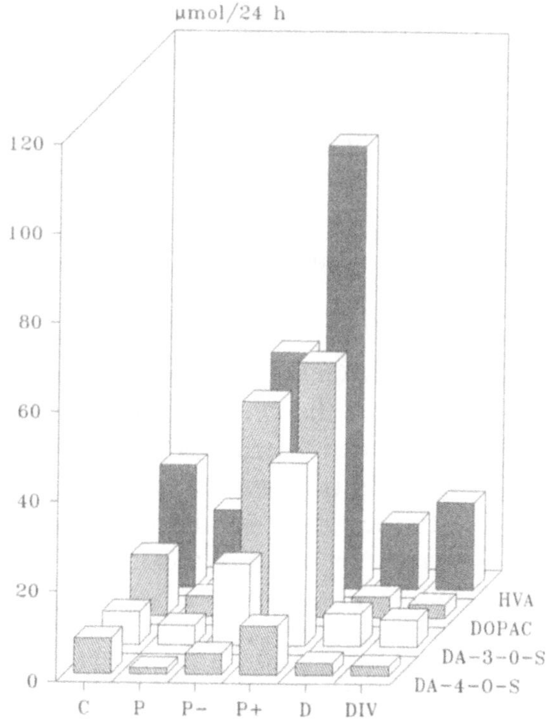

Fig. 3. Summary diagram of the dopamine metabolites in *C* controls, *P* Parkinson's disease (PD) de novo, *P-* PD/no L-DOPA, *P +* PD/+ L-DOPA, *D* depression, *DIV* diverse neurological disorders

excretion was tenfold increased compared to de novo PD patients. In PD patients, treated with amantadine-sulfate in combination with terguride or anticholinergic drugs without L-DOPA, the output of DA-sulfoconjugates was also significantly increased (Table 1, Fig. 3). The predominant isomer of the excreted DA-sulfoconjugate in PD and in all examined neurological disorders is DA-3-O-S.

The ratio DA-4-O-S/DA-3-O-S was calculated to be 1.07 in healthy subjects compared to 0.29 in de novo PD, 0.25 in PD without L-DOPA treatment, 0.26 in PD with L-DOPA treatment, 0.38 in depression and 0.35 in other neurological disorders. Daily fluctuations in the concentrations were observed. In order to assess age effects on conjugated DA, a series of linear correlations were calculated. We found no age dependency in the excretion of DA-sulfates (data not given).

Dopamine metabolites: DOPAC and HVA

Changes in the rate of DA-sulfoconjugation are paralleled with changes in the 24 h levels of DOPAC and HVA. Lowest levels of DOPAC and HVA were observed in de novo PD patients and depression. Also in the different neurological disorders the formation of DOPAC and HVA remained significantly suppressed. MADOPAR®-therapy of PD resulted in a 4-fold increase of DOPAC and HVA excretion. Significant changes were also obtained in the PD group treated with other substances than L-DOPA. In this group the increase of the 24 h DOPAC excretion was 3.5-fold as compared to a 2-fold increase in the 24 h HVA formation.

Discussion

The role of sulfoconjugation of catecholamines in relation to deamination by MAO and O-methylation by COMT in CNS and in the periphery is not clear. Deamination by MAO is the principal pathway for the enzymatic inactivation of DA. Phenol sulfotransferase (PST) contributes up to 15% of total DA inactivation. At concentrations less than 100 μM, soluble COMT contributes less than 5 % to the total catabolism of DA (Rivett et al., 1982). In humans the following pattern is found: Wide distribution of PST in brain, high excretion of DA sulfate in urine of subjects expecially undergoing L-DOPA-therapy and low levels of DA sulfate in brain (Elchisak et al., 1983; Rivett et al., 1984). Moreover, DA-4-O-S is metabolically labile in contrast to DA-3-O-S and can be hydrolyzed by an aryl sulfatase under physiological conditions (Kienzl et al., 1990). In rat brain, increase in DA sulfate was significantly correlated with the degree of MAO inhibition and directly proportional to free DA concentrations (Buu, 1985).

However, the present study reveals that the urinary excretion patterns mirror to some extend the full spectrum of free Da released in individual neurological disorders. There is evidence that central dopaminergic mechanisms appear to be involved in the regulation of the peripheral sympathetic innervation (Jellinger, 1989):

1. It is suggested that PD is a generalized disorder not limited to distinct CNS areas, and that impairment of the dopaminergic nigro-striatal system may involve the tyrosine hydroxylase (TH), the key-enzyme of the catecholaminergic system, also in the adrenal medulla, thus inducing disorders of the peripheral sympathetic system (Barbeau, 1969; Riederer et al., 1978). Animal experiments indicate that an increase in adrenal TH activity is associated with stimulation of central dopamine receptors (Quick and Sourkes, 1977).

2. Bacopoulos (1979) provided an experimental evidence that a significant portion of circulating DA metabolites originates from brain. Alterations in DOPAC and HVA concentrations of dopaminergically innervated brain regions induced by nigrostriatal pathway lesions in rats resulted in parallel changes in plasma concentrations.

3. Drug-induced alterations of DA-concentrations in the periphery may reflect neutrotransmitter mechanisms which regulate the sympathetic activity. Bromocriptine, an ergot alkaloid, may act as a mixed agonist-antagonist at DA-receptors to alter plasma catecholamine concentrations (Van Loon, 1979). This effect is explained by a central mechanisms. Bromocriptine is suggested to act presynaptically on an unknown catecholaminergic neuron. The site is postulated to be located in the medulla oblongata where it regulates the sympathetic outflow (Van Loon, 1980).

4. Administration of amantadine to patients with PD was not followed by increased excretion of monoamine metabolites except in those patients who received additional anticholinergic drugs (Lenman et al., 1977). Nevertheless, it was concluded that the effect of amantadine could be an indirect dopamine-releasing action and a direct agonist effect on dopamine receptors (Bailey and Stone, 1975).

A major alternative interpretation of the spectrum of DA-metabolites in the urinary excretion is that they reflect in a more general way the function of the whole dopaminergic system.

References

Bacopoulos NG, Hattox SE, Roth RH (1979) 3,4-Dihydroxyphenylacetic acid and homovanillic acid in rat plasma: possible indicators of central dopaminergic activity. Eur J Pharmacol 56:225–236

Bailey EV, Stone TW (1975) The mechanism of action of amantadine in parkinsonism – a review. Arch Int Pharmacodyn 216:246–262

Barbeau A (1969) Parkinson's disease as a systemic disorder. In: Gillingham FJ, Sonaldson ML (eds) 3rd Symp. on Parkinson's disease. Livingstone, Edinburg, pp 66–73

Bronaugh RL, Mattox SE, Hoehn MM, Murphy RC, Rutledge CO (1975) The separation and identification of dopamine 3-O-sulfate and dopamine 4-O-sulfate in urine of parkinsonian patients. J Pharmacol Exp Ther 195:441–452

Buu NT (1985) Dopamine sulfoconjugation in the rat brain: regulation by monoamine oxidase. J Neurochem 45(2):470–476

Elchisak MA (1987) Analytical techniques for the determination of phenolic amine neurotransmitter conjugates. Life Sci 41(7):913–916

Hoehn MM and Yahr MD (1967) Parkinsonism: onset, progression and mortality. Neurology (Minneap) 17:427–442

Jellinger K (1989) Pathology of Parkinson's syndrome. In: Calne DB (ed) Handbook of experimental pharmacology, vol 88. Springer, Berlin Heidelberg New York Tokyo, pp 47–112

Jenner NW, Rose FA (1973) Studies on the separation of 3,4-dihydroxyphenylethylamine (dopamine) and related compounds by rat tissues. Biochem J 135:109–114

Johnson CA, Baker CA, Smith RT (1980) Radioenzymatic assay of sulphate conjugates of catecholamines and DOPA in plasma. Life Sci 26:1591–1598

Kienzl E, Eichinger K (1988) Preparation of dopamine 3-O-sulfate and dopamine 4-O-sulfate as reference substances and high performance liquid chromatographic trace determination. J Chromatogr 430:263–269

Kienzl E, Eichinger K, Jellinger K, Kuhn W, Fuchs G, Riederer R (1990) Dopamine-sulfate conjugates in Parkinson's disease. In: New vistas in drug research. Springer, Wien New York (in press)

Lenman JAR, Turnbull MJ, Reid A, Fleming AM (1977) Urinary monoamine metabolite excretion in disorders of movement. J Neurol Sci 32:219–225

Quick H, Sourkes TL (1977) Central dopaminergic and serotoninergic systems in the regulation of adrenal tyrosine hydroxylase. J Neurochem 28:137–147

Riederer P, Rausch WD, Birkmayer W, Jellinger K, Seemann D (1978) CNS modulation of adrenal tyrosine hydroxylase in Parkinson's disease and metabolic cncephalopathies. J Neural Transm [Suppl] 14:121–131

Rivett AJ, Eddy BJ, Roth JA (1982) Contribution of sulfate conjugation, deamination and O-methylation to metabolism of dopamine and norepinephrine in human brain. J Neurochem 39:1009–1016

Rivett AJ, Francis A, Whittemore R, Roth JA (1984) Sulfate conjugation of dopamine in rat brain: regional distribution of activity and evidence for neuronal localisation. J Neurochem 5:1444–1448

Rutledge CD, Hoehn M (1973) Sulfate conjugation and L-dopa treatment of parkinsonian patients. Nature 244:447–450

Scott MC, Elchisak MA (1983) Cardiovascular effects of dopamine sulfate in anesthetized dogs. Fed Proc 42:1363

Sofic E (1986) Thesis, Techn Univ Vienna, Austria

Van Loon GR, Sole MJ, Bain J (1979) Effects of bromocriptine on plasma catecholamines in normal men. Neuroendocrinology 28:425–434

Van Loon GR (1980) Abnormal catecholamine mechanisms in hypothalamic-pituitary disorders. Metabolism 29(11) [Suppl] 1:1198–1202

Westerink BHC, ten Kate N (1986) 24 h excretion patterns of free, conjugated and methylated catecholamines in man. J Clin Chem Clin Biochem 24:513–519

Yamamoto T, Yamatodani A, Nishimura AM (1985) Determination of dopamine-3- and -4-O-sulfate in human plasma and urine by anion-exchange high-performance liquid chromatography with fluorimetric detection. J Chromatogr 342:261–267

Authors' address: Prof. Dr. K. Jellinger, Ludwig Boltzmann Institute of Clinical Neurobiology, Lainz-Hospital, Wolkersbergenstrasse 1, A-1130 Wien, Austria.

J Neural Transm (1990) [Suppl] 32: 481–484

Effect of adrenal-caudate transplantation on CSF levels of salsolinol sulfate in patients with Parkinson's disease

B. A. Faraj, V. M. Camp, and **R. L. Watts**

Departments of Radiology (Division of Nuclear Medicine) and Neurology, Emory University School of Medicine, Atlanta, Georgia, U.S.A.

Summary. The main objective of this study was to determine whether the activation of dopaminergic pathways through adrenal-caudate transplantation stimulates the production of the dopamine cyclic metabolite salsolinol in CSF of patients with Parkinson's disease. Salsolinol sulfate in CSF samples was assayed by radioenzymatic technique. The outcome of this study revealed that the replacement of degenerative nigrostriatal neurons with new dopamine-producing cells by adrenal brain transplants resulted in significant increase in CSF concentration of salsolinol sulfate as compared to preoperative levels.

Introduction

Experimental findings continue to accumulate which support the etiological involvement of biologically active dopamine-aldehyde adducts, salsolinol, and other tetrahydroisoquinoline (TIQs) in the abnormal consumption of alcohol (Collins, 1988). Moreover, it has been proposed that dopamine-containing neurons may be implicated in the complex phenomena underlying alcohol drinking, in terms of not only its rewarding properties, but also its addictive liability (Blum et al., 1990; Pohorecky and Brick, 1988). The possibility exists that the synthesis of dopamine-aldehyde adducts occurs as more alcohol is consumed and that one of the principal actions of the TIQs is an alteration of the dopamine neuronal system in the brain (Minano et al., 1989).

One important issue surrounding the neurochemical significance of salsolinol and other TIQs such as tetrahydropapaveroline is their respective formation endogenously within the central nervous system in the absence of alcohol consumption. We have chosen to study the cerebrospinal fluid (CSF) level of the dopamine cyclic metabolite salsolinol in patients with

Parkinson's disease because Parkinsonism is a pathologic state that results primarily from dopamine deficiency in the basal ganglia. Because Parkinson's disease is caused by progressive loss of nigrostriatal neurons which make dopamine, it is reasonable to consider replacing them with new dopamine-producing cells from other sources. The possible sources of tissue that could be transplanted include adrenal medulla and fetal substantia nigra from a donor. In the Parkinsonian primates that receive fetal nigral implants, there was biochemical as well as behavioral improvement, indicated by an increase in the amount of dopamine metabolites in the CSF (Bakay et al., 1987). Furthermore, Madrazo et al. (1987) reported dramatic clinical improvement in two Parkinson's disease patients following transplantation of autologous adrenal medullary tissue into the head of the right caudate nucleus.

The main objective of this paper, therefore, was to determine whether the implantation of dopamine-producing cells into the brains of Parkinsonian humans for correction of the dopamine deficiency state has an effect on the endogenous production of the dopamine TIQ metabolite salsolinol as measured in the CSF of these patients.

Methods

Subjects

Ten subjects with advanced Parkinson's disease were evaluated. In these patients, adrenal-caudate transplant was achieved following the grafting of autologous adrenal medullary tissue into the right caudate nucleus using a transcortical technique similar to Madrazo et al. (1987). Lumbar CSF samples were collected following a 4-day drug-free period.

Salsolinol sulfate

Salsolinol sulfate in CSF was assayed by radioenzymatic assay according to the procedures of Faraj et al. (1989).

Results

The results of this preliminary study (mean $\pm$ SD) showed that the concentration of salsolinol sulfate in CSF of 3 untreated Parkinson's disease patients was 145 ± 120 pg/ml, respectively (Table 1). Following adrenal-caudate transplantation, there was a significant ($p < 0.05$) increase in CSF concentration of salsolinol sulfate (443 ± 262 pg/ml) at 6–12 months postoperatively.

Table 1. CSF salsolinol sulfate (SAL-S) levels in patients with Parkinson's disease* before and after adrenal-caudate transplantation (ACT)

	CSF (pg/ml)**	
Subject #	Before ACT SAL-S	After ACT SAL-S
1	283	444
2	58	200
3	96	193
4	–	1108
5	–	374
6	–	511
7	–	252
8	–	363
9	–	490
10	–	500
mean	145	443
S.D.	120	262

* Parkinson's disease patients were studied during a 4-day drug-free period.
** Mean $\pm$ SD

Discussion

It has been demonstrated that salsolinol and tetrahydropapaveroline (THP) can modify alcohol intake when these compounds are applied directly to the brain (Myers, 1978). However, reports of the detection of salsolinol in tissues and fluids by mammalian brain in absence of alcohol consumption have been controversial. This questions the importance of salsolinol as an endogenous dopamine-like neuromodulator produced within the CNS. With the availability of a specific and sensitive radioenzymatic assay, we measured the CSF levels of salsolinol in a group of patients with Parkinson's disease before and after adrenal-caudate transplantation. In agreement with our previously reported study of plasma salsolinol levels in humans, we found that the major fraction of salsolinol in CSF of untreated Parkinson's patients who have been off the dopaminergic therapy for a period of four days was represented by salsolinol sulfate and that free salsolinol amounted to less than 5% of total salsolinol.

Interestingly, the replacement of degenerative nigrostriatal neurons with new dopamine-producing cells by autologous transplantation of tissues of adrenal medulla in patients with Parkinson's disease resulted in a 2- to

3-fold increase in the CSF levels of salsolinol sulfate at 6–12 months post-operatively as compared to preoperative levels. Similar increases were obtained in CSF of other dopamine metabolites (Watts et al., 1989).

These observations lend support to the thesis that salsolinol, an endogenous metabolite of dopamine, is produced in vivo along distinct population of dopamine-producing cells within the CNS.

References

Bakay RA, Barrow DL, Fiandaca MS, Iuvone PM, Schiff A, Collins DC (1987) Biochemical and behavioral correction of MPTP Parkinson-like syndrome by fetal cell transplantation. Ann NY Acad Sci 495:623–640

Blum K, Noble EP, Sheridan PJ, Montgomery A, Ritchie T, Jagadeeswaran P, Nogami H, Briggs AH, Cohn JB (1990) Allelic association of human dopamine D_2 receptor gene in alcoholism. JAMA 263:2055–2060

Collins MA (1988) Recent development in alcoholism, vol 6. Plenum Press, New York, p 387

Faraj BA, Camp VM, Davis DC, Lenton JD, Kutner M (1989) Elevation of plasma salsolinol sulfate in chronic alcoholics as compared to non-alcoholics. Alcohol Clin Exp Res 13:155–163

Madrazo I, Drucker-Colin R, Diaz V (1987) Open microsurgical autograft of adrenal medulla to the right caudate nucleus in two patients with intractable Parkinson's disease. N Engl J Med 316:831–834

Minano FJ, McMillen BA, Myers RD (1989) Interaction of tetrahydropapaveroline with inhibition of dopa decarboxylase by Ro 4-4602 in brain: effects on alcoh: drinking in the rat. Alcohol 6:133–137

Myers RD (1978) Psychopharmacology of alcohol. Annu Rev Pharmacol Toxicol 18:125–144

Pohorecky LA, Brick J (1988) Pharmacology of ethanol. Pharmacol Ther 36:335–427

Watts RL, Freeman A, Bakay RA, Iuvone M, Watts N, Graham S (1989) Adrenal-caudate transplantation in patients with Parkinson's disease (PD): 1-year follow-up. Neurology [Suppl 1] 39:127

Authors' address: Dr. B. A. Faraj, Department of Radiology, 467 Woodruff Memorial Building, Emory University School of Medicine, Atlanta, GA 30322, U.S.A.

Subject Index

α-2-Presynaptic receptors 405–412
α-2-Presynaptic receptors, down regulation 410
α- and β-adrenergic receptors 30
α-Adrenoceptor, agonist 457
α-Adrenoceptor, in depression 457–461
α-receptors 30, 31
α_1-receptor 33
α_1-up-regulation 29–37
α_2-adrenergic receptors 29–37
Acetylcholine 4
Acetylcholinesterase 4
Acid metabolites, brain levels 116
Adenylate cyclase 29–37
Adrenal medulla 481–484
Adrenal-caudate transplantation 481–484
Adrenaline 13, 431–435
Adverse events 181
Aging 240
Alcohol 483
Alzheimer's disease 64
Amantadine 269–275
Amantadine, L-deprenyl synergism 273
Amine oxidase activity in breast tumours 325
Amine oxidase, in bovine lung 341–344
Amine oxidase, yeast, peroxisomal 315–318
Amine oxidases, plasma 281, 315–318, 323–326
2-Aminoethylcarboxamide derivatives 165–170
Amphetamine 32
Anticonvulsant 90
Antidepressants, mechanism of action 29–37
Autoradiography of MAO 62
Ascorbic acid 253
Autologous transplantation 483

β1-Adrenoceptor 463
β-Adrenoreceptor 319
β-down regulation 30
β-receptors 29–37
Blood platelets, monoamine oxidase 161–164
Blood pressure 444
Blood vessels 395–404
Brofaromine 31, 95, 172
Brofaromine, depressive disorder 189–195
Brofaromine, estimation of plasma concentration 197–201
Brofaromine, in depression 199

CAKI-1 cells, human 437–440
Calbindin 244
Calcium 239–248
Calcium antagonists 240
Calcium binding protein 244
Calcium channel 244
Calcium homeostasis, intracellular 244
Calcium-mediated cell killing 244
Carbamazepine 31, 33
Catalase 233, 240
Catechol-0-methyltransferase 3–18, 376, 463–467
Catechol-0-methyltransferase, inhibition by CGP 28014 381–391
Catechol-0-methyltransferase, inhibition by RO 40-7592 375, 378, 379
Catechol-0-methyltransferase, membrane bound 369
Catechol-0-methyltransferase, monoclonal antibodies 369–374
Catechol-0-methyltransferase, N-terminal amino acid sequence 366
Catechol-0-methyltransferase, pH-activity profile 364
Catechol-0-methyltransferase, placental 364

Catechol-0-methyltransferase, properties 359–368
Catechol-0-methyltransferase, purification 363
Catechol-0-methyltransferase, soluble 369–374
Catechol-0-methyltransferase, western blot analysis 372
COMT and MAO of the extraneuronal metabolizing system 14, 15
Catecholamine release 405–412
Catecholamines 21
Catecholamines, extraction from plasma 431–435
Catecholamines, supersensitivity 463–467
Catecholamines, systemic fractional extraction 431–435
cDNA clones encode MAO-A 41
cDNA clones encode MAO-B 41
Cerebrospinal fluid 447–456, 481–484
Cerebrospinal fluid, salsolinol sulfate 483
c-fos protooncogene 34
Cheese effect 139–144, 149–154
Clonidine 406, 410, 457
Clorgyline 85, 87, 139–144, 172, 221, 329, 399, 405
Cocain 398, 399
cognitive impairments 185–188
Complex I and III 244
Convulsions 91
Copper-containing enzymes 280
cortical electric activity 139–144
Corticosteroids 4
Cytotoxic oxygen free radicals 240

DATATOP study 217–229, 237
Deamination of branched aliphatic amines 89–92
Dementia of Alzheimer type 61–63, 259–267
^{3}H-deprenyl 61
Deprenyl 55, 58, 62, 76, 85, 87, 96, 171–175, 204, 221, 223, 235, 239–248, 269–275, 407
Deprenyl, amantadine synergism 273
Deprenyl, arterial blood clearance curve 56
Deprenyl, neuroprotective action 245
^{3}H-L-deprenyl, quantitative autoradiography 64
Deprenyl, uptake of ^{11}C in the brain 57
Depression 177–184, 457–461, 475
Desferrioxamine 242, 245
Desimipramine binding 413
Desimipramine binding, effect of thiol and disulfide 414
Desimipramine binding, loss by enzymes 415
Desimipramine binding site, of PC 12 cells 417
Desimipramine binding site, purification 415
Desimipramine binding site, solubilization 415
Desimipramine sensitive processes 431–435, 437–440
Desmethyl-selegiline 131–137, 139–144
Desmethyl-selegiline, EEG frequency 141
Desmethyl-selegiline, EEG pattern 142
Diamine oxidase 315–322, 345–349
Diamine oxidase, inhibition, kinetic analysis 303, 306, 307
Diamine oxidase, intestinal 291–314
Diamine oxidase, use of blockers 310
Differential scanning thermograms 259–267
Dihydropyridine sensitive calcium channel 244, 245
3, 4-Dihydroxymandelic acid 5, 395–404
3, 4-Dihydroxyphenylglycol 5, 8, 10, 395–404, 421–429, 441–445
Dismutase activity 240
DOPA 231, 232, 233, 375–380
DOPA decarboxylase inhibitors 232
DOPA formation 158
DOPA, catabolism
DOPA, plasma concentration 389, 390
DOPA, time course 383
DOPAC 32, 79–84, 124, 155–159, 233, 269–275, 387–391, 475, 476
DOPAC, MAO-A inhibition 157
DOPAC, MAO-B inhibition 157
DOPAC, RO 19-6327 82
Dopamethylester 385
Dopamine 132, 239–248, 269–275, 375, 397, 413

Dopamine in the striatum 234
Dopamine metabolites, in depression 475
Dopamine metabolites, in neurological disorders 475
Dopamine neurons 240
Dopamine sulfates, urinary 471–479
Dopamine, deamination of monoamine oxidase A and B 155–159
Dopamine, MAO-A inhibition 157
Dopamine, MAO-B inhibition 157
Dopamine, RO 41-1049 82
Dopamine, turnover 271
Dopamine-3-0-sulfate 471–479
Dopamine-4-0-sulfate 471–479
Dopamine-β-hydroxylase 406
Dopamine-aldehyd adducts 481

EEG 141
Electron transport system 244
Endoperoxides 286
Endothelium 395
Enzyme inhibition, rate of metabolite formation 9
Epilepsy 123–129
Erythrocytes 250
Essential hypertension 421–429
Ethanol 319–322
Extraneuronal cells 4
Extraneuronal uptake, low affinity$_2$ 3–18

false transmitters 19–27
Ferritin 241
Fluoxetine 31
Flupirtine 143
Fluvoxamine 31

G-protein 33
GABA 123
Glutathione 229–238, 241, 253
Glutathione disulfide 229
Glutathione peroxidase 240
Glutathione reductase 230
Glutathione, oxidized levels 232, 234, 235
glycine 126
Growth hormone 457, 459
Growth hormone, response to clonidine 457, 459, 460

Haloperidol 21, 233, 234
Harmaline 95
Harman 93–102
harman binding 93–102
Harman displacement 95, 96
Harman labels MAO-A 99
Harmine 95
Hepatotoxicity 149
Herbicide 218
High affinity uptake$_1$ 3–18
Histamine 291–314, 319
Histamine methyltransferase 292
Histamine, plasma levels 307
Histaminosis 291–314
Histidine decarboxylase 292
Homovanillic acid 79–84, 233, 269–275, 381–391, 475, 476
Homovanillic acid, CSF 32
Homovanillic acid, RO 41-1049 82
Homovanillic acid, serum vs. striatal 382
5-HT_2 receptors 32
Huntington's disease 61
Hydrogen peroxide 90, 222, 230–232, 240, 249–257, 286
Hydrogen peroxide during normal DA turnover 236
Hydrogen peroxide formation, MAO 235
6-Hydroxydopamine 239
8-Hydroxy-DPAT 32
5-hydroxyindoleacetic acid 30, 171–175
5-Hydroxyindoleacetic acid 269–275
5-hydroxyindoleacetic acid, in plasma 149–154
5-Hydroxytryptamine, oxidation 249–257
5-Hydroxytryptamine, peroxidative products 251
Hypertension, essential 421–429
Hypertension essential, plasma DOPEG 426, 427
Hypertensives, plasma noradrenaline 427

Imaging of catecholaminergic nerve terminals 19–27
Imaging peripheral sympathetic innervation 23
Imaging striatal dopaminergic neurons 21
Imipramine 29

Imipramine binding 447–456
Imipramine binding, effect of human CSF 449
Imipramine, Hamilton Depression Rating scale 180
Immunofluorescence cytochemistry of MAO 67–77
Inhibitor metabolism 168
in situ hybridization histochemistry 51
Interaction, uptake and inside enzyme 12
In vivo voltammetry 85
Iprindole 116
Iproniazide 29
Ipsapirone 32
Iron 239–248
Iron chelator 241
Iron induced lipid peroxidation 240
Iron neurotoxicity and nigro-striatal dopamine 239
Isoprenaline 7, 431
Isotopes 19–27

Kynuramine 93–102

L-amino acid decarboxylase 155–159
Levoprotiline 31
Lipid deposition 335
Lipid peroxidation 244
Lipid peroxidation, induction by iron 242
Localisation of MAO-A & MAO-B 50, 52
LSD 33

Mammary cancer 323–326
MAO 4, 9, 12–16, 21, 29–37, 61, 203, 217-229, 269, 292, 323, 341, 345, 351–356
MAO activity, CNS 149
MAO and oxidative stress 229–238
MAO in bovine eye 327–330
MAO inhibition by substituted tryptamine analogues 103–105
MAO inhibitor phenelzine 113–118
MAO inhibitors 51, 109
MAO inhibitors and dopamine outflow 79–84
MAO inhibitors, reversible chemical structures 166
MAO inhibitors, slow binding 165–170
MAO platelet 149, 161, 163, 164
MAO, chronic inhibition 405–412
MAO, effects of human CSF 449
MAO, ex vivo inhibition by reversible inhibitors 146
MAO, iron-melanin interaction 239–248
MAO, neuronal 411
MAO, neuronal, saturation inside the metabolizing system 11
MAO, partial inhibition, neuronal 10
MAO-A 41, 67–77, 85, 103–105, 139–144, 155–159, 221
MAO-A activities in brain and liver 126
MAO-A activity 149–154
MAO-A activity in rat striatum 81
MAO-A and -B cDNA probes 42
MAO-A and -B inhibitors in plasma enzymatic assay 146
MAO-A cDNA clone 45
MAO-A cDNA subfragments 42
MAO-A genes, human 41–47
MAO-A inhibition in depression 191
MAO-A inhibitors in plasma 145–148
MAO-A inhibition, dopamine 157
MAO-A selective inhibitors 145–148
MAO-A transcripts Northern analysis 45
MAO-A, 5Kb transcript 43
MAO-A, enzymatic detection 147
MAO-A, molecular neuroanatomy 49–53
MAO-A, reversible and selective inhibitors 49, 79–84
MAO-B 41, 61, 67–77, 85, 103–105, 121, 124, 139–144, 155–159, 221
MAO-B activities in brain and liver 126
MAO-B activity in rat striatum 81
MAO-B cDNA subfragments 43
MAO-B genes, human 41–47
MAO-B inhibition by RO 19-6327, 211–214
MAO-B inhibition, dopamine 157
MAO-B inhibition, dose-response curves 213
MAO-B inhibition, time course after RO 19-6327 212
MAO-B inhibitor 79–84

MAO-B inhibitors in plasma 145–148
MAO-B, 3Kb transcript 43
MAO-B, enzymatic detection 147
MAO-B, molecular neuroanatomy 49–53
MAO-B, reversible and selective inhibitors 49
MAO-B, visualization 61–63
MAO-B-activity, kinetic evaluation 131–137
MAO-B-inhibition, duration 134, 136
MAO-B-inhibition, ex vivo 133
MAO-B-inhibition, time course 134
MAO-inhibition and psychometric function 187
MAO-turnover 55–59
Maprotiline 185–188, 457–461
Maprotiline, Hamilton Rating Scale for Depression 191
Maprotiline, side effects 191, 192
MDL 72.974A, pharmacokinetic parameters 207
MDL 72.974A, platelet MAO-B activity 203–209
MDL 72.974A, tolerability 203–209
MDL 72527 351, 353
Melanin 241
Melanin-containing neurons 76
Melatonin 171
Membrane constituents 259–267
Membrane fluidity 259–267
Metabolism, dopamine 85–88
Metabolism, serotonin 85–88
Metabolizing system 4
3-Methoxy-4-hydroxyphenylglycol 30
1-methyl-β-carboline 94
Methylamine oxidase 315–318
3-0-Methyldopa 376, 381–386, 390
3-0-Methyldopa and CGP 28014 384, 387–391
3-0-Methyldopa, serum vs. striatal 384
3-0-Methyldopamine 395–404
5-F-α-methyltryptamine 95
Microtubulae associated proteins 34
Migraine 161–164
Milacemide 89, 123–129, 351–356
Milacemide, anticonvulsant action 123
Milacemide, metabolites 127
Milacemide, urinary metabolism 352
Moclobemide 31, 79–84, 165–170, 172, 183, 457–461
Moclobemide & melatonin synthesis 171–175
Moclobemide compared with imipramine 177–184
Moclobemide, autonomic side effects 182
Moclobemide, cardiovascular tolerability 182
Moclobemide, depressive disorder 189–195
Moclobemide, efficacy 181
Moclobemide, Hamilton Depression Rating scale 180, 191
Moclobemide, psychometric alterations 185–188
Moclobemide, safety 181
Moclobemide, side effects 192
Motorische Leistungs-Serie 186
MPP^+ 218
MPTP 22, 121, 239, 269–275
MPTP analogs and MAO 219, 221
MPTP analogs oxidation by MPTP analogs 121
MPTP bioactivation 217–229
MPTP, changes of amines 271
MPTP, Stylbasole analogs as MAO-substrates 119–122

N-(L-propylpentyl)glycinamide glycine 89
Neuronal uptake 3–18
Neurotoxins 217–229, 239, 280
Neurotransmitter amines, brain levels 116
Nigro-striatal system 477
NMDA receptor 124
NMDA-antagonists 274
Nomifensine 32
Noradrenaline 30, 413
Noradrenaline transporter, affinity labeling 416
Noradrenaline transporter, protein 418
Noradrenaline, fractional release 405
Noradrenaline, leakage 428
Noradrenaline, metabolism 398
Noradrenaline, plasma 441–445
Noradrenaline, plasma clearance 431–435
Noradrenaline, re-uptake 421–429

Noradrenaline, release, endogenous 408, 442
Noradrenaline, synaptic concentration 425, 441–445
Noradrenaline, transporter 413–419
Noradrenaline, uptake 1 437–440
Noradrenaline, uptake 2 437–440, 463–467
Normetanephrine 395–404
Normotensives, plasma DOPEG 427
Normotensives, plasma noradrenaline 427
Norrie disease 42
Nucleus basalis of Meynert 259–267

Ornithine decarboxylase 246
Oxaprotiline 31, 32
Oxidative stress 243, 245
Oxidative stress and monoamine oxidase 229–238
Oxygen radical species 239–248

Pargyline 10, 171–175
Pargyline, effect on extracellular DOPAC 86
Parkinson's disease 61, 203–209, 230, 237, 239–248, 274, 375, 387, 475, 481–484
Parkinsonian patients 217–230
Paroxetine binding 447–456
Paroxetine binding, effect of human CSF 451
Pathogenesis of Parkinsonism 222
Peripheral organs 4
Peroxidase 240, 249–257
Peroxidative stress 230
PET analysis of [^{11}C]-deprenyl binding 51
PET imaging 25
PET imaging of the striatum 22
Pharmaco-EEG changes 139–144
Phenelzine 113–118, 150, 153
Phenelzine, metabolites 115
Phenylethylamines 413
Pheochromocytoma cells (PC 12) 413–419
Phosphatidylinositol system 29–37
Pithed rat model 407
Placenta 359–368
Polyamine oxidase 345–349, 351–356
Polyamines 239–248, 292, 319, 352
Polyamines, N-acetylated 246
Position emission tomography 19–27
Positron emission tomography using ^{11}C-L-deprenyl 55–59
Positron-emitting isotopes 20
Preadipozytes 331–336
Presynaptic autoreceptors 32
Progression of Parkinsonism 223
2-Propyl-1-aminopentane 89–92
Prostaglandin 286
Pump and leak system 3–18
Pump and leak system, enzyme inside 7
Pump and leak system, two enzymes inside 8

Radionuclides 20
Reserpine 15, 231, 233, 234
Respiratory chain activity 239–248
RO 16-6491 165–170
RO 19-6327 52, 79–84, 156, 158, 165–170, 211–214
RO 41-1049 79–84, 156, 158, 165–170
RO 12812 400
Rolipram 33

Salsolinol sulfate 481–484
Second messengers 29–37
selective inhibitors 284
selegiline 131–137, 139–144
Selegiline, EEG frequency 141
Selegiline, EEG pattern 142
Semicarbazide-sensitive amine oxidase 395
Semicarbazide-sensitive amine oxidase, production by preadipocytes 331–336
Semicarbazide-sensitive amine oxidases, in bovine eye 327–330
Semicarbazide-sensitive amine oxidases, in breast tumours 323–326
Semicarbazide-sensitive amine oxidase, in rat aortic smooth muscle 337–339
Semicarbazide-sensitive amine oxidases, pharmacology 279–290
Semicarbazide-sensitive amine oxidases, physiological aspects 282
Semicarbazide-sensitive amine oxidases, tissue bound 281, 282
sensorimotor disturbances 185
Serotonin 30, 149, 270
Serotonin in blood 150

Serotonin in plasma 149–154
Serotonin receptors 31
Serotonin uptake 447–456
Serotonin uptake, transporter 448
Serotonin, effect of human CSF 453
Serotonin, MAO inhibitors 151
Serotonin, removal from the bloodstream 153
Stroke 240
Stylbasole derivatives kinetic parameters 120
Substantia nigra 67–77, 217, 239–248
Sympathetic activity 477
Sympathetic denervation 25
sympathetic innervation in peripheral organs 19–27
sympathetic nerve stimulation 21
Sympathetic nerves 407, 409, 421–429
Sympathomimetic amines 465

t-butyl hydroperoxide 231
Tetradecanoyl phorbol acetate 33
Tetrahydroisoquinoline 481
Tetrahydropapaveroline 481, 483
Thermostability 259–267
Third messenger 33
Transferrin 243
Transferrin receptors 243
Transport mechanismus, interaction 3–18
Transport mechanismus, intracellular enzymes 3–18
Tranylcypromine, analogues as monoamine oxidase inhibitors 107–112
Trauma 240
Tropolone 381
Tryptamine, analogues 103–105

U 47500 A 245
$Uptake_1$ 23, 24
$Uptake_2$ 4, 23
$Uptake_1$, non-neuronal cells 3–18
Uterine artery 401

Valproic acid 90
Valproic acid precursors 89–92
VAS deferens 407, 409, 421
Vascular tissue 395–404

Yohimbine 405

GPSR Compliance

The European Union's (EU) General Product Safety Regulation (GPSR) is a set of rules that requires consumer products to be safe and our obligations to ensure this.

If you have any concerns about our products, you can contact us on ProductSafety@springernature.com

In case Publisher is established outside the EU, the EU authorized representative is:

Springer Nature Customer Service Center GmbH
Europaplatz 3
69115 Heidelberg, Germany

Batch number: 10723890

Printed by Printforce, the Netherlands